Orthobiologics

Giuseppe Filardo · Bert R. Mandelbaum
George F. Muschler · Scott A. Rodeo
Norimasa Nakamura
Editors

Orthobiologics

Injectable Therapies for the Musculoskeletal System

Editors
Giuseppe Filardo
Rizzoli Orthopaedic Institute
Bologna
Italy

Università della Svizzera Italiana
Ente Ospedaliero Cantonale
Lugano
Switzerland

George F. Muschler
Orthopaedic Surgery and Biomedical
Engineering
Cleveland Clinic
Cleveland, OH
USA

Norimasa Nakamura
Institute for Medical Science in Sports
Osaka Health Science University
Osaka
Japan

Bert R. Mandelbaum
Orthopaedic & Sports Medicine Group
Santa Monica Orthopaedic & Sports
Medicine
Santa Monica, CA
USA

Scott A. Rodeo
Hospital for Special Surgery
New York, NY
USA

ISBN 978-3-030-84746-3   ISBN 978-3-030-84744-9   (eBook)
https://doi.org/10.1007/978-3-030-84744-9

This Springer imprint is published by the registered company Springer Nature Switzerland AG
The registered company address is: Gewerbestrasse 11, 6330 Cham, Switzerland

Preface

Orthobiologics are gaining increasing attention in the musculoskeletal field. In particular, the local application of biological substances through injections represents a promising minimally invasive treatment approach. The use of blood derivatives as well as cells derived from various sources and tissues with the goals of improved and faster healing is growing in clinical practice. *Orthobiologics: Injectable Therapies for the Musculoskeletal System* provides an authoritative and timely reference that summarizes the rationale for these different strategies together with the available evidence exploring the role of this approach for lesions ranging from muscle, ligament, tendon, bone, meniscus, cartilage as well as early osteoarthritis.

The first part of the book provides scientific and clinical foundations of this field. A broad range of creative thinking and strategic approaches were inspired by the potential of orthobiologics to meet the clinical demand for more effective therapies. At the same time, the sheer number of potential disease targets, cell sources, cell types, processing options, composition, and associated variables can be overwhelming and confusing for clinicians, patients, as well as the scientists working in the field. The rational design, development, deployment, and rigorous assessment of evolving clinical strategies require unified frameworks and nomenclature on which to build clear communication. To this aim, this textbook seeks to advance the field by bringing a diverse array of current approaches together in a unified conceptual model. The book starts with a chapter specifically dedicated to introducing core themes, biological paradigms, engineering principles, and nomenclature (including a glossary on cell-related terms) that each subsequent chapter builds upon to effectively explore and define the current state and evolving opportunities in cellular injection therapies for musculoskeletal diseases.

The first part of the book also explores the preclinical evidence supporting a diversity of approaches. This includes cell-based as well as non-cell-based strategies. It is important to note that evidence in an evolving field comes preferably through rigorously controlled science. However, insight and evidence can accumulate through anecdotal, uncontrolled, and even controversial observations. Objective findings are necessary to validate new treatments, but legitimate practical and ethical concerns can frustrate exploration based solely on objective clinical data. This diversity is presented and analyzed in the hope of educating the reader, as well as animating and enabling scientific debate in the coming years.

Injectable orthobiologics are fashionable treatments that have a risk of false expectations and over hype. Patients are attracted to the allure of these emerging therapies, and this fact imposes a large psychological component on any objective assessment of biological effects. For this reason, a specific chapter in the second part of the book is dedicated to the concept of the placebo effect. All proposed products will need to demonstrate efficacy in randomized clinical trials where the placebo effect can be controlled. This largest portion of the book systematically examines the available clinical evidence for each approach for specific musculoskeletal tissues. For tissues/diseases where there was larger body of current evidence, the topic has been split into specific chapters focusing on cell-based and non-cell-based treatments. The third part of the book is dedicated to new and emerging orthobiologic options.

The spirit in this book is a tone of cautious optimism sometimes mixed with frank skepticism. Applications in some tissues and diseases have more evidence than others. Overall, the editors and authors of this book broadly agree that, at the present time, biological therapies have not been optimized. The magnitude of improvement and probability of success associated with specific approaches has been rigorously defined only for few indications. Clear indications and contraindications have not been established. As a result, orthobiologic therapies for musculoskeletal conditions remain imperfectly developed and in many settings "unproven." Each of the chapters in this text examines the progress that has been achieved to date, while also pointing out gaps in evidence, current contraindications, and specific opportunities for future improvement. Clinical judgment and ethical principles strongly suggest that the current use of bioactive or cellular injection therapies should only be performed as part of rigorous trials or registries. Only in this way can patient safety be assured, and definitive evidence can emerge to justify a broader use in clinical practice.

The field will not be limited to current formulations or products, the clinical demand for minimally invasive restorative therapies and the rapid ongoing advance of biological and cellular sciences will continue to offer new opportunities. That fact places important responsibilities on us now to define and build clinical systems and tools for rigorous systematic assessment, as well as appropriate protections for patient safety and responsible reporting of outcomes. Clinicians and professional organizations as well as regulatory bodies will need to work assertively to facilitate and enable the establishment of collaborative interinstitutional networks, registries, and cooperative study groups. With all of these goals and principles in mind, we hope that *Orthobiologics: Injectable Therapies for the Musculoskeletal System* will provide the reader with an approachable and rigorous point of entry into the foundational science, clinical needs, and the preclinical and clinical evidence, and provide them with tools to both understand and potentially contribute to the next phase of rigorous ongoing development of biological and cellular injection therapies to serve current and future patients.

We sincerely thank all of the authors for their thorough and insightful chapters, and we are confident that our readers will appreciate the contributions provided by some of the principal thought leaders and experts in the world of orthobiologics.

Bologna, Italy	Giuseppe Filardo
California, USA	Bert R. Mandelbaum
Cleveland, USA	George F. Muschler
New York, USA	Scott A. Rodeo
Osaka, Japan	Norimasa Nakamura

Contents

Part I Injectable Orthobiologics: Formulations and Rationale

1 The Stem and Progenitor Cell Paradigms and Engineering Principles Guiding the Clinical Use of Cells or Cell-Derived Products for Regenerative Medicine 3
George F. Muschler, Hannah Simmons,
Venkata Mantripragada, and Nicolas S. Piuzzi

2 Bone Marrow as a Source of Cells for Musculoskeletal Cellular Therapies. 29
George F. Muschler, Hannah Simmons,
Venkata Mantripragada, and Nicolas S. Piuzzi

3 Adipose-Derived Stem/Stromal Cells, Stromal Vascular Fraction, and Microfragmented Adipose Tissue 47
Enrico Ragni, Marco Viganò, Paola De Luca,
Edoardo Pedrini, and Laura de Girolamo

4 Injections of Synovial Mesenchymal Stromal Cells. 63
Ichiro Sekiya and Nobutake Ozeki

5 Placenta, Umbilical Cord, and Umbilical Cord Blood-Derived Cultured Stromal Cells. 75
Jin-A Kim and Chul-Won Ha

6 Injectable Allogenic Mesenchymal Stromal Cells: Advantages, Disadvantages, and Challenges 89
Lucas K. Keyt, Matthew D. LaPrade, Aaron J. Krych,
and Daniel B. F. Saris

7 Injection of Steroid Hormones. 97
Tristan W. Juhan, Andrew J. Homere, Alexander E. Weber,
George F. Hatch, and Frank A. Petrigliano

8 Intra-articular Hyaluronic Acid Injections 109
Karan Vishwanath and Lawrence J. Bonassar

9 Cytokines, Chemokines, Alpha-2-Macroglobulin, Growth Factors 123
Claire D. Eliasberg and Scott A. Rodeo

10 Platelet-Rich Plasma: Processing and Composition 133
Spencer M. Stein and Bert R. Mandelbaum

11 Placental Tissue Extracts 145
Bogdan A. Matache, Eric J. Strauss, and Jack Farr

12 Secretome, Extracellular Vesicles, Exosomes 155
Florien Jenner and Iris Ribitsch

**Part II Injectable Orthobiologics: Methods
and Results Based on Anatomy and Pathology**

13 Rotator Cuff Tendinopathy: Cell Therapy 169
Philippe Hernigou and Jacques Hernigou

14 Rotator Cuff Tendinopathy: Biologics 181
Pietro Simone Randelli, Chiara Fossati, Marianna Vitale,
Francesca Pedrini, and Alessandra Menon

15 Orthobiologics for the Treatment of Tennis Elbow 191
William D. Murrell, Sharmila Tulpule, Nagib Atallah Yurdi,
Agnes Ezekwesili, Nicola Maffulli, and Gerard A. Malanga

16 Patellar Tendinopathy: Cell Therapy 205
Chris H. Jo and Sanghoon Oh

17 Patellar Tendinopathy: Biologics 215
Rahman Kandil and Jason Dragoo

18 Orthobiologics for the Treatment of Achilles Tendinopathy 221
Joseph D. Lamplot, Cort D. Lawton, and Scott A. Rodeo

19 Orthobiologics for the Treatment of Plantar Fasciitis 237
Filippo Rosati Tarulli, Cristian Aletto, and Nicola Maffulli

20 Ligament Lesions: Cell Therapy 245
Robert S. Dean, Nicholas N. DePhillipo,
and Robert F. LaPrade

21 Ligament Lesions: Biologics 257
David Figueroa, Rodrigo Guiloff, and Francisco Figueroa

22 Meniscal Lesions: Cell Therapy 265
Kazunori Shimomura, David A. Hart,
and Norimasa Nakamura

23 Meniscal Lesions: Biologics 277
Stefano Zaffagnini, Alberto Poggi, Luca Andriolo, Angelo
Boffa, and Giuseppe Filardo

24 Orthobiologics for the Treatment of Muscle Lesions 287
Alberto Grassi, Giacomo Dal Fabbro, and Stefano Zaffagnini

25 Cartilage Lesions and Osteoarthritis: Cell Therapy 301
Tiago Lazzaretti Fernandes, Kazunori Shimomura,
David A. Hart, Angelo Boffa, and Norimasa Nakamura

26 Cartilage Lesions and Osteoarthritis of the Knee: Biologics 315
Giuseppe Filardo, Angelo Boffa, Luca Andriolo,
Alberto Poggi, and Alessandro Di Martino

27 Cartilage Lesions and Osteoarthritis of the Hip and Ankle: Orthobiologics 329
Francesca Vannini, Simone Ottavio Zielli, and Cesare Faldini

28 Injectable Orthobiologics for the Treatment of Subchondral Insufficiency Fractures of the Knee (SIFK) and Related Pathogenic Processes 349
Kyle N. Kunze, Zaamin B. Hussain, Mikel Sánchez,
and Jorge Chahla

29 Injections: Orthobiologics and the Power of Placebo 361
Davide Previtali, Marco Cuzzolin, Giorgio Di Laura Frattura,
Christian Candrian, and Giuseppe Filardo

Part III Future Directions

30 Pluripotent Stem Cells: Embryonic/Fetal Stem Cells and Induced Pluripotent Stem Cells 371
Gun-Il Im

31 Gene Therapy 383
Magali Cucchiarini

32 In Situ Targeting of Stem and Progenitor Cells in Native Tissues 393
Cierra A. Clark, Takeshi Oichi, Joshua M. Abzug,
and Satoru Otsuru

Part I

Injectable Orthobiologics: Formulations and Rationale

The Stem and Progenitor Cell Paradigms and Engineering Principles Guiding the Clinical Use of Cells or Cell-Derived Products for Regenerative Medicine

George F. Muschler, Hannah Simmons, Venkata Mantripragada, and Nicolas S. Piuzzi

1.1 Overview

This chapter introduces core themes, biological paradigms, engineering principles, and nomenclature that each subsequent chapter builds upon to effectively explore and define the current state and evolving opportunities in cellular injection therapies for musculoskeletal disease.

This chapter introduces key concepts, terms, and definitions that are essential to the rigorous design and execution of cellular therapy strategies and to ensuring clarity and rigor in oral and written communication in this rapidly evolving field. These terms, as well as additional terms introduced in subsequent chapters, are collected in the *Glossary of Terms* provided at the end of this chapter.

Regenerative medicine is defined as any intervention that seeks to repair, replace, augment, or regenerate new tissues. In the context of this book, we limit our focus on regenerative medicine to therapies that are nonsurgical and can be delivered by percutaneous injection. Our particular focus is on injection therapies that utilize viable cells or cell-derived materials (e.g., secretory products, extracellular matrix, or exosomes). However, the biological concepts, engineering principles, therapeutic targets, and clinical outcome assessments that are articulated in this text are also applicable to injection of drugs and synthetic bioactive agents that may be developed in the future.

Why develop injectable cell-based therapies? Current musculoskeletal therapies are robust and alleviate or prevent tremendous clinical burdens for individual patients and our broader community. However, many conditions still have imperfect or unpredictable outcomes. The cost and risk of some therapies can be high. Much greater interest is being expressed for preventive therapies and early interventions with less risk that may improve function, reduce cost and risk, and delay or eliminate the need for invasive therapies. Moreover, scientific advances in cell biology and biomaterials have generated a broad array of new therapeutic options. This alignment of clinical demand and biological potential inspires a broad range of creative thinking and strategic approaches.

Why this book? The sheer number of potential disease targets, cell sources, cell types,

G. F. Muschler (✉) · N. S. Piuzzi
Orthopaedic Surgery and Biomedical Engineering, Cleveland Clinic, Cleveland, OH, USA
e-mail: MUSCHLG@ccf.org

H. Simmons · V. Mantripragada
Department of Biomedical Engineering, Cleveland Clinic, Cleveland, OH, USA

G. Filardo et al. (eds.), *Orthobiologics*, https://doi.org/10.1007/978-3-030-84744-9_1

processing options, composition, and associated variables can be overwhelming and confusing for clinicians, patients, as well as the scientists and engineers working in the field.

The rational design, development, deployment, and rigorous assessment of evolving clinical strategies demand unified frameworks and nomenclature on which to build clear communication. Communication in turn leads to clear and rigorous clinical and scientific consensus regarding clinical objectives, tools and methods comparisons between approaches, and ultimately clear evidence-based decision-making about the optimal use of new therapy options.

This textbook seeks to advance the field by bringing a diverse array of current approaches together in a unified conceptual model. In doing so we seek to inspire innovation, quantitative rigor, and objective communication. However, this must be done without oversimplification of the many variables involved. Moreover, if concepts are to be communicated clearly, they must be done with terminology that is precisely defined and used. To that end, this chapter introduces a set of key terms. Each is underlined when introduced. Specific definitions for these terms are provided in a glossary that is provided at the end of the book.

1.2 Domains of Progress

Three essential domains of progress recur throughout each chapter in this book and in the development and testing of any future cellular therapy:

1. Clinical Assessment—Assessment of safety and efficacy in specific clinical settings.
2. Cell Sourcing—Clinical cell harvest and isolation of cell source materials.
3. Cell Processing /Expansion/Fabrication—In vitro processing, expansion and potential modification of cells for optimal therapy.

Each of these activity domains involves a series of generally consistent process steps that can directly influence the outcome. Moreover, drilling deeper, each of the process steps inevita-

bly involves a series of individual elements or variables that must be defined and controlled in order to study its effects, and optimize the repeatability and reproducibility of outcome.

Ishikawa diagrams provided in Figs. 1.1–1.3 provide an efficient conceptual paradigm with which to organize and illustrate key process steps and variables in each of these activity domains. Reading from left to right, diagonal arrows illustrate the order in which key process steps come into play. Along each diagonal arrow is a partial list of the process variables that must be defined so that critical process parameters (CPPs) in each process step that may be controlled and contribute to the ultimate outcome.

On the far right of the diagram for each activity domain is the specific outcome that is achieved. These outcomes represent critical resources of materials or information that are needed to guide the rational development and use of injectable therapies, specifically:

- **Clinical Assessment** (Fig. 1.1)—Clinical assessment provides generalizable evidence of safety and efficacy of a specific therapy in a specific clinical setting or patient population. An essential premise in clinical assessment is that it is performed in a manner that is repeatable and potentially reproducible. Executed appropriately, clinical assessment both informs us about clinical performance and enables the design and execution of future studies with greater precision.
- Prospective randomized controlled trials (RCTs) are the "gold standard" for clinical assessment, where single variables can be isolated and alternative treatment options are very similar. Given the diversity of product and process variables associated with current cellular therapy approaches, comparative effectiveness trials (CETs) may be more commonly employed. The available RCTs and CETs are systematically explored in this text. However, there is also a productive history in musculoskeletal medicine of systematic enrollment in prospective clinical registries (PCRs) (https://www.aaos.org/registries/) [1, 2]. Registries lack power and precision with respect to direct comparisons; however they

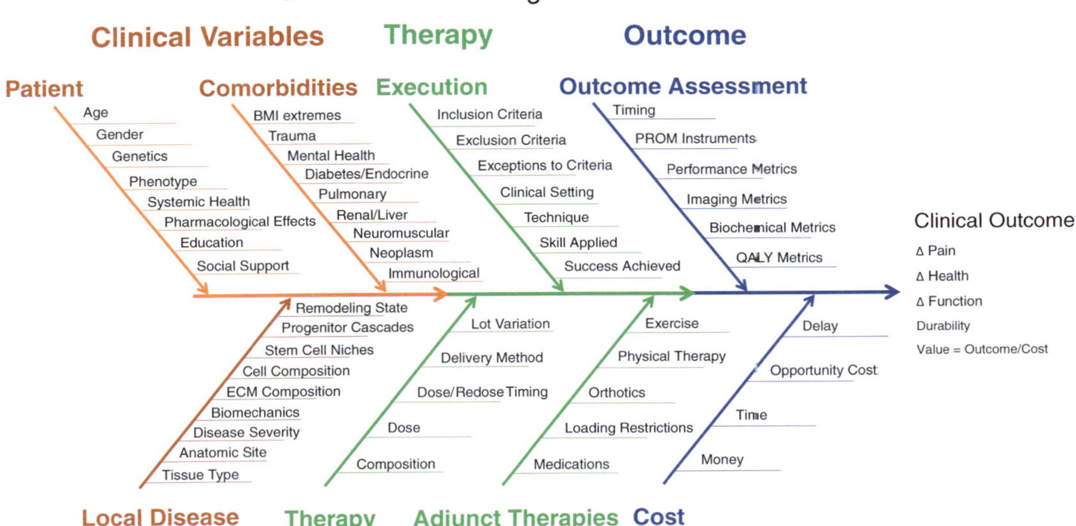

Fig. 1.1 Clinical assessment—Ishikawa diagram illustrating process steps and variables associated with clinical care and assessment of clinical outcome

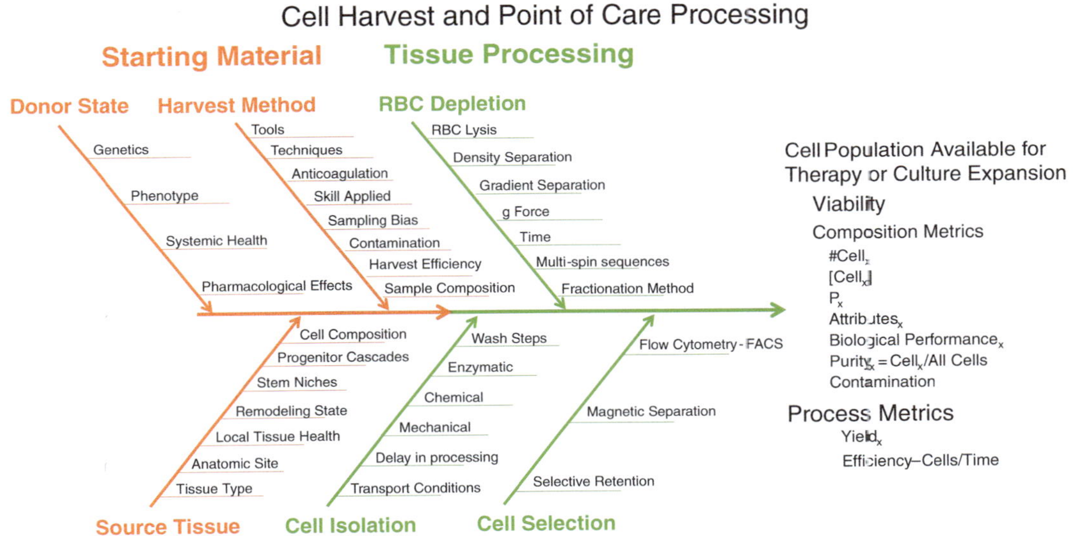

Fig. 1.2 Cell harvest and point of care processing—Ishikawa diagram illustrating process steps and variables associated with cell harvest and rapid point of care processing

have an important role to play in the environment where (a) concerns about safety (the risk of adverse events) are low, (b) <u>equipoise</u> is present (potential benefits equal and potentially outweigh the risks), and (c) diversity of approaches is already available and in clinical use in a definable range of patients and clinical settings, making investment in paired comparisons impractical. The value of registries include (a) engagement of otherwise independent clinical teams in collaborative discussion and analysis; (b) standardization and

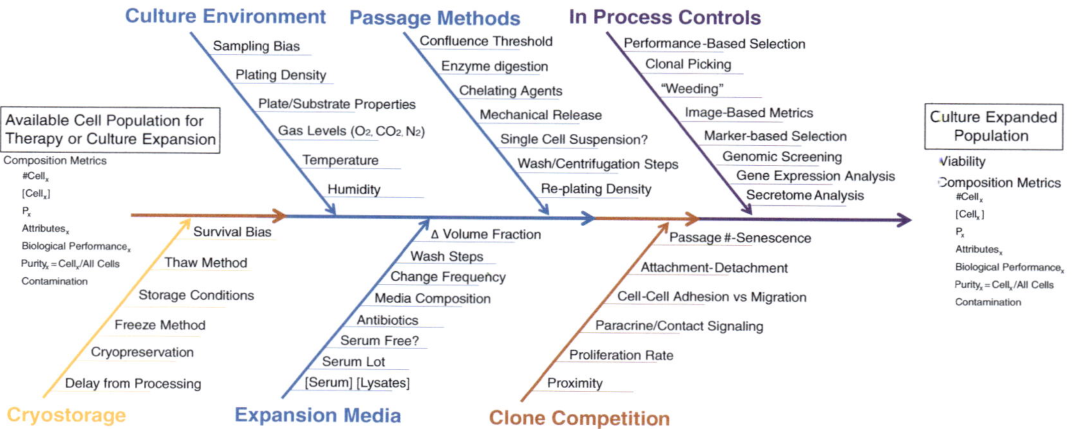

Fig. 1.3 In vitro cell expansion and cell product fabrication—Ishikawa diagram illustrating process steps and variables associated with in vitro cell expansion and the fabrication and release of a culture-expanded cell product

consensus around capture of key clinical variables, key cell product variables, outcome metrics, and time points of outcome assessment; (c) capture of utilization data and trends (patients being treated and approaches being used); (d) high-level assessment of which treatments are enduring and are being abandoned (an early warning system for unsuccessful approaches), and (e) a first-order estimation of the relative magnitude and duration of benefit that may be expected with specific approaches in specific settings. These data can sometimes identify and focus attention on best practices. However, most often, registries provide a sieve that enables the most effective approaches to rise to the top, where more rigorous analysis through prospective RCTs and CETs can be framed and the critical network of aligned clinical teams that are capable of executing those trials can be established.

- The authors of this book generally agree that at the present time, (a) cellular therapies have not been optimized, (b) the magnitude of improvement and probability of success with specific approaches has not been rigorously defined for more than a few indications, and (c) clear indications and contraindications have not been broadly established. In this context, cellular therapies for musculoskeletal conditions remain imperfectly developed and in many settings "unproven." At this time, patient care and professional ethics suggest that all clinical use of cellular injection therapies should be performed as part of RCTs, CETs, or PCRs.

- **Cell Sourcing: Point of Care Processing** (Fig. 1.2)—Cell sourcing provides a freshly isolated cell population, from a specific individual and tissue type and location, with a measurable <u>yield</u> (number of cells harvested) and <u>composition</u> (i.e., concentration, prevalence, and viability of various cell types), each with definable biological potential. Cell sourcing is an essential starting point for any cell-based or cell-derived injectable product. In many current approaches, the output of cell sourcing is, in fact, the actual <u>cellular product</u> [e.g., bone marrow aspirate concentrate (BMAC), platelet-rich plasma (PRP), stromal vascular fraction (SVF) of processed fat]. However, the <u>repeatability</u> and <u>reproducibility</u> of cell sourcing is often compromised by wide variability in <u>clinical patient variables</u>, donor tissue health (or disease), and the <u>harvest efficiency</u> (a function of harvest methods and processing methods). These, in turn, impose variation in total cell <u>yield</u> and composition, even when concerted effort is made to standardize other process steps [3–15]. As a result, when freshly isolated cells are used, the dose (number of cells administered for therapy), cell composition, as well as intrinsic biological potential of the cells may be different in every episode of therapy.

- Moreover, the heterogeneous mixture of cells in given injection may contain effector cells (i.e., cells that directly benefit the desired outcome, either directly by generating new tissue or secretory products or indirectly by synergy with other effector cells); inhibitor cells (i.e., cells that actively block, compete with, or inhibit the desired outcome); and contaminating cells (i.e., cells that are passive "bystanders," playing no active role in outcome, but which may still adversely impact outcome by (a) competing with effector cells for space or local nutrients or (b) by degenerating and creating secondary debris and inflammation) [16]. Extending this paradigm, it can be assumed that there is some definable optimal range of dose, concentration, and composition of effector cells and delivery strategy (e.g., disease state, site preparation, delivery method, carrier medium) for any given clinical problem. Furthermore, it can be assumed that minimizing inhibitor cells and other contaminating cells will further optimize the performance of any population of effector cells that is implanted. However, at present, the optimal dose, composition, and delivery method are not known for any clinical setting. Each of the subsequent chapters explores and defines the state of current progress.

- While it is clear that current cell sourcing methods cannot yet be relied upon to provide a cell population that is repeatable and reproducible, nor a dose or composition that is optimized for a specific clinical application, there is tremendous room for improvement. Moreover, a perfect dose and composition of cells is not a necessary requirement to provide safe and effective clinical benefits. Current use is not based on the premise that any current cellular therapy has been optimized. Decisions for current use are based on expediency (i.e., default to the most practical and available current option) in service to patients seeking help today.

- This can be justified, even if optimization is not yet achieved, provided clinical assessment of specific cellular products (i.e., specific harvest and processing protocols) (a) established clinical therapies are inadequate and do not provide satisfactory clinical benefits established that clinical therapies have not yet achieved satisfactory benefits, (b) appropriate legal and regional guidelines are respected and followed, (c) rigorous clinical judgment is used to assess and limit risks, (d) patients are informed in clear and transparent terms, (e) the therapy utilized is clearly documented, and (f) the outcome of clinical use is evaluated and reported with appropriate institutional (e.g., Institutional Review Boards) and regulatory oversight, either as part of an internal quality control process or through a prospective process of clinical assessment in RCTs, CETs, or PCRs with the intent of providing generalizable recommendations. Each of the chapters in this text examines and reports on the progress that has been achieved to date through responsible clinical development and assessment while also pointing out gaps in evidence, contraindications for current use, and specific opportunities for future improvement.

- It is important to note that, even in a point of care "same day" setting, cell processing to improve the concentration and prevalence of effector cells is not limited to the choice of donor tissue and harvest methods. As Fig. 1.2 illustrates, a variety of options are available for improving cell isolation and yield. The most common current method is density separation (DS) (e.g., use of a centrifuge to separate high density cells from low density cells). DS is most often used to deplete red blood cells (RBCs), the most common and abundant contaminating cell population in most settings, and for separation and concentration of nucleated cells and platelets. Other common methods for cell selection include selective retention (i.e., use of the native tendency of some effector cells to attach preferentially to some surfaces), magnetic separation (i.e., selective capture or removal of cells based on labeling with a ferromagnetic tag allowing cells to be captured or diverted in a strong magnetic field), or fluorescence-activated cell sorting (FACS) (i.e., automated cell selection based on immunofluorescent labeling using one or more surface markers).

- The regulatory environment around each of these strategies will vary from country to country. In the USA, DS processing is currently the only one of these that is compliant with the FDA definition of "minimal manipulation" required for the so called 361 exemption [17].

- **Cell Expansion/Cell Fabrication In Vitro** (Fig. 1.3)—In vitro cell expansion can be used to provide a population of cells that is more numerous and more uniform than the mixed cell populations that are derived from native tissues. In theory, it is possible to utilize expansion as a means of enriching for "effector" cells and depletion of inhibitor and contaminating cells.

- However, if culture expansion is to be effective, the process of in vitro expansion must, (a) preserve the desirable intrinsic biological potential of effector cells, if not enhance this potential in culture, and (b) insure that effector cells outcompete other inhibitor cells or contaminating cell populations during the expansion process. The experience of the past 30 years has shown that these two critical assumptions are not necessarily true.

- The history of culture-expanded cell populations for musculoskeletal applications and the nomenclature that has been used are important to understand and define. The field of culture-expanded bone marrow-derived cells was born out of hematopoietic research [18]. Bone marrow stromal cells were originally identified as adherent fibroblastic cells from bone marrow that were essential to support hematopoietic differentiation in vitro. Friedenstein and Owen, however, demonstrated the capacity of marrow-derived stromal cells to express markers of bone, adipose, or cartilage differentiation in addition to a fibroblastic phenotype (referred to as "trilineage potential") [19, 20]. In 1991, Caplan introduced the concept of a mesenchymal stem cell or

"MSC," which was initially defined as a purified, homogeneous culture-expanded cell population that retained trilineage potential, as a possible product for clinical therapy [21]. It was assumed that culture-expanded MSCs would be capable of survival after transplant and generation of new tissues; however, this assumption proved to be generally incorrect and rarely achieved. Subsequently, Caplan revised his definition of MSC to "medicinal signaling cell," arguing that the biological effects sometimes documented following injection of culture-expanded MSCs must be due to the secretome of these cells, since they did not need to persist to have an effect [22]. This assertion continues to be based on a premise of reproducibility and repeatability that has not been demonstrated, and does not necessarily capture or acknowledge the large variation that is present from batch to batch when MSC populations are expanded from different patients, tissues, or even the same heterogeneous sample of starting cells.

- After a period in which the term "MSC" was requently applied relatively indiscriminately in the literature to any culture-expanded fibroblastic cell, regardless of phenotype or reproducible function, the ISCT proposed redefining the term MSC as "mesenchymal stromal cell" (not "stem" cell) [23]. Moreover, the minimal criteria were defined as necessary before a designation of multipotent MSCs could be used. Specifically, the ISCT limited the use of "mesenchymal stromal cell" to culture-expanded plastic adherent cells with in vitro or in vivo demonstration of "trilineage" differentiation potential (i.e. differentiation into progeny expressing adipocyte, chondrocyte, and osteoblast features) AND cells that expressed CD73, CD90, and CD105 but were negative for hematopoietic and endothelial markers. This definition has subsequently been further refined by ISCT to include other markers [24].

- Unfortunately, however, cells meeting these ISCT criteria at MSCs can easily be obtained from various adult or embryonal tissues. They have been frequently used in clinical trials. Yet, repeatable and reproducible fabrication of

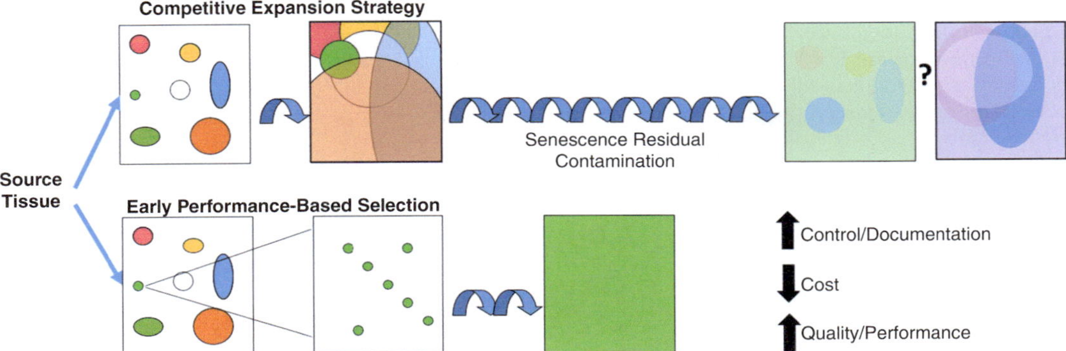

Fig. 1.4 Diagram comparing competitive expansion vs performance-based selection. Competitive expansion (CE) of polyclonal populations inevitably forces clones to compete with one another for space and resources. The outcome of this method is the persistence of the clones with the fastest and longest proliferative potential in vitro. However, outcome will vary depending on time and passage number in culture and small variations in starting conditions. Performance-based selection is defined by the purposeful choice of a preferred clone or clone type, rather than the passive acceptance of the "winner" of CE, but depends upon knowledge of clonal attributes that are associated with preferred future performance. Copyright Cleveland Clinic

MSC populations have been difficult to achieve and to document, based on both in vitro cellular performance and clinical performance criteria.

- Some of this challenge is illustrated in the upper half of Fig. 1.4. The starting population of connective tissue progenitor (CTP) cells in native tissues is inevitably heterogeneous [6, 25, 26]. When this diverse mixture of cells is placed into culture, the attachment and proliferation of individual cells inevitably generates a diversity of clonal populations that are placed into a setting of competition for space and soluble nutrients. The growth, migration, mixing, and interactions in this "competitive expansion (CE)" environment are inevitably associated with early winners and losers, based on proliferation rate. However, as time progresses in culture and subsequent passage, cell senescence and the secretion of paracrine cytokines may favor the evolution of the population and the rise of additional clones over others. Moreover, due to stochastic variation in the initial separation or co-location of various clones, the outcome of CE may be different even when the same sample and a similar diversity of clones is plated in different culture wells.
- This unavoidable variation and evolution in the outcome of CE processing influences the features of MSCs, both batch to batch and over time. As a result, the experimental results and clinical efficacy is often highly variable between batches and lots of product. This recognition of heterogeneity, despite a homogenous appearance under the microscope, and the fact that MSC cultures undergo massive clonal selection over time. This clonal competition and evolution over serial passages during expansion has recently been quantiatively demonstrated using multicolor fluorescence labeling and deep sequencing by Selich et al. as shown in Fig. 1.5 (reprinted from REF) [27].
- An alternative to CE that is now emerging is the use of performance-based selection (PBS) (lower section of Fig. 1.4), in which early single founding cell attributes or clonal colony performance (prior to overgrowth and mixing between clones) may be used to identify preferred "effector" clones or "inhibitor" or "contaminating" clones, which can then be selectively isolated or depleted using automated methods [28, 29].

1.2.1 Choice between Rapid Point of Care Processing and Cell Expansion

The intrinsic dilemma between the use of autologous cellular products derived from rapid point

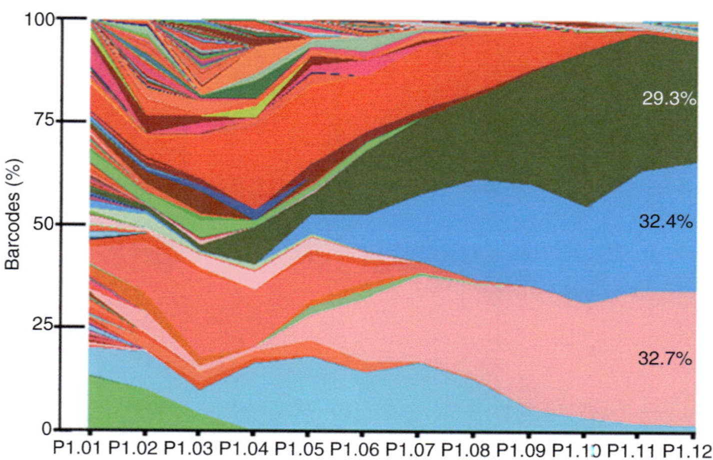

Fig. 1.5 Time-dependent variation in clone dominance during in vitro expansion of polyclonal progenitor-derived cells—This figure, reprinted from Selich et al. [27], illustrates the rise and fall of various clones in prevalence within a culture-expanded population of "MSCs" over time. Each clone is represented as a different color. (Reprinted with permission from "Selich et al from Stem Cell and Translational Medicine, 2016;5:591–601, 2016")

of care (POC) cell processing vs highly processed cellular products that are derived by in vitro cell expansion and modification is worth highlighting here. Neither page space nor clinical data permit rigorous, data-driven comparisons of these approaches. However, a few key and recurring nodes of tension in this comparison are worth pointing out, as they recur throughout discussions in this text and in future development and regulation of cellular therapies. Both strategic options have serious drawbacks.

The inevitable variation among patients and samples has been cited as a reason to discount the use of freshly isolated cells, because the composition of the materials injected cannot be completely controlled or duplicated from one patient to the next, even if the anatomic site, harvest method, and all other downstream process steps and CPPs are controlled. This is a common reflex from a purely traditional regulatory perspective of pharmaceuticals. However, this pharmaceutical mindset may be misguided when it comes to cell populations. Minimally processed autogenous cells have a preferential safety profile (immunological reaction, infection, oncogenic risk) and often greater intrinsic biological potential over more aggressively processed and expanded cell populations. Even if the number of cells and some attributes of processed and expanded allograft cells can be better controlled,

patient to patient variation in immunological responses and cell-cell interactions represent a substantial barrier to allograft cell performance, far beyond the recent recognition of the need for "personnalization" in assessment of responses to otherwise pure and truely homogeneous pharmaceuticals. This suggests that the burden of purity, absolute homogeneity, and precise compositional reproducibility that is applied to pharmaceuticals may be inappropriate, and impose expectations that are incompatible with biological realities of patient to patient variation in genetics, tissue source health, and cell sourcing options. Rather than imposing pharmaceutical chemical standards of purity and quantitative compositional reproducibility, an alternative but no less rigorous focus on reproducible safety and relative clinical impact that accepts unavoidable individual variation in cell source composition and the absolute magnitude of intrisic biological potential of individual autogenous cellular therapy may be required to best serve patients. The potential benefits of a less pharmaceutical approach that would enable alternative metrics for compositional documentation and optimization of autogenous cellular therapies are a potential acceleration of active investigation of autogenous cellular therapies as well as a substantial improvement in the safety, efficacy, cost, and risk associated with current approaches to in vitro expansion strategies.

1.3 Conceptual Paradigm of Stem and Progenitor Biology in the Context of Regenerative Medicine: Cell Composition and Cellular Kinetics

Cells are the foundation of life and medicine. Every new life, every stage of development, every change in health status, and every response to injury or therapy are mediated by the function of cells. As a result, cells are the target of virtually every medical therapy intervention. This includes drugs as well as biological implants and endoprosthetics. Over 200 different cell types are recognized. Therefore, while the concentration and dose of a drug therapy may be defined, the cell composition of the target organ or organ system of each patient may vary widely. The cell composition of any given tissue can be defined by the <u>concentration</u> and <u>prevalence</u> of each cell type. The <u>cell composition</u> of a tissue is directly related to its health and function and will inevitably be changed in the setting of disease.

All cells are transient. They have an origin, as well as a fate. Growth, repair, and remodeling all require the generation of new cells. However, they also require the systematic loss or removal of cells (such as surface shedding from the skin and gut) or apoptosis (in situ cell death). Cell migration also plays a huge role, both in arrival (e.g., inflammatory cells) and departure (e.g., maturing hematopoietic cells leaving bone marrow).

In an adult, a state of health (<u>homeostasis</u>) is defined by the balance of the rate at which cells are added or removed from a given tissue, and the residence time or life span of a given cell or cell type in that tissue [30]. Disease is inevitably associated with (if not caused by) an imbalance in the addition or loss of cells of various types in an involved tissue, and an associated change in the cell composition. Moreover, the success of many medical therapies can be defined as the restoration of normal <u>cellular kinetics</u> (i.e., the balance of proliferation, migration, differentiation, and survival of diverse cell types within a tissue).

In this context, life and health can be viewed as a "river" of cells of every type that flow through each tissue, each at their own rate. The cellular kinetics of any individual cell type is defined on one side by the rate at which new cells are added (e.g., local proliferation and differentiation or migration from another site) and on the other side by the rate at which they are removed (migration or death) and the mean life span for a given cell type in that tissue. Mean life span for cells can vary from hours (in the case of macrophages responding to infection) to decades (in the case of neurons, chondrocytes, or osteocytes). The phenomenon of cellular senescence and the accumulation of scenescent cells, which fail to undergo apoptosis at the end of their functional life, is increasingly recognized as a feature of aging and disease. Senescent cells can further shifts the balance of concentration and prevelence, but also open a new potential therapeutic target cell population, which itself will be highly heterogeneous.

The mathematics and mechanisms of cellular kinetics are often applied to studies of neoplasm or infection. However, these same principles are equally relevant to tissue engineering applications and particularly to cellular therapies. These concepts have been effectively illustrated in the context of bone formation and remodeling and the balance of function of osteoblasts and osteoclasts, and their separate <u>stem/progenitor cell systems</u> that are coresident in bone and bone marrow tissue [30, 31].

1.3.1 Stem and Progenitor Cell Systems and Niches

Stem cells and progenitors are present in virtually all normal tissues [30, 32–35]. The concept of a stem cell and the difference between stem cells and other downstream progenitor cells are particularly important to clarify in the discussion of cellular therapies. For the purposes of this chapter and this book, we bypass the complexity of embryological growth and development, or neoplasm, and focus instead on the stem and progenitor cell systems present in native adult tissues, particularly connective tissues.

Throughout life, the process of generating and sustaining each of the 200+ adult cell types

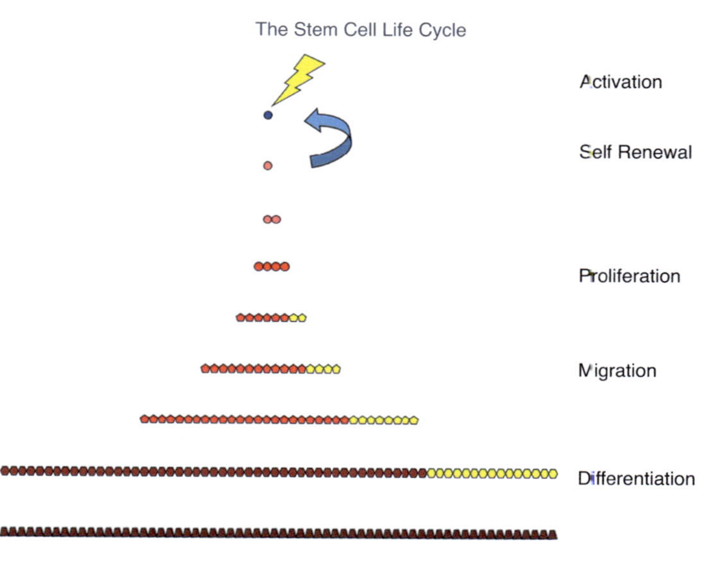

Fig. 1.6 Stem cell and progenitor cell expansion cascade. This simplified model illustrates the attributes of the stem cell (self-renewal without differentiation). It also illustrates the attributes of the downstream proliferating progenitor cell population. Progenitors inevitably vastly outnumber true stem cells (i.e. they have much greater prevelance). They also demonstrate much greater diversity (heterogeneity) through the capacity to differentiate through a diversity of stages and potential differentiation paths and outcomes

begins with a "stem cell." It is estimated that there are at least 30 separate stem cell systems associated with the maintenance of health and homeostasis in adults [36]. The most prominent and best characterized systems include hematopoietic cells in the bone marrow, skin, gut, vascular endothelium, and nervous system. In the adult, each of these systems is biologically separate and distinct, and each system may generate a plurality of mature cell types.

For the purposes of this book, and our focus on musculoskeletal therapy, the most important systems to consider are those that give rise to cells with connective tissue phenotypes, other than the blood, specifically the bone, cartilage, tendon, ligament, fat, synovium, or meniscus/labrum. We refer to systems that generate progeny with the potential to differentiate into one or more connective tissue phenotype as tissue-specific connective tissue stem and progenitor cell systems (CTP systems) and the cells themselves as CTPs (discussed in more detail below).

Stem cells are critically important in biological systems, and the term "stem cell" has come into public awareness through a series of impactful events over three decades, including the success of bone marrow transplantation (hematopoietic stem cell (HSC) transplantation), the discovery and debate over embryonic stem cells (ES cells), and the Nobel Prize in 2011 for the generation of induced pluripotent stem cells (iPSCs). Each represented profound discoveries and advances with transformational effects and potential. However, in recent years, the term "stem cell" has been degraded through widespread misuse as a marketing term, often with unsupported and even "magical" implications that misrepresented the nature and biology of a true stem cell. Misuse of the term "stem cell" has corrupted effective communication between clinicians, scientists, and particularly patients [37]. We have paid particular attention in this textbook to avoid this trap and limit the use of the term "stem cell" to its biological definition and avoid its use as a gratuitous, colloquial, or marketing term.

The core concept of a stem cell system and important differences between stem cells and progenitor cells are illustrated in Fig. 1.6. This illustrates a hypothetical stem cell (green) that is activated to proliferate and give rise to a group of progeny (progenitors) (pink) that progress through a series of cell divisions, and a series of progressive differentiation states, to ultimately provide a population of mature functional cells (amber), all derived from a single activation of a single stem cell, which ultimately die (black). The essential defining feature of a true adult stem cell is that after cell division, one of the two cells retains the properties and remains in the niche of the cell that divided, and preserves the functional potential for further identical cell divisions that may spawn future generations of proliferating progenitors. This unique "stem cell" function and process is referred to as asymmetrical cell division and "self renewal".

Because many adult stem cells are capable of generating progeny that can follow more than one path to terminal differentiation, each stage of proliferation is also associated with a probability that progeny will go down one path or another. This probability is influenced by a combination of local signaling molecules, chemical conditions (e.g., pH and O_2 concentration), and mechanical environment.

In this model, the total cells (N_c) generated following a given stem cell activation event are a function of the number of cell divisions (n) before terminal differentiation, and the cumulative probability (p) of differentiation along specific pathways, where ($N_c = 2^n \times p$) [30].

This diagram is, of course, highly simplified. Even so, it is applicable as a first approximation to virtually every stem cell system in adult tissues and illustrates five key features of stem cell systems that we have laid out in more detail in a prior publication [30]:

1. Stem cells, almost by definition, divide by "asymmetrical" cell division, producing two daughter cells: one daughter cell that goes on to expand and generate progeny that ultimately differentiates into mature cell phenotypes and a second cell that retains the features of the original stem cell, effectively "self-renewing" the original stem cell, leaving it available to divide again in the future. Were this "self-renewal" not to occur, each activation of a stem cell would result in the loss of a stem cell, rapidly depleting the sparse stem cell pool.

2. The number of mature cells formed as a result of each stem cell activation event is determined by the number of cycles of symmetrical cell division during proliferation prior to differentiation, and the probability that cell products of each division cycle will become committed to a specific end cell type (e.g., osteoblast or adipocyte). Note that, in contrast to the diagram in Fig. 1.6, in most systems 10–20 cell divisions occur prior to differentiation, so rather than 64 cells, as illustrated here, each stem cell activation may generate between 1000 and one million new cells. Shifting the kinetics of cell division or the probability of differentiation events can have rapid and profound impact on cell composition within a given tissue.

3. True stem cells will divide rarely, and much more slowly than the downstream progenitor population. Any single stem cell division is magnified immensely by the expansion of downstream progenitors. In adult tissues, under normal conditions a stem cell will be in G_0 (not dividing). The work of expanding cell numbers is done by progenitor cells. This protects the upstream stem cell population from accumulating chromosomal abnormalities damage through transcription errors or senescence.

4. The prevalence of true stem cells in any tissue is always vastly smaller than the number of progenitors (i.e., cells that are active in or capable of proliferation and differentiation into a mature phenotype).

5. The number of mature cells will always be vastly great than the number of progenitors because the life span of cells in the mature cell state is vastly longer (3 days to 30+ years) than the life span of cells at any stage of upstream proliferation (generally 1–5 days); each stem cell in a given tissue may support many layers of mature cells. As a result, at steady state in healthy tissue, the ratio of downstream progenitors to true stem cells will generally be in the range 1000:1 to 100,000:1).

The stem cells that are at the apex of these stem cell systems are almost always located in a particular anatomic site or "stem cell niche" within a tissue. These sites are often characterized by either a unique population of local support cells that provide a specific extracellular matrix or a secretory environment that spatially/directionally orients, supports, and protects long-term stem cell function and asymmetrical cell division. Detailed characterization has been accomplished for many stem cell niches (e.g., hematopoietic stem cells), but niches in connective tissue have been more elusive [36, 38]. Perivascular cells (e.g., pericytes or other cells in the perivascular adventicia) are at least one proposed niche for connective tissue stem or progenitor cells [39]. However, niches are also proposed in the trabecular bone surface [40], synovium [41], and the superficial zone of cartilage [15].

1.3.2 Tissue-Specific Kinetics and Stem/Progenitor Populations

In many stem cell systems, including the connective tissue stem and progenitor cell system, the progeny arising from a given stem cell activation may be induced to differentiate along a number of pathways (e.g., bone, cartilage, fat, fibrous tissue) as part of normal tissue generation and turnover. This may involve a series of commitment events or stages of differentiation before commitment to a particular differentiation fate well beyond the simple diagram in Fig. 1.6 [30, 42]. Moreover, in many systems, progenitors themselves may become arrested for periods of time at a particular stage as "transit expanding populations" or even migrate from one location to another (e.g., through systemic circulation) [30, 43–45] before resuming the process of proliferation. This has the effect of creating multiple sets of progenitor cell types and states that may be induced to "home" to a specific tissue location, or to be activated and proliferate in response to particular therapeutic stimuli.

It is self-evident that the kinetics and functional hierarchy of the stem or progenitor cell populations in any given tissue (marrow, fat, muscle, bone, cartilage, synovium, tendon) will be tissue-type specific and reflect the composition and dynamic state and health of the tissue of origin. Some literature has pointed to the fact that multilineage potential (bone, cartilage, fat, fibrous tissue) can be demonstrated among proliferating progenitors that are isolated in vitro from a variety of connective tissues (marrow, fat, periosteum) [41, 46–48]. However, increasing evidence has accumulated that the stem and progenitor populations from different tissues are distinct with respect to local tissue niches, biological attributes, and performance [46, 48–50]. Biological needs, developmental history, and current state of each tissue type will define different stem cell niches, activation signals, migration pathways, and differentiation options for the cells in that tissue. Populations of stem and progenitor cells isolated from fat tissue will inevitably be different in concentration, prevalence, and biological attributes than cells from marrow [7]. It is therefore most accurate to think and communicate in terms of "tissue-specific" connective tissue stem/progenitor populations [51].

1.3.3 Tissue-Derived Cell Populations: Heterogeneous Mixes of Cells and Biological Potential

As illustrated in Fig. 1.1, native tissues will inevitably provide the starting materials on which any cell therapy may be based. Based on the discussion above, it is also self-evident that any sample of tissue will contain a highly heterogeneous mixture of cells. This includes connective tissues, which are sometimes considered to be less complex and more homogeneous than other tissues. The vast majority of cells found in connective tissues will be mature cells which have long functional life span (residence time) in tissue from the time of maturation to the time when they are removed due to senescence, apoptosis, remodeling activity, or injury. The mean survival time of an osteocyte, for example, can be estimated to be 15–25 years. This is even longer for cartilage tissue. A much smaller fraction of cells will represent progenitor cells which are either in the process of expansion through proliferation or represent quiescent progenitors that can be stimulated to reenter the cell cycle.

Methods for liberation and isolation of cells from connective tissues are varied, but most are designed to generate suspensions of viable single cells. These methods also seek to preserve the intrinsic attributes and biological potential of any progenitor cells that are isolated, but they may not be perfect, and some selection bias may be expected due to loss of some relevant cells.

If cells are placed into culture, under appropriate conditions, any cell that has retained proliferative capacity may grow and generate observable progeny. If the clones derived from individual proliferating cells are sufficiently separated, the progeny of individual cells can be identified as discrete groups of cells or "colonies," where the originators of each colony are defined as "colony-forming units" (CFUs). Moreover, the biological attributes and performance of the cells in each colony provide insight into the intrinsic biological potential of each individual colony founding CFU.

When thinking about the number of CFUs and the diversity of performance among the progeny of each CFU, it is important to remember the complexity and kinetics present in the source tissues. Cells with proliferative potential may come from a diversity of stem cell systems (connective tissue, vascular, neural), even hematopoietic cells, in the case of cancellous bone or bone marrow. Colony founding CFUs will be drawn from every stage of maturation between the stem cell and mature cell phenotype and every possible variation in levels of commitment along potential paths of differentiation. True stem cells will give rise to a very small number of colonies. Moreover, because the extraction process has likely removed the few stem cells that are isolated from their native stem cell niche, the properties of the true stem cell will be unlikely to be preserved. Even so, the progeny of true stem cells may still exhibit and be identifiable based on unique in vitro performance (proliferation, migration, morphology, markers, etc.).

It is also important to recognize that culture conditions strongly influence which cells survive and which cells proliferate in vitro. Therefore, while any given subpopulation of CFUs ("X") will have a true prevalence (tP_{CFUx}), the number of CFU-derived colonies that is observed (oP_{CFUx}) will almost always be lower depending on the efficiency with which colony formation is activated in a given setting (colony forming efficiency (CFE)), where $oP_{CFUx} = tP_{CFUx} \times CFE$, and CFE is always less than 1.0 [5, 52–54].

CFE may also be influenced by interactions between CFUs and their progeny, as well as by other non-proliferating non-progenitors in the cultures. In fact, initially, the secretory products of non-progenitors will dwarf the secretory product or RNA expression profile of any stem or progenitor cell population in the sample. This interaction between the culture-expanded progeny of colony founding progenitor cells as well as the vastly more numerous non-progenitors in tissue-derived cell populations results in a high probability of stochastic variation in the outcome of in vitro isolation and expansion procedures that rely on mixed polyclonal cell cultures.

Recognizing this milieu in vitro also reminds us that when using freshly isolated cells for therapies (e.g., BMAC), it is entirely inappropriate to refer to these therapies as "stem cells" or "stem cell therapy." In fact, if stem cells are present in these preparations, they are the least common cell in the mixture and have very likely been altered from their native state during harvest and isolation.

This complexity does not however preclude a potential benefit from the harvest, processing, and transplantation of the mix cell populations that are derived from native tissues. Rare stem and progenitor cell populations may have the ability to preferentially survive and even to proliferate after transplantation.

Moreover, that fact that this mixed population will inevitably contain cells that may have a positive effect ("effector cells") and cells that may have a negative effect ("inhibitor cells") inspires and catalyzes optimism that further optimization is possible through the application of scientific and engineering principles. This quest defines the purpose and mission of this book.

1.3.4 Tissue-Specific Connective Tissue Progenitors

Among all of the stem and progenitors in native tissues, it is only a subset of those cells that are capable of proliferating and generating progeny capable of forming one or more connective tissue (bone, cartilage, tendon, ligament, meniscus, fat, scar, stroma). This heterogeneous population of stem and progenitors represents a hierarchy of stem and progenitor cells that are collectively referred to as "connective tissue progenitors" or

CTPs [16, 30, 31]. CTP populations are by nature tissue specific and heterogeneous [5, 6, 14, 15, 25, 26, 48, 55, 56]. While other populations of progenitors may be important in musculoskeletal applications (vascular endothelial progenitors, hematopoietic-derived immune cells), the measurement and characterization of CTP population represents the main focus of the remainder of this chapter and subsequent chapters in this text.

1.4 Engineering Principles

Engineering is the discipline of applying the basic knowledge and tools of biology, chemistry, physics, and mathematics to solve practical problems [16]. In the context of cellular therapies, the engineering "mission" can be defined as:

> *Provide clinical benefit by delivery of effective cell populations to the right location in the right way at the right time in the right patient for the right reason (disease state), safely and reproducibly.*

The potential benefits of cellular therapy could be to repair, replace, augment, or regenerate new tissue through long-term cell survival and direct contribution. However, benefits may also involve short-term survival and indirect contribution through modulation of the signaling environment by secretory products or by cell-cell interactions [57, 58].

This is a complex challenge. The pathophysiology of each disease state and each tissue is unique. As a result, the underlying deficits in cell composition, cellular function, and signaling environment will differ. For each clinical setting, the preferred cell(s), cell composition, dose, mode of delivery, and timing of delivery will vary.

The path to successful therapy for any clinical setting must start with a rational clinical mechanism, and a strategy to systematically optimize the identity, composition, and dose (cell number) (i.e., the "product" to be used) as well as the method, location, and timing of delivery.

Systematic and quantitative application of engineering concepts begin and end with measurement and documentation. As outlined in Figs. 1.1–1.3, measurement and control is

required in each domain of clinical cell therapy: clinical assessment, cell harvest and isolation, and (when applicable) in vitro expansion and processing.

This universal theme of disease-specific therapy and application of rigorous measurement and engineering principles is an underlying principle of cellular engineering that is revisited throughout this book. The following provides an overview of the paradigm and nomenclature used for measurement and engineering optimization throughout this text.

1.4.1 Clinical Assessment Measurement

Clinical assessment includes two main domains of assessment—clinical setting and clinical outcomes.

Clinical Setting: Diagnosis, Disease State, and Clinical Context—A specific <u>diagnosis</u> is essential and often recorded in the form of an ICD or WHO code. The stage or <u>severity of disease</u> is equally important. This is most often documented using established disease-specific scoring systems based on physical findings or imaging (e.g., Kellgren-Lawrence scores for osteoarthritis or Ficat stage for osteonecrosis). Other disease-specific clinical variables may also be relevant (e.g., time since onset, laxity, strength, laboratory measures). Finally, the clinical context of care must be defined, including standardized measures of <u>demographics</u> (age, gender, work and social status, education, height, weight, BMI) and disease-defining laboratory measures, as well as the burden of <u>comorbidities</u> (e.g., Charlson, Elixhauser).

<u>Clinical Outcome</u>—Clinical outcome is the most basic objective assessment relevant to the efficacy and value of a given therapy. The <u>efficacy</u> (E) can be defined as the magnitude of improvement (or decline) by documentation of difference between a starting state of health (H1) and an ending state (H2), where E = H2-H1. The <u>value</u> (V) of a given therapy is a function of efficacy per unit cost, where $V = E/C$, and C represents the total cost of the therapy. Cost can be measured from various perspectives,

however: cost to the patient (money, time, and risk), cost to the insurer (immediate cash and probability of future liability), and cost to society (economic impacts of disability).

We focus here on clinical efficacy, which itself can and should be measured in several domains: general health, function, pain, and metrics of tissue health or function.

General Health and Function—These measures are generally provided by using standardized patient-reported outcome metrics of <u>pain</u> (e.g., visual analog scale (VAS)), <u>general health</u> (e.g., SF-36, VR-12, EQ-5D, PHQ-9), as well as disease-specific measures of <u>function</u> (e.g., HOOS, KOOS, WOMAC, DASH).

Target Tissue Health and Function—The status of disease at the local tissue level will have profound effects on any biological therapy. The tissue where cells will be delivered is directly analogous to the "soil, sun, and water" where the "seeds" (cells) will be planted. The biological starting state of a tissue can be directly measured by a combination of anatomy and dynamic physiological parameters: <u>structure</u>, <u>composition</u> (cells and/or extracellular matrix (ECM)), <u>kinetics</u> (metabolic rate, oxygen tension, cell division, cell migration, gene expression), and/or <u>signaling environment</u> (local secretome, including growth factors, cytokines, and ECM).

Similarly, the biological efficacy of therapy can also be measured by the change in structure, composition, kinetics, or environment in the target tissue—when before and after therapy assessment can be performed.

In some settings, the starting state of a tissue may be routinely measured in the course of diagnostic biopsy procedures, where histological and histomorphometric data can be captured. However, with rare exceptions, the starting and ending states of target tissue at a histological level are only available in preclinical animal studies.

Structural status and changes are often measured based on alteration of anatomy (tissue loss or formation) using clinical imaging modalities (e.g., CT, MRI). Some compositional or kinetic tissue attributes may also be measured before and after treatment through noninvasive imaging techniques (e.g., DGEMRIC, T1Rho, PET) [59–61].

Note that the tissue structure, composition, and kinetics may also be relevant to the selection of tissue donor sites for either autogenous or allograft cells. Some starting states are likely to be incompatible with successful cell transplantation. However, to date, little available data addresses this question.

1.4.2 Defining the Cellular Product and Process

A fundamental principle in providing clinical therapy is to know what therapy was provided:

- What is the composition of the cell product?
- What cells are being isolated?
- What is the number, concentration, and prevalence of the cells being used?
- What is the identity of the effector cells being used?
- What are the biological attributes that are used to identify the effector cells?
- What inhibitor cells or contaminating cells are present, and in what prevalence?
- What is the variation in composition, dose (cell number), or biological attributes from patient to patient or batch to batch?
- What methods are used to increase the prevalence of effector cells and reduce the prevalence of inhibiting or contaminating cells?

These questions have been surprisingly difficult to answer. This is particularly true for cellular therapies (products) that are generated through a process of cell harvest and point of care processing, as illustrated in Fig. 1.2. However, they can still be challenging for cellular products that are fabricated through culture expansion, as in Fig. 1.3.

Basic information about the number, concentration, composition, and biological attributes of the cells that are processed and delivered are often missing from clinical publications on injectable cellular therapies for musculoskeletal disease. One reason for this dearth of information is that much of the early phases of investigation involve autogenous cell populations derived from the blood, marrow, or fat, which were processed at the bedside. In this setting, concerns about biocompatibility and sterility were low. Regulatory barriers are low. The threshold for equipoise in potential patient benefit was low for many benign musculoskeletal conditions, particularly when considering care that may delay potential surgery and associated costs and risks. Moreover, the analytical tools for counting and characterizing cells were often not available. As reported in the following chapters of this book, the literature findings point the potential benefit of injection therapies but also highlight to the critical need for the field of cellular therapy in general, and musculoskeletal therapy in particular, to step up to a higher standard of documentation and reporting.

1.4.3   Cell Composition

<u>Cell composition</u> is defined by number, concentration, prevalence, and biological attributes of each cell population in a tissue or cell sample.

The overall composition of all cells in this mixture can be characterized by the number and <u>concentration</u> of each cell type (X), i.e., $N_{Cell\ X}$ and [Cell X]. The composition of the whole mixture of cells is the sum of the number and concentration of all cell types and each cell stage. Concentration can be measured in cells per ml for cell suspensions but can also be represented as cells per mg of tissue.

The most abundant cells in any tissue will always be mature nondividing cells. This will include a diversity of cell phenotypes. For example, bone marrow will include hematopoietic cells, vascular and perivascular cells, fat cells, fibrous stromal cells, and osteocytes embedded in bone fragments, and osteoblasts, osteoclasts, and lining cells on the surface of bone fragments. As in any healthy remodeling tissue, isolated cells will also include a small composition of the native tissue-specific stem cells that are involved in generating the cells that are needed to preserve these individual cell types and tissue compartments and replace cells that are lost to injury or senescence and apoptosis. This will include the full range of downstream progenitors on their developmental journey from a stem cell to a fully differentiated mature cell phenotype. As a result, tissue-derived

cells will include a small population of CTPs and their downstream progeny on their way to become bone, fat, or stromal cells [16, 31].

Each individual cell type "x" has a <u>prevalence (Px)</u> within the whole cell population, defined as: $P_x = N_{Cell\ X}\ /(N_{cells})$. The mean <u>prevalence</u> of colony founding CTPs in cells isolated from bone marrow and from subcutaneous fat is roughly $50/10^6$ cells and $2000/10^6$ cells, for example [6, 7, 56]. However, prevalence can also be represented as a percentage, particularly for more abundant contaminating cells (e.g., lymphocytes or monocytes).

The identification of individual cell types is a point or rapid ongoing development and is not yet standardized. Identification is generally determined by physical attributes (e.g., size, granularity, morphology, adherence, or surface markers), or functional performance attributes (e.g., colony formation, migration, secretory proteome, or the differentiation of downstream progeny). However, functional performance is the most clinically relevant of all of these parameters.

1.4.4   Analytical Method for Cell Composition Analysis

<u>Manual or automated differential cell counts</u> are routinely performed on fresh cell suspensions isolated from bone marrow or other tissues. Anticoagulation is necessary (e.g., heparin, EDTA, acid citrate dextrose). Manual counting is subject to large subjective variation between observers. Auto-analyzers (while standardized, and generaly parse cells based on size and light refraction) are generally calibrated for assay of cells from peripheral blood. As a result, measurement for other cell types must be specifically validated before autoanalyzer data can be trusted. Cells from bone marrow, contain a diversity of immature hematopoietic cells, endothelial cells, and cells from fat or other tissues. Each cell type requires a tissue-specific profiles to quantitatively capture the diversity of cell subtypes.

<u>Flow cytometry</u> can be used to resolve the prevalence of individual cell types by parsing cells into groups based on size, granularity, or the presence of one or more cell surface or

intracellular markers (e.g., CD34, CD146). A sufficient number of cells (e.g., 10^5) must be assayed, however, to allow estimation of populations of cells with a low prevalence, such as CTPs, even if a specific set of CTP-associated markers could be established.

Colony-forming unit assays (i.e., in vitro measurement of the number of cells that are capable of proliferation under specific in vitro growth conditions and characterization of the biological attributes of the progeny of the individual colony founding cells that make up each colony provides) are a powerful analytical tool. CFU assay is the only precise method for the assay of hematopoietic stem cells (HSCs) and downstream hematopoietic progenitors (e.g., BFU-E, CFU-GM, CFU-GEMM) in bone marrow. CFU assay is also the only reliable way to assay the prevalence and heterogeneity of CTPs derived from connective tissue. It is important to note, however, that all colony-forming assays are dependent on media conditions. Change in media and other culture conditions can result in large differences in the observed prevalence of colony formation by a given cell type "x." The efficiency of colony formation (CFE) is dependent on processing and media conditions and that the observed prevalence (oP_x) is always an approximation of the true tP_x, and $oP_x = tP_x \times CFE$. It is tempting to assume that CFE is equal to 1.0 (i.e. 100% conversion of possible colony forming cells into colonies. Howwever, as media conditions, oxygen tension and other variables are refined, incremental advances in CFE continue to be made, refining our analytical systems.

Single-cell RNA sequencing (scRNAseq) is a rapidly emerging tool for assay and analysis of the number of different cell types in a given tissue samples. However, scRNAseq is expensive ($2–4 K per sample) and does not yet provide resolution for rare cell populations (like stem and progenitor cells).

1.4.5 Sampling Bias

Quantitative comparison of cell composition and yield from different tissue sources is also dependent on efficient methods for cell extraction and isolation. Each tissue will require a different method for mechanical, chemical, and often enzymatic processing to optimize the yield, viability, and biological performance of the cells that are isolated. Each isolation method may impose a sampling bias that increases or decreases the probability that a cell of a given type will be preserved in the sample. For example, bone marrow-derived cells are generally isolated and preserved in a single cell suspension, just using an anticoagulant to prevent the formation of fibrin clots. However, BMA samples can also contain fragments of tissue and fat that are buoyant and float to the top of a sample. Any cells that are trapped in a floating layer of tissue may be lost unless processing is added to isolate and release those cells. Similarly, BMA samples inevitably contain some fragments of cancellous bone. This bone is dense and will sink to the bottom of a harvest reservoir and will be discarded in the red cell mass, unless processing is added to specifically capture these fragments. Fat- and bone-associated cells may be particularly relevant to some BMA applications as a significant portion of the CTP population may be resident in tissue fractions on or adjacent to bone or in perivascular tissue associated with intramedullary fat. These elements of BMA harvested materials are often neglected when considering the potential yield of bone marrow aspirate as a source tissue.

1.4.6 Defining and Documenting the Process

The composition of cells that are available from an individual patient will be critically dependent on the patient health status and the health and dynamic state of the tissue of origin. Those two factors will always be variables which have a profound effect on the quality of the starting material that is available at the beginning of a cellular therapy process. In the case of autogenous cellular therapy, the patients and their health are not a variable that can be controlled or standardized. This is a unique aspect of autogenous cellular therapy that is distinct from drug development and testing, where starting materials can be highly standardized.

However, many other aspects of the harvest, isolation, and processing of cells can be standardized. These standardizable process steps comprise most of the content of Figs. 1.2 and 1.3.

It is sometimes said in regulatory circles that "process is product." The assumption is that if each individual process step is defined and performed in a highly regulated and reproducible manner, then the outcome (composition) of any cell sample that is passed through this process will also be repeatable and reproducible.

Substantial effort is invested in defining both rigorous process steps and controls. Systematic comparison of outcome allows process parameters that influence outcome to be defined and controlled. A well-characterized process will include well-defined Critical Process Parmeters (CPPs) (e.g., g force, time, temperature, oxygen tension, media composition) that are particularly important to the quality, repeatability, and reproducibility of outcome. The individual chapters in this text frequently explore and report on the CPPs that have been identified for specific products and clinical applications.

Critical quality attributes (CQAs) are another part of process and defined as attributes that are necessary for confirmation of the quality and potency of an end product. CQAs are used as quality metrics as well as "release criteria" for individual batches of a fabricated product. In the case of cellular products, CQAs are specifically those parameters of composition or biological performance that predict efficacy.

As present, CQAs are generally not known for cellular therapy products. CQAs can be guessed at and used in early stages of production or clinical trial. However, ultimately, CQAs will need to be discovered, and refined for each clinical setting, by linking product composition to measurable clinical outcomes.

Each phase of cell harvest and processing in Figs. 1.2 and 1.3 can be defined by measurement of the effect of that processing step on the number, concentration, prevalence, or biological behavior of cells. Those effects can be measured in terms of defined process metrics:

Yield—The total number of cells isolated.

Yield Efficiency—The fraction of the cells that were present at the beginning of a process that remained at the end.

Enrichment—Increase in the relative prevalence of one cell type over another based on the selective retention or removal of one population over another.

Purification—Increase in prevalence of one cell type to the point of near homogeneous representation in the population.

Repeatability—The capacity to achieve the same result when a process is repeated on the same or similar sample at the same location (e.g., same lab or operating room team)

Reproducibility—The capacity to achieve the same result with a process is repeated on the same or similar sample at a different location (e.g., another lab or another operating room team).

Measurement and refinement of those CPPs that optimize yield, efficiency, enrichment, purity, repeatability, and reproducibility represent a next critical phase of clinical cell sourcing and processing and must be central in our ongoing development of cellular therapies.

Differences among various patients, tissue sources, harvest techniques, or processing methods are each readily identified and compared based on the relative differences in cell number (yield), concentration, and/or the prevalence (purity) of individual cell types of interest.

1.4.7 Defining the Efficacy of Processes and Process Steps

While CPPs and CQAs are the goal, the process of optimizing and refining processes to achieve the best possible outcome involves a series of measurable steps.

1.5 Conclusion

The clinical use of injection therapies and particularly the use of cells and cell-derived products in both nonsurgical and surgical settings have become a highly active area of interest and progress. However, the translation of this promise into safe and effective therapies demands advancement in three key domains of progress:

1. Clinical Assessment—Assessment of safety and efficacy in specific clinical settings.
2. Cell Sourcing—Clinical cell harvest and isolation of cell source materials.
3. Cell Expansion/Cell Fabrication—In vitro expansion and modification of cells for therapy or generation of cell-derived products.

Advancement of the field also demands clear, precise, and accurate understanding and use of unified *conceptual paradigm of stem and progenitor cell biology* and application of rigorous *engineering principles* of quantitative measurement, process analysis and reporting related to cell sourcing, expansion and cell-based fabrication, and particularly clinical assessment.

In addition to traditional physiological and biochemical methods, measurement and reporting will increasingly demand rigorous measure of the cell composition and *biological attributes* of Stem or Progenitor cell that are proposed for use in cellular therapies, defining those cell populations that are presumed to be "effector cells," "inhibitor cells," or "contaminating" cells in any preparation. The outcome of processes of sourcing and expansion (i.e., yield, efficiency, efficacy, purity, repeatability, reproducibility) must be based on quantitative changes in composition and attributes. Processes themselves must be defined and reported with specific CPPs. Clinical outcomes must be linked to measurable and relevant CQAs.

The strategic development and refinement of cell sourcing, in particular, will be dependent upon ongoing advances in our understanding of stem cell niche biology and the kinetics associated with the in vivo proliferation, migration, differentiation, and survival of stem and progenitor cell systems in health and disease states.

Finally, the use of *nomenclature* (*naming conventions*) must be consistent and adopt generalizable consensus-based standards. In particular, the casual, gratuitous, and reflexive use of the term "stem cell" in both scientific publication and clinical marketing needs to be abandoned.

We hope that this introductory chapter and the work that follows contribute directly to achieving these goals, for our patients and for the entire field of musculoskeletal sciences and medicine.

Take-Home Messages

- Translation of injection therapies into safe and effective therapies demands advancement in three key domains: clinical assessment, cell and material sourcing, cell expansion/cell fabrication.
- Advancement of the field demands unified conceptual paradigm of stem and progenitor cell biology and application of rigorous engineering principles of quantitative measurement, process analysis, and reporting.
- Measurement and reporting will increasingly demand rigorous measure of the cell composition and biological attributes of "effector cells," "inhibitor cells," or "contaminating" cells in any preparation.
- Cell sourcing, processing, and expansion steps must be defined quantitatively and compared with respect to yield, efficiency, efficacy, purity, repeatability, and reproducibility.
- Processes must be defined and reported with specific critical process parameters.
- Clinical outcomes must be linked to measurable and relevant critical quality attributes.
- Strategic targeting of cellular therapies will increasingly demand an understanding of the changes in stem cell systems and their kinetics, associated with health and disease states.
- Nomenclature (naming conventions) must be consistent and adopt generalizable consensus-based standards. In particular, the casual gratuitous use of the term "stem cell" in both scientific publication and clinical marketing needs to be abandoned.

Glossary of Terms

Asymmetric cell division A mitotic process that results in two daughter cells with different cellular fates. Is essential to the maintenance of stem cell populations through the process of "self-renewal"

Cell expansion The process in which cells are cultured under controlled conditions in vitro with the intention of increasing their number while preserving their biological potential

Cell fabrication The process in which cells are expanded and potentially modified in vitro to generate a cellular product. In vitro modification may include one or more of the following: In vitro adaptation or selection resulting from competition; exposure to agents that result in reprogramming or change of cell fate via differentiation; or genetic modification that results in purposeful control or optimization of current biological performance or future biological potential

Cell product (or cellular product) A defined composition of viable cells which is intended for use in treatment of a specific condition or as a source material for research or for cell fabrication. Cell products are defined by cell composition, chemical composition, and concentration

Cell sourcing The purposeful choice of a specific donor, tissue, anatomic site, or cell type to provide the source material from which a cellular or cell-based product is fabricated

Cell-derived product A product whose active ingredients are the direct result of cellular metabolic function (e.g., secreted factors, exosomes, etc.). A cell-derived product does not contain viable cells as the active ingredient

Cellular kinetics The mathematical description or approximation describing the number of various cells in a tissue and the rate of flow of cells from one cell state or one cellular compartment to another. Dynamic kinetics may be used to describe the number of cells in each cell compartment or state, and the rate or probability of activation, proliferation, migration, differentiation, survival, and functional life span of cells in each state or compartment

Clinical assessment The process of gathering clinical information about the demographics, diagnosis, disease stage, clinical history, comorbidities, and current status of general health, pain, and function for an individual patient

Clinical efficacy Efficacy is generally measured as a change in health, pain, or functional status from the beginning to the end of a treatment episode. Efficacy is also defined by a magnitude of variation in possible outcomes, and the probability that a given treatment will give the intended or expected result as well as the probability of adverse events compromising outcome

Clinical efficiency Work or outcome accomplished per unit time

Clinical evaluation The evaluation of the safety and efficacy of a given treatment in a specific therapeutic setting. This process can include, but is not limited to, prospective randomized trials, comparative effectiveness trials, and clinical registries

Clinical function The state of physical or mental or emotional performance at a specific point in time. Most often defined using standardized measures or strength or speed (e.g., site to stand or timed walking or stair climbing) or use of standardized patient-reported outcome measures (PROMs) (e.g., HOCS, KOOS, WOMAC, and DASH)

Clinical outcome The ending state of a patient after a given treatment episode

Clinical safety A measure of the presence or absence of side effects and adverse effects caused by a treatment or drug

Colony-forming unit (CFU) The single founding cell that gives rise to an identifiable group of progeny that are its direct decedents. This is commonly used as an assay in tissue-derived cells to estimate the prevalence of biologically potent stem and progenitors of a particular type under specific conditions. However, a CFU assay can be performed on any cell population to provide a measure of the prevalence and diversity of cells in a population that is capable of ongoing proliferation

Comorbidity One or more disease or condition co-occurring with, and generally independent of, another disease or conditions

Comparative effectiveness trials (CETs) Research directly comparing the benefits, risks, and usefulness of two or more treatment approaches that differ in multiple variables (e.g., treatment modality, timing, stage, sequence, dose regime, active agent or intervention, cost). CET is generally distinct from prospective randomized clinical trials, where a limited number of variables are systematically changed, keeping others purposefully constant

Competitive expansion (CE) An environment and process that is created when multiple colony founding cells are placed into a vessel with limited space and nutrients. As a result clones will compete for space and nutrients

Composition The property defined by the concentration, prevalence, and biological potential of each cell type in a mix of cells

Concentration [cell] The number of cells of a particular type per unit volume (e.g., milliliter) or cells per unit of tissue mass (e.g., gram)

Conditioned media Growth medium that has been used to support the viability and growth of a defined cell population and thereby has accumulated in it the secretory products that are derived from that cell population

Connective tissue progenitors (CTPs) Cells resident in native tissue that are capable of proliferation and generation of progeny that can differentiate into one or more connective tissue phenotypes (e.g., bone, cartilage, fat, tendon, muscle, scar, stroma)

Contaminating cells Cells present in cellular product that do not play a direct role in improving or inhibiting a desired outcome

Critical process parameters (CPPs) Key variables in a manufacturing process that must be defined and controlled in order to achieve successful fabrication of an effective repeatable and reproducible end product

Critical quality attributes Measurable features in an end product that, when present, provide evidence that the product has clinical potency, appropriate purity, and minimal risk of adverse events

Delivery strategy The method used to place a cell product or drug product at an intended location where it will be made bioavailable at an appropriate concentration and state

Demographics General personal and sociological information that are commonly used to define the life state and environment in which a patient is seeking care. Demographic data generally includes age, biological sex, employment, social status, education, height, weight, body mass index, home environment, sexual history, and use of drugs, tobacco, or alcohol

Density separation (DS) A class of methods for isolating individual cell populations based on differences in density (i.e., buoyancy or sedimentation rate) in the aqueous environment of serum or culture medium. This is generally accomplished using a centrifuge

Diagnosis A defined disease state and stage of pathology or severity

Dose The prescribed amount of a drug recommended to be taken at a specific time. For cell-based products, dose is defined by the total number of cells of one or more cell types that are delivered with therapeutic intent

Effector cells Cells, which by their metabolic function, actively contribute (directly or indirectly) to the positive or desired therapeutic effect of a cellular product

Equipoise The state of genuine uncertainty within a patient, provider, or the medical community at large regarding whether or not one treatment will be more beneficial than another

Exosome A subset of membrane-bound extracellular vesicles that are 30–150 nm in diameter, produced by a cell in the endosomal compartment, and released upon fusion with the plasma membrane. These vesicles share a common structure, biogenesis, trafficking, and cell-type-associated pathophysiological functions. They contain a diversity of signaling molecules, mRNA, microRNA, and lipids which are cell-type specific and contribute to intercellular communication and modulation

Extracellular vesicle Cell-derived membranous structures of various sizes (30–1000 nm in diameter) surrounded by a phospholipidic bilayer that function as intercellular messengers via receptor-mediated interactions and by ferrying bioactive lipid, protein, and nucleic acid cargo to recipient cells to stimulate regenerative processes and homeostasis

General health The physical, mental, and social state of a patient. Can be measured by metrics like SF-36, VR-12, EQ-5D, and PHQ-9

Harvest efficiency A measure of the yield of a specific cell population per unit time

Homeostasis The state of equilibrium where composition and function of cells and tissue are maintained in a constant state, through balance of the rate of cell and tissue loss and regeneration

Homing The biological process by which cells at one anatomic location or transit state (in circulation) are induced to migrate or become resident in another location through signaling interactions associated with chemical gradients and local adhesion molecules interacting with cell surface receptors (e.g., SDF-1 and CXCR4)

Inhibitor cells Cells, which by their metabolic function, actively inhibit (directly or indirectly) the positive or desired therapeutic effect of a cellular product

Magnetic separation (MS) The process of isolating specific populations of cells by labeling them with a ferromagnetic tag that allows them to be captured or diverted in a strong magnetic field

Mesenchymal stromal cell (MSC) Culture-expanded, plastic-adherent cells that express CD73, CD90, and CD105 but not hematopoietic and endothelial markers (CD11b, CD14, CD19, CD34, CD45, CD79a, and HLA-DR) which possess the capacity for differentiation into osteoblast, chondrocyte, and adipocyte lineages [23, 24]

MicroRNA (miRNA) A class of noncoding RNA that function to posttranscriptionally regulate gene expression within a cell. They do this through base pairing with complementary sequences in mRNA molecules and cause either (1) cleavage of the mRNA, (2) destabilization of the mRNA through shortening the poly-A tail, or (3) interfering with the translation of the mRNA by ribosomes

Performance-based selection (PBS) A method active choice of the starting materials to be used in generating a cellular product, based on specific physical or metabolic performance or acquired attributes that enable prediction of the future performance of the progeny of the selected cells or cell population

Point of care (POC) The time and immediate physical location in which care is provided to a patient

Prevalence of cell type X (P_x) Total number of cells of type X (e.g., CTPs) divided by the total number of cells present in a population

Process steps The sequential series of events, interventions, or decisions that define an overall process or protocol and determine its outcome or end products

Progenitor cells Cells with the capacity to generate progeny that will adapt, evolve, or differentiate into a phenotype state that is different than the current state of the cell. A "stem cell" is a specific subset of progenitor cell that has the capacity to divide asymmetrically, and thereby renew itself, while generating new progenitors, whose progeny will contribute to new tissue formation or the replacement of cells being lost to disease, senescence, or programmed cell death (e.g., apoptosis)

Prospective clinical registries (PCRs) A database containing documentation of patients with consistently definable attributes of disease type and disease state, as well as documentation of the specific treatment that was provided and the outcome that was achieved. Effective registries are designed to provide unbiased and complete prospective documentation of the treatments being applied and the outcome of those treatments so that retrospective can be used to estimate the efficacy or available treatments, identify patient groups or treatment approaches that are not resulting in sufficient benefit (i.e., value), and define areas of opportunity where specific questions of comparative effectiveness or clinical randomization should be addressed in formal clinical trials

Prospective randomized controlled trials (RCTs) A study in which participants are randomly allocated into experimental groups to measure difference in efficacy resulting from the modification of a single or limited defined set of defined well-controlled variable

Repeatability The ability to repeat a protocol or process in a single location (e.g., lab or hospital setting) and achieve the same outcome

Reproducibility The ability to repeat a protocol or process that has been established at one location (e.g., lab or hospital setting) and achieve the same outcome in one or more new locations

Selective retention (SR) The process of isolating specific populations of cells based on the innate natural ability or response of cells to attach to a surface. Note that SR processing may involve surfaces that are flat, porous, fibrous, or granular or bead-like in geometry

Self-renewal The process of cell division that results in offspring that are identical to the founding cell, with no loss of biological potential

Senescence The process in which a cell ages, loses the ability to continue to divide or to provide biological functions of mature cells, but remains alive as a passive non-contributing member of the population. Senescenct cells may also have adverse effects, by competing with functional cells for nutrients or proucing a secretome that mediates undesired effects

Signaling environment The conditions surrounding a given cell of soluble bioactive molecules or molecules embedded in or on the surface of the extracellular matrix that interact with cell surface receptors and thereby modulate the response of the cell. This signaling environment may include growth factors, cytokines, extracellular matrix proteins, small inhibitor RNAs, as well as exosomes. Autocrine, paracrine, or endocrine signals all contribute

Stem and progenitor cell systems Groups of cells that have a definable/discoverable relationship within a biological hierarchy that supports the formation, preservation, and remodeling of diverse tissue and organ systems beginning in embryonic life through adult life

Stem cell niche The anatomic sites and signaling environment that enables the preservation of the stem cell phenotype and the capacity for asymmetric cell division of a stem cell or stem cell type

Stem cells Generally, resting cells that possess the capacity for asymmetric cell division and self-renewal. Stem cells are most often characterized and preserved in their function by residence within a niche of extracelllar matrix or other cells that enable asymmetrical cell division

Symmetrical cell division A mitotic process that results in two daughter cells with equivalent biological potential

Tissue composition The full inventory of cells and/or extracellular matrix components present within a tissue

Value (V) The magnitude of improvement (I) (i.e., change) in health status, pain, or function per unit cost (C), where $V = I/C$. note that cost may be measured from a diversity of perspectives (e.g., patient, payer, employer, government, or society at large)

Yield (Y) The absolute magnitude or material resulting from a process step (e.g., the number of cells of a given type, or the mass of a given chemical or secreted product)

Yield fraction A measure of the efficacy of a process that is defined by the amount of a given material in a final product (e.g., cells of a certain type), divided by the maximum that was theoretically achievable (e.g., the amount of cells of that type that was present in the starting material before a process step)

References

1. Malchau H, Garellick G, Berry D, et al. Arthroplasty implant registries over the past five decades: development, current, and future impact. J Orthop Res. 2018;36(9):2319–30.
2. Delaunay C. Registries in orthopaedics. Orthop Traumatol Surg Res. 2015;101(1):S69–75.
3. Mareschi K, Ferrero I, Rustichelli D, et al. Expansion of mesenchymal stem cells isolated from pediatric and adult donor bone marrow. J Cell Biochem. 2006;97(4):744–54.
4. Siegel G, Kluba T, Hermanutz-Klein U, Bieback K, Northoff H, Schäfer R. Phenotype, donor age and gender affect function of human bone marrow-derived mesenchymal stromal cells. BMC Med. 2013;11:1.

5. Mantripragada VP, Kaplevatsky R, Bova WA, et al. Influence of glucose concentration on colony-forming efficiency and biological performance of primary human tissue-derived progenitor cells. Cartilage. 2020.

6. Mantripragada VP, Piuzzi NS, Bova WA, et al. Donor-matched comparison of chondrogenic progenitors resident in human infrapatellar fat pad, synovium, and periosteum—implications for cartilage repair. Connect Tissue Res. 2019;60(6):597–610.

7. Mantripragada VP, Bova WA, Piuzzi NS, et al. Native-osteoarthritic joint resident stem and progenitor cells for cartilage cell-based therapies: a quantitative comparison with respect to concentration and biological performance. Am J Sports Med. 2019:1–10.

8. de Lima PK, Orellana MD, De Santis GC, et al. Effects of high-dose chemotherapy on bone marrow multipotent mesenchymal stromal cells isolated from lymphoma patients. Exp Hematol. 2010;38(4):292–300.e4.

9. Mindaye ST, Ra M, Lo Surdo JL, Bauer SR, Alterman MA. Global proteomic signature of undifferentiated human bone marrow stromal cells: evidence for donor-to-donor proteome heterogeneity. Stem Cell Res. 2013;11(2):793–805.

10. Muschler GF, Nitto H, Boehm CA, Easley KA. Age- and gender-related changes in the cellularity of human bone marrow and the prevalence of osteoblastic progenitors. J Orthop Res. 2001;19(1):117–25.

11. Mantripragada VP, Boehm C, Bova W, Briskin I, Piuzzi NS, Muschler GF. Patient age and cell concentration influence prevalence and concentraion of progenitors in bone marrow aspirates: an analysis of 436 patients. J Bone Joint Surg. 2021:1–9.

12. Naung NY, Suttapreyasri S, Kamolmatyakul S, Nuntanaranont T. Comparative study of different centrifugation protocols for a density gradient separation media in isolation of osteoprogenitors from bone marrow aspirate. J Oral Biol Craniofac Res. 2014;4(3):160–8.

13. Luangphakdy V, Boehm C, Pan H, Herrick J, Zaveri P, Muschler GF. Assessment of methods for rapid intraoperative concentration and selection of marrow-derived connective tissue progenitors for bone regeneration using the canine femoral multidefect model. Tissue Eng Part A. 2016;22(1–2).

14. Mantripragada VP, Bova WA, Boehm C, et al. Primary cells isolated from human knee cartilage reveal decreased prevalence of progenitor cells but comparable biological potential during osteoarthritic disease progression. J Bone Joint Surg Am. 2018;100(20):1771–80.

15. Mantripragada VP, Bova WA, Boehm C, et al. Progenitor cells from different zones of human cartilage and their correlation with histopathological osteoarthritis progression. J Orthop Res. 2018;36(6):1728–38.

16. Muschler GF, Nakamoto C, Griffith LG. Engineering principles of clinical cell-based tissue engineering. J Bone Joint Surg Ser A. 2004;86(7):1541–58.

17. Regulatory considerations for human cells, tissues, and cellular and tissue-based products: minimal manipulation and homologous use guidance for industry and food and drug administration staff contains nonbinding recommendations regulatory considerations. *Guid Ind Food Drug Adm Staff.* 2020.

18. Dexter TM, Moore AS, Sheridan AP. Maintenance of Hemopoietic stem cells and production of differentiated progeny in allogenic and semiallogenic bone marrow chimeras in vitro. J Exp Med. 1977;145:1612–6.

19. Owen M, Friedenstein AJ. Stromal stem cells: marrow-derived osteogenic precursors. Ciba Found Symp. 1988;136:42–60.

20. Friedenstein AJ, Chailakhyan RK, Gerasimov UV. Bone marrow osteogenic stem cells: in vitro cultivation and transplantation in diffusion chambers. Cell Tissue Kinet. 1987;20:263–72.

21. Caplan AI. Mesenchymal stem cells. J Orthop Res. 1991;9(5):641–50.

22. Caplan AI. Mesenchymal stem cells: time to change the name! Stem Cells Transl Med. 2017;6(6):1445–51.

23. Horwitz EM, Le Blanc K, Dominici M, et al. Clarification of the nomenclature for MSC: the International Society for Cellular Therapy position statement. Cytotherapy. 2005;7(5):393–5.

24. Viswanathan S, Shi Y, Galipeau J, et al. Mesenchymal stem versus stromal cells: International Society for Cell & Gene Therapy (ISCT®) Mesenchymal Stromal Cell committee position statement on nomenclature. Cytotherapy. 2019;21(10):1019–24.

25. Mantripragada VP, Tan K-L, Vasavada S, Bova W, Barnard J, Muschler GF. Characterization of heterogeneous primary human cartilage-derived cell population using non-invasive live-cell phase-contrast time-lapse imaging. Cytotherapy. 2020;000:1–12.

26. Mantripragada VP, Bova WA, Piuzzi NS, et al. Native-osteoarthritic joint resident stem and progenitor cells for cartilage cell-based therapies: a quantitative comparison with respect to concentration and biological performance. Am J Sports Med. 2019;47(14):3521–30.

27. Selich A, Daudert J, Hass R, et al. Massive clonal selection and transiently contributing clones during expansion of mesenchymal stem cell cultures revealed by lentiviral RGB-barcode technology. Stem Cell Transl Med. 2016;5:591–601.

28. Mantripragada VP, Luangphakdy V, Hittle B, Powell K, Muschler GF. Automated in-process characterization and selection of cell-clones for quality and efficient cell manufacturing. Cytotechnology. 2020;72(5):615–27.

29. Kwee E, Herderick EE, Adams T, et al. Integrated colony imaging, analysis, and selection device for regenerative medicine. SLAS Technol. 2017;22(2):217–23.
30. Muschler GF, Midura RJ, Nakamoto C. Practical modeling concepts for connective tissue stem cell and progenitor compartment kinetics. J Biomed Biotechnol. 2003;2003(3):170–93.
31. Muschler GF, Midura RJ. Connective tissue progenitors: practical concepts for clinical applications. Clin Orthop Relat Res. 2002;1(395):66–80.
32. Dua HS, Azuara-Blanco A. Limbal stem cells of the corneal epithelium. Surv Ophthalmol. 2000;44(5):415–25.
33. Sainio K, Raatikainen-Ahokas A. Mesonephric kidney—a stem cell factory? Int J Dev Biol. 1999;43(5):435–9.
34. Rao MS. Multipotent and restricted precursors in the central nervous system. Anat Rec. 1999;257(4):137–48.
35. Yelick PC, Zhang W. Mesenchymal stem cells: biology and potential clinical uses. Tissue Eng Princ Pract. 2012;28(May 1999):10-1-10-14.
36. Morrison SJ, Spradling AC. Stem cells and niches: mechanisms that promote stem cell maintenance throughout life. Cell. 2008;132(4):598–611.
37. Sipp D, Robey PG, Turner L. Clear up this stem-cell mess. Nature. 2018;561(7724):455–7.
38. Moore KA, Lemischka IR. Stem cells and their niches. Biomater Regen Med. 2006:1880–5.
39. Shi S, Gronthos S. Perivascular niche of postnatal mesenchymal stem cells in human bone marrow and dental pulp. J Bone Miner Res. 2003;18(4):696–704.
40. Patterson TE, Boehm C, Nakamoto C, et al. The efficiency of bone marrow aspiration for the harvest of connective tissue progenitors from the human iliac crest. J Bone Joint Surg Am. 2017;99:1673–82.
41. Fernandes TL, Kimura HA, Pinheiro CCG, et al. Human synovial mesenchymal stem cells good manufacturing practices for articular cartilage regeneration. Tissue Eng Part C Methods. 2018;24(12):709–16.
42. Russell KC, Phinney DG, Lacey MR, Barrilleaux BL, Meyertholen KE, O'Connor KC. In vitro high-capacity assay to quantify the clonal heterogeneity in trilineage potential of mesenchymal stem cells reveals a complex hierarchy of lineage commitment. Stem Cells. 2010;28(4):788–98.
43. Kumagai K, Vasanji A, Drazba JA, Butler RS, Muschler GF. Circulating cells with osteogenic potential are physiologically mobilized into the fracture healing site in the parabiotic mice model. J Orthop Res. 2008;26:2.
44. Kumagai K, Takeuchi R, Ishikawa H, et al. Low-intensity pulsed ultrasound accelerates fracture healing by stimulation of recruitment of both local and circulating osteogenic progenitors. J Orthop Res. 2012;30:9.
45. Otsuru S, Tamai K, Yamazaki T, Yoshikawa H, Kaneda Y. Circulating bone marrow-derived osteoblast progenitor cells are recruited to the bone-forming site by the CXCR4/stromal cell-derived Factor-1 pathway. Stem Cells. 2008;26(1):223–34.
46. Danisovic L, Bohac M, Zamborsky R, et al. Comparative analysis of mesenchymal stromal cells from different tissue sources in respect to articular cartilage tissue engineering. Gen Physiol Biophys. 2016;35:207–14.
47. McLain RF, Fleming JE, Boehm CA, Muschler GF. Aspiration of osteoprogenitor cells for augmenting spinal fusion: comparison of progenitor cell concentrations from the vertebral body and iliac crest. J Bone Joint Surg Ser A. 2005;87(12 I):2655–61.
48. Qadan MA, Piuzzi NS, Boehm C, et al. Variation in primary and culture-expanded cells derived from connective tissue progenitors in human bone marrow space, bone trabecular surface and adipose tissue. Cytotherapy. 2018;20(3):343–60.
49. Juhl M, Tratwal J, Follin B, et al. Comparison of clinical grade human platelet lysates for cultivation of mesenchymal stromal cells from bone marrow and adipose tissue. Scand J Clin Lab Invest. 2016;76(2):93–104.
50. Hoogduijn MJ, Betjes MGH, Baan CC. Mesenchymal stromal cells for organ transplantation: different sources and unique characteristics? Curr Opin Organ Transplant. 2014;19(1):41–6.
51. Bianco P, Robey PG, Simmons PJ. Mesenchymal stem cells: revisiting history, concepts, and assays. Cell Stem Cell. 2008;2(4):313–9.
52. Marcantonio NA, Boehm CA, Rozic RJ, et al. The influence of tethered epidermal growth factor on connective tissue progenitor colony formation. Biomaterials. 2009;30(27):4629–38.
53. Villarruel SM, Boehm CA, Pennington M, Bryan JA, Powell KA, Muschler GF. The effect of oxygen tension on the in vitro assay of human osteoblastic connective tissue progenitor cells. J Orthop Res. 2008;26(10):1390–7.
54. Heylman CM, Caralla TN, Boehm CA, Patterson TE, Muschler GF. Slowing the onset of hypoxia increases colony forming efficiency of connective tissue progenitor cells in vitro. J Regen Med Tissue Eng. 2013;2.
55. Mantripragada VP, Boehm C, Bova W, Briskin I, Piuzzi NS, Muschler GF. Patient age and cell concentration influence prevalence and concentration of progenitors in bone marrow aspirates: an analysis of 436 patients. J Bone Joint Surg. 2021:1–9.
56. Patterson TE, Boehm C, Nakamoto C, et al. The efficiency of bone marrow aspiration for the harvest of connective tissue progenitors from the human iliac crest. J Bone Joint Surg Am. 2017;99:19.
57. Yagi H, Soto-Gutierrez A, Farekkadan B, et al. Mesenchymal stem cells: mechanisms of immunomodulation and homing. Cell Transplant. 2010;19(6–7):667–79.

58. Najar M, Raicevic G, Crompot E, et al. The immunomodulatory potential of mesenchymal stromal cells: a story of a regulatory network. J Immunother. 2016;39(2):45–59.
59. Link TM, Li X. Establishing compositional MRI of cartilage as a biomarker for clinical practice. Osteoarthr Cartil. 2018;26(9):1137–9.
60. Wáng Y-XJ, Zhang Q, Li X, Chen W, Ahuja A, Yuan J. T1ρ magnetic resonance: basic physics principles and applications in knee and intervertebral disc imaging. Quant Imaging Med Surg. 2015;5(5):858–85885.
61. Stach CM, Bäuerle M, Englbrecht M, et al. Periarticular bone structure in rheumatoid arthritis patients and healthy individuals assessed by high-resolution computed tomography. Arthritis Rheum. 2010;62(2):330–9.

Bone Marrow as a Source of Cells for Musculoskeletal Cellular Therapies

George F. Muschler, Hannah Simmons, Venkata Mantripragada, and Nicolas S. Piuzzi

2.1 Introduction

This chapter builds upon and extends the quantitative engineering principles and the paradigm of stem and progenitor cell biology that are outlined in the first chapter and applies these tools and tenants to the clinical harvest, rapid point of care processing, and quantitative characterization of human bone marrow-derived cells for use in cellular therapy.

The aims of this chapter are as follows:

1. Define best practices for safe and effective bone marrow aspiration (BMA).
2. Define means for measurement of the composition and quality of a given BMA sample or BMA-derived product prior to injection.
3. Provide an overview of what is known about the outcome of clinical cell sourcing using BMA, including variation between patients, aspiration site, and aspiration methods.
4. Define parameters that can be used effectively to report out and compare the composition, efficacy, efficiency, repeatability, and reproducibility of BMA when used for clinical therapy.

2.2 Bone Marrow as a Cell Source

Autologous bone marrow harvested as a bone marrow aspirate (BMA) is perhaps the most frequently used cell source for musculoskeletal applications [1–4]. *Lindholm and Urist* [5] were among the first to add unprocessed BMA to bone matrix to enhance clinical bone healing in the late 1970s. In the mid-1980s, *Connolly and Shindell* [6, 7] reported successful treatment of tibia nonunion with percutaneous injections of unprocessed BMA. Many subsequent reports describe the use of BMA in osteonecrosis [8–12], nonunion [13–17], bone defects [18–20], surgical fusion procedures [21], augmentation of distraction osteogenesis [22], chondral defects [23–25], osteoarthritis [26–29], tendinopathy [30], and tendon repair augmentation [31].

As outlined in Chap. 1, bone marrow, like all tissues, contains a heterogeneous mixture of mature cells representing the diversity of tissues that are present in the harvest site, as well as a much smaller population of stem and progenitor

G. F. Muschler (✉) · N. S. Piuzzi
Orthopaedic Surgery and Biomedical Engineering, Cleveland Clinic, Cleveland, OH, USA
e-mail: MUSCHLG@ccf.org

H. Simmons · V. Mantripragada
Department of Biomedical Engineering, Cleveland Clinic, Cleveland, OH, USA

G. Filardo et al. (eds.), *Orthobiologics*, https://doi.org/10.1007/978-3-030-84744-9_2

cells that support the preservation of those tissues over time. When bone marrow is harvested from bones such as the pelvis or vertebral body, hematopoietic stem cells and hematopoietic progenitor cells are the most abundant population, approaching 1–2%. Vascular endothelial progenitors are also present. However, a small fraction of the population of cells is derived from the connective tissue stem and progenitor population, defined as *connective tissue progenitors (CTPs)* [16, 32–41]. CTPs include progenitors of bone marrow stromal cells that support hematopoiesis, as well as progenitors capable of bone, cartilage, fat, and other connective tissues. The prevalence of CTPs varies widely from one individual and one bone site to another, ranging from 1 in 5000 to 1 in 100,000 (0.5–0.0001%) [42]. CTPs are defined as a class of heterogeneous stem and progenitor cells that are present in native tissues and capable of proliferation to generate progeny that can differentiate to express one or more connective tissue phenotype. This definition of CTPs as a heterogeneous class of native stem and progenitor populations, by design, establishes a paradigm of communication that stimulates and enables the description and exploration of (a) the tissue-specific hierarchy of CTPs from different tissues, (b) the tissue-specific impact of aging and disease on CTPs, and (c) tissue-specific differences that may impact the sourcing of CTPs for specific clinical applications [33, 41, 43–47].

There is no one phenotype for CTPs that can be defined by cell size or surface markers, though some markers have been associated with some CTPs (e.g., STRO-1, CD146, CD90, PDPN hyaluronan) [48, 49]. However, the concentration and prevalence of CTPs in a given sample can be measured directly based on the primary attribute of this population of cells, which is to survive, attach to a diversity of surfaces (including tissue culture plastic, many ceramics, as well as bone matrix), and then proliferate to form an observable colony of progeny in vitro [50, 51]. Several assays are available using a variety of plating surfaces and media conditions. The grounding assumption of these assays is that each colony that is observed is derived from a single colony founding CTP. When CTPs are assayed in this

way, the assay may include a small subset of cells that were pulled from true stem cell niches in bone marrow tissue. However, the colonies observed will also represent the diversity of cell states that may be present in the tissue among CTPs that are on their way to a variety of future differentiation states [32, 52–54].

The term CTP is not unique to marrow. Every organ system contains connective tissue and therefore will contain CTPs that are associated with preservation of the homeostasis of the connective tissue population in that tissue. However, because the composition and mature connective tissue phenotypes in each tissue are different, the CTP populations that are assayed in cells from each tissue will be "tissue specific."

It is important to note that the term CTP is specific only to the cell that was present in native tissue and then went on to form a colony in vitro. By definition, when CTP populations are allowed to form colonies in vitro, the cells in those colonies are the progeny of CTPs, but they are not CTPs themselves. These progeny can be expanded in vitro into large batches of CTP-derived progeny. If they meet specific criteria, they may be categorized as "marrow stromal cells" or "mesenchymal stromal cells" ("MSCs"). The International Society for Cell & Gene Therapy (ISCT) has defined standardized terminology and minimal criteria for classification of culture-expanded cells as "MSCs." These criteria include (1) adherence to tissue culture plastic in standard culture conditions; (2) the presence of surface markers, including CD73, CD90, and CD105; (3) the absence of hematopoietic markers CD34, CD45, CD14, CD19, and HLA-DR; and (4) the ability to differentiate to osteoblasts, adipocytes, and chondrocytes in vitro (aka "trilineage differentiation") [55, 56].

2.3 Clinical Rationale for Bone Marrow-Derived Cells in Cellular Therapy

As outlined in Chap. 1, for any given clinical application, there are some cells that may contribute to outcome ("effector cells"). There may also be some that may actively inhibit outcome ("inhibitory cells"). Furthermore, many cells may have no direct action pro or con. However, these "contaminating cells" ("bystanders"), just by their presence at the local site, will inevitably compete with effector cells for space and nutrients and thereby reduce any positive effect that may be accomplished by effector cells. Therefore, the clinical goal in using bone marrow-derived cells for therapy is to (1) optimize the delivery of effector cells and (2) minimize the delivery of inhibitory and contaminating cells. The obvious strategy, of course, depends on developing a knowledge of which cells are effector cells (contributing directly or indirectly to outcome), which may be inhibitory, and which are contaminating.

In the current clinical paradigm in the use of bone marrow for cellular therapy, we often assume that CTPs are the primary "effector cell." However, this may not always be true. Vascular endothelial cells or hematopoietic cells, or other populations of cells, may contribute to outcome in some settings. The preferred composition remains an open question in almost every clinical application.

In settings of bone or cartilage defects or deficiencies, it is generally assumed that a loss or deficiency of local stem and progenitor cells is part of the pathophysiology and, therefore, that replacing or supplementing the population of potential tissue-forming cells will improve outcome [9, 57, 58]. Transplantation of cells and matrix from autologous cancellous bone (ACB) was reported by Phemister in the 1920s [59, 60] and further characterized by Burwell [61] as a process dependent on transplanted cells. ACB grafting remains the gold standard for local enhancement of bone regeneration [62, 63]. ACB provides several components that are thought to contribute to clinical efficacy: bone matrix (the extracellular component of ACB); "osteoconductive" properties, i.e., providing a matrix surface that facilitates the attachment; and migration of osteogenic and vascular endothelial cells throughout a given tissue volume. Bone matrix may also provide "osteoinductive" signals that are embedded in the matrix or released from the matrix, which induce proliferation and differentiation of

local progenitors along a bone differentiation pathway. However, the presence of CTPs with the capacity to differentiate into new osteoblasts (aka, connective tissue progenitor - osteogenic – CTP-Os) is generally assumed to be the most important biological component of ACB [63].

To avoid the risk of pain, bleeding, scarring, and infection associated with ACB harvest [64–67], many surgeons have elected the alternative of aspirating bone marrow as a means of obtaining ("sourcing") the CTP-Os that they need. Over the past four decades, this experience with bone marrow aspiration (BMA) has established BMA harvest as a safe and effective means of collecting bone marrow-derived cells [68]. This clinical experience now continues to drive the exploration BMA-derived cells as a cell source for other injectable cellular therapies that are developed throughout this text.

Clinical experience supports the safety and potential efficacy of introducing bone marrow-derived cells into joint, and other connective tissue sites. The technique of microfracture was introduced to clinical practice by Steadman [69] in the 1980s. Microfracture, as a marrow stimulation technique, provides a pathway for CTPs in subchondral bone to migrate from bone marrow into the joint where they may contribute with cells or secreted factors that improve the formation of cartilage or fibrocartilage within a cartilage defect. Subsequently, Osteochondral Autograft Transplantation (OATS) [70], which transfers one or more cylindrical osteochondral autografts into a cartilage defect, also exposed joints to bone marrow-derived cells, including CTPs. All of these strategies follow the paradigm of reintroducing cells into a defect with the intention of using the biological potential of cells to survive, proliferate, and differentiate to form new functional tissue. All of these experiences support the use of cellular constructs in specific settings. The lack of biological complications or clinical risks associated with the introduction of marrow-derived cells to a joint space does not preclude the potential long-term risks, but does mitigate most concerns to date. It is important to note that undesirable consequences can occur when bone marrow is transplanted to some tissue sites, particularly injured muscle. Heterotopic bone is a known complication following hip arthroplasty, complex medial collateral ligament injury, traumatic amputations associated with blast injury [71, 72], and even local soft tissue injections [73].

The use of BMA as a cell source continues to increase as it offers several advantages over alternative cell sources. First, percutaneous aspiration has minimal morbidity (post-procedure pain, scarring, or infection risk). BMA procedures are well tolerated with a local anesthetic and mild sedation and require only a Band-Aid to cover the harvest entry [39, 74]. Second, a BMA sample can be immediately anticoagulated to provide a single-cell suspension that can be further processed without the need for more invasive tissue handling [75, 76] (e.g., mincing, homogenization, or enzymatic digestion). Immediate access to a cell suspension enables a BMA sample to be processed using a variety of point of care, rapid processing methods that can be used to purposefully modify or improve upon the composition of marrow-derived cells that can be used for therapy, particularly increasing the concentration of CTPs ([CTP], usually defined as CTPs per mL). Density separation (DS) using a centrifuge allows the concentration of all nucleated cells, including CTPs, by removal of the lower density plasma and the higher density red cell mass that dilute the samples. DS processing has the least regulatory barriers [77, 78]. It is also possible to increase the prevalence of CTPs (P_{CTP}) (i.e., CTPs per million nucleated cells) through selective retention [79–84], magnetic separation [85–87], or FACS sorting, by preferentially isolating CTPs or preferentially removing non-CTPS or undesirable CTPs ("negative selection").

2.4 Bone Marrow Aspiration Technique

Aspiration technique has been reviewed in detail [68, 88]. Harvest of bone marrow by aspiration involves several key steps: (1) selecting an anatomic site for aspiration; (2) passing a hollow needle with central trocar through cortex into the intramedullary cavity of a bone containing bone marrow; (3) applying sufficient negative pressure (using a syringe or vacuum pump) to induce the flow of material from the marrow space (in addition to inducing flow of the relatively viscous marrow contents, this process also induces rupture of vascular sinusoids in the marrow cavity and a flow of contaminating peripheral blood into the needle); (4) collecting the fluid aspirated material in a reservoir for later use; and (5) using an anticoagulant to prevent clotting (when needed to allow further processing or mixing).

The effectiveness of each BMA harvest procedure will vary widely and can be measured by (a) yield of total nucleated cells, (b) CTP concentration ([CTP]), and (c) CTP prevalence (P_{CTP}). Over 70% of the variance that is present will be due to variation between patients. However, an optimal technique will have a profound impact on these effectiveness parameters, in particular:

Aspiration Site—The intramedullary canal of any bone can be aspirated. The iliac crest and vertebral body provide the highest yield and concentration [89]. Aspiration from the proximal humerus, or the proximal or distal femur or tibia, or the calcaneus is also possible. These sites will contain nucleated cells and CTPs, but at much lower concentration, particularly with advancing age. If sites below mid-thigh are aspirated, the use of a tourniquet is discouraged. While not systematically studied, in the few times that authors tested aspiration with a tourniquet, the resulting aspirates tend to flow less readily, be fatty, and show lower viability in culture.

Either the anterior or posterior iliac crest can be used. The posterior crest requires either prone or lateral decubitus positioning. However, it also provides a larger volume of bone for aspiration, and slightly higher cell concentration. It is essential to have a profound understanding of the iliac crest anatomy and safety zone in order to pursue the aspirations avoiding complications [90, 91].

Needle Type—We found no difference in cell yield between 8 and 11 gauge BMA needles (data not shown). Nonetheless, the larger 8 gauge needle is stiffer adding control and enables the use of a longer needle. A needle length of 7 inches is very helpful when collecting bone marrow samples from obese patient. Traditional aspiration needles have been designed for use by hematologists. A hollow needle with trocar is used whenever advancing the needle into bone to avoid packing the needle tip with bone. Aspiration of marrow through a tip packed with bone will result in excessive shear forces as cells are pulled through small holes between bone spicules, causing cell lysis and death. Several needles have been introduced with a diversity of grip styles. Some have end cutting features that enable "excavation" of marrow and bone to enhance harvest. Others have one or more side holes to avoid bone packing without a trocar. Needles with side holes cannot be used in a lateral approach, as all holes must be buried in bone to be effective aspiration portals. They also have two drawbacks: (a) side holes increase the risk of needle fracture due to bending, and (b) if aspirating through two side holes, equal flow will be unlikely. Flow may favor low viscosity blood through one hole to the exclusion of the other.

Surgical Approach—There are two main surgical approaches to aspiration from the iliac crest: "lateral" or "parallel." (See Figs. 2.1 and 2.2.)

Lateral Approach—Using a #11 blade, a 2 millimeter stab incision is made parallel to Langer's lines in the skin. For the anterior crease, this is made approximately 4–5 cm posterior and lateral to the anterior superior iliac spine. For the posterior crest, this is approximately 4 cm lateral to the posterior superior iliac spine. Using a single skin site access through those incisions allow a needle can be advanced in a fan of projections to engage the lateral cortex of any site. A BMA needle is placed perpendicular to the iliac crest, and the tip of the needle is advanced with the obturator in place using alternating rotation and axial force to position the tip just beneath the lateral

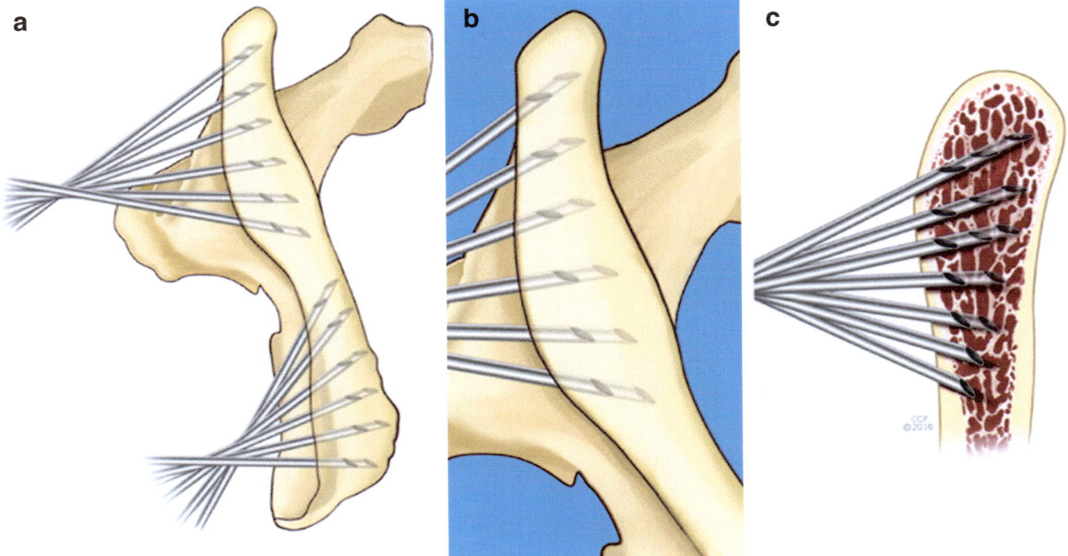

Fig. 2.1 Bone marrow aspiration technique—perpendicular or lateral technique. Bone marrow can be harvested by aspiration of either the anterior or the posterior iliac crest as illustrated (**a**). In most patients, the thickness of the iliac crest allows aspiration from a lateral approach immediately after entry of the needle into the medullary cavity and then advancement of the needle by 5 mm once or twice to obtain two or three aspirates through the same cortical hole (**b** Axial view representation). Only a single site of skin entry but multiple sites of bone entry are required with the lateral approach. (**c** Coronal view representation) (Reproduced, with permission, from: Hernigou J, Alves A, Homma Y, Guissou I, Hernigou P. Anatomy of the ilium for bone marrow aspiration: map of sectors and implication for safe trocar placement. Int Orthop. 2014;38(12):2585–90. Epub 2014 Apr 30. Reprinted with permission, Cleveland Clinic Center for Medical Art & Photography 2016–2018. All Rights Reserved)

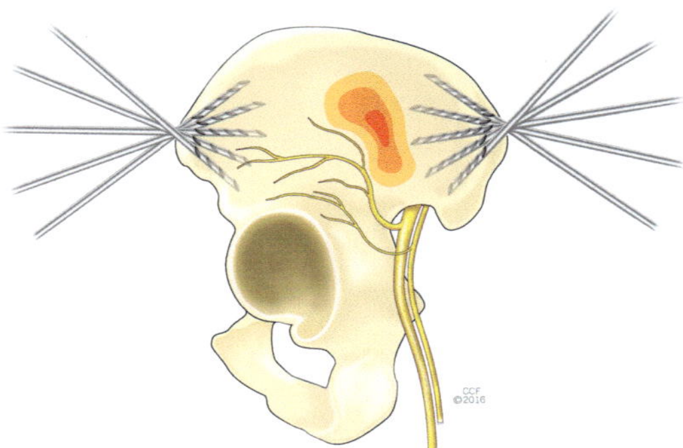

Fig. 2.2 Bone marrow aspiration technique—parallel technique. In this approach the space between the inner and outer table is entered through the anterior superior iliac spine (ASIS) or posterior superior iliac spine (PSIS). The needle can then be advanced using the trocar in 5 mm increments through the flat portion of the anterior or posterior iliac wings often to a depth of 6–8 cm, before the needle path must be redirected to access fresh sites. Multiple, separate aspirates can be obtained by advancing the needle between and parallel to the inner and outer tables of the iliac crest in increments of 5 to 10 mm and by redirecting the needle along various trajectories in a fan-like projection from the entry hole. The color map represents the part of the ilium that should be avoided where the thickness of the bone is <3 mm (yellow) to <1 mm (red) [90]. NOTE: the correct reference for these diagrams is in a Current Concepts paper in JBJA - Piuzzi and Muschler authors

cortex. Following aspiration the needle's trocar is reintroduced, and the needle is advanced 5 mm to aspirate a new site. Advancement is limited by the inner cortex. Advancing just 5 mm in depth at a time and changing cortical entry site by 5–10 mm results in harvest of samples that are essentially independent [39] (Fig. 2.1).

Parallel approach—In this approach the aspiration needle is passed in the space between the inner and outer table. Entry sites to the pelvis are either through the anterior superior iliac spine (ASIS) or posterior superior iliac spine (PSIS) (Fig. 2.2). A needle with a single distal hole can be advanced using a trocar in 5 mm increments through the flat portion of the anterior or posterior iliac wings often to a depth of 6–8 cm, before the needle path must be redirected to access fresh sites. Aspiration from the iliac crest has been reported to provide an increase in cellularity and CTP yield compared to the anterior crest

[92]. The practice of fully advancing the needle and then backing up, rather than advancing with each new site, in theory, would increase the opportunity for contamination with blood leaking into the needle from previous aspiration sites along the needle track. When a needle with multiple side holes is used, one must advance the needle sufficiently to position all of the side holes in the bone before aspiration. Repositioning of a side hole needle by varying direction and depth or rotation will impact the yield and composition of an aspirate, but specific methods for use of a side hole needle have not been systematically studied.

Aspiration Volume—Limiting the volume of aspiration at any individual needle position is very important to reduce dilution and contamination of marrow-derived cells with peripheral blood. Using a traditional needle with a single distal hole, limiting aspiration volume to 1–2 ccs will increase the yield of nucleate cells and CTPs per cc [93]. Data indicate that approximately 85% of the marrow-derived cells that can be harvested from a given site have been harvested in the first 2 ccs and that aspiration beyond 4 ccs has no value, and only adds contaminating blood. Limiting aspiration volume to 2 ccs will provide a sample with CTP concentration that is fourfold higher than a 10 cc aspirate, before any further processing [39] (Fig. 2.3). Aspiration of more than 4 cc from a single needle position predominantly adds peripheral blood to the sample, diluting marrow-derived cells [39, 68].

Applying Negative Pressure During Aspiration—The use of a 10 mL syringe fixed to the needle provides excellent control of the magnitude of negative pressure during aspiration and limits the volume of aspiration [68, 94]. Achieving an air-tight seal of the skin, soft tissue, and cortical bone around the aspirating needle is important. This can be difficult in sites, such as the tibia or calcaneus where overlying soft tissue is limited. Failure of this seal will result in air leak and prevent effective aspiration. Drawing the plunger of an empty 10 mL syringe back to 10 cubic centimeter marker will create a vacuum of −441 torr (mm Hg) [95]. Although larger syringe size can generate more maximum vacuum, all syringe

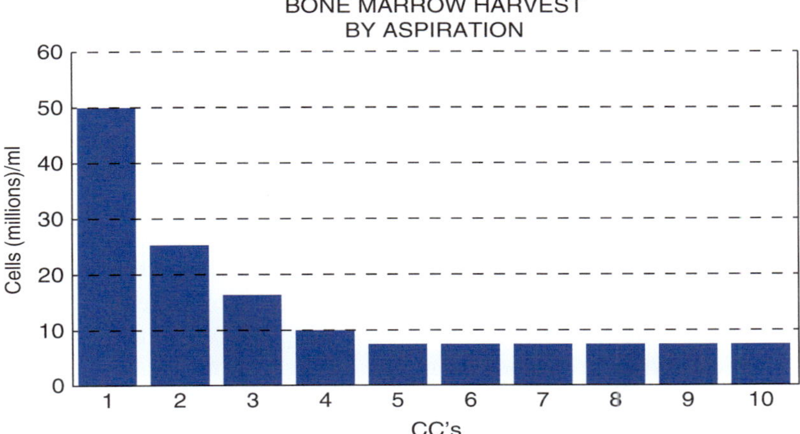

Fig. 2.3 Influence of aspiration volume on the harvest of marrow-derived cells and progenitors. A rapid dilution of bone marrow-derived cells by peripheral blood occured as aspiration volumes were increased from 1 cc (in which we found a mean of about 50 million nucleated cells) to 4 ccs. Also, about 85% of all of the marrow-derived cells available from a given site were collected in just a 2 cc aspirate [96]

devices, 1 mL through 20 mL, generate approximately identical vacuum at the same volume displacement of the plunger. A 20 mL syringe (maximum vacuum −517 torr) requires almost twice the force as the 10 mL to achieve maximum vacuum, which makes it increasingly difficult to control [95]. The magnitude of force that is needed to move marrow into the syringe will vary at each site and in each patient, depending on the histological composition of the marrow space. Hematopoietic marrow sites and regions of osteopenia/osteoporosis tend to decrease the pressure required. Regions of myelofibrosis will increase the force required, and the rate of flow. However, marrow-derived cells and CTPs can be harvested by aspiration from any cancellous bone site. In theory, rotating the bevel of the needle 360° while aspirating might increase the sweep of the needle tip and improve the aspiration efficiency; however this was not found to be significant (unpublished data).

2.5 Aspirate Yield, Composition, and Efficiency

The yield and composition of each BMA sample will vary from patient to patient and aspirate to aspirate [42, 96]. The success of aspiration can be measured by a number of metrics: yield and composition of nucleated cells (N_{Cell}), concentration of nucleated cells ([Cell]), yield of CTPs (N_{CTP}), concentration of CTPs ([CTP]), and the prevalence of CTPs (P_{CTP}) [88]. The number, concentration, and prevalence of other relevant cell types could also be included (e.g., hematopoietic stem cells (HSCs), endothelial progenitor cells (EPCs), platelets).

Mean Cell Concentration [Cell]—When performed using optimal methods, the mean [Cell] could be 15–35 million cells per mL. Age has a strong effect on marrow cellularity [46]. Mean [Cell] in a 20-year-old may be close to 40–60 million cells/mL, while an 80-year-old may be closer to 5–10 million cells/mL. Lower numbers of cellularity over a series of patients imply that the technique of aspiration or processing may be flawed [97].

Mean CTP prevalence (P_{CTP}) (CTPs per million nucleated cells) could be 20–60 CTPs/million cells, a mean ratio of approximately 1 CTP per 25,000 cells. CTP prevalence remains relatively constant with advancing age but will decrease in patients with low turnover senile osteoporosis and increase in high turnover states, such as sickle cell anemia [42, 93]. Note that BMA is sometimes criticized when compared to adipose tissue as a source of CTPs, because the prevalence of CTPs in adipose tissue is higher, in the range of 1 CTP per 2–4 K cells [33]. This is a fair criticism but needs to be balanced against the increased morbidity of harvesting 60 ccs of BMA versus 60 cc of adipose tissue and the fact that the CTP population derived from adipose tissue has a difference composition of biological starting states and biological potential than CTPs from bone marrow.

Mean CTP concentration ([CTP]) will be approximately 850 CTPs/mL (range 400–1500). [CTP] is a calculated number (i.e., the product of [Cell] and P_{CTP}) and therefore not an independent variable [46, 97].

Magnitude of Dilution—Patterson et al. directly compared the concentration and prevalence of CTPs in native cancellous bone with the concentration and prevalence of CTPs in BMA samples in 33 patients [97]. This comparison included both cells and CTPs in the marrow space, as well as cells and CTPs that are intimately attached to the surface of cancellous bone. This assessment demonstrated that the concentration of CTPs in a BMA sample is three- to fourfold lower than the concentration that is present in the native bone tissue from which it is aspirated. Dilution was not constant across all patients, but this is the best estimation of the magnitude of dilution that results from contamination of marrow-derived cells from serum, nucleated cells, and RBCs from peripheral blood. It also demonstrates that the two- to sixfold increase in cell concentration that can be accomplished using density separation (centrifuge processing) is sufficient to provide a preparation of cells and CTPs that has concentration of CTPs that is comparable to autogenous cancellous bone graft [97].

Therapeutic Value—The therapeutic value (i.e., quality) of a given BMA sample is not known. There are no established metrics that can link BMA composition directly to outcome. However, if the CTP population is assumed to be the primary effector population, then CTP concentration [CTP] and CTP prevalence (P_{CTP}) can be assumed to be the most important metrics. [CTP] defines the number of CTPs that can be delivered into a given tissue sites, which is always limited by volume. Similarly, P_{CTF} defines the competition environment into which CTPs are placed. If prevalence is 1 in 25 K, then each CTP will be competing with 25 K non-CTPs for space and for nutrients. Oxygen in particular is a limiting resource for the survival and performance of transplanted cells, including CTPs. This fact drives much of the considerations that are discussed below for further processing of BMA-derived cells to increase [CTP] and P_{CTP}, by removing RBCs, serum, potential "inhibitors," and other "contaminating" cells and non-CTP progenitors.

Harvest Efficiency—Efficiency can also be measured in units of the number of cells or CTPs that are collected per unit time or per unit of expense. Improving the efficiency of bone marrow aspiration or cell harvest procedures is not discussed in this chapter, but is a valid area of future development, particularly once the effector cells for a given application are known and settings are defined where the number of effector cells must be increased beyond the level available in a traditional aspirate. Bone marrow transplantation procedures which target the transplantation of 10^7 CD34+ cells or 10^9 total marrow cells for hematopoietic engraftment may be the first target for improved harvest efficiency [98].

Bone Marrow Aspiration Safety—BMA from the iliac crest has been reported to be a safe technique with a low rate of complication, especially when performed in the orthopedic field [99]. Some serious adverse events and even death have been reported related to BMA in the hematological literature, but these complications may be related more to the patient's morbidity (e.g., thrombocytopenia, myeloproliferative disorder) than to the procedure itself [100–102]. Hernigou et al. [91] reported an experience of BMA in 1800 orthopedic patients using primarily the parallel technique, including 4 hematomas requiring transfusion, 1 transient sciatic palsy, and 3 transient episodes of numbness in the area of the lateral femoral cutaneous nerve. The primary author (GFM) has had clinical experience in more than 1500 patients undergoing elective orthopedic procedures using the lateral approach from both the anterior and posterior crest over

30 years. BMA was performed for a mix of therapeutic and research purposes, either under a spinal or general anesthetic. In this experience the median BMA volume was 40 cc (range 16–300 mL). In this time period, there were only two reported bruises and one symptomatic hematoma, no infections, and no chronic pain at the aspiration site [39, 103]. On direct questioning, most patients report no discomfort at the aspiration site on the day following aspiration, and aspiration did not delay discharge from the hospital.

2.6 Processing Options for BMA

Density Separation (DS) (i.e., Centrifuge Processing)—Most methods described in this text will utilize density separation as the primary method for bone marrow processing. Traditional manual processing of bone marrow used "buffy coat" isolation. An anticoagulated marrow sample (acid citrate dextrose or sodium heparin most common) is placed in a test tube and centrifuged for 10–15 min at 400 g. Following this, serum and most platelets rise to the top. RBCs sink to the bottom, along with some dense granulocytes. However, the vast majority of nucleated cells, at intermediate density below 1.025 gm/mL, will layer in a fluffy white sheet that can be gently aspirated from the surface of the RBCs. With skill, this will generate a sample where the concentration of nucleated cells [Cell] is comparable to the concentration of continuing RBCs [RBCs]. This separation can be made more cleanly by adding a density gradient layer (e.g., sucrose or Ficoll™). However, these agents must be removed before injection and may harm nucleated cell and CTP performance, and are not part of current practice.

A variety of commercial centrifuges is available. Each provides disposable tubing and sometimes a unique apparatus to enable separation of nucleated cells from serum and RBCs using cutoffs defined by density floats or optical sensors. There is variation among devices with respect to speed, ease of use, priming volume (marrow volume required), and cost. Almost all will reduce [RBC] by tenfold or more and increase the concentrate of a given marrow sample by three- to sixfold over that of the raw anticoagulated marrow aspirate sample. None of the centrifuge devices increases the prevalence of CTPs above that in the starting marrow sample. All lose some cells and CTPs in the process. In many cases, CTPs in the initial marrow samples can be lost by attachment to tubing or container surfaces, likely due to the innate attribute of CTPs to attach to surfaces based on charge or surface bound adhesion molecules in serum that may coat tubing surfaces (e.g., fibronectin, vitronectin, laminin). A limited number of studies directly compared the yield of different centrifuge products for marrow processing with respect to cell, RBC, and CTP yield and loss [104, 105]. However, process efficiency, as well as optimization of [CTP] and P_{CTP}, is an ongoing area of interest and a point of competition between available centrifuge devices used for bone marrow processing.

Selective Retention (SR)—SR processing is a method that involves the use of the intrinsic ability for CTPs to attach to tissue culture plastic, extracellular matrix molecules (e.g., fibronectin, vitronectin, laminin), and some ceramic materials to selectively retain CTPs in an immobilized state and remove them from other non-adherent cells. This method was designed and tested for use in preparation of a bone graft material and can concentrate CTPs as much as 20-fold in a porous material with appropriate pore size and surface area and increase the P_{CTP} 2–six-fold [51, 82, 106, 107].

SR processing could be used as a method for preparation of materials for injection, but this would require that CTPs that are retained be subsequently released and then deconcentrated prior to injection. In contrast to centrifuge processing, the US FDA considers any attachment and release process to an external substrate to be more than "minimal manipulation," even if this is a normal property of the cell, and not an antibody-mediated process. As a result, the cost of bringing such a therapy to market through a premarket approval (PMA) process as a Class III product is much higher, and little investment has been made in developing this rational alternative for bone grafting as well as for injection therapy. SR processing may face less regulatory restriction in some settings, and this remains a rational low-cost opportunity for marrow processing and CTP isolation in the future.

Magnetic Separation (MS)—MS processing refers to the generalizable capability of labeling cells that express a discriminating surface marker using either an antibody to that marker or some other selectively binding molecule and using that attachment to tag selected cells with a bead or other magnetic feature [48, 108]. Once bound, if the mixture of magnetically tagged cells and non-magnetic cells is placed in a magnetic field, tagged cells will be accelerated in the magnetic field to flow in a predicted path or to become retained upon a limiting surface. MS processing is commonly used clinically for processing large batches of marrow-derived cells to select for CD34+ cells or remove NK cells.

MS processing has been applied successfully to processing of marrow-derived cells using hyaluronan (hyaluronic acid – HA) as a marker for positive selection, using the naturally occurring HA binding peptide in aggrecan and linking it to a magnetic bead. This resulted in rapid isolation of a subset of marrow-derived CTPs that retained HA on their surface, presumably part of their native niche within the marrow tissue. Those HA-positive CTPs were found to proliferate more rapidly and to express an osteogenic phenotype in vitro [48]. Moreover, transplantation of HA+ cells isolated using MS processing improved the quality of bone formation in canine bone defects [108]. MS processing using HA or other putative CTPs markers is an option for rapid processing for injection therapies. However, again, the extra cost and regulatory burden of developing such a therapy must be balanced against the cost and efficacy of DS processing of bone morrow that is already clinically available and alternative injectable product options.

Fluorescence-Activated Cell Sorting (FACS)—FACS processing is a rational means of isolating a specific population or purging an undesirable population, provided specific cell surface markers are known. Despite widespread availability as processing tool for research and for clinical diagnosis, FACS is intrinsically limited as a tool for marrow cell processing for injection therapies by (1) the expense of antibodies required for both positive and negative selection, (2) complexity and time of manual processing for labeling, (3) throughput limits (~25 M cells per hour), (4) dilution of selected cells requiring reconcentration, and (5) trauma to cells during processing, resulting in loss of viability or change in performance. Freshly isolated CTPs in particular are sensitive to FACS processing. Colony forming efficiency of fresh BMA samples is reduced by as much as 90–100% after passing through the fluid dynamics associated with FACS processing, with or without addition of cell separation agents.

2.7 Standardized Measurement and Report of BMA Dose and Composition

As outlined in Chap. 1, understanding the relationship between BMA dose and composition is critically needed in our field. Developing this understanding requires the development of consensus regarding what should be measured and reported, and then collaboration across the discipline, between practitioners, authors, journal editors, medical societies, as well as regulators to collect, share, and use this information. Only by systematic measurement of the composition of BMA or BMA-derived products and linking this to clinical outcomes will our literature support the process of systematic reviews and comparative effectiveness analysis needed to define indications and contraindications and adequately inform patients and clinicians regarding the risks, benefits, and roles of cellular therapies in individual patients and clinical settings. As in other domains of orthopedics and musculoskeletal medicine, through patient registries and reports on cohorts, attention can gradually focus on the critical quality attributes that need to be compared and optimized in randomized trials.

Several recent publications have outlined these issues and provided specific recommendations for reporting, including the source of cells (tissue and anatomic location); methods of harvest, processing, and delivery; as well as dose and composition (volume, concentration of cells, RBCs, platelets, and CTPs) [1, 109]. These principles and reporting parameters are outlined in Chap. 1, Figs. 2.1–2.3.

Assay of cells, RBCs, and platelets can be obtained using standardized automated cell counting systems. Automated cell counters are generally optimized for counting blood samples, but some counters can also be calibrated to provide metrics of bone marrow-derived cells.

Assay of CTPs based on colony formation, while not complex, is not a skill set and resource that is universally available in clinical practices. However, CTP assay methods have been defined and could be performed in any cell biology laboratory, using standard culture technique [33, 42, 110–113]. Marrow samples can therefore be sent to a collaborating or reference laboratory for culture and analysis. Automated methods for CTP colony analysis have been defined and can be adopted [43, 45, 114–116], as shown in Fig. 2.3.

Flow cytometry (FC) and fluorescence-activated cell sorting (FACS) are robust methods for analysis of the composition of a mix cell suspension. Cells can be parsed based on size and granularity, as well as the presence or absence of various surface markers. Generally, 10^5 or 10^6 cells are used in each analysis. Unfortunately, FC and FACS have little value as yet for analysis of CTPs in marrow, since no specific set of markers is yet available that defines CTPs or separates CTPs from non-CTPs. Even putative markers like STRO-1, CD146, CD73, CD90, and PDPN are present on far more non-CTPs than on CTPs in a given marrow population [117]. Moreover, since the prevalence of CTPs is only about 1 in 25 K cells, even a population of 10^6 cells will have only 50 CTPs, which will be buried in the data among a vastly larger number of non-CTPs. Some progress is ongoing, however, the search for CTP-specific markers or marker combinations continues, and FC and FACS may have a greater role to play in the future [49].

Assay of Bioactive Soluble Factors in BMA—There are also opportunities to look at other metrics of potential importance, such as the composition of growth factors, cytokines, and exosomes in the final products. Practical methods for assay of these components exist, but are time and materials intensive, making them practical only in a setting of robust funding. It is important to note that the composition of soluble molecules in a BMA sample is not high on the list of probably effectors of the clinical effect of a given BMA product. Most of the composition of soluble factors in a BMA sample will be derived from contaminating serum. Some will represent the immediate secretory product of the harvested cells in response to the trauma of being ripped from bone marrow. It is most likely that the clinical outcome will be far more dependent on the cellular composition and then the biological output and secretome that those harvested cells will generate in the future, than the early chemical composition of a BMA that will be clinically important.

Biobanking—The development of low-cost biobanking of isolated cell and tissue samples is opening the opportunity for prospective banking of BMA samples, which then become available for future targeted analysis of cell, CTP, and cytokines using standardized methods with appropriate clinical controls. For example, if 10 patients out of 100 go on to fail a BMA treatment and 10 patients out of 100 go on to dramatic clinical success, those 20 samples could be selective pulled out for comparison providing optimal power while avoiding the cost of analysis for all samples. Biobanking is likely to be a common feature of many prospective collaborative clinical registries and many prospective clinical trials, as the field evolves [118, 119].

RNAseq is a new robust methodology that may become valuable for bone marrow analysis and characterization. In particular, single-cell RNAseq (scRNAseq) is a method that can be used to map the diversity of cell populations (e.g., 10,000 cells at a time) in a mixture based on gene expression profile states. This opens a window for unbiased search for BMA population attributes that may be associated with either donors or BMP product composition that can be associated with outcome. As in FC or FACS analysis, the low abundance of CTPs in a given population keeps the CTP population and its heterogeneity well below the radar of scRNAseq. scRNAseq is expensive, but the cost is falling rapidly. Furthermore, methods are being developed that may enable biobanking of cells or isolated intact nuclei allowing later selective analysis [120–122].

2.8 Putting it all Together

Optimizing the use of bone marrow and bone marrow-derived cells for use in cellular therapy continues to represent an area of great challenge and great opportunity. Progress requires understanding the nature and diversity of bone marrow and bone marrow-derived cells. Optimizing care requires the combination of the right patient, disease, disease state, cell source, processing method, composition, and delivery system in order to achieve outcomes that are repeatable and reproducible and know that the outcome is achieved at lowest possible risk and at an acceptable cost.

Linking clinical data (demographics diagnosis, disease stage, comorbidities, and functional status), BMA harvest and composition data, and rigorously documented clinical outcomes is the first step in this process. To date, most of the available data has been generated in a diversity of studies, where the vast majority have not provided a full set of these data. The remaining chapters in this book systematically unveil the current state of the field.

We are grateful for the work of innovators and pioneers who have patiently and thoughtfully brought us and our patients to this promising threshold. We look forward to a next era in the systematic and rational development of rigorously characterized cellular therapies, including bone marrow as a cell source. We expect this era to be dominated by collaborative networks of clinicians and scientists that establish the infrastructure to capture, share, and analyze data using robust standardized methodology through prospective registries and rigorously designed prospective clinical trials.

Take-Home Messages

- Bone marrow aspiration (BMA) provides a diversity of stem and progenitor cells, including connective tissue progenitors (CTPs), hematopoietic stem and progenitor cells (HSCs), and progenitors of vascular endothelium (EPCs).
- BMA can be harvested percutaneously with minimal morbidity and provides a readily processed cell suspension using appropriate anticoagulation.
- BMA can be provided as a point of care product using a centrifuge.
- Processed marrow can be used also to provide the starting materials needed for in vitro culture expansion and fabrication of a diversity of cellular and cell-derived products.
- Yield of CTPs and other marrow-derived cells can be strongly influenced by patient selection, aspiration site, aspiration technique, and subsequent processing.
- BMA is most frequently and effectively harvested from the iliac crest.
- A diversity of processing methods can be used to increase the concentration and prevalence of specific cell types: density separation, selective retention, magnetic separation, and FACS.
- Consensus standards are needed for defining BMA and BMA-derived product composition and the efficacy of processing methods (yield, concentration, prevalence, etc.) and will accelerate the development and comparison of novel BMA-derived therapies and the objective comparison of bone marrow and alternative cell sources.
- Specific clinical indications and optimal processing methods and composition for marrow-derived therapies have yet to be determined and must be defined through rigorous prospective clinical registries and clinical trials.

References

1. Piuzzi NS, Hussain ZB, Chahla J, et al. Variability in the preparation, reporting, and use of bone marrow aspirate concentrate in musculoskeletal disorders: a systematic review of the clinical orthopaedic literature. J Bone Joint Surg Am Vol. 2018;100(6):517–25.
2. Piuzzi NS, Khlopas A, Newman JM, et al. Bone marrow cellular therapies: novel therapy for knee osteoarthritis. J Knee Surg. 2018;31(1):22–6.
3. Piuzzi NS, Chahla J, Jiandong H, et al. Analysis of cell therapies used in clinical trials for the treatment of osteonecrosis of the femoral head: a systematic review of the literature. J Arthroplast. 2017;32(8):2612–8.
4. Marcucio RS, Nauth A, Giannoudis PV, et al. Stem cell therapies in orthopaedic trauma. J Orthop Trauma. 2015;29:S24–7.
5. Lindholm TS, Urist MR. A quantitative analysis of new bone formation by induction in compositive grafts of bone marrow and bone matrix. Clin Orthop Relat Res. 1980;150:288–300.
6. Connolly JF, Shindell R. Percutaneous marrow injection for an ununited tibia. Nebr Med J. 1986;71(4):105–7.
7. Connolly JF, Guse R, Tiedeman J, Dehne R. Autologous marrow injection as a substitute for operative grafting of tibial nonunions. Clin Orthop Relat Res. 1991;266:259–70.
8. Gangji V, De Maertelaer V, Hauzeur J-P. Autologous bone marrow cell implantation in the treatment of non-traumatic osteonecrosis of the femoral head: five year follow-up of a prospective controlled study. Bone. 2011;49(5):1005–9.
9. Hernigou P, Flouzat-Lachaniette CH, Delambre J, et al. Osteonecrosis repair with bone marrow cell therapies: state of the clinical art. Bone. 2014;70:102–9.
10. Yoshioka T, Mishima H, Akaogi H, Sakai S, Li M, Ochiai N. Concentrated autologous bone marrow aspirate transplantation treatment for corticosteroid-induced osteonecrosis of the femoral head in systemic lupus erythematosus. Int Orthop. 2011;35(6):823–9.
11. Piuzzi NS, Chahla J, Schrock JB, et al. Evidence for the use of cell-based therapy for the treatment of osteonecrosis of the femoral head: a systematic review of the literature. J Arthroplast. 2017;32(5):1698–708.
12. Piuzzi NS, Chahla J, Jiandong H, et al. Analysis of cell therapies used in clinical trials for the treatment of osteonecrosis of the femoral head: a systematic review of the literature. J Arthroplast. 2017;32(8):2612–8.
13. Hernigou P, Mathieu G, Poignard A, Manicom O, Beaujean F, Rouard H. Percutaneous autologous bone-marrow grafting for nonunions. Surgical technique. J Bone Joint Surg. 2006;88-A(Supplement 1, Part 2):322–7.
14. Sugaya H, Mishima H, Aoto K, et al. Percutaneous autologous concentrated bone marrow grafting in the treatment for nonunion. Eur J Orthop Surg Traumatol. 2013;24(5):671–8.
15. Fernandez-Bances I, Perez-Basterrechea M, Perez-Lopez S, et al. Repair of long-bone pseudoarthrosis with autologous bone marrow mononuclear cells combined with allogenic bone graft. Cytotherapy. 2013;15(5):571–7.
16. Garg NK, Gaur S. Percutaneous autogenous bone-marrow grafting in congenital tibial pseudarthrosis. J Bone Joint Surg Br. 1995;77(5):830–1.
17. Goel A, Sangwan SS, Siwach RC, Ali AM. Percutaneous bone marrow grafting for the treatment of tibial non-union. Injury. 2005;36(1):203–6.
18. Jäger M, Herten M, Fochtmann U, et al. Bridging the gap: bone marrow aspiration concentrate reduces autologous bone grafting in osseous defects. J Orthop Res. 2011;29(2):173–80.
19. Petri M, Namazian A, Wilke F, et al. Repair of segmental long-bone defects by stem cell concentrate augmented scaffolds: a clinical and positron emission tomography—computed tomography analysis. Int Orthop. 2013;37(11):2231–7.
20. Gessmann J, Köller M, Godry H, Schildhauer TA, Seybold D. Regenerate augmentation with bone marrow concentrate after traumatic bone loss. Orthop Rev (Pavia). 2012;4:1
21. Khashan M, Inoue S, Berven SH. Cell based therapies as compared to autologous bone grafts for spinal arthrodesis. Spine (Phila Pa 1976). 2013;38(21):1885–91.
22. Lee DH, Ryu KJ, Kim JW, Kang KC, Choi YR. Bone marrow aspirate concentrate and platelet-rich plasma enhanced bone healing in distraction osteogenesis of the tibia. Clin Orthop Relat Res. Published online. 2014:1–9.
23. Gobbi A, Karnatzikos G, Sankineani SR. One-step surgery with multipotent stem cells for the treatment of large full-thickness chondral defects of the knee. Am J Sports Med. 2014;42(3):648–57.
24. Buda R, Vannini F, Castagnini F, et al. Regenerative treatment in osteochondral lesions of the talus: autologous chondrocyte implantation versus one-step bone marrow derived cells transplantation. Int Orthop. Published online. 2015:893–900.
25. Giannini S, Buda R, Battaglia M, et al. One-step repair in talar osteochondral lesions: 4-year clinical results and T2-mapping capability in outcome prediction. Am J Sports Med. 2012;41(3):511–8.
26. Do KJ, Lee GW, Jung GH, et al. Clinical outcome of autologous bone marrow aspirates concentrate (BMAC) injection in degenerative arthritis of the knee. Eur J Orthop Surg Traumatol. Published online. 2014:1–7.
27. Wong KL, Lee KBL, Tai BC, Law P, Lee EH, Hui JHP. Injectable cultured bone marrow–derived mesenchymal stem cells in Varus knees with cartilage defects undergoing high tibial osteotomy: a prospective, randomized controlled clinical trial with

2 years' follow-up. Arthrosc J Arthrosc Relat Surg. 2013;29(12):2020–8.

28. Chahla J, Piuzzi NS, Mitchell JJ, et al. Intra-articular cellular therapy for osteoarthritis and focal cartilage defect of the knee. J Bone Joint Surg. 2016;98:1511–21.

29. Piuzzi NS, Khlopas A, Newman JM, et al. Bone marrow cellular therapies: novel therapy for knee osteoarthritis. J Knee Surg. 2018;31(1):22–6.

30. Pascual-Garrido C, Rolón A, Makino A. Treatment of chronic patellar tendinopathy with autologous bone marrow stem cells: a 5-year-followup. Stem Cells Int. 2012:953510.

31. Stein BE, Stroh DA, Schon LC. Outcomes of acute Achilles tendon rupture repair with bone marrow aspirate concentrate augmentation. Int Orthop. 2015;39:901–5.

32. Muschler GF, Midura RJ. Connective tissue progenitors: practical concepts for clinical applications. Clin Orthop Relat Res. 2002;1(395):66–80.

33. Mantripragada VP, Bova WA, Piuzzi NS, et al. Native-osteoarthritic joint resident stem and progenitor cells for cartilage cell-based therapies: a quantitative comparison with respect to concentration and biological performance. Am J Sports Med. Published online. 2019:1–10.

34. Brighton CT, Lorich DG, Kupcha R, Reilly TM, Jones AR, Woodbury RA. The pericyte as a possible osteoblast progenitor cell. Clin Orthop Relat Res. 1992;275:287–99.

35. Connolly J, Guse R, Lippiello L, Dehne R. Development of an osteogenic bone-marrow preparation. J Bone Joint Surg Am. 1989;71(5):684–91.

36. Gimble JM, Robinson CE, Wu X, Kelly KA. The function of adipocytes in the bone marrow stroma: an update. Bone. 1996;19(5):421–8.

37. Huard C, Moisset PA, Dicaire A, et al. Transplantation of dermal fibroblasts expressing MyoD1 in mouse muscles. Biochem Biophys Res Commun. 1998;248(3):648–54.

38. O'Driscoll SW. Articular cartilage regeneration using periosteum. Clin Orthop Relat Res. 1999;367(Suppl):S186–203.

39. Muschler GF, Boehm C, Easley K. Aspiration to obtain osteoblast progenitor cells from human bone marrow: the influence of aspiration volume. J Bone Joint Surg. 1997;79-A:1699–709.

40. Mantripragada VP, Bova WA, Boehm C, et al. Primary cells isolated from human knee cartilage reveal decreased prevalence of progenitor cells but comparable biological potential during osteoarthritic disease progression. J Bone Joint Surg. 2018;100:1771–80.

41. Mantripragada VP, Bova WA, Boehm C, et al. Progenitor cells from different zones of human cartilage and their correlation with histopathological osteoarthritis progression. J Orthop Res. 2018;36(6):1728–38.

42. Mantripragada VP, Boehm C, Bova W, Briskin I, Piuzzi NS, Muschler GF. Patient age and cell con-centration influence prevalence and concentration of progenitors in bone marrow aspirates: an analysis of 436 patients. J Bone Joint Surg. Published online. 2021:1–9.

43. Qadan MA, Piuzzi NS, Boehm C, et al. Variation in primary and culture-expanded cells derived from connective tissue progenitors in human bone marrow space, bone trabecular surface and adipose tissue. Cytotherapy. 2018;20(3):343–60.

44. Mantripragada VP, Bova W, Boehm C, et al. Primary cells isolated from human knee cartilage reveal decreased prevalence of progenitors cells but comparable biological potential during osteoarthritic disease progression. J Bone Joint Surg Am Vol. Published online Accepted. 2018;100(20):1771–80.

45. Mantripragada VP, Piuzzi NS, Bova WA, et al. Donor-matched comparison of chondrogenic progenitors resident in human infrapatellar fat pad, synovium, and periosteum - implications for cartilage repair. Connect Tissue Res. 2019;60(6): 597–610.

46. Mantripragada VP, Boehm C, Bova W, Briskin I, Piuzzi NS, Muschler GF. Patient age and cell concentration influence prevalence and concentration of progenitors in bone marrow aspirates: an analysis of 436 patients. J Bone Joint Surg. Published online. 2021:1–9.

47. Mantripragada VP, Tan K-L, Vasavada S, Bova W, Barnard J, Muschler GF. Characterization of heterogeneous primary human cartilage-derived cell population using non-invasive live-cell phase-contrast time-lapse imaging. Cytotherapy. 2020:1–12.

48. Caralla T, Boehm C, Hascall V, Muschler G. Hyaluronan as a novel marker for rapid selection of connective tissue progenitors. Ann Biomed Eng. 2012;40(12):2559–67.

49. Chan CKF, Gulati GS, Sinha R, et al. Identification of the human skeletal stem cell. Cell. 2018;175(1): 43–56.e21.

50. Kim EJ, Fleischman AJ, Muschler GF, Roy S. Response of bone marrow derived connective tissue progenitor cell morphology and proliferation on geometrically modulated micro-textured substrates. Biomed Microdevices. 2013;15(3):385–96.

51. Muschler GF, Matsukura Y, Nitto H, et al. Selective retention of bone marrow-derived cells to enhance spinal fusion. Clin Orthop Relat Res. 2005;432:242–51.

52. McLain RF, Fleming JE, Boehm CA, Muschler GF. Aspiration of osteoprogenitor cells for augmenting spinal fusion: comparison of progenitor cell concentrations from the vertebral body and iliac crest. J Bone Joint Surg Am. 2005;87(12):2655–61.

53. Muschler GF, Nakamoto C, Griffith LG. Engineering principles of clinical cell-based tissue engineering. J Bone Joint Surg Ser A. 2004;86(7):1541–58.

54. Muschler GF, Midura RJ, Nakamoto C. Practical modeling concepts for connective tissue stem cell

and progenitor compartment kinetics. J Biomed Biotechnol. 2003;2003(3):170–93.

55. Dominici M, Le Blanc K, Mueller I, et al. Minimal criteria for defining multipotent mesenchymal stromal cells. The International Society for Cellular Therapy position statement. Cytotherapy. 2006;8(4):315–7.

56. Krampera M, Galipeau J, Shi Y, Tarte K, Sensebe L. Immunological characterization of multipotent mesenchymal stromal cells-the international society for cellular therapy (ISCT) working proposal. Cytotherapy. 2013;15(9):1054–61.

57. Hernigou P, Poignard A, Rouard H. Percutaneous autologous bone-marrow grafting for nonunions: influence of the number and concentration of progenitor cells. J Bone Joint Surg. 2005;87-A(7):1430–7.

58. Piuzzi N, Ng M, Chughtai M, et al. The stem-cell market for the treatment of knee osteoarthritis: a patient perspective. J Knee Surg. 2018;31(6):551–6.

59. Phemister DB. The fate of transplanted bone and regenerative power of its various constituents. Surg Gynecol Obstetr. 1914;19:303–14.

60. Phemister DB. Treatment of the necrotic head of the femur in adults. J Bone Joint Surg. 1949;31(1):55–66.

61. BURWELL RG. Studies in the transplantation of bone. V. the capacity of fresh and treated homografts of bone to evoke transplantation immunity. J Bone Joint Surg Br. 1963;45 B:386–401.

62. Mauffrey C, Barlow BT, Smith W. Management of segmental bone defects. J Am Acad Orthop Surg. 2015;23(3):143–53.

63. Pape HC, Evans A, Kobbe P. Autologous bone graft: properties and techniques. J Orthop Trauma. 2010;24(SUPPL. 1):36–40.

64. Younger EM, Chapman MW. Morbidity at bone graft donor sites. J Orthop Trauma. 1989;3(3):192–5.

65. Robertson PA, Wray AC. Natural history of posterior iliac crest bone graft donation for spinal surgery: a prospective analysis of morbidity. Spine (Phila Pa 1976). 2001;26(13):1473–6.

66. Banwart JC, Asher MA, Hassanein RS. Iliac crest bone graft harvest donor site morbidity. A statistical evaluation. Spine (Phila Pa 1976). 1995;20(9):1055–60.

67. Urgery S, Ncorporated I. Prospective observational study of donor-site. J Bone Joint Surg. 2012;94(18):1649–54.

68. Piuzzi NS, Mantripragada VP, Kwee E, et al. Bone marrow-derived cellular therapies in orthopaedics part 1: recommendations for bone marrow aspiration technique and safety. JBJS Rev. 2018;6(11):e5.

69. Steadman RJ, Rodkey WG, Rodrigo JJ. Microfracture: surgical technique and rehabilitation to treat chondral defects. Clin Orthop Relat Res. 2001;391S:S362–9.

70. Cole BBJ, Pascual-garrido C, Grumet RC. Surgical Management of Articular Cartilage Defects in the knee. J Bone Joint Surg. 2009;91(7):1778–90.

71. Hoyt BW, Pavey GJ, Potter BK, Forsberg JA. Heterotopic ossification and lessons learned from fifteen years at war: a review of therapy, novel research, and future directions for military and civilian orthopaedic trauma. Bone. 2018;109:3–11.

72. Forsberg JA, Pepek JM, Wagner S, et al. Heterotopic ossification in high-energy wartime extremity injuries: prevalence and risk factors. J Bone Joint Surg Ser A. 2009;91(5):1084–91.

73. Shankar A, Daniel RT, Walter N, Chandy MJ. Heterotopic ossification in the orbit. Surg Neurol. 2002;58(6):421–3.

74. Jäger M, Hernigou P, Zilkens C, et al. Cell therapy in bone healing disorders. Orthop Rev (Pavia). 2010;2(2):e20.

75. Papathanasopoulos A, Kouroupis D, Henshaw K, McGonagle D, Jones EA, Giannoudis PV. Effects of antithrombotic drugs fondaparinux and tinzaparin on in vitro proliferation and osteogenic and chondrogenic differentiation of bone-derived mesenchymal stem cells. J Orthop Res. 2011;29(9):1327–35.

76. Do Amaral RJFC, Da Silva NP, Haddad NF, et al. Platelet-rich plasma obtained with different anticoagulants and their effect on platelet numbers and mesenchymal stromal cells behavior in vitro. Stem Cells Int. 2016:7414036.

77. Luangphakdy V, Boehm C, Pan H, Herrick J, Zaveri P, Muschler GF. Assessment of methods for rapid intraoperative concentration and selection of marrow-derived connective tissue progenitors for bone regeneration using the canine femoral multidefect model. Tissue Eng Part A. 2016;22(1–2):17–30.

78. Meppelink AM, Wang XH, Bradica G, et al. Rapid isolation of bone marrow mesenchymal stromal cells using integrated centrifuge-based technology. Cytotherapy. 2016;18(6):729–39.

79. Muschler GF, Matsukura Y, Nitto H, et al. Selective retention of bone marrow-derived cells to. Clin Orthop Relat Res. 2005;432:242–51.

80. Chu W, Zhuang Y, Gan Y, Wang X, Tang T, Dai K. Comparison and characterization of enriched mesenchymal stem cells obtained by the repeated filtration of autologous bone marrow through porous biomaterials. J Transl Med. 2019;17(1):1–16.

81. Luo K, Gao X, Gao Y, et al. Multiple integrin ligands provide a highly adhesive and osteoinductive surface that improves selective cell retention technology. Acta Biomater. 2019;85:106–16.

82. Luangphakdy V, Boehm C, Pan H, Herrick J, Zaveri P, Muschler GF. Assessment of methods for rapid intraoperative concentration and selection of marrow-derived connective tissue progenitors for bone regeneration using the canine femoral multidefect model. Tissue Eng Part A. 2016;22(1–2).

83. Yousef MAA, La Maida GA, Misaggi B. Long-term radiological and clinical outcomes after using bone marrow mesenchymal stem cells concentrate obtained with selective retention cell technology in posterolateral spinal fusion. Spine (Phila Pa 1976). 2017;42(24):1871–9.

84. Yang P, Xing J, Chen B, et al. The clinical use of the enriched bone marrow obtained by selective cell retention technology in treating adolescent idiopathic scoliosis. J Orthop Transl. 2021;27(September 2019):146–52.

85. Caralla T, Joshi P, Fleury S. et al, In vivo transplantation of autogenous marrow-derived cells following rapid intraoperative magnetic separation based on hyaluronan to augment bone regeneration.

86. Jia Z, Liang Y, Xu X, et al. Isolation and characterization of human mesenchymal stem cells derived from synovial fluid by magnetic-activated cell sorting (MACS). Cell Biol Int. 2018;42(3):262–71.

87. Petters O, Schmidt C, Thuemmler C, et al. Point-of-care treatment of focal cartilage defects with selected chondrogenic mesenchymal stromal cells—an in vitro proof-of-concept study. J Tissue Eng Regen Med. 2018;12(7):1717–27.

88. Piuzzi NS, Mantripragada VP, Kwee E, et al. Bone marrow-derived cellular therapies in orthopaedics part 2: recommendations for reporting the quality of bone marrow-derived cell populations. JBJS Rev. 2018;6(11):e5.

89. McLain RF, Fleming JE, Boehm CA, Muschler GF. Aspiration of osteoprogenitor cells for augmenting spinal fusion: comparison of progenitor cell concentrations from the vertebral body and iliac crest. J Bone Joint Surg Ser A. 2005;87(12 I):2655–61.

90. Hernigou J, Alves A, Homma Y, Guissou I, Hernigou P. Anatomy of the ilium for bone marrow aspiration: map of sectors and implication for safe trocar placement. Int Orthop. 2014;38(12):2585–90.

91. Hernigou J, Picard L, Alves A, Silvera J, Homma Y, Hernigou P. Understanding bone safety zones during bone marrow aspiration from the iliac crest: the sector rule. Int Orthop. 2014;38(11):2377–84.

92. Pierini M, Di Bella C, Dozza B, et al. The posterior iliac crest outperforms the anterior iliac crest when obtaining mesenchymal stem cells from bone marrow. J Bone Joint Surg Am. 2013;95(12):1101–7.

93. Patterson TE, Boehm C, Nakamoto C, et al. The efficiency of bone marrow aspiration for the harvest of connective tissue progenitors from the human iliac crest. J Bone Joint Surg Am. 2017;99:19.

94. Hernigou P, Homma Y, Flouzat Lachaniette CH, et al. Benefits of small volume and small syringe for bone marrow aspirations of mesenchymal stem cells. Int Orthop. 2013;37(11):2279–87.

95. Haseler LJ, Sibbitt RR, Sibbitt WL, et al. Syringe and needle size, syringe type, vacuum generation, and needle control in aspiration procedures. Cardiovasc Interv Radiol. 2011;34(3):590–600.

96. Muschler GF, Boehm C, Easley K. Aspiration to obtain osteoblast progenitor cells from human bone marrow: the influence of aspiration volume. J Bone Joint Surg Ser A. 1997;79(11):1699–709.

97. Patterson TE, Boehm C, Nakamoto C, et al. The efficiency of bone marrow aspiration for the harvest of connective tissue progenitors from the human iliac crest. J Bone Joint Surg Am. 2017;99:1673–82.

98. Jillella AP, Ustun C. What is the optimum number of CD34+ peripheral blood stem cells for an autologous transplant? Stem Cells Dev. 2004;13(6):598–606.

99. Hernigou P, Desroches A, Queinnec S, et al. Morbidity of graft harvesting versus bone marrow aspiration in cell regenerative therapy. Int Orthop. 2014;38(9):1855–60.

100. Bain BJ. Bone marrow biopsy morbidity and mortality. Br J Haematol. 2003;121(6):949–51.

101. Bain BJ. Bone marrow biopsy morbidity and mortality: 2002 data. Clin Lab Haematol. 2004;26(5):315–8.

102. Bain BJ. Bone marrow biopsy morbidity: review of 2003. J Clin Pathol. 2005;58(4):406–8.

103. Majors AK, Boehm CA, Nitto H, Midura RJ, Muschler GF. Characterization of human bone marrow stromal cells with respect to osteoblastic differentiation. J Orthop Res. 1997;15(4):546–57.

104. Gaul F, Bugbee WD, Hoenecke HR, D'Lima DD. A review of commercially available point-of-care devices to concentrate bone marrow for the treatment of osteoarthritis and focal cartilage lesions. Cartilage. 2019;10(4):387–94.

105. Dragoo JL, Guzman RA. Evaluation of the consistency and composition of commercially available bone marrow aspirate concentrate systems. Orthop J Sport Med. 2020;8(1):1–8.

106. Muschler GF. Methods of preparing a composite bone graft. Published online 1998.

107. Muschler GF. Apparatus and methods for preparing an implantable graft. Published online 2000.

108. Caralla T, Joshi P, Fleury S, et al. In vivo transplantation of autogenous marrow-derived cells following rapid intraoperative magnetic separation based on hyaluronan to augment bone regeneration. Tissue Eng Part A. 2013;19(1–2):125–34.

109. Murray I, Chahla J, Safran M, et al. International expert consensus on a cell therapy communication tool: DOSES. J Bone Joint Surg. 2019;101:904–11.

110. Mantripragada VP, Piuzzi NS, George J, et al. Reliable assessment of bone marrow and bone marrow concentrates using automated hematology analyzer. Regen Med. 2021;14(7):639–46.

111. Bidula J, Boehm C, Powell K, et al. Osteogenic progenitors in bone marrow aspirates from smokers and nonsmokers. Clin Orthop Relat Res. 2006;442:252–9.

112. Muschler GF, Boehm C, Easley K. Aspiration to obtain osteoblast progenitor cells from human bone marrow: the influence of aspiration volume. J Bone Joint Surg Am. 1997;79(11):1699–709.

113. McLain RF, Boehm CA, Rufo-Smith C, Muschler GF. Transpedicular aspiration of osteoprogenitor cells from the vertebral body: progenitor cell concentrations affected by serial aspiration. Spine J. 2009;9(12):995–1002.

114. ASTM F2944 - 20. Standard practice for automated colony forming unit (CFU) assays—image

acquisition and analysis method for enumerating and characterizing cells and colonies in culture. ASTM Int. Published online. 2020.

115. Mantripragada VR, Luagphakdy V, Kwee E, Piuzzi N, Powell K, Muschler G. Automated imaging and analysis of colony founding stem and progenitor cells—correlation of early quality attributes with future biological performance. Cytotherapy. 2017;19(5):S119.

116. Powell K, Nakamoto C, Villarruel S, Boahm C, Muschler G. Quantitative image analysis of connective tissue progenitors. Anal Quant Cytol Histol. Published online. 2007:112–21.

117. Kwee E, Saidel G, Powell K, Heylman C, Boehm C, Muschler G. Quantifying proliferative and surface marker heterogeneity in colony-founding connective tissue progenitors and their progeny using time-lapse microscopy. J Tissue Eng Regen Med. 2019;13(2):203–16.

118. Piuzzi NS, Dominici M, Long M, et al. Proceedings of the signature series symposium "cellular therapies for orthopaedics and musculoskeletal disease proven and unproven therapies—promise, facts and fantasy," international society for cellular therapies, Montreal, Canada, may 2, 2018. Cytotherapy. 2018;20(11):1381–400.

119. Chu CR, Rodeo S, Bhutani N, et al. Optimizing clinical use of biologics in orthopaedic surgery: consensus recommendations from the 2018 AAOS/NIH U-13 conference. J Am Acad Orthop Surg. 2019;27(2):E50–63.

120. Freeman BT, Jung JP, Ogle EM. Single-cell RNA-Seq of bone marrow-derived mesenchymal stem cells reveals unique profiles of lineage priming. PLoS One. 2015;10:9.

121. Wolock SL, Krishnan I, Tenen DE, et al. Mapping distinct bone marrow niche populations and their differentiation paths. Cell Rep. 2019;28(2):302–11.e5.

122. Tikhonova AN, Dolgalev I, Hu H, et al. The bone marrow microenvironment at single-cell resolution. Nature. 2019;569(7755):222–3.

Adipose-Derived Stem/Stromal Cells, Stromal Vascular Fraction, and Microfragmented Adipose Tissue

Enrico Ragni, Marco Viganò, Paola De Luca, Edoardo Pedrini, and Laura de Girolamo

3.1 Introduction

Adipose tissue, for a long time, has been considered merely a storage of excess energy, but more recent evidence has helped shed some light on its role [1], comprising energy balance storage [2], as well as regulating bone metabolism, hematopoiesis, and the inflammatory response [3].

Adipose is a highly vascularized structure, composed of a heterogeneous mixture of cell populations, primarily derived from interlobular and perivascular connective tissues, consisting of mature adipocytes, preadipocytes, fibroblasts, vascular smooth muscle cells, endothelial cells, resident monocytes/macrophages, and lymphocytes, as well as progenitor cells and mesenchymal stem/stromal cells (MSCs). The presence of MSCs within this tissue (ASCs, adipose-derived stem/stromal cells) has recently drawn significant clinical attention due to their purported paracrine effects and multipotent differentiation capacity [4]. To date, its use as source of pro-

regenerative cells has been successfully reported in a variety of preclinical and clinical applications, including musculoskeletal conditions, cardiac diseases, ischemia, amyotrophic lateral sclerosis, diabetes, and Alzheimer's and Parkinson's diseases [5]. Considering the promising results achieved so far, a wide array of lab-driven technologies is actively studied to undergo the process of a more efficient translation into the clinical setting.

Adipose tissue either can be used to isolate ASCs or can be processed at the point of care to obtain adipose-derived products. In the former case, ASCs are efficiently isolated by tissue enzymatic digestion and then culture expanded as adherent monolayers. In this setting, ASCs are generally consistent with the International Society for Cellular Therapy (ISCT) accepted attributes mesenchymal stromal cell populations (MSCs). Differently, adipose tissue can be processed at the point of care into cell suspensions or microfragments that have been commonly referred to as stromal vascular fraction (SVF) or microfragmented adipose tissue (microfat), respectively [6].

Both these strategies for the use of adipose-derived therapeutic cellular products have advantages and pitfalls. The approach based on cultured ASCs provides a standardized cell population of stem/stromal cells, compared to the use of SVF or microfat, in which different cell types (i.e.,

E. Ragni · M. Viganò · P. De Luca · E. Pedrini
L. de Girolamo (✉)
Orthopaedic Biotechnology Laboratory, IRCCS
Istituto Ortopedico Galeazzi, Milan, Italy
e-mail: enrico.ragni@grupposandonato.it;
marco.vigano@grupposandonato.it;
deluca.paola@grupposandonato.it;
laura.degirolamo@grupposandonato.it

G. Filardo et al. (eds.), *Orthobiologics*, https://doi.org/10.1007/978-3-030-84744-9_3

endothelial cells, progenitor cells, and leukocytes) are represented together with mesenchymal stem/stromal cells [7]. On the other hand, the use of microfat or SVF has the theoretical and practical advantages of providing a point of care therapy that does not imply the cost and risk of in vitro culture expansion. Moreover, preparation of SVF and microfat may preserve the tissue native niche, which is composed by different cell types including stem and progenitor cells.

Still many controversial points animate the debate on the most effective procedure. To shed some light, in the next paragraphs, a more in-depth description of both cells and techniques as well as applications will be discussed, with a final focus on orthopedic-related tissues and diseases.

3.2 Adipose-Derived Stem/ Stromal Cells and Adipose-Derived Products: Two Sides of the Same Moon

3.2.1 SVF and Microfat

Although they have some similarities, including being prepared at the point of care and the characteristic of preserving the tissue niche, SVF and microfat also present some substantial differences.

The adipose tissue SVF is defined as a heterogeneous population of freshly isolated cells comprising all the different types of cells residing in the tissue such as fibroblasts, preadipocytes, vascular smooth muscle cells, endothelial cells, resident monocytes/macrophages, and lymphocytes, except mature adipocytes. The process to obtain SVF may exceed the definition of "minimal manipulation" as it is frequently based on enzymatic tissue digestion. However mechanical dissociation, albeit less efficient in terms of cell recovery, is currently favored mainly for regulatory reasons. In contrast, microfat, obtained by mechanical processing only, is composed of clusters of blood- and lipids-free adipose tissue ranging from tens to few hundred micrometers in diameter, containing all the adipose tissue cells, including adipocytes, within their native niche [8,

9]. Moreover, microfat, beyond preserving the cell composition, also preserves the tissue microarchitecture [10]. Borrowing the concept from the world of bone marrow and bone marrow concentrate (BMAC), it is quite common to refer to these adipose-derived products as "cell concentrates." Actually, this is improper since, especially for microfat, the production process is not designed to concentrate any population type, but rather to eliminate blood and lipid residuals known to be pro-inflammatory agents [10]. Both the SVF and microfat have similar nucleated cell number per gram of product, as well as similar proliferation abilities and the expression of the typical MSC marker CD90; nevertheless, the proportion of cells positive for CD34 and CD45 appears to be higher in SVF compared to microfat [11, 12], underlying the higher blood contamination in SVF.

Both products have shown anti-inflammatory and immunomodulatory potential, and reparative effects in vivo [13], and safety in a growing number of clinical trials [14–16], including musculoskeletal diseases. Moreover, the undisputable practical advantages associated to the use of SVF and microfat over culture-expanded ASCs have made them very popular among the orthopedic community, as revealed by the increasing number of publications reporting the results of their application [17].

3.2.2 Culture-Expanded Adipose-Derived Stem Cells (ASCs)

Within the SVF, not all the cells are likely to have a therapeutic effect [18]. Among them ASCs have a role of paramount importance in regenerative medicine, and therefore many therapeutic approaches are based on the use of these cells only. A small fraction of the adipose tissue is in fact represented by ASCs that can be isolated and induced to proliferate in culture to generate expanded populations. The process starts with the enzymatic isolation of the SVF, and then it further proceeds with in vitro expansion in appropriate culture media leading to the loss of the native adipose structure and the achievement of a homogeneous population of expanded cells that can be rigorously characterized in terms of cell markers, morphology, and secretory profiles. Interestingly, adipose tissue contains up to 3% of MSCs, whereas in bone marrow it is reported between 0.002% and 0.02% [19]. The identification of the heterogeneous stem/stromal cell types and native phenotypes in their environment is still a matter of debate [20]. There is growing evidence supporting the hypothesis that these cells and more in general MSCs reside in a perivascular location. Consistently, the ability of MSCs to stabilize blood vessels and contribute to tissue homeostasis in both physiological and injury conditions has also led some authors to propose that MSCs are a subpopulation of pericytes [21].

Culture-expanded ASCs match the criteria reported in the ISCT guidelines aimed to standardize the concept and metrics used for culture-expanded products and the appropriate use of the term MSCs. The definition and required attributes for MSC included the adherence to plastic support, the capacity for tri-lineage differentiation (adipocyte, chondroblast, osteoblast) in vitro, the expression of cell surface markers (CD73, CD90, and CD105), but the lack of cell surface markers associated with hematopoietic stem cells and progenitors (CD45, CD34, CD14 or CD11b, CD79a or CD19, and HLA-DR) [22]. More recently, other potentially useful markers have been proposed, like positivity for CD13, CD29, and CD44 and absence of CD31 and CD235a [23]. Further, cell size and granularity, telomere length, senescence status, trophic factor secretion, and immunomodulation ability [24, 25] can also be evaluated. The opportunity to characterize ASCs, in theory, should lead to more reproducible product assessment and outcomes [26]. However, since the techniques of expansion can affect the relative proportion and features of the expanded cell populations [27, 28], individual batches of ASCs can vary significantly with respect to these metrics. All the aforementioned are attributes that must be considered as predictive of the potency of any culture-expanded cell population that may be used in regenerative medicine [29]. Therefore, being able to optimize the

population attributes, including the secretion of soluble factors, might allow the development of tailored cell-based protocols to achieve the desired result.

However, this strategy requires a GMP facility and a minimum of two procedures (harvest and administration) to complete the treatment, increasing the cost for both patients and NHS or other payors.

3.3 Influence of Patient-Specific Factors on Adipose-Derived Cells and Products

The tissue source selection, processing methods, injection techniques, cell composition, and cell dose have been extensively studied for years, and the efforts of researchers are still aimed at their standardization. Nevertheless, the variability in terms of outcome suggests the presence of patient-specific factors such as age, body mass index (BMI), gender, and harvest sites as confounding variables in the evaluation.

Studies have shown a slight decrease in the overall yield of nucleated cells with increasing age [30], as well as a significant decrease in the proliferative and differentiation capacities of culture-expanded ASCs [31]. This result is in keeping with studies on bone marrow-derived expanded MSCs where age is negatively correlated with cell viability and overall potential [32]. Nevertheless, despite a lower yield of pro-regenerative cells per gram of tissue, the autologous transplantation of ASCs seems to be still a feasible option for elderly patients [33].

A higher BMI has been associated with a reduced number of viable mature adipocytes per gram of tissue, a lower differentiation capacity of the culture-expanded ASCs, reduced capacity of cell migration, and angiogenic and proliferative abilities [30], probably due to the low oxygen condition and inflammatory conditions observed in adipose tissue of obese patients. Interestingly, the effect of BMI on cell performance can be reverted. Bariatric surgery and diet-induced long-term calorie restriction could improve cultured ASCs profile, with reduced DNA damage, improved viability, and extended replicative life span [34]. This evidence is in line with studies reporting a positive connection between weight loss and reduced inflammation [35].

The role of gender and donor site is still controversial. Some studies on human ASCs isolation failed to show any difference in adipose tissue native stem/stromal cell concentration, prevalence, or yield by gender. However, another study suggested that men might have a higher yield compared to women [36]. Likewise, the ideal donor site for fat harvest is yet to be defined.

Some studies [37–39] showed that fat from the lower abdomen and medial thighs has higher yield compared to the upper abdomen, trochanteric region, knees, and flanks but similar differentiation potential. However, previous studies suggest that the choice of donor site has little effect on fat graft outcomes and the choice should be based on ease and safety of access to the tissue [30]. Other parameters, such as diet, lifestyle, drug consumption, and smoke and alcohol habit, should be also investigated to identify a possible influence on the pre-regenerative properties of adipose-derived cells or products.

3.4 The Rationale for Using Injections of Culture-Expanded ASCs or Adipose-Derived Products

Both ASCs and adipose-derived products can be delivered mainly with two approaches, which imply different mechanisms of action. The first one relies on the seeding of cells/SVF or microfat on scaffolds to generate tissue and organs, and it is typically used in association with surgery, such as repair of focal chondral lesion or tendon rupture, as well as treatment of critical bone defects. Cells/SVF or microfat are seeded on a support (scaffold) and can exert their function by both paracrine regulations of the microenvironment and direct differentiation into tissue-specific cells, albeit not complete. The second approach relies on the direct delivery of cells/SVF or microfat to damaged sites, typically by injections or infusions. In this case, many findings suggest that, despite still being a valid model in different appli-

cations [40], the direct cell trans-differentiation mechanism would not be the main responsible for the benefits observed after MSCs transplantation, but rather the therapeutic effect is related to the secretion of soluble factors able to regulate the cross-talk with resident cells [41]. However, in the absence of adequate support for attachment, cells alone after injection on the site are generally stressed, sometimes leading to a rapid death [42]. In this view the delivery of cells within their niche, as it happens with SVF and even more with microfat, could protect them from this phenomenon. Nevertheless, the initiation of the restoration process is guaranteed by the initial cross-talk between stem/stromal cells and resident cells, and therefore their long term-survival at the site of injection is not a strict requirement for their functioning. The low engraftment rate documented in lung injury models or cardiac infarcts after MSCs infusion [43, 44], and studies demonstrating similar or even improved organ function upon infusion of MSC-derived conditioned medium (MSC-CM) with respect to whole MSCs [45], are all supporting a paracrine role of MSCs. Therefore, the research interest is also shifting on the characterization of secreted factors, collectively termed as the "secretome."

3.4.1 Paracrine Potential (Soluble Mediators and Exosomes/Microvesicles)

The term secretome refers to the wide array of secreted factors, such as cytokines and chemokines or lipids with trophic and immunomodulatory activities [46]. Since Caplan's description of MSCs as "drugstores," i.e., elements that recognize injury signals and became activated in order to release bioactive molecules able to modulate local immune response and to establish a regenerative microenvironment [47], a number of elements, such as trophic (anti-scarring, anti-apoptotic, mitogenic, angiogenic), immunomodulatory, and also antimicrobic factors, were identified in MSCs secretome [48]. Therefore, the traditional paradigm of MSCs as a "cell replacement tool" has been now enriched by a new vision of MSCs as "sensing cells" that interact with tissue progenitor cells through a paracrine action, which stimulates the innate potential of the tissue in the repair and modulation of inflammatory and immune reactions. These features have defined the rationale behind the use of MSCs as therapeutic tool in treating joint diseases like osteoarthritis. Accordingly, MSCs were shown to modulate the function of the immune system typically dysregulated during joint inflammation, by suppressing B cells and inhibiting T cells proliferation, together with attracting regulatory T cells and promoting the release of anti-inflammatory factors [49]. Even more importantly, MSCs were reported to promote in macrophages the transition from pro-inflammatory M1 to anti-inflammatory M2 polarization, inhibiting the release of pro-inflammatory cytokines (TNF-α and IL-1β), and augmenting the secretion of anti-inflammatory molecules (IL-10) [50]. As a consequence, polarization switch may reduce the cartilage degeneration mediated by inflammatory macrophages [51]. Consistently, the effectiveness of native or culture-expanded ASCs (and related products) paracrine action was demonstrated on chondrocytes and tenocytes exposed to pathological conditions, with results suggesting a restoration of tissue homeostasis [52, 53]. Then, a significant amount of research explored the possibility of modulating these factors through the adoption of different culturing conditions, paving the way for the development of acellular therapeutic interventions for autoimmune, inflammatory, and malignant diseases and tissue regeneration from cellular secretions derived from MSCs (Fig. 3.1).

Among all the components of the secretome, extracellular vesicles (EVs) were also identified as active entities [54]. EVs embed different type of molecules (DNA, mRNAs, miRNAs, pre-miRNAs, ncRNAs, and proteins), can be found in different biological fluids, and are secreted by a wide range of cell types including MSCs [55]. The recent advent of omics techniques allowed a better characterization of these vesicles and fostered research on their involvement in the regulation of different biological processes [56]. Consistently, EVs from MSCs showed an immunosuppressive role on many types of immune cells [57]. In specific, treatment of T cells in vitro resulted in a marked decrease in proliferation and downregulation of IFN-γ and TNF-α secretion [58], with inflammation efficiently suppressed in vivo [59]. Moreover, EVs from cultured ASCs had positive effects in skin regeneration and cardiac, liver, and neuroprotection [60] with strong attractive properties as potential therapeutic candidates also in the orthopedic settings since the reported attenuation of the inflammatory response and the degeneration after both tendon or cartilage injury [61, 62].

Overall, although further studies involving the safety and duration of EVs therapeutic effect are needed, MSC-derived EVs are the most promising candidates for a rational design of next-generation cell-free MSC-based therapeutics mainly derived from adipose tissue. In fact, the use of EVs avoid potential safety concerns typical of cell-based approaches (i.e., tumorigenicity and undesired spontaneous differentiation). Considering their natural biogenesis process, EVs are generated with high biocompatibility, enhanced stability, and limited immunogenicity, which provide multiple advantages as drug delivery systems over traditional synthetic methods. In this context, EVs can penetrate the tissues and be bioengineered to enhance the targetability, avoiding off-target effects. In comparison with cell-based approaches, their manufacturing is also more competitive in terms of cost-effectiveness. In this perspective, few clinical trials of Phase I, II, and III have been opened in the last years, covering diseases such as macular holes (NCT03437759) or diabetes mellitus type 1 (NCT02138331) or ischemic stroke (NCT03384433) [63]. Rational and potential of extracellular vesicles—exosomes are reported more in detail in Chap. 11.

3.5   In Vitro and Preclinical Findings

As already mentioned, the interest in the use of ASCs and adipose-derived products such as SVF and microfat in musculoskeletal applications is dramatically increasing over the last years. In the following paragraph, we will comment on the most relevant findings of in vitro preclinical

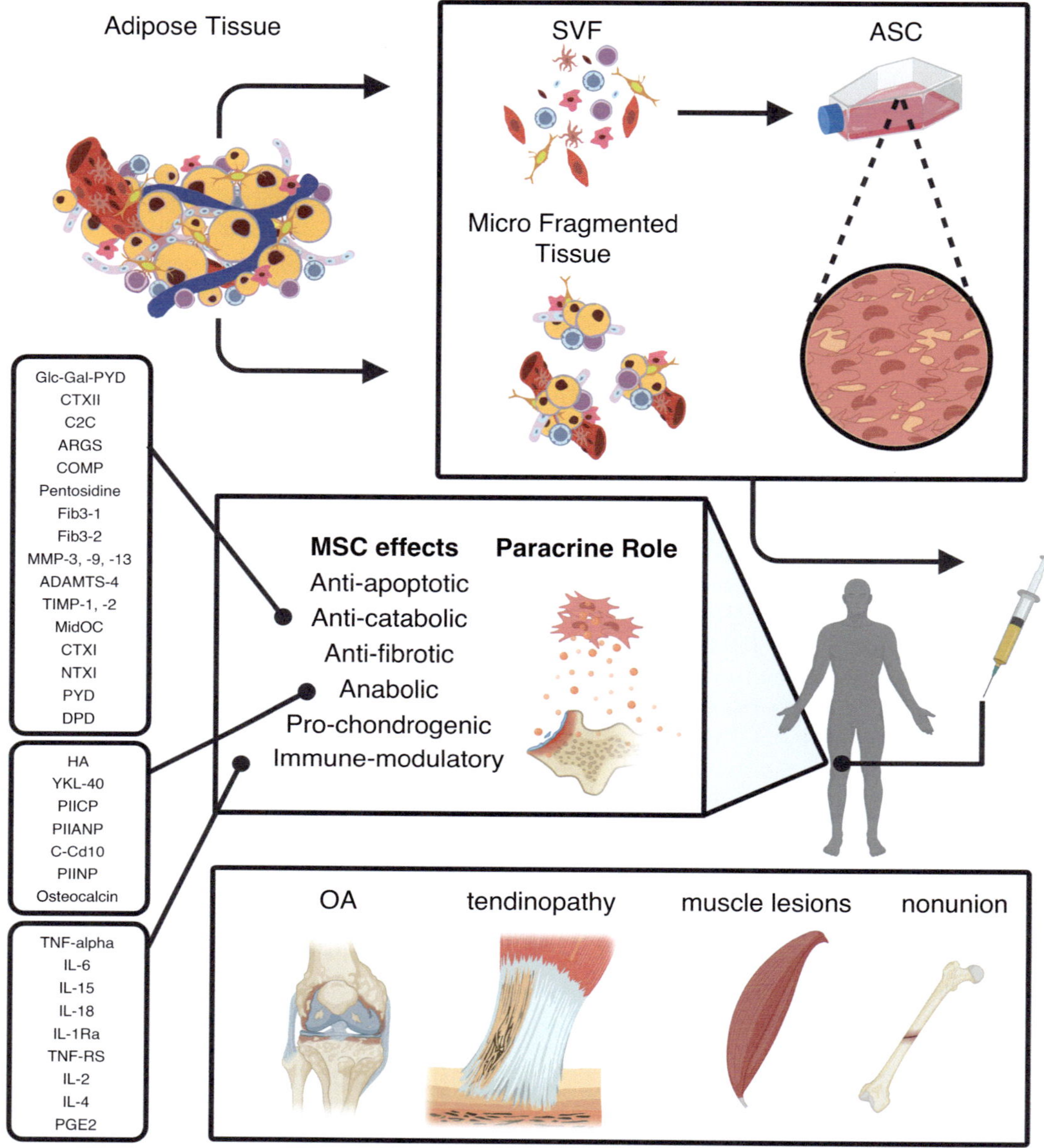

Fig. 3.1 Adipose-derived products: applications and properties

studies published so far for the treatment of joint lesions/degeneration, tendon and bone repair, as well as muscle lesions, to give the readers significant insights about their mechanisms of action. Given the preclinical settings, most of the studies show the results of the use of culture-expanded ASCs, although some results are about the unprocessed products. Up-to-date reviews and meta-analysis can also provide the readers with the most recent papers about the clinical applications of both ASCs and SVF and microfat [17, 64–66].

3.5.1 Focus on Culture-Expanded ASCs and SVF/Microfat in Joint Degeneration

Articular cartilage degeneration eventually gives rise to osteoarthritis (OA), the main cause of disability in developed countries [67]. The current conservative options may relieve symptoms but are ineffective in the restoration of the damaged tissues. Recently, innovative therapies for cartilage regeneration showed efficacy [68, 69], with particular regard to MSCs thanks to their immunomodulatory and pro-regenerative potential [70].

Pivotal in vitro studies reported the ability of culture-expanded ASCs to induce chondrocyte proliferation and extracellular matrix production, through their paracrine activity with anti-inflammatory, anti-apoptotic, and chondrogenic properties [71].

Also, the potential of autologous ASCs infusion for osteochondral defects treatment has been assessed in numerous animal models [72, 73]. Interestingly, the successful regeneration of cartilage has also been reported with an allogeneic transplant of ASCs in a sheep OA model [74]. Similar results have been observed in a rabbit model, where ASCs infusion promoted histological healing [75]. Single intra-articular injections of ASCs have been tested in dogs with hip OA. ASCs-treated animals were reported to have improved their condition [76] with improved limb function within 3 months from the procedure [77]. Conversely, the intravenous injection of ASCs in dogs with elbow OA failed to significantly improve the animals' conditions [78].

For what concern adipose-derived products, in a model of goat osteochondral defect, the application of SVF showed higher regeneration compared to the controls. SVF-treated animals exhibited more extensive collagen type II, hyaline-like cartilage, and more tissue native-like content of glycosaminoglycan in the cartilaginous layer. Moreover, in the defect regions, it has been observed more intense collagen type I staining [79]. Similar results have been obtained in a rat model of full-thickness cartilage defect treated with native stem/stromal cell-enriched microfat where it was able to effectively restore cartilage tissue [80]. A very interesting paper reports a direct comparison of cultured ASCs, SVF, and microfat for the treatment of OA in a rabbit model of bilateral transection of the anterior cruciate ligament. The rabbits were either left untreated or injected with culture-expanded ASCs or SVF or 300µl of microfat. The analysis conducted at 2- and 4-month follow-ups showed no macroscopic differences among the groups. However, at both experimental times, microfat showed the most promising results with a more uniform cartilage staining and a smoother cartilage surface than the untreated group [81].

3.5.2 Focus on Culture-Expanded ASCs and SVF/Microfat in Tendon Repair

Tendon tissue has poor healing potential, given by the limited cellular content and vascularization. Thus, the response to treatment is generally low, and prolonged recovery is needed [82]. In addition, spontaneous tendon repair often fails in adequately restoring the structural and molecular composition of the tissue, often resulting in scar tissue rich in collagen type III, more vulnerable to injuries and relapses [83]. Surgical repair also showed frequent relapses. Conservative treatments were able to improve symptoms, but none of them provided a long-term solution [84], and therefore, the application of ASCs or adipose tissue-derived products has been explored for tendon regeneration.

In vitro models demonstrated that the co-culture of primary tenocytes and ASCs could drive the differentiation of the latter into tenocytes in vitro [85, 86]. In vivo, in a mice tendon repair model, the local administration of ASCs has been reported to accelerate the tendon healing process through differentiation of ASCs into tenocytes, and by increasing the expression of angiogenic growth factors [87]. Similar results were obtained on a rabbit calcaneal tendon injury model, which showed that the application of ASCs associated with platelet-rich plasma increased the resistance of tendons as well as the amount of collagen type I, VEGF, and FGF [88]. More recently, using a rat tendinopathy model, the application of ASCs significantly improved the pathological picture [89]. ASCs have also been used on racehorses suffering from superficial flexor digitorum longus tendon (SFDLT) lesions. The injection of ASCs significantly improved healing, with treated horses showing shorter periods of lameness and better organization of collagen fibers in the injured tendon [90]. Similarly, in a horse model of collagenase-induced SFDLT lesions, the administration of ASCs resulted in a better organization of collagen fibers and a reduction of the inflammatory infiltrate. Besides, the ultrasound evaluation showed a lack of lesion progression compared to the control group [91].

Analyzing the effect of uncultured adipose tissue products, some authors reported that in vitro microfat significantly increased the proliferation rate of tendon progenitor cells as well as the expression of VEGF, which is crucial for the neovascularization of the tissue during the healing process [92]. In a similar experimental model, it was also demonstrated that microfat was effectively able to counteract the detrimental effect of experimentally induced inflammation in co-cultures with autologous tenocytes [53]. Likewise, in a rotator cuff tear model in rabbits, the application of native stem/stromal cell-enriched SVF caused a significant improvement in few physiological parameters, and it accelerated the transformation of collagen fibers from type III to type I, the crucial step of repaired tissue maturation [93].

3.5.3 Focus on Culture-Expanded ASCs and SVF/Microfat in Bone Repair

Bone fractures and segmental bone defects are a significant source of patient morbidity and place a substantial economic burden on the healthcare system. Generally, after damage, bone can regenerate itself, but in the case of significant loss of tissue, surgery with bone grafts or bone substitutes is required. These approaches may be characterized by long immobilization periods, donor site morbidity (in case of autologous graft), muscular atrophy, and potential complications such as infection, pain, or hemorrhage [94, 95] that may lead to incorrect graft integration, resorption, and eventually relapses [96]. Therefore, potential applications of ASCs in this context have then been explored [95, 97, 98].

In vitro studies have reported, under specific stimuli, the ability of ASCs to differentiate into osteocytes, unequivocally showing markers of the mature tissue [99, 100]. Interestingly, it has been reported that osteogenic induction might not be mandatory as the primary function of adhesion, migration, proliferation, and differentiation can also be achieved using native ASCs [101, 102]. Animal models mainly relied on the use of scaffolds populated by ASCs, with few applications involving ASCs injection. Some studies explored the use of ASCs and osteocyte-induced ASCs in the context of distraction osteogenesis (DO) [94]. In a rabbit model of tibial defect, the authors reported a shorter consolidation period using osteo-differentiated or undifferentiated stem/stromal cells compared to the control, but osteo-differentiated ASCs seem to perform better in terms of tissue density and quality [103]. Similarly, in a rat model of DO, the authors demonstrated that the injection of ASCs resulted in a significantly higher density and fracture strength after 6 weeks, supported by molecular evidence as ASCs' derived tissue expressed osteogenic markers [104].

For what concern the uncultured adipose tissue product, mechanical generated-SVF (mSVF) and enzymatic generated-SVF (eSVF) were compared to test whether the mechanical approach influences the biological features and functions of SVF. Albeit less efficient in terms of cell recovery and CFU-F than eSVF (five times less), mSVF preserved the functions of cell populations within the adipose tissue, with similar osteo-differentiation commitment and similar release of VEGF, HGF, IGF-1, and PDGF-bb, involved in pathways mediating osteochondral repair and cell migration, and of the anti-inflammatory cytokine IL-10 [105].

3.5.4 Focus on Culture-Expanded ASCs and SVF/Microfat in Muscle Repair

Among musculoskeletal tissues, the muscle is more prone to regenerate after injury, thanks to the presence of satellite cells, a subpopulation with stem cell-like properties [106, 107]. Although these cells are able to regenerate muscle tissue after strains, tears, or lacerations, they fail to resolve conditions of greater damage with significant muscle tissue loss, indicated as volumetric muscle loss injuries [108].

As per the other tissues, the use of ASCs for muscle regeneration and repair may rely on direct differentiation or on the release of paracrine effectors. Indeed, ASCs are able to differentiate in vitro into skeletal myoblasts and myotubes, and they maintain myogenic potential also after expansion [109], but if properly stimulated using dedicated scaffold, they may also differentiate in vivo [110].

ASCs with specific myogenic properties, and able of homing to the injured muscle tissues, have been obtained [111] and used in a mice model of Duchenne muscular dystrophy, with promising results [112].

The potential of ASCs to regenerate the skeletal muscle showed to be comparable to muscle-derived progenitor cells in a volumetric muscle loss injury murine model employing tissue-engineered muscle repair (TEMR) construct [113].

Cultured homologous ASCs injected into injured soleus muscles showed an acceleration of skeletal muscle repair in rat [114].

Similar results were obtained when human ASCs were implanted in a model of murine hind limb ischemia: an improvement in the functionality of the damaged limb occurred faster than in the control mice. In this work, the authors hypothesize a paracrine action of IL-6 released from ASCs, leading to stimulation of M2 macrophages and inducing muscle repair through neovascularization [115].

The paracrine activity of ASCs for muscle regeneration has been investigated specifically in different animal models. The conditioned media of ASCs have been suggested to improve muscle tissue healing in a rabbit model of critical limb ischemia [116]. The effects of ASC whole secretome or isolated extracellular vesicle fraction were evaluated in an in vivo cardiotoxin-induced skeletal muscle injury model, and this study demonstrated that both extracellular vesicles and soluble molecules released in the ASC secretome promote muscle regeneration acting in synergistic manner [117].

Interestingly, the rat ASCs paracrine activity for muscle regeneration can be improved by pre-treatment of stem/stromal cells with IL-4 and SDF-1. Indeed, ASCs treated with these factors were able to improve muscle structure and function and decrease fibrosis in a rat model of skeletal muscle injury [118].

In an attempt to determine the importance of the direct use of ASCs, ASCs and ASC-conditioned medium were used in type I collagen hydrogel, and the action of these constructs were directly compared in volumetric muscle loss rat model. The results indicated that hydrogels bearing ASCs or conditioned medium only were able to induce similar increase of angiogenesis and myogenesis, as well as M2 stimulation, suggesting that both elements retain an immunomodulatory role on macrophages transition. A decrease of inflammation and collagen deposition was also observed, resulting in improved muscle repair [119], confirming once more the pivotal ASCs paracrine role.

3.6 Conclusions

The rationale for the use of adipose-derived stem/stromal cells and adipose-derived products such as SVF and microfat, as well as their safety profile, for the treatment of several musculoskeletal conditions is strong and well documented in both in vitro and preclinical studies. The possibility of local survival and differentiation of tissue-derived cells and the formation of new tissues is theoretically appealing but as yet unproven. Moreover, this effect could be mainly observed when the adipose-derived cells or products are associated with surgery and delivered locally at the injury/defect site. Paracrine action mediated by soluble factors as well as by exosomes and microvesicles may play a key role in ASCs-based therapies by modulating the microenvironment, especially in a setting of injury or degeneration. In some cases, ASCs or the adipose tissue-derived products may act not only on symptoms relief but also as disease-modifying agents, possibly reverting the pathological progression. The current efforts of the scientific community are aimed to improve the knowledge of the most effective strategies to improve the therapeutic effects of these approaches. In particular, cell priming, that is the modulation of the secretory ability of cells through the use of cytokines and growth factors, hypoxia, pharmacological drugs, biomaterials, or different culture conditions, has been indicated as one of the most promising ones. In fact, an appropriate priming can modulate the cell secretory profile so that the molecule cargo is able to exert a specific therapeutic effect for each different pathology. Regardless of the mechanism of action, the optimization of dose and delivery strategies to achieve both predictable and durable positive effects needs to be further evaluated in high-quality clinical studies. While ASCs have the undisputable advantage of being homogeneous and therefore more controlled, SVF and microfat are easier to use and do not have to follow strict regulatory pathways. Overall, both are associated with pros and cons, and only further research studies will allow to identify the best approach for the different musculoskeletal pathologies and the different type of patient.

> **Take-Home Messages**
> - The presence of ASCs within the adipose tissue has drawn significant clinical attention due to their purported paracrine effects and multipotent differentiation capacity. Practical methods to exploit the properties of adipose tissue at the point of care, such as SVF and microfat, have been developed to promote an efficient use into the clinical setting.
> - ASCs, SVF, or microfat can be delivered in association with surgery for the treatment of local defects or through injection to damaged sites to treat wider areas of degeneration.
> - The main therapeutic effect of ASCs and adipose-derived products is mainly mediated by the release of soluble factors as well as by exosomes that interact with the resident cells creating a pro-regenerative microenvironment.
> - The efficacy of ASCs, SVF, and microfat for the treatment of several musculoskeletal conditions, as well as their safety profile, is well documented in both in vitro and preclinical studies. Therefore, there is a high potential of the individual fat component to be used in regenerative medicine.
> - While ASCs have the undisputable advantage of being homogeneous and therefore more controlled, SVF and microfat are easier to use and do not have to follow strict regulatory pathways. Further research studies will allow to identify the best approach for the different musculoskeletal pathologies and type of patient.

References

1. Schäffler A, Büchler C. Concise review: adipose tissue-derived stromal cells-basic and clinical implications for novel cell-based therapies. Stem Cells Wiley. 2007;25:818–27.
2. Casteilla L, Dani C. Adipose tissue-derived cells: From physiology to regenerative medicine. Diabetes Metab. Elsevier Masson SAS. 2006:393–401.
3. Li Y, Meng Y, Yu X. The unique metabolic characteristics of bone marrow adipose tissue. Front Endocrinol. Frontiers Media S.A. 2019:69.
4. Mazini L, Rochette L, Amine M, Malka G. Regenerative capacity of adipose derived stem cells (ADSCs), comparison with mesenchymal stem cells (MSCs). Int J Mol Sci. MDPI AG. 2019;20:2523.
5. Chu D-T, Nguyen Thi Phuong T, Tien NLB, Tran DK, Minh LB, Van Thanh V, et al. Adipose tissue stem cells for therapy: an update on the progress of isolation, culture, storage, and clinical application. J Clin Med. MDPI AG. 2019;8:917.

6. Casteilla L. Adipose-derived stromal cells: their identity and uses in clinical trials, an update. World J Stem Cells. Baishideng Publishing Group Inc. 2011;3:25.

7. Polancec D, Zenic L, Hudetz D, Boric I, Jelec Z, Rod E, et al. Immunophenotyping of a stromal vascular fraction from microfragmented Lipoaspirate used in osteoarthritis cartilage treatment and its lipoaspirate counterpart. Genes (Basel). 2019;10(6):474.

8. Aronowitz JA, Lockhart RA, Hakakian CS. Mechanical versus enzymatic isolation of stromal vascular fraction cells from adipose tissue. SpringerPlus. 2015;4:1–9.

9. Shah FS, Wu X, Dietrich M, Rood J, Gimble JM. A non-enzymatic method for isolating human adipose tissue-derived stromal stem cells. Cytotherapy. Elsevier Inc. 2013;15:979–85.

10. Vezzani B, Shaw I, Lesme H, Yong L, Khan N, Tremolada C, et al. Higher pericyte content and secretory activity of microfragmented human adipose tissue compared to enzymatically derived stromal vascular fraction. Stem Cells Transl Med. 2018;7:876–86.

11. Carelli S, Messaggio F, Canazza A, Hebda DM, Caremoli F, Latorre E, et al. Characteristics and properties of mesenchymal stem cells derived from microfragmented adipose tissue. Cell Transplant. SAGE Publications Inc. 2015;24: 1233–52.

12. Rodriguez J, Pratta A-S, Abbassi N, Fabre H, Rodriguez F, Debard C, et al. Evaluation of three devices for the isolation of the stromal vascular fraction from adipose tissue and for ASC culture: a comparative study [internet]. Stem Cells Int. 2017:e9289213.

13. Leto Barone AA, Khalifian S, Lee WPA, Brandacher G. Immunomodulatory effects of adipose-derived stem cells: fact or fiction? Biomed Res Int. 2013;2013:383685.

14. Koh YG, Choi YJ. Infrapatellar fat pad-derived mesenchymal stem cell therapy for knee osteoarthritis. Knee. 2012;19:902–7.

15. Koh YG, Choi YJ, Kwon SK, Kim YS, Yeo JE. Clinical results and second-look arthroscopic findings after treatment with adipose-derived stem cells for knee osteoarthritis. Knee Surg Sports Traumatol Arthrosc. 2015;23:1308–16.

16. Comella K, Blas JAP, Ichim T, Lopez J, Limon J, Moreno RC. Autologous stromal vascular fraction in the intravenous treatment of end-stage chronic obstructive pulmonary disease: a phase I trial of safety and tolerability. J Clin Med Res. Elmer Press, Inc. 2017;9:701–8.

17. Lopa S, Colombini A, Moretti M, de Girolamo L. Injective mesenchymal stem cell-based treatments for knee osteoarthritis: from mechanisms of action to current clinical evidences. Knee Surg Sports Traumatol Arthrosc. 2019;27:2003–20.

18. Nielsen FM, Riis SE, Andersen JI, Lesage R, Fink T, Pennisi CP, et al. Discrete adipose-derived stem cell subpopulations may display differential functionality after in vitro expansion despite convergence to a common phenotype distribution. Stem Cell Res Ther. BioMed Central Ltd. 2016;7:1–13.

19. Alvarez-Viejo M, Menendez-Menendez Y, Blanco-Gelaz MA, Ferrero-Gutierrez A, Fernandez-Rodriguez MA, Gala J, et al. Quantifying mesenchymal stem cells in the mononuclear cell fraction of bone marrow samples obtained for cell therapy. Transplant Proc. 2013;45:434–9.

20. Gomez-Salazar M, Gonzalez-Galofre ZN, Casamitjana J, Crisan M, James AW, Péault B. Five decades later, are mesenchymal stem cells still relevant? Front Bioeng Biotechnol [Internet] Frontiers. 2020;8:148.

21. Crisan M, Yap S, Casteilla L, Chen C-W, Corselli M, Park TS, et al. A perivascular origin for mesenchymal stem cells in multiple human organs. Cell Stem Cell. 2008;3:301–13.

22. Dominici M, Le Blanc K, Mueller I, Slaper-Cortenbach I, Marini FC, Krause DS, et al. Minimal criteria for defining multipotent mesenchymal stromal cells. The International Society for Cellular Therapy position statement. Cytotherapy. 2006;8:315–7.

23. Bourin P, Bunnel BA, Casteilla L, Dominici M, Katz AJ, March KL, et al. Stromal cells from the adipose tissue-derived stromal vascular fraction and culture expanded adipose tissue-derived stromal/stem cells: a joint statement of the International Federation for Adipose Therapeutics and Science (IFATS) and the International Society for Cellular Therapy (ISCT). Cytotherapy. Elsevier. 2013;15:641–8.

24. Domergue S, Bony C, Maumus M, Toupet K, Frouin E, Rigau V, et al. Comparison between stromal vascular fraction and adipose mesenchymal stem cells in remodeling hypertrophic scars. PLoS One. 2016;11. Public Library of Science.

25. Nyberg E, Farris A, O'Sullivan A, Rodriguez R, Grayson W. Comparison of stromal vascular fraction and passaged adipose-derived stromal/stem cells as point-of-care agents for bone regeneration. Tissue Eng Part A. Mary Ann Liebert Inc. 2019;25:1459–69.

26. Pittenger MF, Discher DE, Péault BM, Phinney DG, Hare JM, Caplan AI. Mesenchymal stem cell perspective: cell biology to clinical progress. Npj Regen Med Nat Res. 2019:1–15.

27. Peng Q, Alipour H, Porsborg S, Fink T, Zachar V. Evolution of ASC Immunophenotypical subsets during expansion in vitro. Int J Mol Sci. Multidisciplinary Digital Publishing Institute. 2020;21:1408.

28. Dominici M, Paolucci F, Conte P, Horwitz EM. Heterogeneity of multipotent mesenchymal stromal cells: from stromal cells to stem cells and vice versa. Transplantation. 2009;87:S36–42.

29. Samsonraj RM, Raghunath M, Nurcombe V, Hui JH, van Wijnen AJ, Cool SM. Concise review: multifaceted characterization of human mesenchymal stem cells for use in regenerative medicine.

Stem Cells Transl Med. John Wiley and Sons Ltd. 2017;6:2173–85.

30. Varghese J, Griffin M, Mosahebi A, Butler P. Systematic review of patient factors affecting adipose stem cell viability and function: implications for regenerative therapy. Stem Cell Res Ther. 2017;8:45.

31. De Girolamo L, Lopa S, Arrigoni E, Sartori MF, Baruffaldi Preis FW, Brini AT. Human adipose-derived stem cells isolated from young and elderly women: their differentiation potential and scaffold interaction during in vitro osteoblastic differentiation. Cytotherapy. Elsevier Inc. 2009;11:793–803.

32. D'Ippolito G, Schiller PC, Ricordi C, Roos BA, Howard GA. Age-related osteogenic potential of mesenchymal stromal stem cells from human vertebral bone marrow. J Bone Miner Res. 1999;14:1115–22.

33. Gonzalez-Garza MT, Cruz-Vega DE. Regenerative capacity of autologous stem cell transplantation in elderly: a report of biomedical outcomes. Regen Med. Future Medicine Ltd. 2017;12:169–78.

34. Mitterberger MC, Mattesich M, Zwerschke W. Bariatric surgery and diet-induced long-term caloric restriction protect subcutaneous adipose-derived stromal/progenitor cells and prolong their life span in formerly obese humans. Exp Gerontol. Elsevier Inc. 2014;56:106–13.

35. Moschen AR, Molnar C, Geiger S, Graziadei I, Ebenbichler CF, Weiss H, et al. Anti-inflammatory effects of excessive weight loss: potent suppression of adipose interleukin 6 and tumour necrosis factor a expression. Gut. BMJ Publishing Group. 2010;59:1259–64.

36. Faustini M, Bucco M, Chlapanidas T, Lucconi G, Marazzi M, Tosca MC, et al. Nonexpanded mesenchymal stem cells for regenerative medicine: yield in stromal vascular fraction from adipose tissues. Tissue Eng Part C Methods. 2010;16:1515–21.

37. Padoin AV, Braga-Silva J, Martins P, Rezende K, Rezende ARDR, Grechi B, et al. Sources of processed lipoaspirate cells: influence of donor site on cell concentration. Plast Reconstr Surg. 2008;122:614–8.

38. Geissler PJ, Davis K, Roostaeian J, Unger J, Huang J, Rohrich RJ. Improving fat transfer viability: the role of aging, body mass index, and harvest site. Plast Reconstr Surg. Lippincott Williams and Wilkins. 2014;134:227–32.

39. Jurgens WJFM, Oedayrajsingh-Varma MJ, Helder MN, ZandiehDoulabi B, Schouten TE, Kuik DJ, et al. Effect of tissue-harvesting site on yield of stem cells derived from adipose tissue: implications for cell-based therapies. Cell Tissue Res. 2008;332:415–26.

40. Yorukoglu AC, Kiter AE, Akkaya S, Satiroglu-Tufan NL, Tufan AC. A concise review on the use of mesenchymal stem cells in cell sheet-based tissue engineering with special emphasis on bone tissue regeneration. Stem Cells Int. 2017;2017:2374161.

41. Si Z, Wang X, Sun C, Kang Y, Xu J, Wang X, et al. Adipose-derived stem cells: sources, potency, and implications for regenerative therapies. Biomed Pharmacother. Elsevier Masson SAS. 2019;114:108765.

42. Hutton DL, Grayson WL. Hypoxia inhibits de novo vascular assembly of adipose-derived stromal/stem cell populations, but promotes growth of preformed vessels. Tissue Eng Part A. 2016;22:161–9.

43. Jae WL, Gupta N, Serikov V, Matthay MA. Potential application of mesenchymal stem cells in acute lung injury. Expert Opin Biol Ther. NIH Public Access. 2009;9:1259–70.

44. Basu J, Ludlow JW. Developments in tissue engineered and regenerative medicine products: a practical approach. Developments in tissue engineered and regenerative medicine products: a practical approach. Elsevier Ltd; 2012.

45. Gnecchi M, Danieli P, Malpasso G, Ciuffreda MC. Paracrine mechanisms of mesenchymal stem cells in tissue repair. Methods Mol Biol. Humana Press Inc. 2016;1416:123–46.

46. Madrigal M, Rao KS, Riordan NH. A review of therapeutic effects of mesenchymal stem cell secretions and induction of secretory modification by different culture methods. J Transl Med. BioMed Central Ltd. 2014;12:260.

47. Caplan AI, Correa D. The MSC: an injury drugstore. Cell Stem Cell. Cell Press. 2011;9:11–5.

48. Murphy MB, Moncivais K, Caplan AI. Mesenchymal stem cells: environmentally responsive therapeutics for regenerative medicine. Exp Mol Med. Nature Publishing Group. 2013;45:e54.

49. Rawat S, Gupta S, Mohanty S. Mesenchymal stem cells modulate the immune system in developing therapeutic interventions. Immune response activation and immunomodulation. IntechOpen; 2019.

50. Vasandan AB, Jahnavi S, Shashank C, Prasad P, Kumar A, Jyothi PS. Human mesenchymal stem cells program macrophage plasticity by altering their metabolic status via a PGE 2 -dependent mechanism. Sci Rep. Nature Publishing Group. 2016;6.

51. Fernandes TL, Gomoll AH, Lattermann C, Hernandez AJ, Bueno DF, Amano MT. Macrophage: a potential target on cartilage regeneration. Front Immunol. Frontiers Media S.A. 2020;11:111.

52. Denkovskij J, Bagdonas E, Kusleviciute I, Mackiewicz Z, Unguryte A, Porvaneckas N, et al. Paracrine Potential of the Human Adipose Tissue-Derived Stem Cells to Modulate Balance between Matrix Metalloproteinases and Their Inhibitors in the Osteoarthritic Cartilage In Vitro. hindawi.com. 2017.

53. Viganò M, Lugano G, Orfei CP, Menon A, Ragni E, Colombini A, et al. Autologous microfragmented adipose tissue reduces the catabolic and fibrosis response in an in vitro model of tendon cell inflammation. Stem Cells Int. Hindawi Publishing Corporation. 2019;2019:5620286.

54. Théry C, Witwer KW, Aikawa E, Alcaraz MJ, Anderson JD, Andriantsitohaina R, et al. Minimal

information for studies of extracellular vesicles 2018 (MISEV2018): a position statement of the International Society for Extracellular Vesicles and update of the MISEV2014 guidelines. J Extracell Vesicles. Taylor and Francis Ltd. 2018;7:1535750.

55. Hong P, Yang H, Wu Y, Li K, Tang Z. The functions and clinical application potential of exosomes derived from adipose mesenchymal stem cells: a comprehensive review. Stem Cell Res Ther. BioMed Central Ltd. 2019:1–12.

56. Zhang M, Zhang F, Sun J, Sun Y, Xu L, Zhang D, et al. The condition medium of mesenchymal stem cells promotes proliferation, adhesion and neuronal differentiation of retinal progenitor cells. Neurosci Lett. Elsevier Ireland Ltd. 2017;657:62–8.

57. Burrello J, Monticone S, Gai C, Gomez Y, Kholia S, Camussi G. Stem cell-derived extracellular vesicles and immune-modulation. Front Cell Dev Biol. Frontiers Media S.A. 2016.

58. van den Akker F, Vrijsen KR, Deddens JC, Buikema JW, Mokry M, van Laake LW, et al. Suppression of T cells by mesenchymal and cardiac progenitor cells is partly mediated via extracellular vesicles. Heliyon. Elsevier Ltd. 2018;4:e00642.

59. Cosenza S, Toupet K, Maumus M, Luz-Crawford P, Blanc-Brude O, Jorgensen C, et al. Mesenchymal stem cells-derived exosomes are more immunosuppressive than microparticles in inflammatory arthritis. Theranostics. Ivyspring International Publisher. 2018;8:1399–410.

60. Lee M, Ban JJ, Yang S, Im W, Kim M. The exosome of adipose-derived stem cells reduces β-amyloid pathology and apoptosis of neuronal cells derived from the transgenic mouse model of Alzheimer's disease. Brain Res. Elsevier B.V. 2018;1691:87–93.

61. Shen H, Yoneda S, Abu-Amer Y, Guilak F, Gelberman RH. Stem cell-derived extracellular vesicles attenuate the early inflammatory response after tendon injury and repair. J Orthop Res. John Wiley and Sons Inc. 2020;38:117–27.

62. Mianehsaz E, Mirzaei HR, Mahjoubin-Tehran M, Rezaee A, Sahebnasagh R, Pourhanifeh MH, et al. Mesenchymal stem cell-derived exosomes: a new therapeutic approach to osteoarthritis? Stem Cell Res Ther. BioMed Central Ltd. 2019.

63. Yin K, Wang S, Zhao RC. Exosomes from mesenchymal stem/stromal cells: a new therapeutic paradigm. Biomark Res. BioMed Central Ltd. 2019;7:8.

64. Vasiliadis AV, Galanis N. Effectiveness of AD-MSCs injections for the treatment of knee osteoarthritis: analysis of the current literature. J Stem Cells Regen Med. 2020;16:3–9.

65. Di Matteo B, Vandenbulcke F, Vitale ND, Iacono F, Ashmore K, Marcacci M, et al. Minimally manipulated mesenchymal stem cells for the treatment of knee osteoarthritis: a systematic review of clinical evidence. Stem Cells Int. 2019;2019:1735242.

66. Robinson DM, Eng C, Makovitch S, Rothenberg JB, DeLuca S, Douglas S, et al. Non-operative orthobiologic use for rotator cuff disorders and glenohumeral osteoarthritis: a systematic review. J Back Musculoskelet Rehabil. 2020;34:17.

67. Hiligsmann M, Cooper C, Arden N, Boers M, Branco JC, Luisa Brandi M, et al. Health economics in the field of osteoarthritis: an Expert's consensus paper from the European Society for Clinical and Economic Aspects of Osteoporosis and Osteoarthritis (ESCEO). Semin Arthritis Rheum. 2013;43:303–13.

68. Bendich I, Rubenstein WJ, Cole BJ, Ma CB, Feeley BT, Lansdown DA. What is the appropriate Price for PRP injections for knee osteoarthritis? A cost-effectiveness analysis based on evidence from level 1 randomized controlled trials. Arthroscopy. Elsevier BV. 2020.

69. Mehranfar S, Abdi Rad I, Mostafavi E, Akbarzadeh A. The use of stromal vascular fraction (SVF), platelet-rich plasma (PRP) and stem cells in the treatment of osteoarthritis: an overview of clinical trials. Artif Cells Nanomed Biotechnol. Taylor and Francis Ltd. 2019;47:882–90.

70. Colombini A, Perucca Orfei C, Kouroupis D, Ragni E, De Luca P, ViganÒ M, et al. Mesenchymal stem cells in the treatment of articular cartilage degeneration: new biological insights for an old-timer cell. Cytotherapy. 2019;21:1179–97.

71. Torres-Torrillas M, Rubio M, Damia E, Cuervo B, Del Romero A, Peláez P, et al. Adipose-derived mesenchymal stem cells: a promising tool in the treatment of musculoskeletal diseases. Int J Mol Sci. MDPI AG. 2019;20:3105.

72. De Girolamo L, Niada S, Arrigoni E, Di Giancamillo A, Domeneghini C, Dadsetan M, et al. Repair of osteochondral defects in the minipig model by OPF hydrogel loaded with adipose-derived mesenchymal stem cells. Regen Med. Future Medicine Ltd. 2015;10:135–51.

73. Hsu YK, Sheu SY, Wang CY, Chuang MH, Chung PC, Luo YS, et al. The effect of adipose-derived mesenchymal stem cells and chondrocytes with platelet-rich fibrin releasates augmentation by intra-articular injection on acute osteochondral defects in a rabbit model. Knee. Elsevier B.V. 2018;25:1181–91.

74. Feng C, Luo X, He N, Xia H, Lv X, Zhang X, et al. Efficacy and persistence of allogeneic adipose-derived mesenchymal stem cells combined with hyaluronic acid in osteoarthritis after intra-articular injection in a sheep model. Tissue Eng Part A. 2018;24:219–33.

75. Oshima T, Nakase J, Toratani T, Numata H, Takata Y, Nakayama K, et al. A scaffold-free allogeneic construct from adipose-derived stem cells regenerates an osteochondral defect in a rabbit model. Arthroscopy. W.B. Saunders 2019;35:583–93.

76. Rubio M, Sopena J, Carrillo JM, Cugat R, Dominguez JM, Vilar J, et al. Hip osteoarthritis in dogs: a randomized study using mesenchymal stem cells from adipose tissue and plasma rich in growth factors. Int J Mol Sci. MDPI AG. 2014;15:13437–60.

77. Vilar JM, Batista M, Morales M, Santana A, Cuervo B, Rubio M, et al. Assessment of the effect

of intraarticular injection of autologous adipose-derived mesenchymal stem cells in osteoarthritic dogs using a double blinded force platform analysis. BMC Vet Res. BioMed Central Ltd. 2014;10:143.

78. Olsen A, Johnson V, Webb T, Santangelo KS, Dow S, Duerr FM. Evaluation of intravenously delivered allogeneic mesenchymal stem cells for treatment of elbow osteoarthritis in dogs: a pilot study. Vet Comp Orthop Traumatol. Georg Thieme Verlag. 2019;32:173–81.

79. Jurgens WJFM, Kroeze RJ, Zandieh-Doulabi B, van Dijk A, Renders GAP, Smit TH, et al. One-step surgical procedure for the treatment of osteochondral defects with adipose-derived stem cells in a caprine knee defect: a pilot study. Biores Open Access. Mary Ann Liebert Inc. 2013;2:315–25.

80. Xu T, Yu X, Yang Q, Liu X, Fang J, Dai X. Autologous micro-fragmented adipose tissue as stem cell-based natural scaffold for cartilage defect repair. Cell Transplant. SAGE Publications Ltd. 2019;28:1709–20.

81. Filardo G, Tschon M, Perdisa F, Brogini S, Cavallo C, Desando G, et al. Micro-fragmentation is a valid alternative to cell expansion and enzymatic digestion of adipose tissue for the treatment of knee osteoarthritis: a comparative preclinical study. Knee Surg Sports Traumatol Arthrosc. [Internet]. 2021.

82. Ahmad Z, Wardale J, Brooks R, Henson F, Noorani A, Rushton N. Exploring the application of stem cells in tendon repair and regeneration. Arthroscopy. W.B. Saunders. 2012;28:1018–29.

83. Ni M, Lui PPY, Rui YF, Lee WYW, Lee WYW, Tan Q, et al. Tendon-derived stem cells (TDSCs) promote tendon repair in a rat patellar tendon window defect model. J Orthop Res. John Wiley and Sons Inc. 2012;30:613–9.

84. Schneider M, Angele P, Järvinen TAH, Docheva D. Rescue plan for Achilles: therapeutics steering the fate and functions of stem cells in tendon wound healing. Adv Drug Deliv Rev. Elsevier B.V. 2018;129:352–75.

85. de Aro A, Carneiro G, Teodoro L, da Veiga F, Ferrucci D, Simões G, et al. Injured Achilles tendons treated with adipose-derived stem cells transplantation and GDF-5. Cell. MDPI AG. 2018;7:127.

86. Schneider PRA, Buhrmann C, Mobasheri A, Matis U, Shakibaei M. Three-dimensional high-density co-culture with primary tenocytes induces tenogenic differentiation in mesenchymal stem cells. J Orthop Res. 2011;29:1351–60.

87. Kokubu S, Inaki R, Hoshi K, Hikita A. Adipose-derived stem cells improve tendon repair and prevent ectopic ossification in tendinopathy by inhibiting inflammation and inducing neovascularization in the early stage of tendon healing. Regen Ther. Japanese Society of Regenerative Medicine. 2020;14:103–10.

88. Uysal CA, Tobita M, Hyakusoku H, Mizuno H. Adipose-derived stem cells enhance primary tendon repair: biomechanical and immunohistochemical evaluation. J Plast Reconstr Aesthet Surg. 2012;65:1712–9.

89. Oshita T, Tobita M, Tajima S, Mizuno H. Adipose-derived stem cells improve collagenase-induced tendinopathy in a rat model. Am J Sports Med. SAGE Publications Inc. 2016;44:1983–9.

90. Skutella T. Autologous adipose tissue-derived mesenchymal stem cells affect the regeneration of equine tendon lesions. Ommega Int. 2016;1:1–8.

91. Carvalho ADM, Badial PR, Álvarez LEC, Yamada ALM, Borges AS, Deffune E, et al. Equine tendonitis therapy using mesenchymal stem cells and platelet concentrates: a randomized controlled trial. Stem Cell Res Ther. 2013;4:85.

92. Randelli P, Menon A, Ragone V, Creo P, Bergante S, Randelli F, et al. Lipogems product treatment increases the proliferation rate of human tendon stem cells without affecting their Stemness and differentiation capability. Stem Cells Int. 2016;2016:4373410.

93. Lu LY, Ma M, Cai JF, Yuan F, Zhou W, Luo SL, et al. Effects of local application of adipose-derived stromal vascular fraction on tendon-bone healing after rotator cuff tear in rabbits. Chin Med J. Wolters Kluwer Medknow Publications. 2018;131:2620–2.

94. Morcos MW, Al-Jallad H, Hamdy R. Comprehensive review of adipose stem cells and their implication in distraction osteogenesis and bone regeneration. Biomed Res Int. 2015;2015:842975.

95. Mousaei Ghasroldasht M, Matin MM, Kazemi Mehrjerdi H, Naderi-Meshkin H, Moradi A, Rajabioun M, et al. Application of mesenchymal stem cells to enhance non-union bone fracture healing. J Biomed Mater Res A. John Wiley and Sons Inc. 2019;107:301–11.

96. Sohn HS, Oh JK. Review of bone graft and bone substitutes with an emphasis on fracture surgeries. Biomater Res. BioMed Central Ltd. 2019;9.

97. Yoon D, Kang BJ, Kim Y, Lee SH, Rhew D, Kim WH, et al. Effect of serum-derived albumin scaffold and canine adipose tissue-derived mesenchymal stem cells on osteogenesis in canine segmental bone defect model. J Vet Sci. Korean Society of Veterinary Science. 2015;16:397–404.

98. Dozza B, Salamanna F, Baleani M, Giavaresi G, Parrilli A, Zani L, et al. Nonunion fracture healing: evaluation of effectiveness of demineralized bone matrix and mesenchymal stem cells in a novel sheep bone nonunion model. J Tissue Eng Regener Med. 2018;12:1972–85.

99. Gimble JM, Guilak F. Adipose-derived adult stem cells: isolation, characterization, and differentiation potential. Cytotherapy. Elsevier Inc. 2003;5:362–9.

100. Halvorsen YDC, Franklin D, Bond AL, Hitt DC, Auchter C, Boskey AL, et al. Extracellular matrix mineralization and osteoblast gene expression by human adipose tissue-derived stromal cells. Tissue Eng. 2001;7:729–41.

101. Jeon O, Rhie JW, Kwon IK, Kim JH, Kim BS, Lee SH. In vivo bone formation following transplantation of human adipose-derived stromal cells that are not differentiated osteogenically. Tissue Eng A. Mary Ann Liebert Inc. 2008;14:1285–94.
102. Li X, Yao J, Wu L, Jing W, Tang W, Lin Y, et al. Osteogenic induction of adipose-derived stromal cells: not a requirement for bone formation in vivo. Artif Organs. 2010;34:46–54.
103. Sunay O, Can G, Cakir Z, Denek Z, Kozanoglu I, Erbil G, et al. Autologous rabbit adipose tissue-derived mesenchymal stromal cells for the treatment of bone injuries with distraction osteogenesis. Cytotherapy. 2013;15:690–702.
104. Nomura I, Watanabe K, Matsubara H, Hayashi K, Sugimoto N, Tsuchiya H. Uncultured autogenous adipose-derived regenerative cells promote bone formation during distraction osteogenesis in rats. Clin Orthop Relat Res. 2014;472:3798–806.
105. Desando G, Bartolotti I, Cattini L, Tschon M, Martini L, Fini M, et al. Prospects on the potential in vitro regenerative features of mechanically treated-adipose tissue for osteoarthritis care. Stem Cell Rev Rep. 2021;17(4):1362–73.
106. Mauro A. Satellite cell of skeletal muscle FIBERS. J Biophys Biochem Cytol. 1961;9:493–5.
107. Anderson JE. The satellite cell as a companion in skeletal muscle plasticity: currency, conveyance, clue, connector and colander. J Exp Biol. 2006;209:2276–92.
108. Grogan BF, Hsu JR. Skeletal trauma research consortium. Volumetric muscle loss. J Am Acad Orthop Surg. 2011;19(Suppl 1):S35–7.
109. Vieira NM, Brandalise V, Zucconi E, Jazedje T, Secco M, Nunes VA, et al. Human multipotent adipose-derived stem cells restore dystrophin expression of Duchenne skeletal-muscle cells in vitro. Biol Cell. 2008;100:231–41.
110. Desiderio V, De Francesco F, Schiraldi C, De Rosa A, La Gatta A, Paino F, et al. Human Ng2+ adipose stem cells loaded in vivo on a new crosslinked hyaluronic acid-Lys scaffold fabricate a skeletal muscle tissue. J Cell Physiol. 2013;228:1762–73.
111. Milner DJ, Bionaz M, Monaco E, Cameron JA, Wheeler MB. Myogenic potential of mesenchymal stem cells isolated from porcine adipose tissue. Cell Tissue Res. 2018;372:507–22.
112. Liu Y, Yan X, Sun Z, Chen B, Han Q, Li J, et al. Flk-1+ adipose-derived mesenchymal stem cells differentiate into skeletal muscle satellite cells and ameliorate muscular dystrophy in mdx mice. Stem Cells Dev. 2007;16:695–706.
113. Kesireddy V. Evaluation of adipose-derived stem cells for tissue-engineered muscle repair construct-mediated repair of a murine model of volumetric muscle loss injury Int J Nanomedicine. 2016;11:1461–73.
114. Peçanha R, de Bagno LLES, Ribeiro MB, Robottom Ferreira AB, Moraes MO, Zapata-Sudo G, et al. Adipose-derived stem-cell treatment of skeletal muscle injury. J Bone Joint Surg Am. 2012;94:609–17.
115. Pilny E, Smolarczyk R, Jarosz-Biej M, Hadyk A, Skorupa A, Ciszek M, et al. Human ADSC xenograft through IL-6 secretion activates M2 macrophages responsible for the repair of damaged muscle tissue. Stem Cell Res Ther. 2019;10:93.
116. Tauber Z, Cizkova K, Janikova M, Jurcikova J, Vitkova K, Pavliska L, et al. Serum C-peptide level correlates with the course of muscle tissue healing in the rabbit model of critical limb ischemia. Biomed Pap Med Fac Univ Palacky Olomouc Czech Repub. 2019;163:132–40.
117. Mitchell R, Mellows B, Sheard J, Antonioli M, Kretz O, Chambers D, et al. Secretome of adipose-derived mesenchymal stem cells promotes skeletal muscle regeneration through synergistic action of extracellular vesicle cargo and soluble proteins. Stem Cell Res Ther. 2019;10:116.
118. Zimowska M, Archacka K, Brzoska E, Bem J, Czerwinska AM, Grabowska I, et al. IL-4 and SDF-1 increase adipose tissue-derived stromal cell ability to improve rat skeletal muscle regeneration. Int J Mol Sci. 2020;21.
119. Huang H, Liu J, Hao H, Chen D, Zhizhong L, Li M, et al. Preferred M2 polarization by ASC-based hydrogel accelerated angiogenesis and myogenesis in volumetric muscle loss rats. Stem Cells Int. 2017;2017:2896874.

Injections of Synovial Mesenchymal Stromal Cells

4

Ichiro Sekiya and Nobutake Ozeki

4.1 Introduction

This chapter provides a brief introduction into the conceptual paradigm, basic, preclinical, and clinical research and process optimization that has led to the initiation of an ongoing clinical study evaluating the injection of culture-expanded mesenchymal stromal cells (MSCs) into the human knee for treatment of osteoarthritis (OA).

Culture-expanded MSC populations can be derived from the bone marrow, synovium, adipose tissue, and other connective tissues. MSCs are highly proliferative in early passage and, by definition, must have the potential to differentiate into bone, cartilage, and/or adipose tissue in vitro. The position paper of the International Society for Cellular Therapy (ISCT) identifies a characteristic surface antigen profile in a proposed definition of MSCs, including HLA-DR(−) in combination with CD105(+), CD73(+), CD90(+), CD45(−), CD34(−), CD14(−), CD11b(−), CD79a(−), or CD19(−) [1]. The most frequently studied MSCs have been bone marrow-derived MSCs, but numerous studies since 2000 have reported the isolation of MSCs from a range of connective tissues other than bone marrow, such as adipose tissue and muscle. MSCs share characteristics that are independent of the original tissue; however, they also have characteristics uniquely associated with the original tissue.

4.2 Generation of MSCs Starting with Heterogeneous Mixtures of Colony-Forming Progenitors from Synovial Tissues

Native connective tissues contain a heterogeneous population of cells, which includes a small number of stem and progenitor cells [2, 3]. The prevalence of stem and progenitor cells can be estimated by plating tissue-derived cells in culture and measuring the number of colonies formed per hundred cells plated. Each colony formed in this process is theoretically derived from a single founding stem or progenitor cell, commonly referred to as a colony-forming unit (CFU). Each colony formed differs from others. These differences can be seen in differences in proliferation rate, morphology, and gene expression. The number and heterogeneity of colonies formed from a given tissue sample can be used to estimate the concentration and prevalence of stem and progenitor cells in the original tissue sample. The heterogeneous population of stem and progenitor cells in native connective tissues has been defined as and is commonly referred to as tissue-specific connective tissue progenitors (CTPs) [4]. The prevalence of CTPs in synovium and adipose tissue is particularly high, when

I. Sekiya (✉) · N. Ozeki
Center for Stem Cell and Regenerative Medicine,
Tokyo Medical and Dental University (TMDU),
Tokyo, Japan
e-mail: sekiya.arm@tmd.ac.jp; ozeki.arm@tmd.ac.jp

© ISAKOS 2022
G. Filardo et al. (eds.), *Orthobiologics*, https://doi.org/10.1007/978-3-030-84744-9_4

compared to bone marrow and cartilage. For example, the CTP prevalence can be 100-fold higher in the synovium, anterior cruciate ligament (ACL), and adipose tissue than bone marrow [5].

It is possible to isolate culture-expanded populations that can be defined as MSCs, based on ISCT criteria, from virtually any connective tissue source. However, the outcome of expansion of the heterogeneous starting materials in each tissue source and in individual patient samples may vary. Therefore, comparison of individual tissue source requires the cells to be cultured under the same conditions. In addition, the plating density and culture period can also affect their proliferative and differentiation potentials [6]. Individual colonies derived from a single stem or progenitor cell inevitably compete with one another during in vitro expansion [7]. A high plating density results in a small size of the colonies due to colony-to-colony contact inhibition. Conversely, a low plating density yields colonies of larger size, but the yields per dish can be low due to a higher growth burden from a smaller number of founding cells. Competition among the clones of individual colony-forming cells continues during in vitro expansion and passage. As a result, the relatively homogeneous population of culture-expanded MSCs represent a highly selected population of cells that may be evident during early colony formation. In other words, MSCs cultured at a low plating density acquire attributes and properties of biological potential that differs widely from the mixed starting population of heterogeneous CTPs.

In this light, maximum yields of culture-expanded MSCs can be achieved by choosing the optimum cell density that limits contacts between colonies. Accordingly, our comparisons of MSCs derived from several kinds of connective tissues have been conducted on MSCs cultured at the initial plating density that is optimal for the original MSC source, showing that proliferation ability was lost at passage 4 in muscle-derived MSCs and at passage 7 in adipose tissue-derived MSCs. The proliferative ability of synovium-, periosteum-, and bone marrow-derived MSCs was retained even at passage 10 [8].

4.3 In Vitro Chondrogenic Potential of Synovial MSCs

Cartilage differentiation was achieved by placing passage 1 2.5×10^5 MSCs cultured at a clonal density on dishes coated for cell culture in a 15-mL polypropylene tube, centrifuging for 10 min to form a cell pellet and then culturing the pellet in chondrogenic differentiation medium containing TGF-β, dexamethasone, and BMP. The medium was changed twice a week. The cell pellet becomes rounder and larger over time and differentiates into cartilage tissue after 3 weeks. The increase in size in the cultured pellet during chondrogenesis was mainly due to the production of cartilage matrix and not due to MSC proliferation [9].

The size and weight of the cartilage pellet reflect the number of chondroprogenitor cells and their ability to produce cartilage matrix, which is an indicator of the chondrogenic differentiation potential of the MSC population. When we prepared and cultured MSCs from the synovium, bone marrow, periosteum, adipose tissue, and muscle from the same donor, the cartilage pellets formed in pellet culture were larger and had more abundant glycosaminoglycans when derived from MSCs from synovium and bone marrow aspirate than from adipose tissue or muscle [8] (Fig. 4.1). Similar results have been obtained in rats [10], rabbits [11], and pigs [12]. This indicates that the synovium and bone marrow are excellent MSC sources for regenerative medicine for cartilage and meniscus.

4.4 Culture of MSCs with Autologous Serum

Serum components are required for MSC expansion; therefore, we believe that the use of autologous serum is preferable for clinical application of MSCs to avoid virus infections and immune reactions. Synovium and bone marrow MSCs are useful for cartilage regenerative medicine because of their high chondrogenic differentiation potential; therefore, we examined whether these MSCs could be cultured with autologous

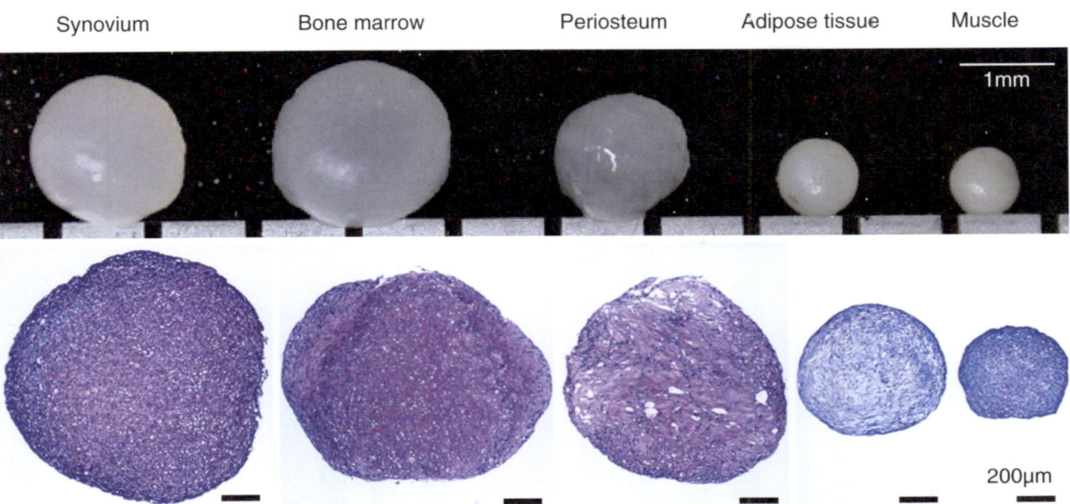

Fig. 4.1 Comparison of the chondrogenic differentiation potential of MSCs from five connective tissues. (Top) Macroscopic images. (Bottom) Histological sections stained with toluidine blue (From Sakaguchi [8] et al.)

serum to obtain sufficient numbers of cells. The synovium was digested with collagenase and the digested cells were filtered through a 70-µm nylon filter. Nucleated cells from the bone marrow were isolated with a density gradient. Nucleated cells were plated into 60-cm^2 dishes at clonal density, which was 10^3 or 10^4 per 60-cm^2 dish for synovial cells and 10^3 or 10^4 per 60-cm^2 dish for bone marrow cells [8]. The cells were cultured with 10% autologous human serum to determine the number of cells at passage 0 ($n = 3$). We calculated number of MSCs harvested after 14-day incubation with 100 mL of autologous serum from 200 mg of synovial tissue and 2 mL of bone marrow from nucleated cell number per synovium weight, nucleated cell number per bone marrow volume, and yields of MSCs. We found that a 14-day culture with 10% autologous serum generated more than ten million synovial MSCs from each of nine patients. By contrast, only about one million bone marrow MSCs were generated only from two of the nine patients (Fig. 4.2).

Next, synovial MSCs at passage 1 and bone marrow MSCs at passage 1 were plated at 50 cells/cm^2 and cultured with 10% autologous serum or 20% fetal bovine serum for 14 days. At 14 days more synovial MSCs were generated with autologous serum, while more bone marrow MSCs were generated with fetal bovine serum. We found that human serum was enriched in the AB isoform of PDGF that binds to PDGFα receptors and that PDGFα receptors were expressed at a higher rate by synovial MSCs than by bone marrow MSCs. These differences in expression led to differences in proliferation with autologous serum [13].

We summarize that the higher prevalence of synovial founding cells represents an advantage of synovial tissue as a source tissue in order to achieve a target number of expanded MSCs since the burden of doubling needed for any individual CTP-derived clone is reduced. This represents an advantage for synovial tissue as a cell source over bone marrow since the reduced doubling and passaging may lower the risk of chromosomal abnormalities. Our digital karyotyping analysis has shown that the proportion of trisomy 7 in synovial MSCs increased from passage 0 to 15 in some donors.

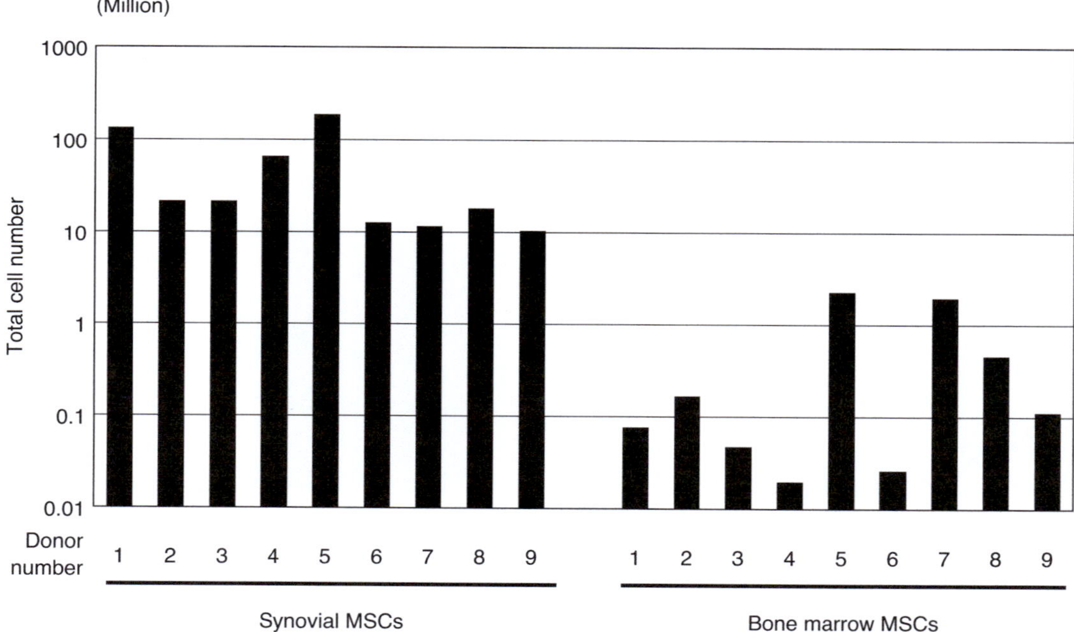

Fig. 4.2 Number of MSCs harvested after 2 weeks of incubation with 100 mL of autologous serum from 200 mg of synovial tissue and 2 mL of bone marrow

4.5 Intra-Articular Injections of Synovial MSCs in a Rat OA Model

ACL in the knee of rats was transected to create an OA model. Three groups were set up: the control group, which received weekly intra-articular injection of PBS; the one-shot group, which received a single intra-articular injection of 1×10^6 rat synovial MSCs; and the weekly group, which received weekly intra-articular injections of 1×10^6 rat synovial MSCs, 12 times over a maximum of 12 weeks. Histological examination of the medial femoral cartilage in the control group showed a reduction in proteoglycan staining at 4 weeks, bone loss at 8 weeks, and widespread bone loss at 12 weeks (Fig. 4.3). By contrast, the one-shot group showed decreased proteoglycan staining in the superficial layer of cartilage at 8 weeks and extensive cartilage loss at 12 weeks. However, the weekly group showed maintenance of the cartilage matrix even after 12 weeks [14].

4.6 Localization of Synovial MSCs after Injection

The localization of MSCs was determined by intra-articular injection of rat synovial MSCs expressing LacZ and evaluation 1 day later. MSCs, which stained blue with X-gal, were extensively observed in the synovium (Fig. 4.4) but not in the cartilage or meniscus. We also investigated the activity of the injected cells and their migration out of the joint by intra-articular injection of rat synovial MSCs expressing luciferase into the knee joint with an in vivo imaging system. The luminescence was more intense in the ACL-transected knee injected once than in the intact knee injected once, but the luminescence was no longer detectable in either knee after 14 days (Fig. 4.5). By contrast, the luminescence was maintained after 14 days in the group that received weekly intra-articular injections. Weekly injections of synovial MSCs maintained high cell activity, and the injected MSCs did not migrate out of the knee joint.

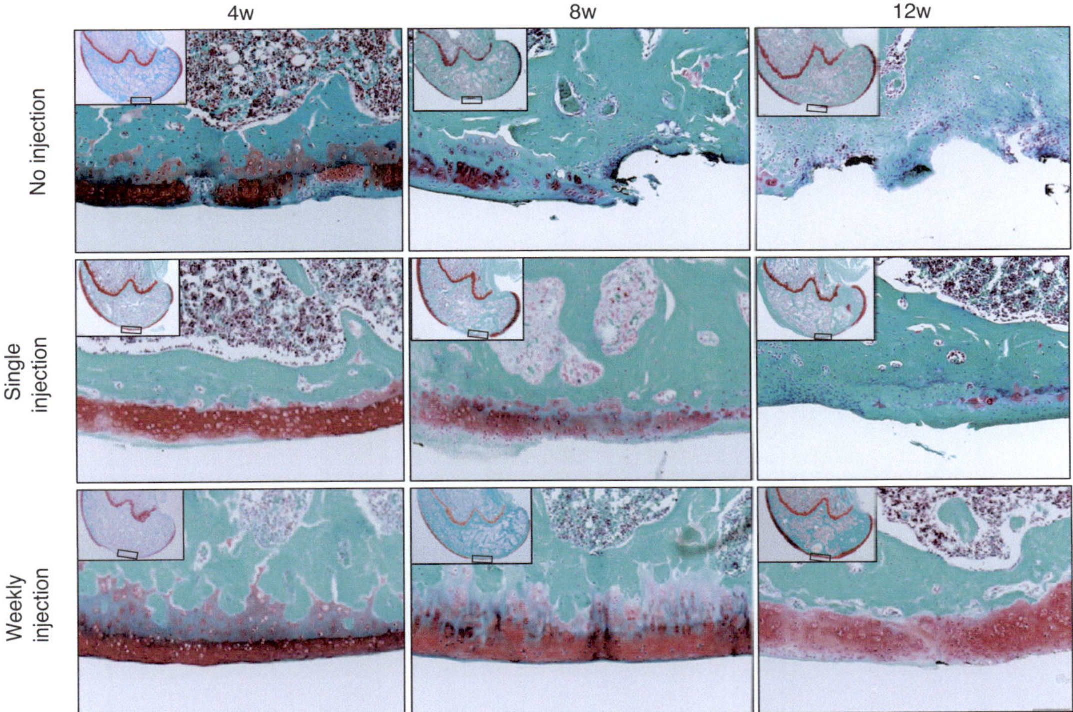

Fig. 4.3 The effect of intra-articular injection of synovial MSCs in ACL-transected rats. Synovial MSCs were injected once or weekly. Histological sections of femoral cartilage stained with Safranin O are shown (From Ozeki [14] et al.)

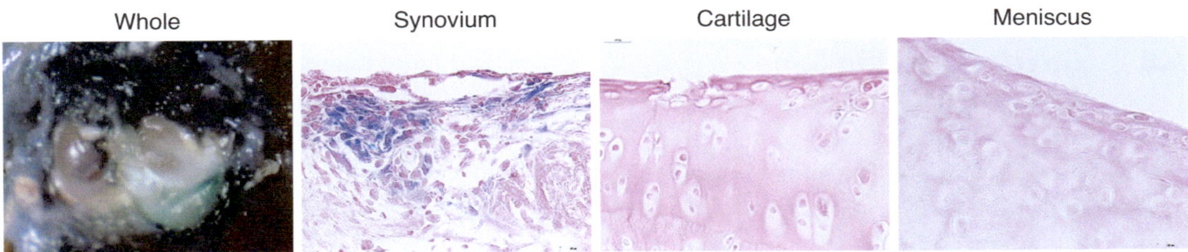

Fig. 4.4 Location of synovial MSCs expressing the LacZ gene 1 day after intra-articular injection into a rat knee. The whole knee joint and histological sections stained with X-gal are shown (From Ozeki [14] et al.)

4.7 Properties of Synovial MSCs after Migration to the Synovium

We also used GFP-expressing MSCs (GFP$^+$ MSCs) to determine whether the injected cells would retain their undifferentiated features. We used flow cytometry to sort the GFP$^+$ MSCs engrafted within the synovium (Fig. 4.6a). The ratio of GFP+ cells against total live cells gradually decreased with time (Fig. 4.6b), but the ratio of CD90 positive cells within the GFP+ cells was maintained at approximately 90% even at 28 days after injection (Fig. 4.6c). The sorted GFP$^+$ cells differentiated into chondrocytes and adipocytes and were calcified in vitro (Fig. 4.6d). These findings confirmed that the injected cells maintained their MSC properties after migration to the synovium.

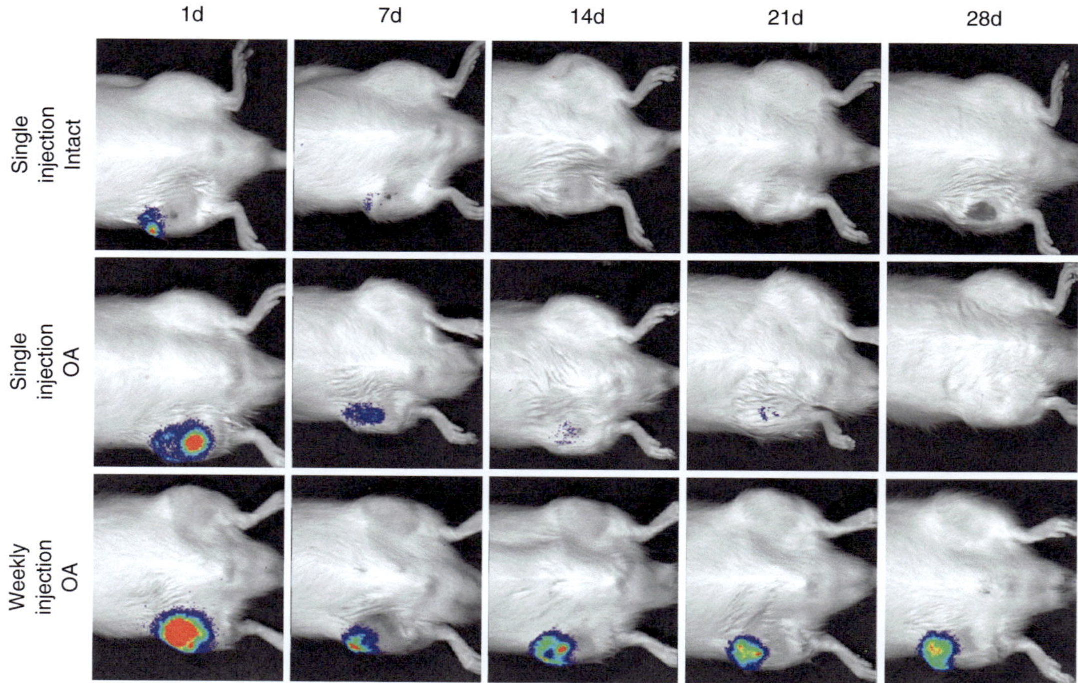

Fig. 4.5 In vivo imaging after intra-articular injection of synovial MSCs expressing the luciferase gene into rat knees. Synovial MSCs were injected once into an intact knee, once into an ACL-transected knee, or weekly into an ACL-transected knee (From Ozeki [14] et al.)

4.8 Species-Specific Gene Expression Analysis

We also used species-specific gene expression to analyze the gene expression changes in synovial MSCs that migrated to the rat synovium after intra-articular injection (Fig. 4.7). We found that human synovial MSCs (hMSCs) detected in the rat synovium accounted for 1% of the total hMSCs 1 day after an intra-articular injection of 1×10^6 hMSCs. We then evaluated the human transcriptomes in the synovium injected with 1×10^6 hMSCs 1 day after injection, using rat synovium mixed with 1×10^4 hMSCs as a control. Microarray analysis for human mRNA revealed that human mRNA increased more than 100-fold in 5 genes, more than 50-fold in 21 genes, and more than ten-fold in 255 genes after the hMSCs migrated into the rat synovial membrane. The ten most highly upregulated human transcripts included hPRG-4 and hBMP-2 (Table 4.1). Further analysis of the expression of human mRNA by RT-PCR revealed significant increases in the expression of hPRG4, hBMP-2, hBMP-6, and hTSG-6 (Fig. 4.8). PRG-4, also known as lubricin, is normally produced by synovial cells or superficial zone chondrocytes and plays an important role in the homeostasis and maintenance of cartilage [15]. BMP-2 and BMP-6 are critical for chondrocyte differentiation, cartilage matrix synthesis, and cartilage protection [16]. TSG-6 has been reported to be secreted by engrafted MSCs to suppress inflammation [17].

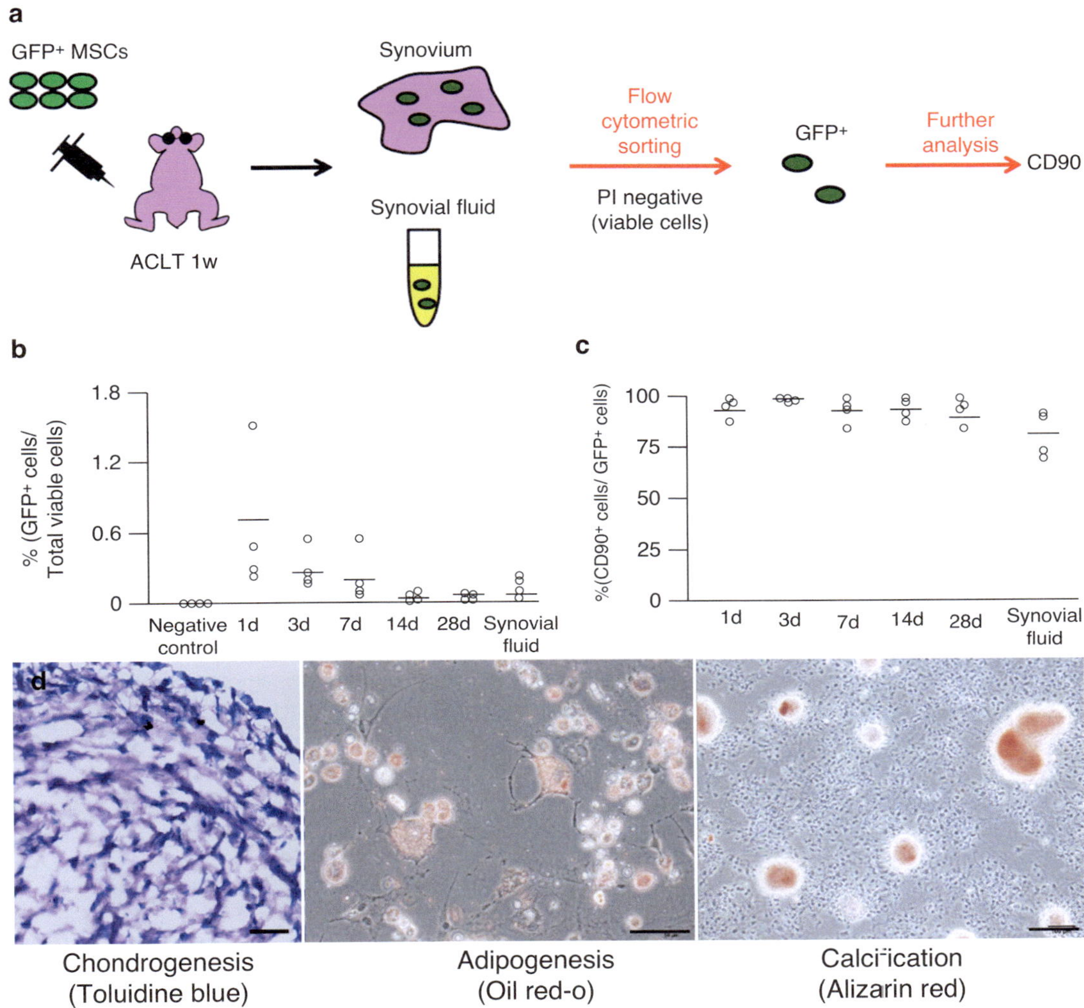

Fig. 4.6 Properties of synovial MSCs after migration to synovium. (**a**) Schema for the flow cytometric assay. ACLT, ACL transection; PI, propidium iodide. (**b**) Sequential ratio of GFP+ MSCs per total live cells in the synovium. Each value is plotted and the average is shown as a crossbar ($n = 4$). (**c**) Ratio of CD90 positive cells in the GFP+ cells ($n = 4$). (**d**) Differentiation potential of the sorted GFP+ cells 23 days after injection (From Ozeki [18] et al.)

4.9 Mechanism by which Injections of Synovial MSCs Delay OA Progression

Most of the synovial MSCs injected into the knee joint migrate into the synovium, and the surviving cells maintain their MSC properties without differentiating into other lineages (Fig. 4.9). The MSCs produce PRG-4 and BMPs for cartilage homeostasis and TSG-6 for anti-inflammation. Two major features of OA include the degeneration of articular cartilage and synovitis; therefore, PRG-4 and BMPs could be useful being chondroprotective for the articular cartilage, while TSG-6 could delay secondary cartilage degeneration by attenuating synovitis.

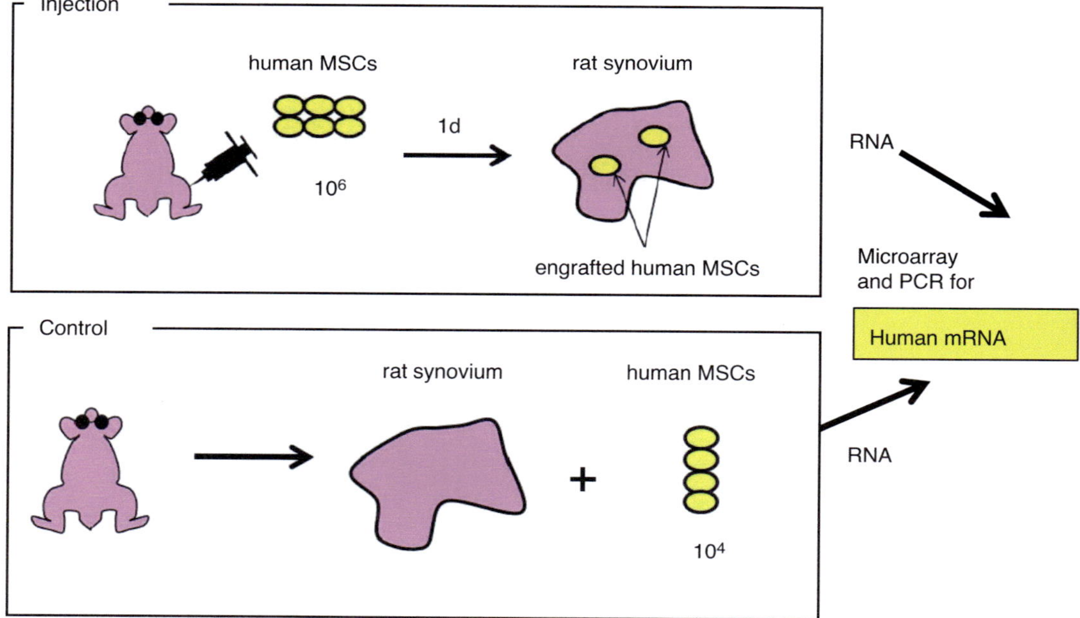

Fig. 4.7 Scheme for species-specific gene expression analysis. A sample containing 1×10^6 human synovial MSCs was injected into an ACL-transected knee of a rat. One day later, the rat synovium was harvested and total RNA was prepared for microarray and RT-PCR. As a control, a rat synovium without MSC injection was harvested, and 1×10^4 human synovial MSCs were mixed with the sample prior to extraction of total RNA

Table 4.1 The top ten human transcripts upregulated in the human MSCs that migrated within the synovium (From Ozeki [14] et al.)

Gene symbol	Gene title	Fold change
TFPI2	Tissue factor pathway inhibitor 2	252.6
PRG4	Proteoglycan 4	162.3
PTHLH	Parathyroid hormone-like hormone	130.5
T.L	Transcribed locus	107.5
LOC285359///PDCL3	Phosducin-like 3 pseudogene///phosducin-like 3	102.6
PLA2G4A	Phospholipase A2, group IVA (cytosolic, calcium-dependent)	92.5
BMP2	Bone morphogenetic protein 2	87.4
RPS4Y1	Ribosomal protein S4, Y-linked 1	75.5
CACNA1D	Calcium channel, voltage-dependent, L type, alpha 1D subunit	63.6
COL15A1	Collagen, type XV, alpha 1	63.3

4.10 Clinical Study of Synovial MSC Injections into OA Knees

We have previously shown that transplantation of synovial MSCs into cartilage defects enhanced cartilage repair in rabbits [18] and pigs [12] and that MRI findings and clinical scores were improved in a human clinical study [19]. We have also reported that transplantation of synovial MSCs promoted meniscus regeneration in rats [20], rabbits [21], pigs [22], and monkeys [23] and that clinical scores [24] and second-look arthroscopy were improved in human clinical studies. In this section, we describe the properties of synovial MSCs and the basic findings follow-

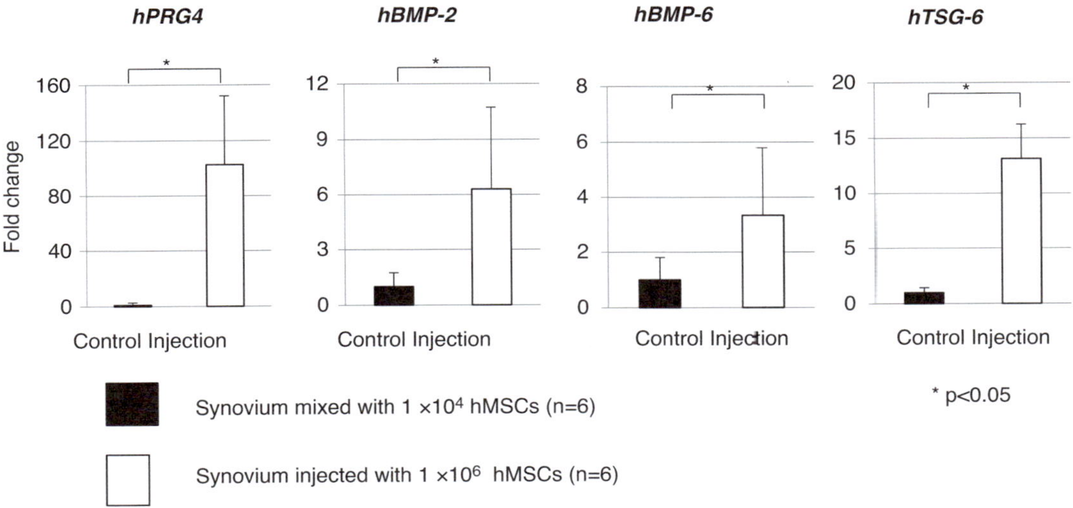

Fig. 4.8 Human-specific gene expressions in the synovium of rat at 1 day, as determined by RT-PCR (From Ozeki [14] et al.)

Fig. 4.9 Possible mechanism for the delay in the progression of cartilage degeneration by injections of synovial MSCs in a rat OA model. Injected synovial MSCs migrate into the synovium, where they express factors such as PRG4, BMP-2, and BMP-6 that prevent cartilage degradation, as well as TSG-6 that inhibits inflammation

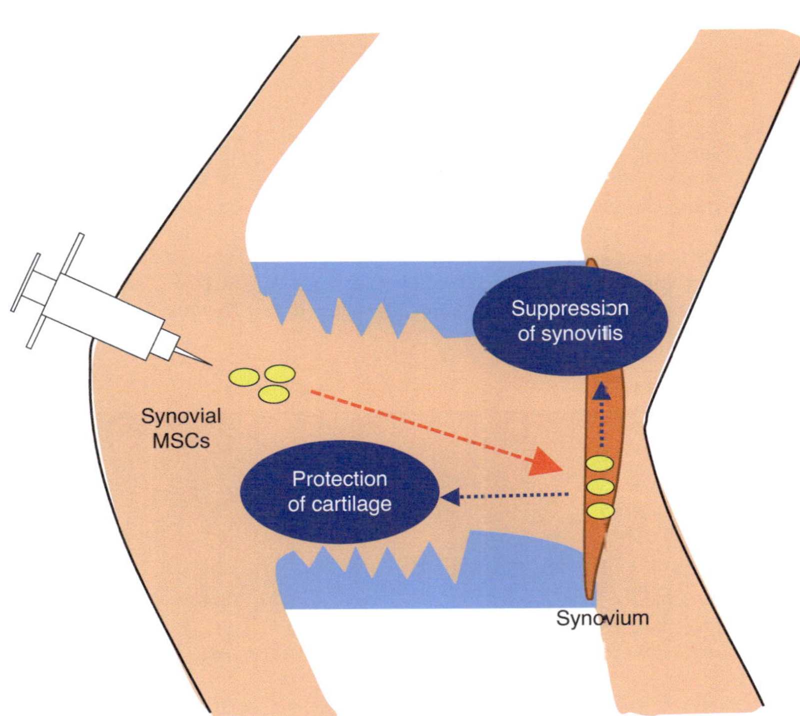

ing intra-articular injections for the treatment of OA of the knee. In addition, we introduce our clinical application of intra-articular injections of synovial MSCs as evaluated by three-dimensional (3D) MRI analysis.

The results from our basic and preclinical animal research have led to the initiation of a clinical study of synovial MSC injections into the OA knees of human patients. The name of the study is "Intraarticular injections of synovial stem cells for osteoarthritis of the knee," registered in Japan (UMIN 000026732), and 14 patients were enrolled. The primary endpoint was the detection of an inhibition of cartilage loss in the OA knee after multiple injections of synovial MSCs. This clinical study included only those patients who had cartilage loss detected by 3D MRI analysis during the first 15 weeks. We harvested synovial tissue arthroscopically from each individual under local anesthesia, cultured the synovial MSCs with autologous serum, and injected two million cells into the knee of that individual twice at 15-week intervals (Fig. 4.10).

Overall, 3D MRI is an attractive analytical method for the quantification of cartilage [25]. We have developed a software for auto-

matic segmentation of cartilage using deep neural networks. The software works by projecting the 3D-reconstructed femoral [26] and tibial cartilage vertically onto the 2D plane [27].

In this clinical study, the last patient's treatment has already been completed, and we are currently in the process of analyzing MRI and clinical outcomes from 30 weeks prior to the first injection to 30 weeks after the first injection. No serious adverse effects have occurred. The preliminary results showed that some participants had decreased cartilage thickness in the posteromedial region of the femoral cartilage prior to injection, but this thickness increased after injection (Fig. 4.11).

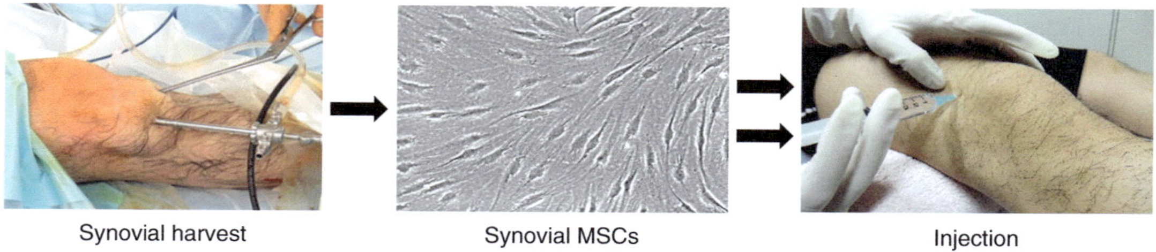

Fig. 4.10 Scheme of a clinical study on autologous synovial MSC injections into the OA knees of human patients. The synovium was harvested and enzymatically digested. The resulting synovial MSCs were expanded with autologous human serum and then injected into the knee twice at a 15-week interval

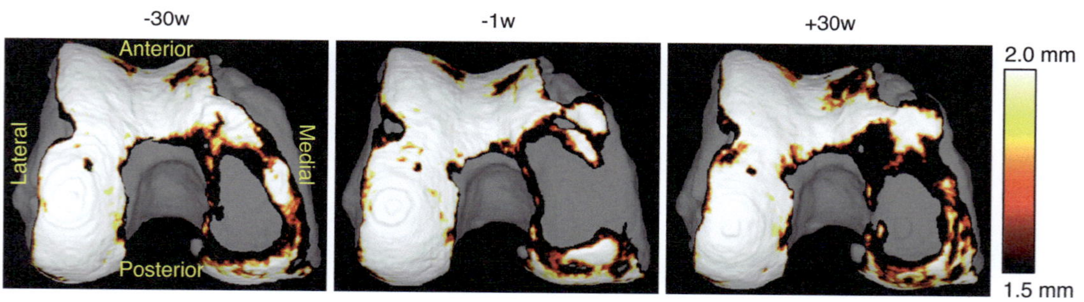

Fig. 4.11 Representative patient case who underwent synovial MSC injections into a knee at time 0 and 15 weeks. MRI examinations were performed at −30, −1, and + 30 weeks. Using the software we developed, the cartilage area was automatically extracted and visualized in three dimensions. Cartilage thickness mappings are shown. Femoral cartilage at the posteromedial region at −30 weeks decreased at −1 week but increased again at +30 weeks

4.11 Conclusions

Synovial MSCs have a high chondrogenic and proliferative potential. Most of the synovial MSCs injected into the knee joint migrate into the synovium, maintain their MSC properties, and produce PRG-4 and BMPs for cartilage homeostasis and TSG-6 for anti-inflammation. We have started a clinical study in which synovial MSCs are injected twice into OA knees and the changes in cartilage measurements are automatically assessed by 3D MRI analysis.

Take-Home Messages
- Maximum yields of culture-expanded mesenchymal stromal cells (MSCs) can be achieved by choosing the optimum cell density that limits contacts between colonies.
- The weight of the cartilage pellet cultured in vitro reflects the number of chondroprogenitor cells and their ability to produce cartilage matrix, which is an indicator of the chondrogenic differentiation potential of the MSC population.
- MSCs isolated from the synovium have a high chondrogenic and proliferative potential.
- Periodic injections of synovial MSCs inhibited osteoarthritis (OA) progression and attenuated synovitis in a rat OA model.
- Most of the synovial MSCs injected into the knee joint migrated into the synovium and increased gene expressions for anti-inflammation, lubrication, and cartilage matrix synthesis.

References

1. Dominici M, Le Blanc K, Mueller I, Slaper-Cortenbach I, Marini F, Krause D, et al. Minimal criteria for defining multipotent mesenchymal stromal cells. The International Society for Cellular Therapy position statement. Cytotherapy. 2006;8(4):315–7.
2. Muschler GF, Midura RJ. Connective tissue progenitors: practical concepts for clinical applications. Clin Orthop Relat Res. 2002;395:66–80.
3. Muschler GF, Midura RJ, Nakamoto C. Practical modeling concepts for connective tissue stem cell and progenitor compartment kinetics. J Biomed Biotechnol. 2003;2003(3):170–93.
4. Mantripragada VP, Piuzzi NS, Bova WA, Boehm C, Obuchowski NA, Lefebvre V, et al. Donor-matched comparison of chondrogenic progenitors resident in human infrapatellar fat pad, synovium, and periosteum—implications for cartilage repair. Connect Tissue Res. 2019;60(6):597–610.
5. Segawa Y, Muneta T, Makino H, Nimura A, Mochizuki T, Ju YJ, et al. Mesenchymal stem cells derived from synovium, meniscus, anterior cruciate ligament, and articular chondrocytes share similar gene expression profiles. J Orthop Res. 2009;27(4):435–41.
6. Sekiya I, Larson BL, Smith JR, Pochampally R, Cui JG, Prockop DJ. Expansion of human adult stem cells from bone marrow stroma: conditions that maximize the yields of early progenitors and evaluate their quality. Stem Cells. 2002;20(6):530–41.
7. Nakamura K, Tsuji K, Mizuno M, Koga H, Muneta T, Sekiya I. Initial cell plating density affects properties of human primary synovial mesenchymal stem cells. J Orthop Res. 2019;37(6):1358–67.
8. Sakaguchi Y, Sekiya I, Yagishita K, Muneta T. Comparison of human stem cells derived from various mesenchymal tissues: superiority of synovium as a cell source. Arthritis Rheum. 2005;52(8):2521–9.
9. Sekiya I, Vuoristo JT, Larson BL, Prockop DJ. In vitro cartilage formation by human adult stem cells from bone marrow stroma defines the sequence of cellular and molecular events during chondrogenesis. Proc Natl Acad Sci U S A. 2002;99(7):4397–402.
10. Yoshimura H, Muneta T, Nimura A, Yokoyama A, Koga H, Sekiya I. Comparison of rat mesenchymal stem cells derived from bone marrow, synovium, periosteum, adipose tissue, and muscle. Cell Tissue Res. 2007;327(3):449–62.
11. Koga H, Muneta T, Nagase T, Nimura A, Ju YJ, Mochizuki T, et al. Comparison of mesenchymal tissues-derived stem cells for in vivo chondrogenesis: suitable conditions for cell therapy of cartilage defects in rabbit. Cell Tissue Res. 2008;333(2):207–15.
12. Nakamura T, Sekiya I, Muneta T, Hatsushika D, Horie M, Tsuji K, et al. Arthroscopic, histological and MRI analyses of cartilage repair after a minimally invasive method of transplantation of allogeneic synovial mesenchymal stromal cells into cartilage defects in pigs. Cytotherapy. 2012;14(3):327–38.
13. Nimura A, Muneta T, Koga H, Mochizuki T, Suzuki K, Makino H, et al. Increased proliferation of human synovial mesenchymal stem cells with autologous human serum: comparisons with bone marrow mesenchymal stem cells and with fetal bovine serum. Arthritis Rheum. 2008;58(2):501–10.

14. Ozeki N, Muneta T, Koga H, Nakagawa Y, Mizuno M, Tsuji K, et al. Not single but periodic injections of synovial mesenchymal stem cells maintain viable cells in knees and inhibit osteoarthritis progression in rats. Osteoarthr Cartil. 2016;24(6):1061–70.
15. Rhee DK, Marcelino J, Baker M, Gong Y, Smits P, Lefebvre V, et al. The secreted glycoprotein lubricin protects cartilage surfaces and inhibits synovial cell overgrowth. J Clin Invest. 2005;115(3):622–31.
16. Sekiya I, Colter DC, Prockop DJ. BMP-6 enhances chondrogenesis in a subpopulation of human marrow stromal cells. Biochem Biophys Res Commun. 2001;284(2):411–8.
17. Lee RH, Pulin AA, Seo MJ, Kota DJ, Ylostalo J, Larson BL, et al. Intravenous hMSCs improve myocardial infarction in mice because cells embolized in lung are activated to secrete the anti-inflammatory protein TSG-6. Cell Stem Cell. 2009;5(1):54–63.
18. Koga H, Muneta T, Ju YJ, Nagase T, Nimura A, Mochizuki T, et al. Synovial stem cells are regionally specified according to local microenvironments after implantation for cartilage regeneration. Stem Cells. 2007;25(3):689–96.
19. Sekiya I, Muneta T, Horie M, Koga H. Arthroscopic transplantation of synovial stem cells improves clinical outcomes in knees with cartilage defects. Clin Orthop Relat Res. 2015;473(7):2316–26.
20. Horie M, Sekiya I, Muneta T, Ichinose S, Matsumoto K, Saito H, et al. Intra-articular injected synovial stem cells differentiate into meniscal cells directly and promote meniscal regeneration without mobilization to distant organs in rat massive meniscal defect. Stem Cells. 2009;27(4):878–87.
21. Hatsushika D, Muneta T, Horie M, Koga H, Tsuji K, Sekiya I. Intraarticular injection of synovial stem cells promotes meniscal regeneration in a rabbit massive meniscal defect model. J Orthop Res. 2013;31(9):1354–9.
22. Hatsushika D, Muneta T, Nakamura T, Horie M, Koga H, Nakagawa Y, et al. Repetitive allogeneic intraarticular injections of synovial mesenchymal stem cells promote meniscus regeneration in a porcine massive meniscus defect model. Osteoarthr Cartil. 2014;22(7):941–50.
23. Kondo S, Muneta T, Nakagawa Y, Koga H, Watanabe T, Tsuji K, et al. Transplantation of autologous synovial mesenchymal stem cells promotes meniscus regeneration in aged primates. J Orthop Res. 2017;35(6):1274–82.
24. Sekiya I, Koga H, Otabe K, Nakagawa Y, Katano H, Ozeki N, et al. Additional use of synovial mesenchymal stem cell transplantation following surgical repair of a complex degenerative tear of the medial meniscus of the knee: a case report. Cell Transplant. 2019;28(11):1445–54.
25. Shakoor D, Guermazi A, Kijowski R, Fritz J, Jalali-Farahani S, Mohajer B, et al. Diagnostic performance of three-dimensional MRI for depicting cartilage defects in the knee: a meta-analysis. Radiology. 2018;289(1):71–82.
26. Hyodo A, Ozeki N, Kohno Y, Suzuki S, Mizuno M, Otabe K, et al. Projected cartilage area ratio determined by 3-dimensional MRI analysis: validation of a novel technique to evaluate articular cartilage. JB JS Open Access. 2019;4(4):e0010.
27. Aoki H, Ozeki N, Katano H, Hyodo A, Miura Y, Matsuda J, et al. Relationship between medial meniscus extrusion and cartilage measurements in the knee by fully automatic three-dimensional MRI analysis. BMC Musculoskelet Disord. 2020;21(1):742.

Placenta, Umbilical Cord, and Umbilical Cord Blood-Derived Cultured Stromal Cells

5

Jin-A Kim and Chul-Won Ha

5.1 Structure and Function of the Placenta, Umbilical Cord, and Umbilical Vessels

The fetal adnexa are composed of the placenta, fetal membranes, and umbilical cord. The placenta is discoid in shape with a diameter of 15–20 cm and a thickness of 3–4 cm. From the margins of the chorionic disc extend the fetal membranes, amnion, and chorion, which enclose the fetus in the amniotic cavity, and the endometrial decidua [1] (Fig. 5.1).

Maternal blood enters the chorionic layer via the maternal spiral arteries. Nutrients and gases are exchanged in the chorionic layer between the chorionic villi. Oxygenated blood is transported to the fetus through the vessels in the chorionic villi and then through the umbilical vein. Deoxygenated fetal blood is carried back to the placenta via the umbilical arteries to reach the vessels in the chorionic villi. Nutrients and gases are then exchanged in the chorionic layer through the chorionic villi. Deoxygenated blood is then transported to the maternal circulation via the maternal veins (Fig. 5.1).

5.2 Placenta-Derived Stromal Cells

Human placenta is well known to not only play a fundamental and essential role in fetal development, nutrition, and tolerance but also function as a bank of multipotent stromal cells [2]. Placental tissue can be easily obtained as medical waste. The human placenta is a fetomaternal entity that consists of a fetal component (the chorionic plate) and a maternal component (the deciduae) [3]. The placenta has a complex structure, consisting of layers of amnion epithelium (AE), amnion (AM), chorionic membrane (CM), chorionic trophoblast (CT), chorionic villi (CV), intervillous space, and deciduous membrane (DC) [1, 4, 5]. It is known that the stromal cells isolated from each of these layers differ in proliferative and differentiation capacity [5, 6].

J.-A. Kim
Department of Health Sciences and Technology, SAIHST, Sungkyunkwan University, Seoul, South Korea

Stem Cell & Regenerative Medicine Research Institute, Samsung Medical Center, Seoul, South Korea

C.-W. Ha (✉)
Department of Health Sciences and Technology, SAIHST, Sungkyunkwan University, Seoul, South Korea

Stem Cell & Regenerative Medicine Research Institute, Samsung Medical Center, Seoul, South Korea

Department of Orthopedic Surgery, Samsung Medical Center, Sungkyunkwan University School of Medicine, Seoul, South Korea

© ISAKOS 2022
G. Filardo et al. (eds.), *Orthobiologics*, https://doi.org/10.1007/978-3-030-84744-9_5

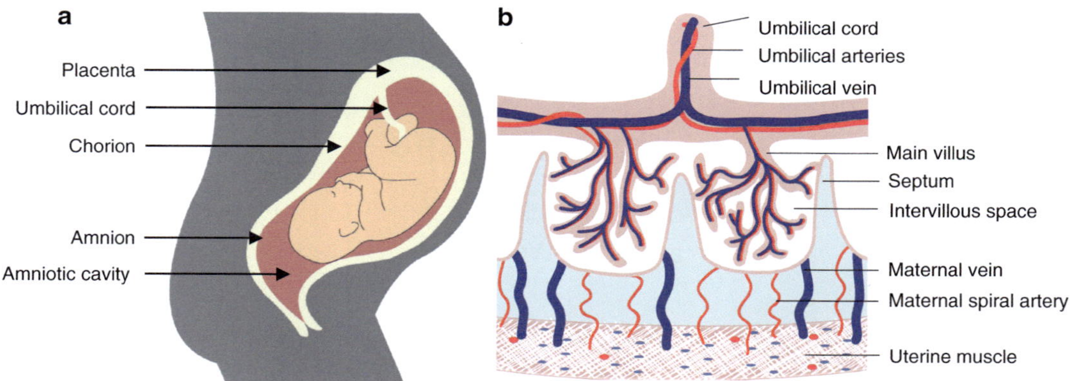

Fig. 5.1 Placental structure and vessels at full term. (**a**) Overview of the uterus, placenta, and a fetus. (**b**) Detailed diagram of the placenta showing the placental structure and fetal and maternal vessels

5.2.1 Formulation

5.2.1.1 Collection

Collection of placental tissue occurs immediately after childbirth, immediately after the umbilical cord and umbilical cord blood are taken from the placenta. The placenta should be obtained with informed consent and following ethical approval guidelines for use in orthobiologics. For use in orthobiologics, the way of collecting the placental tissue should be sterile. Under aseptic surgical conditions, the placenta is collected via cesarean section, and the placental tissue is collected in a sterile container and transferred to the laboratory, or the placenta tissue is divided and collected in a dedicated kit and transferred to the placental tissue bank. It is safe to store the placenta at room temperature or 4 °C during transportation and before dissection.

5.2.1.2 Isolation and Expansion

As the placenta contains a lot of blood cells, the commonly used and efficient method for removing the blood cells is washing as much as possible with Dulbecco's phosphate-buffered saline (DPBS) without calcium or magnesium to remove blood. Since the placenta is made up of different layers, there are many ways to separate the target layer if the investigator prefers to use a specific part of the placenta rather than the whole placenta, before processing with enzyme. The methods of isolating

stromal cells derived from different layers of the placental tissue and commonly used enzymes in many previous studies are listed in Table 5.1. After the enzyme treatment, impurities are filtered out using a cell strainer [5, 7]. Then, the cells are seeded on the culture plate. Cell culture medium is changed 48 h after the initial separation. After incubation for an additional 24 h, the cells are washed twice with DPBS to remove debris and red blood cells. The medium is changed twice a week and cells are incubated until the cell monolayer is 80–90% confluent. This initial expansion can take 4–14 days depending on the quality of the tissue, the amount of starting material, the culture efficiency, and the incubation time with the digestion solution [7].

5.2.2 In Vitro and in Vivo Effects

Placenta-derived stromal cells are known to have more important immunomodulatory effects than bone marrow-derived mesenchymal stromal cells (MSCs) by shifting the differentiation of monocytes from M1 to M2-like macrophages [29]. It has also been shown to inhibit T lymphocyte proliferation and cytokine production and regulate T cell differentiation [30–32]. PGE2, a bioactive lipid synthesized from arachidonic acid by COX-1 and COX-2 enzymes, inhibits T cell proliferation and regulates dendritic cell matura-

Table 5.1 Isolation methods for stromal cells from human placenta

Stromal cell source	Isolation method	References
Whole placenta-derived stromal cells	Collagenase plus trypsin Trypsin plus collagenase II plus dispase II Trypsin-EDTA Collagenase I Collagenase P	[8] [9] [10–12] [13] [14]
Amnion-derived stromal cells	Collagenase II Explant culture	[15] [16]
Chorion-derived stromal cells	Dispase II plus collagenase II or collagenase A Collagenase II Explant culture	[28, 31] [15] [16]
AF-derived stromal cells	Mesh filtering	[16, 17]
AE-derived stromal cells	Trypsin Trypsin-EDTA Explant culture	[18, 19] [20–22] [9]
AM-derived stromal cells	Collagenase plus/or DNase Collagenase V	[18, 19, 22] [5, 21]
CM-derived stromal cells	Collagenase II Collagenase type I plus DNase I plus dispase Trypsin-EDTA plus collagenase I	[9, 19] [7] [23]
CT-derived stromal cells	Trypsin and DNase I plus collagenase DNase I plus collagenase I	[24] [25]
CV-derived stromal cells	Collagenase II plus dispase II Collagenase type I plus DNase I plus dispase Explant culture	[9] [7] [26, 27]
Intervillous space-derived stromal cells	Collagenase I	[5]
DC-derived stromal cells	Collagenase I plus DNase I plus dispase Collagenase plus hyaluronidase plus pronase	[7] [28]

tion and antigen presentation functions [33, 34]. TGF-β is a potent immunomodulatory protein that controls the differentiation, proliferation, and activation of various immune cells [35].

IL-10 is a well-known anti-inflammatory cytokine that regulates the growth and activation of anti-inflammatory cells [36]. All these three important secretory factors were found to be significantly increased in placenta-derived stromal cells cultured with mixed lymphocyte reaction/bead T cell reaction (MLR/BTR) [36–38]. The addition of blockers or neutralizers to PGE2, TGF-β, or IL-10 partially reversed and impaired the inhibitory effect of placenta-derived stromal cells on T cell proliferation [32, 39]. Stimulation of placenta-derived stromal cells by IFN-γ significantly upregulated the release of TGF-β and IL-10 resistant cytokines [32, 40]. These results suggest that placenta-derived stromal cells can be used for the clinical situations where an anti-inflammatory effect is desired [41].

Implantation of allogeneic amniotic membranes or amniotic epithelial cells without immunosuppression is known not to induce acute immune rejection [42–44]. In vitro studies have shown that cells isolated from the amnion and chorion do not trigger an allogeneic or heterologous immune response, but actively inhibit lymphocyte proliferation [22, 45, 46]. Human amniotic membranes and amniotic epithelial cells have been shown to survive long-term in immunocompetent animals, including rabbits [47], rats [48], guinea pigs [49], and bonnet monkeys [50]. After injecting heterogeneous human amniotic membrane cells into neonatal pigs and mice, human microchimerism was detected in the bone marrow, brain, lungs, and thymus, suggesting active migration and integration into specific organs and showing active tolerance of xenogeneic cells [46]. Parolini et al. [51] characterized the effect of human amniotic membrane-derived stromal cells on antigen-specific T cell responses in rheumatoid arthritis (RA) patients and evaluated their therapeutic potential in a preclinical experimental model of RA. Treatment with placenta-derived stromal cells suppressed synovial inflammatory responses and antigen-specific Th1/Th17 activation in cells isolated from RA patients. Moreover, placenta-derived stromal cells stimulated the generation of human CD4+CD25+FoxP3+ Treg cells with a capacity to suppress collagen-specific T cell responses.

Systemic infusion of placenta-derived stromal cells significantly reduced the incidence and severity of collagen-induced arthritis (CIA) by downregulating the two deleterious components of disease: Th1-driven autoimmunity and inflammation. In mice with CIA, placenta-derived stromal cells treatment decreased the production of various inflammatory cytokines and chemokines in the joints, impaired antigen-specific Th1/Th17 cell expansion in the lymph nodes, and generated peripheral antigen-specific Treg cells. The results suggest that the immunosuppressive action of placenta-derived stromal cells is not major histocompatibility complex restricted and that the infused placenta-derived stromal cells are immunologically tolerated by the host, which would be very convenient for a potential clinical application of these cells in RA in the future. The in vitro and in vivo data describe a broad array of immune modulation functions suggesting that placenta-derived stromal cells have immune modulatory and immune tolerance inducing properties [51].

Li et al. [33] investigated the ability of placenta-derived stromal cells to grow on silk fibroin (SF) biomaterials. Placenta-derived stromal cells were maintained in vitro in an allogeneic mixed lymphocytic response (MLR) system to investigate the inhibitory effect on T cell proliferation and in a total of 12 healthy adult New Zealand rabbits after articular cartilage defects of the knee femoral condyle were established. The placenta-derived stromal cells biomaterial complex was implanted, and the articular cartilage defects were observed. The results of MLR indicate that placenta-derived stromal cells inhibit rabbit T cell responses. Cartilage damage was recovered by newly formed free cartilage, no degeneration or infiltration by lymphocytes or leukocytes at 12 weeks, and no silk fibroin biomaterial residues were found. Zhang et al. [26] explored the possibility of using placenta-derived stromal cells for cartilage regeneration. Preinduced placenta-derived stromal cells embedded in a collagen sponge were implanted in the osteochondral defects of nude rats. Coverage with stiff reparative tissue, which was white and had a smooth surface, was shown at 6 weeks after surgery. Histological analysis showed hyaline-like regenerative tissue. The results of their studies suggest that placenta-derived stromal cells can be one of the possible sources of allogeneic cells for cartilage tissue engineering.

5.3 Umbilical Cord-Derived Stromal Cells

The umbilical cord (UC) is an essential part of the placenta, contributing to fetal development by ensuring the blood flow between mother and fetus. The UC is formed within the first weeks of gestation by the enclosure of the vessels (one vein and two arteries) into a bulk of mucous connective tissue, named Wharton's jelly (WJ) and lined by the umbilical epithelium [20].

5.3.1 Formulation

5.3.1.1 Collection

UC tissue collection occurs immediately after the collection of UC blood from the UC and placenta after childbirth. The acquisition of UC tissue is easy and noninvasive. Currently, most hospitals treat UC as medical waste, and the collection procedure is noninvasive. Thus, collecting UC tissue itself does not incur ethical issues. However, UC tissue should be obtained with informed consent and following ethical approval guidelines for use in orthobiologics. WJ-derived stromal cells are abundant in origin, easy to collect, and known to have no adverse effects on the donors [52, 53]. For use in orthobiologics, the way of collecting the UC tissue should be sterile. UC tissue is collected in a sterile container and transported to the laboratory, or a portion of the umbilical cord is divided into several tubes and assembled to transport to the UC tissue bank. It is safe to store the UC tissue at room temperature or 4 °C during transportation and before dissection.

5.3.1.2 Isolation and Expansion

UC-derived stromal cells are usually isolated according to the following method [54]. To isolate UC-derived stromal cells, UC tissue pieces are obtained aseptically after cesarean sections. Depending on the investigator's requirement, 1–3 inches of UC tissue is sufficient to begin culturing UC-derived stromal cells. UC-derived stromal cells are obtained either by enzymatic digestion [55, 56] or by explant culture techniques [57–62]. Most laboratories prefer explant culture methods to isolate UC-derived stromal cells from UC tissue because they are inexpensive and can provide a pure UC-derived stromal cell population [61, 63–65]. However, UC tissue fragments are often separated and floated in the medium, resulting in a small number of cells because the suspended fragments do not provide stromal cells. Two representative methods of obtaining stromal cells from UC tissue include the following steps:

(i) In the explant tissue culture method of stromal cells isolation [57–62]:
   (a) UC tissue is cut into 1–2 mm³ small pieces that are seeded into tissue culture treated flasks or dishes.
   (b) Once the tissue pieces are attached to the plastic surface of the tissue culture flask, add culture medium (5–10% FBS and MEM supplemented with 1% nonessential amino acids) slowly to prevent separation of the pieces.
   (c) The growth of cells in tissue fragments can be observed within a week after incubation.
   (d) Once a sufficient number of cells have been obtained, the tissue fragments are removed, and fresh medium is added to allow these cells to proliferate for a few more days.
   (e) The media is changed twice a week and cells are incubated until the cell monolayer is 80–90% confluent. This initial expansion takes 4–14 days, depending on the quality of the tissue, the amount of starting material, the culture efficiency, and the incubation time with the digestion solution [7].
(ii) The other method is to apply enzymes to the UC tissue [55, 56, 59, 60, 66].
   (a) The enzymes used in this digestion are collagenases (e.g., collagenase type IV), dispase, hyaluronidase, or a mixture of collagenase and trypsin.
   (b) In this method, cord tissue is first cut into small pieces and digested for 30–60 min at 37 °C using collagenase or collagenase and trypsin.
   (c) The solution is filtered through a 70 μm or 100 μm cell strainer and centrifuged at 1000 rpm to pellet the cells.
   (d) Cell pellets are incubated in tissue culture flasks at 37 °C, 5% CO_2 in a humid environment.
   (e) The UC-derived stromal cell medium is replaced with fresh medium within 24 h to remove nonplastic-adherent cells.
   (f) The media is changed twice a week and the cell monolayer is incubated until it is 80–90% confluent. This initial expansion takes 4–14 days, depending on the quality of the tissue, the amount of starting material, the culture efficiency, and the incubation time with the digestion solution [7].

5.3.2 In Vitro and in Vivo Effects

UC-derived stromal cells are not controversial compared to embryonic stem cells and induced pluripotent stem cells [67, 68]. Moreover, UC-derived stromal cells have a higher proliferative potential than multipotent stromal cells from other sources [69]. UC-derived stromal cells have excellent freeze-thaw properties and can be frozen for long periods in liquid nitrogen Dewar bottles and thawed when needed. This feature is convenient for basic experimental work on UC-derived stromal cells and provides a good theoretical basis for the establishment of a clinical resource bank in the future [70].

UC-derived stromal cells have been known to be superior to stromal cells from other sources such as bone marrow in terms of osteogenic and chondrogenic differentiation ability [71]. UC-derived stromal cells exhibit more type II collagen synthesis than bone marrow (BM) stromal cells [72]. Because of the relative difficulty in obtaining bone marrow stromal cells, UC-derived stromal cells have better clinical application prospects. In addition, UC-derived stromal cells are considered to have advantages of high availability, large expansion capacity, no teratoma or tumor formation, and strong immunomodulatory capacity [73]. They show low expression of major histocompatibility complex (MHC) class I molecules and do not express MHC class II molecule and co-stimulatory molecules required for T cell activation. Thus, allogeneic UC-derived stromal cells do not induce T cell proliferation responses [74]. MSCs are known to be self-protected from immunological defenses when used as allografts due to the expression of MHC class I, but not II [75]. The immune system has excellent tolerance to UC-derived stromal cells [72] because of the low immunogenicity of UC-derived stromal cells.

Moreover, pretreatment with pro-inflammatory cytokines may improve the immune regulation of UC-derived stromal cells [76]. In addition, UC-derived stromal cells did not induce the proliferation of xenogeneic and allogeneic immune cells. The expression of immunosuppressive human leukocyte antigens HLA-G6, IL-6, and VEGF and the absence of the costimulatory molecules CD40, CD80, and CD86 further support the immunomodulatory properties of UC-derived stromal cells [77]. Thus, UC-derived stromal cells are potentially one of the best cell types for clinical application.

UC-derived stromal cells are also known to have a paracrine effect. They can secrete a variety of biologically active factors that affect specific biological functions [78]. Exosomes from UC-derived stromal cells have been known to contain cell-associated miRNAs, mRNAs, and proteins secreted by UC-derived stromal cells. UC-derived stromal cell exosomes can inhibit the expression and function of Th22 cells. Th22 are novel CD4+ helper T cells that secrete interleukin 22 and tumor necrosis factor-alpha (TNF-alpha) and have an inflammatory effect in many diseases such as tumors and rheumatoid arthritis [79]. Thus, UC-derived stromal cell exosomes can play an important role in inhibiting inflammation.

Kim et al. [80] investigated anti-inflammatory and tissue regeneration effects after treatment with UC-derived stromal cells in a temporal mandibular joint (TMJ) osteoarthritis (OA) rabbit model. Compared to the untreated control, the potential regeneration outcome and anti-inflammatory effects of UC-derived stromal cells were confirmed in TMJ-OA-induced rabbits. UC-derived stromal cells have shown remarkable cartilage protective effects and additional cartilage regeneration potential. This effect occurred through upregulated expression of growth factors, extracellular matrix markers, and anti-inflammatory cytokines, as well as reduced expression of pro-inflammatory cytokines. The anti-inflammatory effect of UC-derived stromal cells was similar to that of dexamethasone (DEX). In the pathogenesis of OA, the RELA gene is involved in cartilage degradation through MMP-13, and RELA, a member of the NF-κβ gene family, regulates the inflammatory response and activates pro-inflammatory cytokines. The study of synovial tissue in OA patients who underwent total knee replacement (TKR) surgery consisted of six groups who received four repeated treatments [81]. Groups I and II (control) consisted of synovial cells of OA cultured for 24 and 48 h, respectively. Groups III and IV consisted of UC-derived stromal cells cultured for 24 and 48 h, respectively. Groups V and VI were co-culture of synovial cells-WJ-MSCs cultured for 24 and 48 h, respectively. Expression of the MMP-13 and RELA genes in each group was detected by qPCR. The results showed that UC-derived stromal cells reduced MMP-13 gene expression after co-culture for 24 and 48 h in OA synovial cells. Thus, it suggests that UC-derived stromal cells may play an important role in slowing the progression of arthritis.

UC-derived stromal cells were shown to be more chondrogenic than BM-derived stromal cells and resulted in more vitreous cartilage tissue formation [82]. Wu et al. [83] investigated the therapeutic effect of implanting UC-derived stromal cells and hyaluronic acid (HA) hydrogel in a minipig OA preclinical model. Compared with the untreated control group, regeneration of hyaline cartilage was confirmed in gross and histological evaluation. The International Cartilage Repair Society (ICRS) histological score was much higher in the experimental group than in the control group. Zhang et al. [84] combined UC-derived stromal cells with a cell-free chondrocyte extracellular matrix (ECM) oriented scaffold to investigate cartilage regeneration after transplantation in a goat knee full-thickness cartilage defect. Morphologically, the size of cartilage defects 3 months after transplantation was significantly smaller in the experimental group than in the control group. New cartilage-like tissue covered the subchondral bone. The defect was well integrated with the edges of the normal cartilage. Six months after transplantation, the cartilage defect was completely covered with new cartilage tissue. As a result of using various staining methods histologically, the treatment group showed better treatment results than the control group. Quantitative analysis of glycosaminoglycans showed significantly higher levels in the experimental group than in the control group. Lin et al. [85] showed that UC-derived stromal cells seeded on PLGA scaffolds promote cartilage regeneration in a rabbit model with cartilage defects. In a study that used a hydrogel scaffold to encapsulate UC-derived stromal cells and added appropriate cell culture medium after gelation, Alcian Blue and Safranin O staining showed that UC-derived stromal cells encapsulated in hydrogels produced large amounts of extracellular matrix with abundant proteoglycans [70]. Expression of collagen II and aggregated proteoglycans was increased in cultures containing chondrogenic medium, indicating that UC-derived stromal cells have a strong ability to differentiate into chondrocytes under these conditions. These results provided a good rationale for the future clinical application of UC-derived stromal cells in the treatment of cartilage lesions or OA.

Wu et al. [86] explored the clinical and histopathological effects of intra-articular injection of UC-derived stromal cells in a collagen-induced arthritis (CIA) model. In this study, intra-articular injection of UC-derived stromal cells was ineffective in CIA mice and accelerated the progression of arthritis in the presence of TNF-alpha. In order to confirm the role of TNF-alpha, a combination of UC-derived stromal cells and a TNF inhibitor was injected, and it was confirmed that it reduced disease symptoms in CIA mice. Upon exposure to TNF-alpha, there was a significantly reduced expression of CD90 and HLA-G and the level of IL-10 in vitro and in vivo. This showed that TNF-alpha blocks the immunosuppressive effect of human UC-derived stromal cells and that inhibition of TNF-alpha reduces cartilage destruction by inhibiting the immunogenicity of UC-derived stromal cells. Injecting both a TNF inhibitor and UCB-derived stromal cells could be a potentially effective treatment to improve the disease. Santos et al. [87] investigated their treated stromal cells of human umbilical cord tissue (UCX® cells) in induced autoimmune inflammatory arthritis to investigate their immunosuppressive ability. UCX® cells have been shown to inhibit T cell activation and promote the expansion of Tregs better than BM-MSCs. Thus, in a model of acute carrageenan-induced arthritis, administration of heterologous UCX® has shown that human UCX® cells can more efficiently reduce foot edema in vivo than BM-MSCs. Finally, animals treated with intra-articular and intraperitoneal injections of UCX® in a chronic adjuvant-induced arthritis model showed faster relief of local and systemic arthritis symptoms.

Liu et al. [88] investigated the potential immunosuppressive effects of UC-derived stromal cells in RA. Systemic injection of UC-derived stromal cells reduced the severity of CIA in a mouse model. Consistently, levels of pro-inflammatory cytokines and chemokines (TNF-α, IL-6, and monocyte chemoattractant protein-1) were reduced, and anti-inflammatory/regulatory cytokine (IL-10) levels in the serum of UC-derived stromal cells-treated mice were increased. Moreover, these treatments shifted the Th1/Th2-type response and induced Tregs in the CIA. UC-derived stromal cells were also effective in the treatment of RA, especially when cultured in a 3D environment [89].

Therefore, many studies have shown that transplantation of human UC-derived stromal cells in animals has certain therapeutic effects on cartilage lesions, OA, and RA. Currently, the FDA has registered dozens of clinical trials on the transplantation of UC-derived stromal cells for treating refractory diseases, such as knee OA or RA [67, 90].

5.4 Umbilical Cord Blood-Derived Stromal Cells

Umbilical cord blood (UCB) is the blood that remains in the placenta and umbilical cord following the birth of a baby. It is rich in blood stem cells (hematopoietic stem cells) similar to those found in bone marrow, and these cells have already been used to treat many different cancers, immune deficiencies, and genetic disorders. In addition to the hematopoietic stem cells, UCB contains numerous cell types, including a population of stromal cells with the ability to differentiate and generate progeny. UCB-derived stromal cells have emerged as an alternative source for cell therapy because they have plentiful cell banking systems already established with noninvasive collection, immediate transplantation, and hypo-immunogenic properties [91, 92].

5.4.1 Formulation

5.4.1.1 Collection

UCB collection takes place in the cord tissue. UCB is taken from the umbilical vessel immediately after birth and before placental delivery. UCB should be obtained with informed consent and ethical approval for use in orthobiologics. When a baby is born, a medical professional tightens and cuts the umbilical cord. After cleanly disinfecting the cord, blood is collected by inserting a needle into the umbilical cord. Cord blood can still be collected even if there is a delay in cord clamping. The UCB is collected in a sterile syringe or in a bag with citrate phosphate dextrose adenine (CPDA) anticoagulant and delivered to a laboratory or cord blood bank. It is safe to store the UCB at room temperature or 4 °C during transportation and before dissection.

5.4.1.2 Isolation and Expansion

It is known that UCB-derived stromal cells are generally separated by the following method [93, 94]. UCB is obtained aseptically and a minimum volume of approximately 100 cc is required for further processing [95]. UCB-derived stromal cells are obtained by separating the mononuclear cell (MNC) fraction using Ficoll-Hypaque density gradient centrifugation. The steps to obtain stromal cells from UCB are as follows:

(i) Isolation methods of stromal cells from human UCB [16, 96, 97].
  (a) Each UCB unit is diluted 1:1 with phosphate-buffered saline (PBS)/2 Mm ethylenediaminetetraacetic acid (EDTA).
  (b) Carefully loaded onto Ficoll-Hypaque solution.
  (c) Density gradient centrifugation is applied at ×435 g for 30 min at room temperature.
  (d) MNCs are removed from the interphase.
  (e) MNCs are washed two to three times with PBS/EDTA.
  (f) Cell pellets are incubated in tissue culture flasks at 37 °C, 5% CO_2 in a humid environment.
  (g) The stromal cell medium is replaced with fresh medium within 24 h to remove non-plastic adhesion cells.
  (h) The medium is changed twice a week and the cell monolayer is incubated until it is 80–90% confluent. This initial expansion takes 4–14 days, depending on the quality of the tissue, the amount of starting material, the culture efficiency, and the incubation time with the digestion solution [7].

5.4.2 In Vitro and in Vivo Effects

UCB-derived stromal cells are known to be relatively non-immunogenic because they have low expression levels of human leukocyte antigen-MHC class I and lack the MHC class II molecules that induce immune rejection in allogeneic transplantation [98]. In addition, a recent study that used UCB-derived stromal cells in a rabbit model showed no evidence of immune rejection [99]. UCB-derived stromal cells are known to be easily induced to differentiate into chondrocytes [100, 101] and show higher chondrogenic differentiation potential compared to BM-derived stromal cells and adipose-derived stromal cells. As UCB-derived stromal cells are known to have chondrogenic potential and immunomodulatory function, the UCB-derived stromal cells have been considered a potential therapeutic option for arthritis diseases due to their key role in the inflammatory process and related articular cartilage degradation [102].

Compared to other sources of multipotent stromal cells such as BM, UCB-derived stromal cells showed superior secretion of anti-inflammatory cytokines (IL-10 and IL-6) and a superior ability to restore cartilage matrix production in three-dimensional (3D) cartilage structures [103]. This result was confirmed in a recent study showing that injection of allogeneic UCB-derived stromal cells in horses lowered the inflammatory response and decreased cartilage degradation compared to allogeneic and even autologous BM-MSCs [104]. Transplanted into damaged tissue, UCB-derived stromal cells produced a secretome and extracellular vesicles (EVs), stimulating the regenerative process of joint tissue. Jeong et al. [105] demonstrated that UCB-derived stromal cells did not differentiate directly into a chondrocyte phenotype, but exert their action through the secretion of paracrine factors. The secretome of UCB-derived stromal cells treated with OA synovial fluid promoted the differentiation of chondroprogenitor cells into chondrocytes. The authors also reported that thrombospondin-2, a glycoprotein that mediates cell-to-cell interactions, was a key component of this process as it was able to activate various signaling pathways involved in cartilage formation and cartilage in the recipient cell.

Ha et al. [106] implanted the HA hydrogel complex UCB-derived stromal cells into a minipig model to explore its consistent regenerative potential. Minipigs were sacrificed 12 weeks after surgery, and the degree of cartilage regeneration was evaluated by gross and histological analysis, and the transplanted knees resulted in superior cartilage regeneration compared to the control knee.

Park et al. [107] investigated the feasibility of implanting UCB-derived stromal cells and HA hydrogel complexes to repair articular cartilage defects in a rabbit model. The UCB-derived stromal cells and HA composite transplant resulted in an overall superior cartilage repair tissue with better quality than HA alone or without treatment. The cellular structure and collagen arrangement at week 16 were similar to the surrounding normal articular cartilage tissue. Histological scores also showed that cartilage repair in the experimental knee was better than that of the control knee. Zheng et al. [108] investigated a rabbit model of osteochondral regeneration using 3D printed poly-caprolactone-hydroxyapatite (PCL-HA) scaffolds coated with UCB-derived stromal cells and chondrocytes. Mean ICRS scores for the UCB-derived stromal cells and chondrocyte-seeded PCL-HA scaffolds (group A) were significantly higher than the normal unseeded control (NC) PCL-HA scaffold group (group B) ($P < 0.05$). Histology with Safranin O and fast-green staining showed that the UCB-derived stromal cells-seeded PCL-HA scaffolds significantly promoted bone and cartilage regeneration.

Kwon et al. [109] investigated the therapeutic effects and optimal dose of UCB-derived stromal cells injection in a chronic full-thickness rotator cuff tendon tear, and UCB-derived stromal cells injection under ultrasound guidance showed regeneration in a rabbit model, although there were no differences in the regenerative effects between high and low doses of the UCB-derived stromal cells. Lim et al. [110] investigated the effect of allogeneic UCB-derived stromal cells and recombinant methionyl human granulocyte colony-stimulating factor (rmhGCSF) on a canine spinal cord injury model after balloon compression at the first lumbar level. Two weeks after transplantation, the UCB-derived stromal cells groups and UCB-derived stromal cells + rmhGCSF groups had a significantly higher Olby score than the control group. The nerve conduction rate was significantly improved based on the somatosensory evoked potential. In addition, distinct structural consistency of neuronal cell bodies was observed in spinal cord lesions of the UCB-derived stromal cells groups and UCB-derived stromal cells + rmhGCSF groups.

5.5 Conclusions

In conclusion, the placenta, umbilical cord, and umbilical cord blood have emerged as an alternative source to obtain stromal cells. This chapter provided an overview of the cell sourcing options for stromal cells derived from placental tissue, umbilical cord, and umbilical cord blood. There have been various methods of collection, isolation, and culture expansion of the stromal cells obtained from these complex tissues to be used for future cell therapies. Detailed information regarding these methods obtained from the literature are summarized in this chapter. The preclinical data supports the ongoing exploration of these neonatal sources as opportunities for cellular therapies in the musculoskeletal system. The stromal cells isolated and culture-expanded from these neonatal tissues have shown low immunogenicity, significant immunomodulatory effects, and anti-inflammatory effects. These cells have also shown prominent osteogenic and chondrogenic differentiation, as well as paracrine effects that support the regenerative effect in musculoskeletal tissue healing such as cartilage and tendons. Therefore, the use of stromal cells obtained from the neonatal tissue sources seems to be promising in future cell therapy and regenerative medicine.

> **Take-Home Messages**
>
> - The stromal cells isolated and culture-expanded from these neonatal tissues have low immunogenicity, significant immunomodulatory effects, and anti-inflammatory effects.
> - The stromal cells obtained from the placenta, umbilical cord, and umbilical cord blood have also shown prominent osteogenic and chondrogenic differentiation.
> - Paracrine effects of these cells support the regenerative effect in musculoskeletal tissue healing such as cartilage and tendons.

References

1. Parolini O, et al. Concise review: isolation and characterization of cells from human term placenta: outcome of the first international Workshop on Placenta Derived Stem Cells. Stem Cells. 2008;26(2):300–11.
2. Wu M, et al. Comparison of the biological characteristics of mesenchymal stem cells derived from the human placenta and umbilical cord. Sci Rep. 2018;8(1):5014.
3. Parolini O. Human Placenta: a Source of Progenitor/Stem Cells? J Reproduktionsmed Endokrinologie. 2006;3(2):117–26.
4. Hass R, et al. Different populations and sources of human mesenchymal stem cells (MSC): a comparison of adult and neonatal tissue-derived MSC. Cell Commun Signal. 2011;9:12.
5. Choi YS, et al. Different characteristics of mesenchymal stem cells isolated from different layers of full term placenta. PLoS One. 2017;12(2):e0172642.
6. Indumathi S, et al. Comparison of feto-maternal organ derived stem cells in facets of immunophenotype, proliferation and differentiation. Tissue Cell. 2013;45(6):434–42.
7. Pelekanos RA, et al. Isolation and expansion of mesenchymal stem/stromal cells derived from human placenta tissue. J Vis Exp. 2016:112.
8. Genbacev O, et al. Serum-free derivation of human embryonic stem cell lines on human placental fibroblast feeders. Fertil Steril. 2005;83(5):1517–29.
9. Portmann-Lanz CB, et al. Placental mesenchymal stem cells as potential autologous graft for pre- and perinatal neuroregeneration. Am J Obstet Gynecol. 2006;194(3):664–73.
10. Chien CC, et al. In vitro differentiation of human placenta-derived multipotent cells into hepatocyte-like cells. Stem Cells. 2006;24(7):1759–68.
11. Miao Z, et al. Isolation of mesenchymal stem cells from human placenta: comparison with human bone marrow mesenchymal stem cells. Cell Biol Int. 2006;30(9):681–7.
12. Yen BL, et al. Isolation of multipotent cells from human term placenta. Stem Cells. 2005;23(1):3–9.
13. Barlow S, et al. Comparison of human placenta- and bone marrow-derived multipotent mesenchymal stem cells. Stem Cells Dev. 2008;17(6):1095–107.
14. Chang CM, et al. Placenta-derived multipotent stem cells induced to differentiate into insulin-positive cells. Biochem Biophys Res Commun. 2007;357(2):414–20.
15. Yamahara K, et al. Comparison of angiogenic, cytoprotective, and immunosuppressive properties of human amnion- and chorion-derived mesenchymal stem cells. PLoS One. 2014;9(2):e88319.
16. Klein C, et al. Ex vivo expansion of hematopoietic stem- and progenitor cells from cord blood in coculture with mesenchymal stroma cells from amnion, chorion, Wharton's jelly, amniotic fluid,

cord blood, and bone marrow. Tissue Eng Part A. 2013;19(23–24):2577–85.

17. Moraghebi R, et al. Term amniotic fluid: an unexploited reserve of mesenchymal stromal cells for reprogramming and potential cell therapy applications. Stem Cell Res Ther. 2017;8(1):190.

18. Sakuragawa N, et al. Human amnion mesenchyme cells express phenotypes of neuroglial progenitor cells. J Neurosci Res. 2004;78(2):208–14.

19. Soncini M, et al. Isolation and characterization of mesenchymal cells from human fetal membranes. J Tissue Eng Regen Med. 2007;1(4):296–305.

20. Corrao S, et al. Umbilical cord revisited: from Wharton's jelly myofibroblasts to mesenchymal stem cells. Histol Histopathol. 2013;28(10):1235–44.

21. Moore RM, Silver RJ, Moore JJ. Physiological apoptotic agents have different effects upon human amnion epithelial and mesenchymal cells. Placenta. 2003;24(2–3):173–80.

22. Wolbank S, et al. Dose-dependent immunomodulatory effect of human stem cells from amniotic membrane: a comparison with human mesenchymal stem cells from adipose tissue. Tissue Eng. 2007;13(6):1173–83.

23. Koo BK, et al. Isolation and characterization of chorionic mesenchymal stromal cells from human full term placenta. J Korean Med Sci. 2012;27(8):857–63.

24. Semenov OV, et al. Multipotent mesenchymal stem cells from human placenta: critical parameters for isolation and maintenance of stemness after isolation. Am J Obstet Gynecol. 2010;202(2):193.e1–193.e13.

25. Brooke G, et al. Manufacturing of human placenta-derived mesenchymal stem cells for clinical trials. Br J Haematol. 2009;144(4):571–9.

26. Zhang X, et al. Mesenchymal progenitor cells derived from chorionic villi of human placenta for cartilage tissue engineering. Biochem Biophys Res Commun. 2006;340(3):944–52.

27. Igura K, et al. Isolation and characterization of mesenchymal progenitor cells from chorionic villi of human placenta. Cytotherapy. 2004;6(6):543–53.

28. Strakova Z, et al. Multipotent properties of myofibroblast cells derived from human placenta. Cell Tissue Res. 2008;332(3):479–88.

29. Abumaree MH, et al. Human placental mesenchymal stem cells (pMSCs) play a role as immune suppressive cells by shifting macrophage differentiation from inflammatory M1 to anti-inflammatory M2 macrophages. Stem Cell Rev Rep. 2013;9(5):620–41.

30. Liu W, et al. Human placenta-derived adherent cells induce tolerogenic immune responses. Clin Transl Immunol. 2014;3(5):e14.

31. Abumaree MH, et al. Immunomodulatory properties of human placental mesenchymal stem/stromal cells. Placenta. 2017;59:87–95.

32. Chang CJ, et al. Placenta-derived multipotent cells exhibit immunosuppressive properties that are enhanced in the presence of interferon-gamma. Stem Cells. 2006;24(11):2466–77.

33. Li F, et al. Human placenta-derived mesenchymal stem cells with silk fibroin biomaterial in the repair of articular cartilage defects. Cell Reprogram. 2012;14(4):334–41.

34. Yañez R, et al. Prostaglandin E2 plays a key role in the immunosuppressive properties of adipose and bone marrow tissue-derived mesenchymal stromal cells. Exp Cell Res. 2010;316(19):3109–23.

35. Letterio JJ, Roberts AB. Regulation of immune responses by TGF-beta. Annu Rev Immunol. 1998;16:137–61.

36. Kang JW, et al. Immunomodulatory effects of human amniotic membrane-derived mesenchymal stem cells. J Vet Sci. 2012;13(1):23–31.

37. Mareschi K, et al. Immunoregulatory effects on T lymphocytes by human mesenchymal stromal cells isolated from bone marrow, amniotic fluid, and placenta. Exp Hematol. 2016;44(2):138–50.e1.

38. Castro-Manrreza ME, et al. Human mesenchymal stromal cells from adult and neonatal sources: a comparative in vitro analysis of their immunosuppressive properties against T cells. Stem Cells Dev. 2014;23(11):1217–32.

39. Erkers T, et al. Decidual stromal cells promote regulatory T cells and suppress alloreactivity in a cell contact-dependent manner. Stem Cells Dev. 2013;22(19):2596–605.

40. Deuse T, et al. Immunogenicity and immunomodulatory properties of umbilical cord lining mesenchymal stem cells. Cell Transplant. 2011;20(5):655–67.

41. Maxson S, et al. Concise review: role of mesenchymal stem cells in wound repair. Stem Cells Transl Med. 2012;1(2):142–9.

42. Akle CA, et al. Immunogenicity of human amniotic epithelial cells after transplantation into volunteers. Lancet. 1981;2(8254):1003–5.

43. Tylki-Szymańska A, et al. Amniotic tissue transplantation as a trial of treatment in some lysosomal storage diseases. J Inherit Metab Dis. 1985;8(3):101–4.

44. Yeager AM, et al. A therapeutic trial of amniotic epithelial cell implantation in patients with lysosomal storage diseases. Am J Med Genet. 1985;22(2):347–55.

45. Li H, et al. Immunosuppressive factors secreted by human amniotic epithelial cells. Invest Ophthalmol Vis Sci. 2005;46(3):900–7.

46. Bailo M, et al. Engraftment potential of human amnion and chorion cells derived from term placenta. Transplantation. 2004;78(10):1439–48.

47. Avila M, et al. Reconstruction of ocular surface with heterologous limbal epithelium and amniotic membrane in a rabbit model. Cornea. 2001;20(4):414–20.

48. Kubo M, et al. Immunogenicity of human amniotic membrane in experimental xenotransplantation. Invest Ophthalmol Vis Sci. 2001;42(7):1539–46.

49. Yuge I, et al. Transplanted human amniotic epithelial cells express connexin 26 and Na-K-adenosine triphosphatase in the inner ear. Transplantation. 2004;77(9):1452–4.

50. Sankar V, Muthusamy R. Role of human amniotic epithelial cell transplantation in spinal cord injury repair research. Neuroscience. 2003;118(1):11–7.
51. Parolini O, et al. Therapeutic effect of human amniotic membrane-derived cells on experimental arthritis and other inflammatory disorders. Arthritis Rheumatol. 2014;66(2):327–39.
52. Watson N, et al. Discarded Wharton jelly of the human umbilical cord: a viable source for mesenchymal stromal cells. Cytotherapy. 2015;17(1):18–24.
53. Weiss ML, Troyer DL. Stem cells in the umbilical cord. Stem Cell Rev. 2006;2(2):155–62.
54. Mahmood R, Shaukat M, Choudhery MS. Biological properties of mesenchymal stem cells derived from adipose tissue, umbilical cord tissue and bone marrow. AIMS Cell Tissue Eng. 2018;2(2):78–90.
55. Han YF, et al. Optimization of human umbilical cord mesenchymal stem cell isolation and culture methods. Cytotechnology. 2013;65(5):819–27.
56. Semenova E, E.K.M.a.T.O. Comparison of characteristics of mesenchymal stem cells obtained mechanically and enzymatically from placenta and umbilical cord. J Cell Sci Ther. 2017;8(2):262.
57. Choudhery MS, et al. Comparison of human mesenchymal stem cells derived from adipose and cord tissue. Cytotherapy. 2013;15(3):330–43.
58. Mahmood R, et al. In vitro differentiation potential of human placenta derived cells into skin cells. Stem Cells Int. 2015;2015:841062.
59. Hassan G, et al. A simple method to isolate and expand human umbilical cord derived mesenchymal stem cells: using explant method and umbilical cord blood serum. Int J Stem Cells. 2017;10(2):184–92.
60. Hassan G, et al. Platelet lysate induces chondrogenic differentiation of umbilical cord-derived mesenchymal stem cells. Cell Mol Biol Lett. 2018;23:11.
61. Majore I, et al. Growth and differentiation properties of mesenchymal stromal cell populations derived from whole human umbilical cord. Stem Cell Rev Rep. 2011;7(1):17–31.
62. Choudhery MS, et al. Utility of cryopreserved umbilical cord tissue for regenerative medicine. Curr Stem Cell Res Ther. 2013;8(5):370–80.
63. Marmotti A, et al. Minced umbilical cord fragments as a source of cells for orthopaedic tissue engineering: an in vitro study. Stem Cells Int. 2012;2012:326813.
64. Hendijani F, Sadeghi-Aliabadi H, Haghjooy Javanmard S. Comparison of human mesenchymal stem cells isolated by explant culture method from entire umbilical cord and Wharton's jelly matrix. Cell Tissue Bank. 2014;15(4):555–65.
65. Capelli C, et al. Minimally manipulated whole human umbilical cord is a rich source of clinical-grade human mesenchymal stromal cells expanded in human platelet lysate. Cytotherapy. 2011;13(7):786–801.
66. Mennan C, et al. Isolation and characterisation of mesenchymal stem cells from different regions of the human umbilical cord. Biomed Res Int. 2013;2013:916136.
67. Ding DC, et al. Human umbilical cord mesenchymal stem cells: a new era for stem cell therapy. Cell Transplant. 2015;24(3):339–47.
68. Gore A, et al. Somatic coding mutations in human induced pluripotent stem cells. Nature. 2011;471(7336):63–7.
69. Arutyunyan I, et al. Umbilical cord as prospective source for mesenchymal stem cell-based therapy. Stem Cells Int. 2016;2016:6901286.
70. Harris DT. Umbilical cord tissue mesenchymal stem cells: characterization and clinical applications. Curr Stem Cell Res Ther. 2013;8(5):394–9.
71. Reppel L, et al. Chondrogenic induction of mesenchymal stromal/stem cells from Wharton's jelly embedded in alginate hydrogel and without added growth factor: an alternative stem cell source for cartilage tissue engineering. Stem Cell Res Ther. 2015;6:260.
72. Li X, et al. Comprehensive characterization of four different populations of human mesenchymal stem cells as regards their immune properties, proliferation and differentiation. Int J Mol Med. 2014;34(3):695–704.
73. Fong CY, et al. Human Wharton's jelly stem cells have unique transcriptome profiles compared to human embryonic stem cells and other mesenchymal stem cells. Stem Cell Rev Rep. 2011;7(1):1–16.
74. Bassi EJ, Aita CA, Câmara NO. Immune regulatory properties of multipotent mesenchymal stromal cells: where do we stand? World J Stem Cells. 2011;3(1):1–8.
75. Ryan JM, et al. Mesenchymal stem cells avoid allogeneic rejection. J Inflamm (Lond). 2005;2:8.
76. Donders R, et al. Human Wharton's jelly-derived stem cells display immunomodulatory properties and transiently improve rat experimental autoimmune encephalomyelitis. Cell Transplant. 2015;24(10):2077–98.
77. Weiss ML, et al. Immune properties of human umbilical cord Wharton's jelly-derived cells. Stem Cells. 2008;26(11):2865–74.
78. Bai L, et al. Bioactive molecules derived from umbilical cord mesenchymal stem cells. Acta Histochem. 2016;118(8):761–9.
79. Wolk K, et al. IL-22 regulates the expression of genes responsible for antimicrobial defense, cellular differentiation, and mobility in keratinocytes: a potential role in psoriasis. Eur J Immunol. 2006;36(5):1309–23.
80. Kim H, et al. Therapeutic effect of mesenchymal stem cells derived from human umbilical cord in rabbit temporomandibular joint model of osteoarthritis. Sci Rep. 2019;9(1):13854.
81. Sofia V, et al. The influence of Wharton jelly mesenchymal stem cell toward matrix Metalloproteinase-13 and RELA Synoviocyte gene expression on osteoarthritis. Open Access Maced J Med Sci. 2019;7(5):701–6.
82. Wang L, et al. A comparison of human bone marrow-derived mesenchymal stem cells and human umbilical cord-derived mesenchymal stromal cells

for cartilage tissue engineering. Tissue Eng Part A. 2009;15(8):2259–66.

83. Wu KC, et al. Transplanting human umbilical cord mesenchymal stem cells and hyaluronate hydrogel repairs cartilage of osteoarthritis in the minipig model. Ci Ji Yi Xue Za Zhi. 2019;31(1):11–9.

84. Zhang YLS, Guo WM, et al. Experimental study on repairing full-thickness cartilage defect of goat knee joint with human umbilical cord mesenchymal stem cells and acellular chondrocyte extracellular matrix oriented scaffold. Chin Med Biotechnol. 2016;6:32–9.

85. Lin YX, et al. In vitro and in vivo evaluation of the developed PLGA/HAp/Zein scaffolds for bone-cartilage interface regeneration. Biomed Environ Sci. 2015;28(1):1–12.

86. Wu CC, et al. TNF-α inhibitor reverse the effects of human umbilical cord-derived stem cells on experimental arthritis by increasing immunosuppression. Cell Immunol. 2012;273(1):30–40.

87. Santos JM, et al. The role of human umbilical cord tissue-derived mesenchymal stromal cells (UCX®) in the treatment of inflammatory arthritis. J Transl Med. 2013;11:18.

88. Liu Y, et al. Therapeutic potential of human umbilical cord mesenchymal stem cells in the treatment of rheumatoid arthritis. Arthritis Res Ther. 2010;12(6):R210.

89. Miranda JP, et al. The Secretome derived from 3D-cultured umbilical cord tissue MSCs counteracts manifestations typifying rheumatoid arthritis. Front Immunol. 2019;10:18.

90. Fan CG, Zhang QJ, Zhou JR. Therapeutic potentials of mesenchymal stem cells derived from human umbilical cord. Stem Cell Rev Rep. 2011;7(1):195–207.

91. Flynn A, Barry F, O'Brien T. UC blood-derived mesenchymal stromal cells: an overview. Cytotherapy. 2007;9(8):717–26.

92. Kim SM, et al. Gene therapy using TRAIL-secreting human umbilical cord blood-derived mesenchymal stem cells against intracranial glioma. Cancer Res. 2008;68(23):9614–23.

93. Sousa T, et al. Umbilical cord blood processing: volume reduction and recovery of CD34+ cells. Bone Marrow Transplant. 1997;19(4):311–3.

94. U-pratya Y, et al. Collection and processing of umbilical cord blood for cryopreservation. J Med Assoc Thai. 2003;86(11):1055–62.

95. Bieback K, et al. Critical parameters for the isolation of mesenchymal stem cells from umbilical cord blood. Stem Cells. 2004;22(4):625–34.

96. Jin HJ, et al. Comparative analysis of human mesenchymal stem cells from bone marrow, adipose tissue, and umbilical cord blood as sources of cell therapy. Int J Mol Sci. 2013;14(9):17986–8001.

97. Pievani A, et al. Comparative analysis of multilineage properties of mesenchymal stromal cells derived from fetal sources shows an advantage of mesenchymal stromal cells isolated from cord blood in chondrogenic differentiation potential. Cytotherapy. 2014;16(7):893–905.

98. Tempelhof S, Rupp S, Seil R. Age-related prevalence of rotator cuff tears in asymptomatic shoulders. J Shoulder Elb Surg. 1999;8(4):296–9.

99. Jang KM, et al. Efficacy and safety of human umbilical cord blood-derived mesenchymal stem cells in anterior cruciate ligament reconstruction of a rabbit model: new strategy to enhance tendon graft healing. Arthroscopy. 2015;31(8):1530–9.

100. Desancé M, et al. Chondrogenic differentiation of defined equine mesenchymal stem cells derived from umbilical cord blood for use in cartilage repair therapy. Int J Mol Sci. 2018;19:2.

101. Ibrahim AM, et al. Chondrogenic differentiation of human umbilical cord blood-derived mesenchymal stem cells in vitro. Microsc Res Tech. 2015;78(8):667–75.

102. Arrigoni C, et al. Umbilical cord MSCs and their Secretome in the therapy of arthritic diseases: a research and industrial perspective. Cell. 2020;9:6.

103. Lo WC, et al. Preferential therapy for osteoarthritis by cord blood MSCs through regulation of chondrogenic cytokines. Biomaterials. 2013;34(20):4739–48.

104. Bertoni L, et al. Intra-articular injection of 2 different dosages of autologous and allogeneic bone marrow- and umbilical cord-derived mesenchymal stem cells triggers a variable inflammatory response of the fetlock joint on 12 sound experimental horses. Stem Cells Int. 2019;2019:9431894.

105. Jeong SY, et al. Thrombospondin-2 secreted by human umbilical cord blood-derived mesenchymal stem cells promotes chondrogenic differentiation. Stem Cells. 2013;31(10):2136–48.

106. Ha CW, et al. Cartilage repair using composites of human umbilical cord blood-derived mesenchymal stem cells and hyaluronic acid hydrogel in a Minipig model. Stem Cells Transl Med. 2015;4(9):1044–51.

107. Park YB, et al. Single-stage cell-based cartilage repair in a rabbit model: cell tracking and in vivo chondrogenesis of human umbilical cord blood-derived mesenchymal stem cells and hyaluronic acid hydrogel composite. Osteoarthr Cartil. 2017;25(4):570–80.

108. Zheng P, et al. A rabbit model of osteochondral regeneration using three-dimensional printed polycaprolactone-hydroxyapatite scaffolds coated with umbilical cord blood mesenchymal stem cells and chondrocytes. Med Sci Monit. 2019;25:7361–9.

109. Kwon DR, Park GY, Lee SC. Regenerative effects of mesenchymal stem cells by dosage in a chronic rotator cuff tendon tear in a rabbit model. Regen Med. 2019;14(11):1001–12.

110. Lim JH, et al. Transplantation of canine umbilical cord blood-derived mesenchymal stem cells in experimentally induced spinal cord injured dogs. J Vet Sci. 2007;8(3):275–82.

Injectable Allogenic Mesenchymal Stromal Cells: Advantages, Disadvantages, and Challenges

6

Lucas K. Keyt, Matthew D. LaPrade, Aaron J. Krych, and Daniel B. F. Saris

6.1 Introduction

Human stem cell biology presents remarkable untapped potential for further understanding human physiology as well as developing novel treatments across a multitude of medical subspecialties [1]. However, the study of human embryonic or oocyte-derived stem cells is fraught with ethical and political controversy and is currently restricted in a majority of countries [2].

Isolation of adult culture-expanded mesenchymal stromal cell populations starting from mixed populations of tissue-specific connective tissue progenitors from bone marrow, perivascular cells, or adipose tissue represents a fertile alternative for derivation of standardized cellular therapies that avoids the ethical and political dilemmas encountered with fetal stem cells.

Mesenchymal stromal cells (MSCs) [3], historically also variably described as marrow stromal cells, mesenchymal stem cells, mesenchymal progenitor cells, and recently as medicinal signaling cells [4], are currently under investigation for use in both clinical and research settings. Culture-expanded MSC populations can be generated by isolation of mixed populations of native tissue-derived cells and placing cells into culture.

Native connective tissues contain a heterogeneous population of cells, which includes a small number of stem and progenitor cells. The prevalence of stem and progenitor cells in a given tissue sample can be estimated by planting tissue-derived cells in culture and measuring the number of colonies formed per million cells plated. Each colony formed in this process is derived, in theory, from a single founding stem or progenitor cell. Each formed colony differs from others. These differences can be appreciated in proliferation rate, morphology, and gene expression. Differences in performance between individual colonies are interpreted to reflect the diversity of potential niches that cells capable of proliferation in a given tissue may also have in vivo. The number and heterogeneity of colonies formed from a given tissue sample can be used to estimate the concentration and prevalence of stem and progenitor cells in the original tissue sample. The number of progenitor cells reflected in a colony forming assay is always much greater than the number of true underlying stem cells. This heterogeneous population of stem and progenitor cells in native connective tissues has been defined as "tissue-specific connective tissue progenitors" (CTPs).

L. K. Keyt · M. D. LaPrade · A. J. Krych
Department of Orthopedic Surgery, Division of Sports Medicine, Mayo Clinic, Rochester, MN, USA
e-mail: Krych.Aaron@mayo.edu

D. B. F. Saris (✉)
Department of Orthopedic Surgery, Division of Sports Medicine, Mayo Clinic, Rochester, MN, USA

Department of Orthopedics, University Medical Center, Utrecht, The Netherlands
e-mail: Saris.Daniel@mayo.edu

© ISAKOS 2022
G. Filardo et al. (eds.), *Orthobiologics*, https://doi.org/10.1007/978-3-030-84744-9_6

CTPs in native tissues are involved in the formation and remodeling of new tissues, and in response to injury [5, 6]. The concentration and prevalence of CTPs is therefore influenced by the age, systemic health, and local tissue health of the source materials.

Culture-expanded cells, derived from competitive expansion of the mixture of clones that are obtained from any tissue, have been shown to differentiate in vitro into adipocytes, chondrocytes, osteoblasts, myoblasts, and possibly neuron-like cells [1]. This potential to fabricate large numbers of cells with plasticity to form multiple connective tissues has made culture-expanded MSCs an attractive option for tissue restoration in orthopedics and beyond.

6.2 Theoretical Advantages of Allogeneic MSCs

One of the most readily apparent advantages of allogeneic MSCs is avoiding the work and cost of harvesting cells directly from each individual patient undergoing treatment. Generation of autologous MSCs typically requires collection from either bone marrow or adipose tissue. The tissues must be processed to preserve the viability of a heterogeneous stem and progenitor population and then placed into culture where clones of adherent fibroblastic cells can be expanded and compete with one another under conditions that preserve the biological potential of the highly selected clone or clones that persist in the final expanded population. Only after expansion and rigorous characterization of each batch of MSCs with respect to known quality attributes can the cells then be reintroduced into the patient [7–9]. This process is invasive and requires a second procedure, leading to donor site morbidity and increased opportunity for harvest complications. Additionally, this process is time-consuming, requiring a significant time interval between cell harvest and reintroduction of the culture-expanded MSCs, so point of care treatment is not an option. Lastly, individual patients vary profoundly from one to another with respect to the concentration and prevalence and ultimate yield

of CTP-derived clones in culture [10]. This variation from patient to patient and tissue source to tissue source results in large variation in performance from one MSC batch to another. As a result, a culture expanded MSC cellular product could be far better controlled when using allogeneic MSCs, saving patients from increased morbidity, procedural risks, treatment variability, and lost time. Allograft MSC preparation also allows the generation and rigorous assessment of multiple lots of MSCs with respect to key quality attributes that are known to predict future clinical efficacy to be used.

In vitro culture expansion of allogeneic MSCs on a large scale would confer a number of benefits that would not be possible with the use of autologous MSCs. Firstly, rather than harvesting and culture-expanding cells on a patient-by-patient basis, as is the case with autologous MSCs, allogeneic MSCs could be produced in large quantities for use in multiple patients, leading to an overall reduced production cost and theoretically lower market price for the consumer. This may provide access to treatment for patients where MSC therapies would otherwise have been too expensive. Furthermore, patients would not be limited by the cell populations they innately possess. MSCs have unique characteristics [11–13], which may make one specific subset of MSCs more fit than another for a particular task. Furthermore, culture expansion of subsets of MSCs would be possible based on specific cell surface markers [11]. Subculturing of these allogeneic cells would allow for manipulation of MSCs and editing of surface markers to become more effective than wild-type MSCs. Thus, there would be potential for selecting and delivering allogeneic MSCs possessing the specific traits that would be most fit for patient's particular needs, rather than relying on the patient's autologous cells. The use of a consistent allogeneic cellular product would be more akin to the administration of a manufactured drug rather than the typical heterogeneity of biologic injections. This would be particularly advantageous when assessing clinical outcomes and consistency across multiple studies.

Recently, there has been interest in bone marrow aspirate (BMA) and bone marrow aspirate with concentration (BMAC) for the treatment of degenerative changes in the knee. Autogenous BMA and BMAC have been used to augment osteochondral allograft transplantation and integration by potentially improving osseous healing of the graft and incorporation into the knee [14–16].

It is important to remember that the cell population that is present in a bone marrow aspiration and in a BMAC preparation after removal of high density RBCs is distinctly different than the cell population used in a culture-expanded MSC preparation. Native populations of cells isolated from bone marrow contain a heterogeneous mixture of cells from marrow and bone tissue. This includes mature and immature hematopoietic cells from marrow and contaminating blood. These hematopoietic cells represent the highest fraction of cells in terms of the composition of BMAC. BMAC will contain varying numbers of colony founding CTPs (range of 100–5000 per ml of aspirate, with a mean of 1000–2000). These clones will be heterogeneous in performance. Progenitor cells will vastly outnumber true stem cells. As a result, of all of the cells in a BMAC preparation, the least abundant cell will be a stem cell [10]. This yield of CTPs in an aspirate sample is highly sensitive to aspiration technique, and particularly to keeping aspiration volume low from each aspiration site, to limit contamination with peripheral blood [17].

The prevalence of colony founding CTPs in BMA has been reported to vary from 0.001% to 0.02% of all nucleated cells harvested from bone marrow [18–20]. BMAC processing using density separation in a centrifuge can provide a 1.6- to five-fold increase in the concentration of nucleated cells overall, with minimal loss of colony founding CTPs [21–24]. However, CTPs are still a vanishingly small portion of the overall composition of a BMAC preparation. Therefore, while BMAC injection can be considered to be a cellular therapy, it is inappropriate to promote or represent BMAC as a "stem cell" or "progenitor cell" therapy.

Conversely, in vitro culturing of MSCs allows for delivery of allogeneic MSCs at much higher concentrations, where virtually 100% of the injected cells can be characterized with respect to specific markers or attributes. Theoretically, a higher concentration of expanded allograft MSCs could lead to improved control over outcomes. However, rigorous safety controls must also be in place to ensure that expanded cells have not acquired undesired attributes through mutation or clonal selection.

Another potential advantage with the use of allogeneic MSCs is that they can function as "immunomodulators," regulating their local environment by reducing overall immune and inflammatory responses [25–27]. Autologous MSCs are reported to survive long enough to impact a local cell environment in recipient tissue and promote or enable differentiation of desired specialized types of cells, such as chondrocytes, osteoblasts, or myocytes. Conversely, recent studies show allogeneic MSCs are not likely to integrate into the recipient's tissues, and thus function primarily as immunomodulators via paracrine signaling to attract potentially induce the host's own tissue to fill the defects before being identified and removed by the recipient's immune defenses [28]. Allogeneic MSCs have been shown to suppress activity of NK cells [29], dendritic cells [30], neutrophils, macrophages, and B cells [31] possibly due to increased prostaglandin E_2 levels. Allogeneic MSCs may also be effective in the management of systemic inflammatory conditions and with suppressing immune responses after tissue transplants, such as allografts [32–34] via induction of regulatory immunosuppressive lymphocytes [35], such as Tregs. Furthermore, in vivo studies of extracellular vesicles released by allogeneic MSCs show suppression of pro-inflammatory processes and reduction of oxidative stress and fibrosis, creating an environment allowing for the newly recruited, endogenous MSCs to repair damaged tissues [36, 37]. The immunoregulatory capabilities of MSCs are not fully understood; however, current evidence suggests that the manipulation of the recipient's immune response may confer benefit in the treatment of a number of diseases.

The study of injectable allogeneic MSCs in the field of orthopedics provides a number of benefits compared to the study of allogeneic MSCs in other medical specialties. Access into joint spaces has already been well established within orthopedics, making delivery of MSCs intraoperatively or via intra-articular injection relatively straightforward. This also makes biopsy at the site of implementation to measure progress quite feasible in most orthopedic sites, such as the knee or hip, compared to other organs like the heart or nervous system. Intra-articular administration of MSCs is preferred compared to intravenous or other non-localized methods [38] due to restricted biodistribution. Intravenously delivered MSCs often collect in the lung parenchyma and are subsequently eliminated through the kidneys [39]. Due to the relatively impermeable capsules surrounding joints, MSCs introduced intra-articularly primarily remain within the joint.

6.3 Theoretical Disadvantages of Allogeneic MSCs

Culture-expanded allogeneic MSCs may have many future applications and benefits throughout medicine, but there are also a number of potential drawbacks. Though allogeneic MSCs have been said to be "immunopriviledged," [26] the differentiated progeny do induce an immune response and are removed. Introduction of any foreign material into human tissue is never without risk. General medical risks, including infection and allergic reaction, remain potential complications with allogeneic MSCs, as with any other injectable. While the immunomodulating effect of allogeneic MSCs has been shown to confer benefit, immune-mediated adverse events still remain a possibility with injection of allogeneic MSCs. In a horse model, intra-articular injections of autologous, allogeneic, and xenogeneic MSCs each led to an immune response, as indicated by synovial cell hyperplasia and perivascular lymphocytic proliferation; however, only xenogeneic

MSCs induced a persistent immune-modulated response with increased CD4+ cells [40]. A number of additional animal studies have shown similar in vivo activations of both humoral and cell-mediated immune responses when MHC haplotypes are mismatched between donors and recipients [41–43]. Co-administration of monoclonal antibodies specific to MPC surface antigens has been shown to reduce immune responses to MSC injections and may provide benefit as an adjunct therapy.

The use of allogeneic MSCs is on the forefront of modern therapy, and there is still a significant amount of unknown information necessitating further investigation. While use of allogeneic MSCs is cheaper than autologous counterparts, at this point, the scientific evidence backing allogeneic MSCs remains inadequate to justify the current cost. Therapies must be grounded on evidence supporting both safety and efficacy, yet in many new MSC therapies, there is no such basis. For example, there is a lack of scientific evidence suggesting that non-hematopoietic MSCs are able to detect and respond in a preferred manner to the surrounding environment into which they have been introduced, yet much of scientific community holds this belief [44]. Assumptions such as this can cause unrealistic expectations and can lead to patient injury [45–47], ranging from headache and superficial infection to pulmonary embolism and cardiac arrest, indicating a need for oversight to prevent unsubstantiated claims that may mislead patients [48]. Thus, until more is known about allogeneic and autologous MSCs and they can be regulated properly, their use in patients should likely remain limited.

6.4 Challenges for the Future

Currently, two clinical trials in both the United States and in Europe, RECLAIM and IMPACT, respectively, are studying articular cartilage repair in a single-stage treatment of cartilage defects using a mixture of allogeneic MSCs derived from bone marrow (IMPACT) or adipose tissue (RECLAIM) and recycled autologous chondrons [49]. Recent phase I results are promising and demonstrate allogeneic MSCs are safe and efficacious as a supplement to autogenous MSC delivery. These clinical trials represent only a small portion of the ongoing investigation of the capabilities and limitations of allogeneic MSCs.

There are a number of challenges facing further development of allogeneic MSC therapies. The present cost of allogeneic MSCs is unaffordable for the average patient, and the lack of insurance approval for the majority of MSC therapies further compounds this issue. Insurance approval is largely contingent upon adequate scientific evidence proving both safety and efficacy. Currently, according to data presented by the US National Institutes of Health (NIH), there are more than 800 active randomized controlled trials (RCTs) worldwide involving allogeneic MSCs [50] (Fig. 6.1—world map of ongoing RCTs) (Fig. 6.2—pie chart of RCTs by subspecialty). The results of these ongoing and future RCTs will be crucial not only to validate new therapies but also to facilitate patient access.

The regulatory environment has historically been a barrier to MSC research and may pose challenges for the advancement of allogeneic MSC research depending on geographic region. In the United States, the Food and Drug Administration has adopted a new position allowing for expedited approval of therapies designated as regenerative medicine advanced therapy (RMAT) [51], opening an avenue which would allow drug approval to match the rate of novel research. Specifically, this may allow for therapies utilizing allogeneic MSCs to be approved and introduced into the pharmaceutical market faster than historical rates. Standardization in quality, dosing, and scientific reporting on the composition, cellular attributes, and biological reproducibility will be critical and also an enormous challenge to employ on an international scale. However, standardization is essential to ensuring the safety, efficacy, consistency,

and future optimization of allogeneic MSC therapies [52].

6.5 Conclusions

The importance of future regulation and standardization of cellular therapies cannot be understated. There is already a significant amount of misinformation and unsubstantiated therapies available to the public, which can ultimately lead to patient harm, either physically or financially. Scientific nomenclature and terms are at the core of scientific communication, as well as communication in the wider world. The term MSC has been widely and often indiscriminately used to describe a plurality of cell populations. It is essential for use to adopt a more precise system of nomenclature that underscores the fundamental differences between complex heterogeneous populations of cells in native tissues the unique attributes acquired by cell populations in the process of the selective environment of in vitro expansion. This will require adaptation to standards of communication that more consistently and accurately describe population attributes and differences between culture-expanded "MSC" populations that depend on the sources of native connective tissue stem/progenitor cells (CTPs), harvest methods, processing methods, expansion conditions and criteria, characterization of population composition and attributes, transplantation methods and environment, clinical setting (diagnosis and severity at the starting point of therapy), and clinical outcome assessment.

All physicians must, "first, do no harm," and it is crucial that regulation of both safety and efficacy of MSC therapies maintains pace with the development of new treatments. In conclusion, there are a number of limitations that must be overcome prior to the widespread use of allogeneic MSCs, but we are entering a promising period where allogeneic MSCs could revolutionize minimally invasive treatments for musculoskeletal injuries.

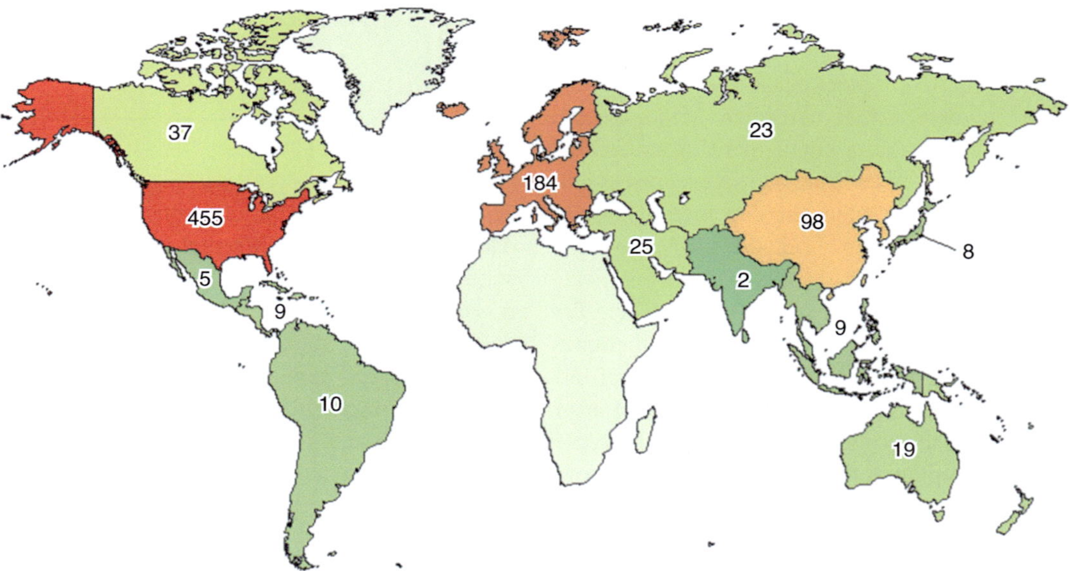

Fig. 6.1 Geographic distribution of ongoing randomized controlled trials involving allogeneic medicinal signaling cells (MSCs) as reported by the National Institutes of Health (NIH, www.clinicaltrials.gov)

Fig. 6.2 Medical subspecialty distribution of ongoing randomized controlled trials involving allogeneic medicinal signaling cells (MSCs) as reported by the National Institutes of Health (NIH, www.clinicaltrials.gov). The subcategory "Other" includes maternal-fetal medicine (7%), infectious disease (6%), endocrinology (6%), cardiology (5%), dermatology (4%), pulmonology (4%), gastroenterology (3%), psychiatry (3%), obstetrics and gynecology (3%), ophthalmology (2%), dentistry (1%), and ear, nose, and throat (<1%)

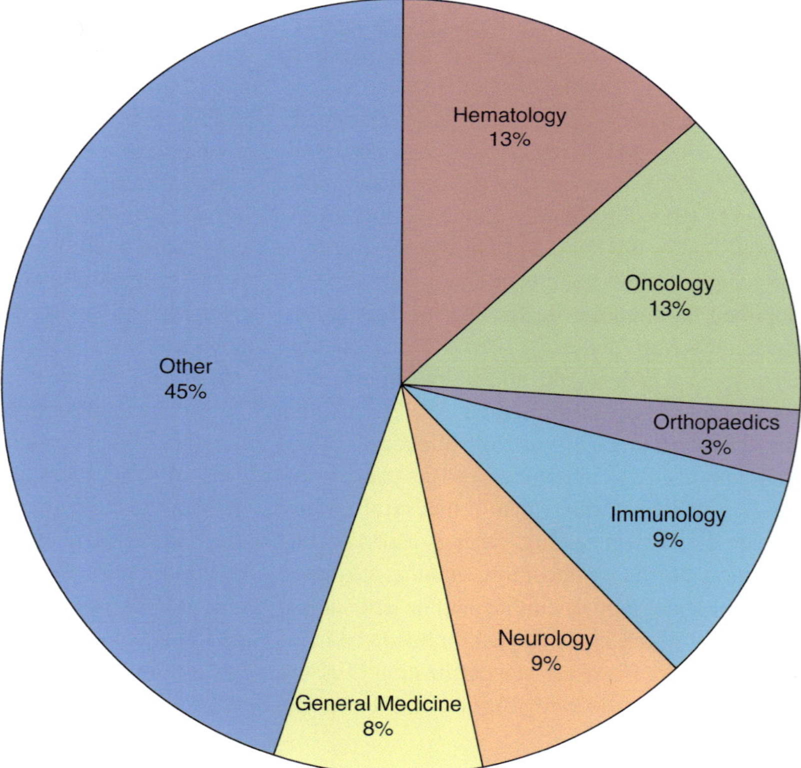

Take-Home Messages

- The use of allogeneic MSCs confers a number of theoretical benefits compared to alternative sourcing of MSCs, including reduced cost and labor and reduced patient and donor site morbidity.
- Mass production of high-quality MSCs, with potential improvement over individual patient's MSCs, is another potential advantage.
- Theoretical disadvantages range from the lack of benefit to the risk of patient injury through the immune response to foreign material.
- More research is needed prior to the widespread use of allogeneic MSCs, but we are entering a promising period where allogeneic MSCs could revolutionize minimally invasive treatments for musculoskeletal injuries.

References

1. Zuk PA, Zhu M, Ashjian P, et al. Human adipose tissue is a source of multipotent stem cells. Mol Biol Cell. 2002;13(12):4279–95.
2. Lo B, Parham L. Ethical issues in stem cell research. Endocr Rev. 2009;30(3):204–13.
3. Viswanathan S, Shi Y, Galipeau J, et al. Mesenchymal stem versus stromal cells: International Society for Cell & gene therapy (ISCT®) mesenchymal stromal cell committee position statement on nomenclature. Cytotherapy. 2019;21(10):1019–24.
4. Caplan AI. Mesenchymal stem cells: time to change the name! Stem Cells Transl Med. 2017;6(6):1445–51.
5. Muschler GF, Midura RJ, Nakamoto C. Practical modeling concepts for connective tissue stem cell and progenitor compartment kinetics. J Biomed Biotechnol. 2003;2003(3):170–93.
6. Muschler GF, Midura RJ. Connective tissue progenitors: practical concepts for clinical applications. Clin Orthop Relat Res. 2002;395:66–80.
7. Armitage JO. Bone marrow transplantation. N Engl J Med. 1994;330(12):827–38.
8. Hequet O. Hematopoietic stem and progenitor cell harvesting: technical advances and clinical utility. J Blood Med. 2015;6:55–67.
9. Korbling M, Freireich EJ. Twenty-five years of peripheral blood stem cell transplantation. Blood. 2011;117(24):6411–6.
10. Patterson TE, Boehm C, Nakamoto C, et al. The efficiency of bone marrow aspiration for the harvest of connective tissue progenitors from the human iliac crest. J Bone Joint Surg Am. 2017;99(19):1673–82.
11. Dominici M, Le Blanc K, Mueller I, et al. Minimal criteria for defining multipotent mesenchymal stromal cells. The International Society for Cellular Therapy position statement. Cytotherapy. 2006;8(4):315–7.
12. Strioga M, Viswanathan S, Darinskas A, Slaby O, Michalek J. Same or not the same? Comparison of adipose tissue-derived versus bone marrow-derived mesenchymal stem and stromal cells. Stem Cells Dev. 2012;21(14):2724–52.
13. Al-Nbaheen M, Vishnubalaji R, Ali D, et al. Human stromal (mesenchymal) stem cells from bone marrow, adipose tissue and skin exhibit differences in molecular phenotype and differentiation potential. Stem Cell Rev Rep. 2013;9(1):32–43.
14. Ackermann J, Mestriner AB, Shah N, Gomoll AH. Effect of autogenous bone marrow aspirate treatment on magnetic resonance imaging integration of osteochondral allografts in the knee: a matched comparative imaging analysis. Arthroscopy. 2019;35(8):2436–44.
15. Wang D, Lin KM, Burge AJ, Balazs GC, Williams RJ. Bone marrow aspirate concentrate does not improve osseous integration of osteochondral allografts for the treatment of chondral defects in the knee at 6 and 12 months: a comparative magnetic resonance imaging analysis. Am J Sports Med. 2019;47(2):339–46.
16. Stoker AM, Baumann CA, Stannard JP, Cook JL. Bone marrow aspirate concentrate versus platelet rich plasma to enhance osseous integration potential for osteochondral allografts. J Knee Surg. 2018;31(4):314–20.
17. Muschler GF, Boehm C, Easley K. Aspiration to obtain osteoblast progenitor cells from human bone marrow: the influence of aspiration volume. J Bone Joint Surg Am. 1997;79(11):1699–709.
18. Hyer CF, Berlet GC, Bussewitz BW, Hankins T, Ziegler HL, Philbin TM. Quantitative assessment of the yield of osteoblastic connective tissue progenitors in bone marrow aspirate from the iliac crest, tibia, and calcaneus. J Bone Joint Surg Am. 2013;95(14):1312–6.
19. Pittenger MF, Mackay AM, Beck SC, et al. Multilineage potential of adult human mesenchymal stem cells. Science. 1999;284(5411):143–7.
20. Veyrat-Masson R, Boiret-Dupré N, Rapatel C, et al. Mesenchymal content of fresh bone marrow: a proposed quality control method for cell therapy. Br J Haematol. 2007;139(2):312–20.
21. Hermann PC, Huber SL, Herrler T, et al. Concentration of bone marrow total nucleated cells by a point-of-care device provides a high yield and preserves their functional activity. Cell Transplant 2008;16(10):1059–69.
22. Jager M, Herten M, Fochtmann U, et al. Bridging the gap: bone marrow aspiration concentrate reduces

autologous bone grafting in osseous defects. J Orthop Res. 2011;29(2):173–80.

23. Betsch M, Schneppendahl J, Thuns S, et al. Bone marrow aspiration concentrate and platelet rich plasma for osteochondral repair in a porcine osteochondral defect model. PLoS One. 2013;8(8):–e71602.

24. Hakimi M, Grassmann JP, Betsch M, et al. The composite of bone marrow concentrate and PRP as an alternative to autologous bone grafting. PLoS One. 2014;9(6):e100143.

25. Liang X, Ding Y, Zhang Y, Tse H-F, Lian Q. Paracrine mechanisms of mesenchymal stem cell-based therapy: current status and perspectives. Cell Transplant. 2014;23(9):1045–59.

26. Berglund AK, Fortier LA, Antczak DF, Schnabel LV. Immunoprivileged no more: measuring the immunogenicity of allogeneic adult mesenchymal stem cells. Stem Cell Res Ther. 2017;8(1):288.

27. Liechty KW, MacKenzie TC, Shaaban AF, et al. Human mesenchymal stem cells engraft and demonstrate site-specific differentiation after in utero transplantation in sheep. Nat Med. 2000;6(11):1282–6.

28. Melick G, Hayman N, Landsman AS. Mesenchymal stem cell applications for joints in the foot and ankle. Clin Podiatr Med Surg. 2018;35(3):323–30.

29. Spaggiari GM, Capobianco A, Abdelrazik H, Becchetti F, Mingari MC, Moretta L. Mesenchymal stem cells inhibit natural killer–cell proliferation, cytotoxicity, and cytokine production: role of indoleamine 2,3-dioxygenase and prostaglandin E2. Blood. 2008;111(3):1327–33.

30. Jiang X-X, Zhang Y, Liu B, et al. Human mesenchymal stem cells inhibit differentiation and function of monocyte-derived dendritic cells. Blood. 2005;105(10):4120–6.

31. Asari S, Itakura S, Ferreri K, et al. Mesenchymal stem cells suppress B-cell terminal differentiation. Exp Hematol. 2009;37(5):604–15.

32. Smith B, Sigal IR, Grande DA. Immunology and cartilage regeneration. Immunol Res. 2015;63(1):181–6.

33. Maumus M, Guérit D, Toupet K, Jorgensen C, Noël D. Mesenchymal stem cell-based therapies in regenerative medicine: applications in rheumatology. Stem Cell Res Ther. 2011;2(2):14.

34. Krampera M, Glennie S, Dyson J, et al. Bone marrow mesenchymal stem cells inhibit the response of naive and memory antigen-specific T cells to their cognate peptide. Blood. 2003;101(9):3722–9.

35. Nasef A, Mathieu N, Chapel A, et al. Immunosuppressive effects of mesenchymal stem cells: involvement of HLA-G. Transplantation. 2007;84(2):231–7.

36. Borger V, Bremer M, Ferrer-Tur R, et al. Mesenchymal Stem/Stromal Cell-Derived Extracellular Vesicles and Their Potential as Novel Immunomodulatory Therapeutic Agents. Int J Mol Sci. 2017;18:7.

37. Yu B, Zhang X, Li X. Exosomes derived from mesenchymal stem cells. Int J Mol Sci. 2014;15(3):4142–57.

38. Lamo-Espinosa JM, Mora G, Blanco JF, et al. Intra-articular injection of two different doses of autologous bone marrow mesenchymal stem cells versus hyaluronic acid in the treatment of knee osteoarthritis: multicenter randomized controlled clinical trial (phase I/II). J Transl Med. 2016;14(1):246.

39. Detante O, Moisan A, Dimastromatteo J, et al. Intravenous administration of 99mTc-HMPAO-Labeled human mesenchymal stem cells after stroke: in vivo imaging and biodistribution. Cell Transplant. 2009;18(12):1369–79.

40. Pigott JH, Ishihara A, Wellman ML, Russell DS, Bertone AL. Investigation of the immune response to autologous, allogeneic, and xenogeneic mesenchymal stem cells after intra-articular injection in horses. Vet Immunol Immunopathol. 2013;156(1–2):99–106.

41. Eliopoulos N, Stagg J, Lejeune L, Pommey S, Galipeau J. Allogeneic marrow stromal cells are immune rejected by MHC class I- and class II-mismatched recipient mice. Blood. 2005;106(13):4057–65.

42. Nauta AJ, Westerhuis G, Kruisselbrink AB, Lurvink EG, Willemze R, Fibbe WE. Donor-derived mesenchymal stem cells are immunogenic in an allogeneic host and stimulate donor graft rejection in a nonmyeloablative setting. Blood. 2006;108(6):2114–20.

43. Zangi L, Margalit R, Reich-Zeliger S, et al. Direct imaging of immune rejection and memory induction by allogeneic mesenchymal stromal cells. Stem Cells. 2009;27(11):2865–74.

44. Marks PW, Witten CM, Califf RM. Clarifying stem-cell therapy's benefits and risks. N Engl J Med. 2016;376(11):1007–9.

45. Kuriyan AE, Albini TA, Townsend JH, et al. Vision loss after intravitreal injection of autologous "stem cells" for AMD. N Engl J Med. 2017;376(11):1047–53.

46. Berkowitz AL, Miller MB, Mir SA, et al. Glioproliferative lesion of the spinal cord as a complication of "stem-cell tourism". N Engl J Med. 2016;375(2):196–8.

47. Bauer G, Elsallab M, Abou-El-Enein M. Concise review: a comprehensive analysis of reported adverse events in patients receiving unproven stem cell-based interventions. Stem Cells Transl Med. 2018;7(9):676–85.

48. Murray IR, Chahla J, Safran MR, et al. International expert consensus on a cell therapy communication tool: DOSES. J Bone Joint Surg Am. 2019;101(10):904–11.

49. de Windt TS, Vonk LA, Slaper-Cortenbach ICM, Nizak R, van Rijen MHP, Saris DBF. Allogeneic MSCs and recycled autologous chondrons mixed in a one-stage cartilage cell transplantation: a first-in-man trial in 35 patients. Stem Cells. 2017;35(8):1984–93.

50. https://www.clinicaltrials.gov/. Published 2020. Accessed 2/21/2020, 2020.

51. Marks P, Gottlieb S. Balancing safety and innovation for cell-based regenerative medicine. N Engl J Med. 2018;378(10):954–9.

52. Murray IR, Murray AD, Geeslin AG, et al. Infographic: we need minimum reporting standards for biologics. Br J Sports Med. 2019;53(15):974–5.

Injection of Steroid Hormones

Tristan W. Juhan, Andrew J. Homere,
Alexander E. Weber, George F. Hatch,
and Frank A. Petrigliano

7.1 Introduction

Corticosteroids, cholesterol derivatives with strong anti-inflammatory properties, have been routinely applied for many orthopaedic conditions. Androgenic-anabolic steroids (AAS) are synthetic testosterone derivatives that have been also used therapeutically for a wide range of conditions. While they may be known to the general public as performance enhancers used to add muscle mass in athletes and bodybuilders, their effects are applicable to the treatment of a wide range of medical conditions. Within medicine, AAS have traditionally been used for hormone imbalance disorders. It has also been shown that their anabolic effects can be used to prevent muscle wasting in a number of chronic conditions such as COPD, HIV, and muscular dystrophy [1]. AAS supplementation has also shown promise for lean body mass loss in patients recovering from severe burns [2].

While AAS appear to show promise in the treatment of certain conditions, there are harmful side effects associated with their use. The negative side effects are more prevalent when used at supraphysiologic doses and when these agents are administered in stacked regimens. Reproductive infertility, cardiomyopathy, atrial fibrillation, and hepatic dysfunction are common side effects that have been well documented in the literature [3]. One of the less-studied side effects of AAS is tendon pathology/rupture, and the implications in repair as well as recovery from injury. The majority of published studies pertaining to the aforementioned issues have been conducted in animal models [4–6]. The anabolic potential of AAS naturally lends itself to a range of orthopaedic investigations. However, human studies investigating the use of AAS as a biologic adjuvant in recovery from musculoskeletal injury are in their early stages. The goal of this chapter is to highlight previous, current, and future work discussing the therapeutic use of AAS for the treatment of orthopaedic injuries.

7.2 Corticosteroids in Orthopaedics

Corticosteroids are derivatives of cholesterol and work by inhibiting both the cyclooxygenase and lipoxygenase pathways of inflammation. In detail, corticosteroids inhibit phospholipase A2, preventing the breakdown of phospholipid to arachidonic acid. Arachidonic acid is the precursor for prostaglandins, thromboxanes, and leukotrienes, which are mediators of inflammation. This

T. W. Juhan · A. J. Homere · A. E. Weber · G. F. Hatch
F. A. Petrigliano (✉)
Department of Orthopaedic Surgery, Keck School of Medicine of USC, University of Southern California (USC), Los Angeles, CA, USA
e-mail: Tristan.juhan@med.usc.edu; Homere@usc.edu; Alexander.weber@med.usc.edu; ghatch@med.usc.edu; frank.petrigliano@med.usc.edu

G. Filardo et al. (eds.), *Orthobiologics*, https://doi.org/10.1007/978-3-030-84744-9_7

allows for inhibition of both the cyclooxygenase and lipoxygenase pathways of inflammation, in contrast to nonsteroidal anti-inflammatory drugs (NSAIDs) which only inhibit the cyclooxygenase pathway. Corticosteroids also work by reducing vascular permeability as well as inhibiting cell migration to areas of injury, ultimately inhibiting oedema, erythema, and pain [7]. Their proven benefit in preventing swelling and pain is particularly attractive for treating orthopaedic injuries, but potential harmful side effects have been a cause for concern.

While corticosteroids are prescribed as injections, pills, and ointments, the most commonly used type in orthopaedics is injectable corticosteroids. Commonly used injectable corticosteroids include dexamethasone (Decadron), methylprednisolone/prednisolone (Depo-Medrol), hydrocortisone, betamethasone (Celestone), and triamcinolone (Kenalog). Selection of which corticosteroid to use is typically based on provider preference, and the amount often depends on the size of the joint injected [7].

Corticosteroids have been used in patients with osteoarthritis to alleviate pain and increase range of motion. A 2006 Cochrane review confirmed the short-term efficacy of corticosteroids in knee osteoarthritis [8]. Additionally, a 2018 Cochrane review of 26 trials on intraarticular corticosteroids for knee osteoarthritis found cortisone injections to be more beneficial than placebo with respect to pain reduction and functional improvement [9]. Improvements in pain were relatively short lived (<6 months), and effects decreased over time. With that said, there is no current evidence of long-term benefits of corticosteroids for osteoarthritis.

Corticosteroid use in the setting of tendonitis is inconclusive, as conflicting literature exists. This has been studied in shoulder/rotator cuff tendinitis, epicondylitis, and Achilles tendinopathy. One systematic review on the efficacy and risks of steroid injections for tendinopathy showed evidence that corticosteroid injections are beneficial in the short term for the treatment of tendinopathy but are worse than other treatments in the intermediate and long terms [10]. With that said, more than three injections around a single tendon are not recommended due to potential collagen degeneration and tendon rupture [7]. This is similar in patients with ligament sprains, in which steroid injections are sometimes used. There is minimal literature supporting their efficacy, and the risk of ligament/tendon rupture is still present. After arthroscopic surgery, corticosteroids have been used for pain relief. Triamcinolone acetonide has been proven to reduce pain levels postoperatively in patients who had arthroscopic knee surgery [7].

Intraarticular steroid injections are a common intervention for frozen shoulder, and a 2017 meta-analysis of 8 randomized controlled trials including 416 patients found that those who received an intraarticular steroid injection had significantly reduced Visual Analog Scale (VAS) pain scores at 4–6, 12–16, and 24–26 weeks post injection when compared to controls [11].

Bursae are commonly injected sites in sports medicine, and corticosteroids have been shown to improve bursitis. Prepatellar bursitis, semimembranosus insertion syndrome, pes anserine bursitis, olecranon bursitis, and shoulder bursitis have all been effectively treated with corticosteroids with minimal side effects [7].

7.3 Side Effects

One primary concern in patients injected with corticosteroids is tendon degeneration and rupture. This is of particular concern in high demand patients who require repeat injections. It has been suggested that injection into the tendon may weaken its structure and increase the risk of rupture. With that said, a systematic review by Coombes et al. suggests this may be an acceptable risk, as they noted a low frequency of tendon rupture [10].

Chondrotoxicity secondary to corticosteroid administration has been noted in the literature as well. A systematic review by Wernecke et al. shows beneficial effects of intraarticular corticosteroid occurring at low doses and durations, while the deleterious chondrotoxic effects occurred at high doses and durations of use [12]. Dragoo et al. note that even a single injection of betamethasone sodium phosphate and betamethasone acetate solution showed significant chondrotoxicity using a physiologically relevant in vitro model, and they noted such injections should be used with caution [13].

Infection is also potential consequence of corticosteroid injections, and this has been seen in patients receiving pre- or postoperative injections. Bhattacharjee et al. reviewed a large database and found that corticosteroid injections within the 2 weeks prior to shoulder arthroscopy may increase the risk of postoperative infections [14]. On the other hand, Kew et al. have demonstrated a significant association between intraarticular corticosteroid injections administered 1 month postoperatively and an increased rate of postoperative infection [15].

A Weber et al. study in 2019 suggested a correlation between pre-operative shoulder injections and an increase in revision rotator cuff repair rates. They also noted a frequency and time dependence to these findings, where more frequent injections and administration of the injection closer to the time of surgery were both independently associated with higher rates of subsequent revision rotator cuff repairs [16].

7.4   AAS Physiology

Androgenic-anabolic steroids are most well-known for use as performance-enhancing drugs in athletes, and their use has been documented as early as the 1950s [17]. While AAS increase muscle mass and strength under certain conditions in healthy adults, there is limited evidence to support that they enhance athletic performance [18–21]. Additionally, they were added to the list of Schedule III Controlled Substances in 1990. Even so, it is estimated that among Americans currently aged 13 to 50 years, 2.9–4.0 million have used AAS. Within this group, roughly one million may have experienced AAS dependence [22].

While the short-term side effects of AAS are generally mild and reversible, long-term, high-dose use is associated with more severe, irreversible cardiovascular disease [23]. Additionally, studies in rodents have shown that exposure to AAS at doses that mimic levels observed with human abuse elicits significant changes in aggression, anxiety, and sexual behaviours [24].

AAS have both anabolic and androgenic effects, and these compounds appear to use three common pathways to exert these effects. Primarily, AAS activate androgen receptors (ARs) to induce a steroid-receptor complex in the cell nucleus inducing a high degree of transcription [25]. A secondary pathway targets 5-alpha hydroxylase, converting AAS into dihydrotestosterone (DHT), a more active form of testosterone. This then stimulates increased protein synthesis by interaction with RNA and DNA. It is thought that this pathway may play a larger role in promoting the androgenic effects of AAS, due to higher 5-alpha reductase activity in male accessory sex glands and lower 5-alpha reductase activity in areas such as skeletal muscle [3]. Lastly, the aromatase pathway is responsible for the conversion of AAS into female sex hormones, such as estradiol and estrone. Female sex hormones bind to oestrogen receptors and form oestrogen-receptor complexes that exert effects on fat tissue, Leydig and Sertoli tissue, and some nuclei in the central nervous system (CNS) [3].

AAS are synthetic derivatives of testosterone that utilize structural modifications to change the relative anabolic-androgenic potency, slow the rate of degradation, change the pattern of metabolism, or decrease the aromatization to estradiol [26]. Most orally administered preparations are 17α-alkylated derivatives of testosterone that are relatively resistant to hepatic first-pass metabolism degradation. Parenteral preparations use esterification of the 17-βhydroxyl group to make the molecule more soluble in lipid vehicles used for injection. Once in the body, blood esterases hydrolyse the esters to yield the active compound. The rate of absorption is dependent on chain length of the acid moiety; in general, a longer chain length means slower absorption, thus a prolonged duration of action [27, 28]. Despite these modifications, all AAS formulations have both anabolic and androgenic activity; none is completely selective. For example, testosterone has an anabolic-androgenic ratio of 1, whereas the ratio for nandrolone is 10 [29].

AAS have been studied in vitro as well as in vivo, and the potency of endogenous androgens/synthetic anabolic steroids was thought to correlate with their affinity in vitro to the androgen receptor, although Feldkoren et al. found that AAS with low affinity to the androgen receptor in vitro (stanozolol, methandienone) are able to act on receptors in vivo to cause biological responses via classical transcriptional mechanisms [30].

Most studies to date have utilized the intramuscular steroid nandrolone decanoate [4–6, 31–44]. Orally administered oxandrolone has also been found to aid in tissue healing and recovery while exhibiting limited side effects. In a prospective randomized controlled trial, Wolf et al. demonstrated that oxandrolone improved protein net balance and lean mass in the severely burned that was associated with increased gene expression for functional muscle proteins [20]. The efficacy, as well as a positive safety profile, was highlighted in a meta-analysis of severely burned patients. This included patients who either received oxandrolone or did not receive this medication during hospital stay (catabolic phase) and recovery phases. Oxandrolone use resulted in decreased hospital stay, reduced weight loss, and greater gain of lean body mass compared to controls. Importantly, oxandrolone administration in this last study did not increase the risk of adverse events [2].

7.5 The Effects of AAS on Musculoskeletal Tissues

The anabolic effects of AAS are mediated primarily by ARs in skeletal muscle. The downstream effect is an increase in the transcription of target genes that may control the accumulation of DNA required for muscle growth [45]. Previously it was thought that ARs become saturated at physiologic levels of testosterone and that providing supraphysiologic doses of testosterone conferred no additional benefit. Recent studies, however, have shown that ARs can be upregulated by exposure to AAS [43, 46] and that the number of ARs is increased by strength training

[47]. This shows the possible mechanism by which we see the complementary effects of AAS administration and exercise.

The increase in muscle size seen with AAS administration is due to an increase in cross-sectional areas of both type I and type II muscle fibres and an increase in myonuclear number [48]. The strength increase seen is from both the muscle fibre hypertrophy and also the change in muscle architecture. Testosterone-treated muscles show an increase in pennation—a finding that is often associated with high force low-velocity contractions [49]. AAS are also associated with increased exercise tolerance by several mechanisms including increasing the rate of protein synthesis during recovery [42].

7.6 AAS Applications in the Treatment of Human Disease

Anabolic agents have been used to improve the net protein balance in patients in catabolic and cachectic states such as burns, wound healing, COPD, HIV, and muscular dystrophy [1].

A meta-analysis of Li et al. to evaluate the efficacy and safety of using oxandrolone in patients with severe burns found that oxandrolone therapy decreased length of rehabilitation stay during the catabolic and rehabilitative phase while leading to additional gains of lean body mass after 6 months when compared to control. Additionally, they found that oxandrolone therapy did not affect mortality, infection, or hepatic function when compared to control [2].

Cachectic patients appear to have a similar benefit, with nutrition-resistant, wasting HIV patients seeing increases in weight and lean body mass after 16 weeks of nandrolone decanoate administration [36]. It has also been shown that testosterone + megestrol acetate reverses the trajectory of involuntary weight loss and increased lean mass in cachectic COPD patients [50].

AAS use has also been studied in spinal cord injuries (SCI), where studies in rodents have suggested AAS can reduce muscle atrophy and reduce bone loss [51, 52]. Loss of lean muscle and bone density seen in SCI is similar to muscle loss seen in other muscle-wasting conditions due to prolonged immobilization [53], and the uses of AAS in disciplines outside of orthopaedics have been shown to benefit patients in muscle-wasting, catabolic states.

7.7 AAS in Orthopaedics
(Table 7.1)

7.7.1 Rotator Cuff Repair

Within the field of orthopaedics, AAS use has been most studied as a treatment adjunct in rotator cuff injuries. Rotator cuff disease is one of the most common musculoskeletal disorders and is estimated to account for 200,000–300,000 orthopaedic procedures per year [54]. After a tear, it has been shown that rotator cuff musculotendinous units typically undergo three phases: retraction, atrophy, and fatty degeneration. It has also been shown that lower tendon retraction lengths and lower Goutallier scores corresponded to decreased repair failure rate [55]. Additionally, it has been demonstrated that muscle atrophy and fatty degeneration of the rotator cuff muscles play a significant role in the functional outcome following rotator cuff repair [54].

Studies looking at the impact of AAS on retraction and fatty degeneration have been conducted in animal models. Gerber et al. used a rabbit model, comparing a group with no intervention (Group 1), a group with local and systemic AAS administration (Group 2), and a group with only systemic AAS administration (Group 3). Mean supraspinatus retraction was highest in Group 1 compared to Groups 2 and 3. Additionally, no fatty degeneration was measured in either Group 2 or 3, while it was measured in Group 1. This study suggests that AAS administration post-rotator cuff injury may partially prevent tendon retraction and fatty degeneration of the rotator cuff musculature [4]. Gerber et al. also evaluated the effects of AAS and insulin-like growth factor (IGF) in re-lengthening of the rotator cuff tendon in sheep starting 16 weeks post tenotomy. They concluded that neither anabolic steroids nor IGF

contribute to the regeneration of the muscle once degenerative changes are established [5]. Looking at these two studies in tandem points to a possible limit of AAS application. AAS may prevent fatty degeneration from progressing but cannot reverse the established degeneration.

The sheep model was again used by Gerber et al. to specifically evaluate AAS on fatty infiltration post-rotator cuff injury. They were able to demonstrate that AAS application at the time of tendon release in sheep significantly reduced fatty degeneration after 16 and 22 weeks when compared to control. AAS also prevented further atrophy in both the AAS after release group and the AAS at time of repair group. This suggests that further muscle fatty degeneration can be prevented with the application of AAS immediately after tendon repair. Additionally, it reinforced his rabbit model findings that fatty muscle degeneration can largely be prevented if AAS are applied immediately at the time of tendon injury [6].

This is consistent with another study by Fluck et al. Their study compared a control group (no intervention) to a group with nandrolone administration immediately after tendon release and a group with nandrolone administration after tendon repair in sheep animal models. Compared to control, nandrolone administration starting immediately after tendon release prevented the increase in area percentage of fat (23% vs. 277% and −1% vs. 398%, respectively) and mitigated the reduction in area percentage of muscle fibres after tendon release (12% vs. 30% and 16% vs. 35%, respectively). This did not affect the changes in muscle volume and muscle composition [33]. Figure 7.1 illustrates nandrolones mitigating effects of muscle to fat transformation via numerous downstream cellular effects in sheep infraspinatus muscles. Overall, these studies have generally shown benefit to using AAS post tendon injury to prevent fatty degeneration of the tendon.

However, contradicting studies have been published, as Papaspiliopoulos et al. examined rabbit's post-rotator cuff repair utilizing four groups: nonsteroid immobilization (Group 1), nonsteroid non-immobilization (Group 2), steroid immobilization (Group 3), and steroid non-immobilization (Group 4). They found that the mean stress at failure was

Table 7.1 Studies on AAS in orthopaedics

Study	Animal	Target tissue	Conclusion
Sloan et al. (1992)	Human	Hip/body	AAS can be safely given to frail elderly subjects with hip fractures
Amory et al. (2002)	Human	Knee	Preoperative supraphysiological testosterone in older men may lead to improvements in some measures of postoperative recovery
Hedström et al. (2002)	Human	Hip/bone	Low doses of AAS + alphacalcidol and calcium have a positive effect on body composition, bone mineral density, and clinical function in elderly women after hip fracture
Tidermark et al. (2004)	Human	Hip	Protein-rich liquid supplementation in combination with nandrolone given to elderly women after femoral neck fracture may improve lean body mass, activities of daily living, and health-related quality of life
Tengstrand et al. (2007)	Human	Bone	Protein-rich supplementation given to elderly female hip fracture patients increased total bone mineral density
Hohmann et al. (2010)	Human	Quadriceps	The use of anabolic steroids results in an improved knee outcome post-TKA and significantly increases extensor strength
Papaspiliopoulos et al. (2010)	Rabbit	Rotator cuff	Better healing and more tendon strength in those that did not receive AAS post-rotator cuff repair
Gerber et al. (2011)	Rabbit	Rotator cuff	AAS administration post-rotator cuff injury may partially prevent rotator cuff retraction and fatty degeneration
Gerber et al. (2012)	Sheep	Rotator cuff	Muscle cells lose reactiveness to anabolic steroids and IGF once retraction has led to fatty infiltration and atrophy of muscle
Seynnes et al. (2013)	Human	Patellar tendon	Tendon adaptations to resistance training and AAS may be different, suggesting differences in collagen remodelling
Mukhopadhyay et al. (2014)	Human	Bone	Use of low-dose AAS results in a significant gain in bone mass and improvement in quality of life in geriatric population
Gerber et al. (2015)	Sheep	Rotator cuff	Further muscle fatty degeneration may be prevented with AAS application immediately following tendon repair
Romani et al. (2016)	Rat	ACL	Rats with normal testosterone levels had higher ACL load to failure and ultimate stress when compared to testosterone-deficient castrated rats
Fluck et al. (2017)	Sheep	Rotator cuff	Prevention of lipid synthesis starting immediately after tendon release is associated with upregulated androgen receptor protein and RNA expression
Wu et al. (2017)	Human	ACL	Perioperative testosterone supplementation increased lean muscle mass 6 weeks after ACL reconstruction

0.1994 N/mm² in Group 1, significantly higher than Group 3 0.1150 N/mm². In the non-immobilization groups, the mean stress at failure was 0.1973 N/mm² in Group 2, again significantly higher than Group 4, 0.0977 N/mm². These results demonstrated better healing and more tendon strength in groups that did not receive AAS, which is therefore detrimental to rotator cuff repair as it relates to tendon healing [35].

7.7.2 Anterior Cruciate Ligament Reconstruction

Although less numerous, several studies have evaluated the use of AAS/testosterone administration in orthopaedic applications related to anterior cruciate ligament reconstruction (ACL-R). It is thought that circulating estradiol may be associated with ligament strength and injury rate, but the role of AAS in ACLs is less understood [56].

Biomechanical properties have been compared between testosterone-deficient castrated rats and normal control rats, which demonstrated that rats with normal testosterone levels had a higher ACL load to failure and ultimate stress. This suggests that physiologic levels of androgens may be important for ligament strength [56].

Wu et al. conducted a study looking at testosterone supplementation after ACL reconstruction. Patients scheduled for ACL reconstruction were randomized into a testosterone or placebo group

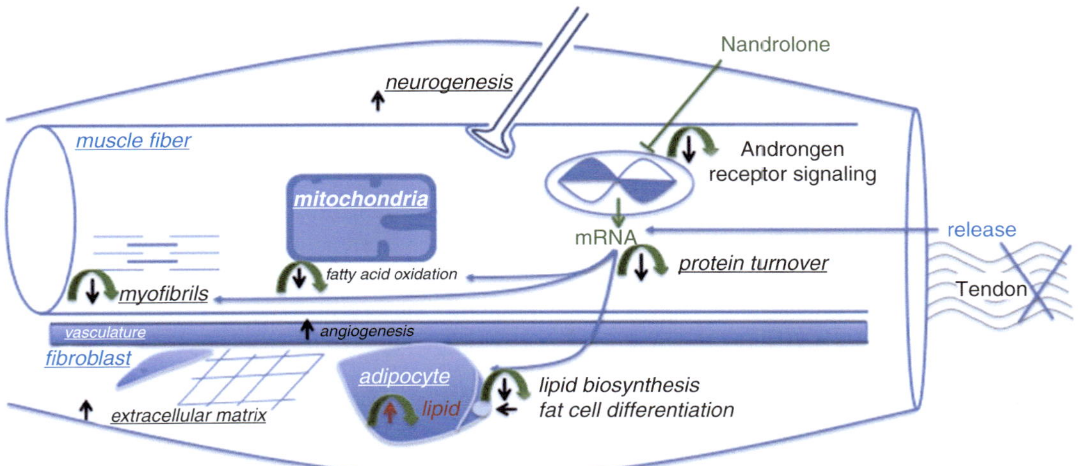

Summary of changes to sheep infraspinatus muscle subsequent to 16 weeks of tendon release. Direction of effects on the transcriptome (black), lipidome (red) and structure (blue/white) of muscle are indicated by arrows. Influence of nandrolone are indicated in green color with curved arrows denoting the inversion of effects.

Fig. 7.1 Influence of nandrolone on sheep infraspinatus muscle subsequent to 16 weeks of tendon release [33] (Reproduced from Flück M, Ruoss S, Möhl CB, et al. (2017) Genomic and lipidomic actions of nandrolone on detached rotator cuff muscle in sheep. J Steroid Biochem Mol Biol 165:382–395 with permission from Elsevier)

for 8 weeks, starting 2 weeks before surgery. They found that perioperative testosterone supplementation increased lean mass 6 weeks after ACL reconstruction (increase of 2.7 ± 1.7 kg compared to a decrease of 0.1 ± 1.5 kg in the control group) while also finding no negative effects of testosterone use. This study suggests that this treatment may help minimize the effects of muscle atrophy associated with ACL injuries and repair [57].

7.7.3   Patellar Tendon

The effects of AAS on in vivo human tendons have also been studied. Seynnes et al. looked at the patella tendons of subjects with resistance training and AAS abuse (RTS), resistance-trained nonsteroid users (RT), and untrained nonsteroid users (CTRL). They found that patellar tendons were stiffer in the RTS group (26%, $p < 0.05$). Additionally, maximal stress was higher in RTS than in RT (15%, $p < 0.05$). They concluded that these findings indicate different tendon adaptations between resistance training and AAS abuse, suggesting differences in collagen remodelling [58].

7.7.4   Total Knee Arthroplasty

The use of AAS in orthopaedic surgery has also been explored in joint reconstruction. Patients with administered nandrolone post-TKA demonstrated greater quadriceps muscle strength at 3, 6, and 12 months postoperatively when compared to placebo, as well as better Knee Society Scores at 6 weeks, 6 months, and 12 months postoperatively [41]. When used for 4 weeks preoperatively, testosterone administration has been shown to improve a patient's ability to stand, climb stairs, and walk while in the hospital postoperatively [59]. Improving mobility and independence are goals following joint reconstruction, and AAS may be important in promoting early strength and ambulation postoperatively.

7.7.5   Hip Fractures

AAS have also been investigated in the postoperative recovery of patients who sustain a hip fracture. Multiple randomized control trials have demonstrated higher Harris hip scores, functional scores, and positive effects on lean body mass in patients treated with AAS postoperatively when compared to nonsteroid groups [37, 38, 60]. With that said, other studies have found no difference in time to mobilization and bone loss in patients treated with AAS [39, 40]. Evidence on the efficacy of AAS in hip fractures is currently inconclusive and should be further explored in future research.

7.8 Potential Side Effects

The use of AAS has been associated with dysplasia of collagen fibrils. This process changes the tendon's crimp morphology, altering the rupturing strain of tendons and the normal biomechanics of the extremities, which significantly decreases tensile strength [61]. Additionally, other studies have shown high-dose AAS have an adverse effect on the biosynthesis of collagen in tendons [62]. Tendon stiffness is also a concern, and one study demonstrated that a stacked anabolic regimen in combination with physical training increased Achilles tendon stiffness in rats, which caused the tendons to fail with less elongation. This did not result in significant differences in the ultimate force at failure, but the energy at the time of tendon failure, toe-limit elongation, and elongation at the time of first failure were all significantly affected [63]. Kanayama et al. demonstrated this risk of tendon rupture clinically. They reported that in AAS abusers, compared to matched non-AAS using bodybuilders, there was a significantly increased risk of tendon ruptures, particularly in the upper body [64]. With that said, differences in tendon stiffness, strength, and energy absorption among AAS-treated tendons at 6 weeks showed reversibility with discontinuation of AAS [65].

7.9 Future Directions

Future studies will likely look to address the biologic sequelae of tendon injuries and subsequent positive potential of AAS in humans further. A currently ongoing clinical trial by Hatch et al. (NCT03091075) aims to examine the potential role of oral oxandrolone to facilitate the healing of repaired rotator cuff tendons and to improve the functional outcomes in patients with chronic, degenerative rotator cuff tears who undergo arthroscopic rotator cuff repair.

While most of the current orthopaedic AAS tendon research involves rotator cuff injury, we anticipate future studies will broaden significantly to evaluate the possible beneficial implications of the use of AAS as a biologic adjuvant therapy for a wide range of musculoskeletal injuries.

Of peak importance in light of some of the discussed studies is determining the optimal timing and duration of treatment, as several studies showed AAS can prevent but not reverse fatty degeneration of muscle [4, 6, 33]. In addition, the benefits of AAS on muscle mass and muscle health have to be weighed against the negative effects of tendon healing, which still has yet to be fully defined in the literature. Also unanswered is what effect AAS and other similar derivatives have on the tendon-bone interface as well as the musculotendinous junction.

Additionally, future work will need to evaluate the dose and drug-dependent responses of various testosterone derivatives as well as the effects of stacking AAS regimens. This will allow surgeons to determine the maximally beneficial treatment for these new potential adjuvant biologics versus the known and still unknown possible negative effects.

7.10 Conclusion

Corticosteroids, utilizing their anti-inflammatory effects, have widespread use for many orthopaedic conditions. AAS use also dates back many decades, but research regarding its therapeutic use as a biologic is not well understood. It has also been demonstrated that AAS affects tendon morphology, leading to decreased tendon strength and increased tendon stiffness. With that said, there seems to be a beneficial role for AAS in the healing of tendon injuries.

In normal rotator cuff tears, fatty degeneration, retraction, and muscle atrophy have been shown to influence disease progression and functional outcomes following repair. There appears to be a temporal relationship, with studies to date mostly demonstrating the prevention of fatty degeneration and atrophy with AAS administration immediately following tendon injury in animal models. These studies have shown that AAS do not appear to be able to reverse fatty degeneration in chronic tendon injuries.

There is also existing evidence that AAS can be effective in minimizing muscle atrophy in ACL injury and repair, as well as other orthopaedic conditions, although further research is needed to determine what this anabolic muscle effect has on recovery time and long-term outcomes.

As we move forward with AAS studies in humans, there needs to be continuous evaluation of the balance between their positive effects on muscle mass and fatty degeneration and the potential negative effects on tendons, remodelling, and other organs. This chapter highlights both the exciting potential of AAS to augment the treatment of musculoskeletal injuries and the need for additional studies to support the use of AAS as a biologic. These studies must be appropriately applied to best protect our patients and help them achieve optimal musculoskeletal recovery.

Take-Home Messages

- Corticosteroids and androgenic-anabolic steroids (AAS) are injectable steroids that use different mechanisms of action to exert therapeutic effects in a number of orthopaedic conditions.
- Corticosteroid injections are commonly used to decrease inflammation and pain by blocking both pathways of inflammation, but one primary concern is tendon degeneration and rupture.
- AAS are known to the general public as performance enhancers but have shown a beneficial role in the healing of tendon

injuries by preventing fatty degeneration and atrophy in animal models.

- AAS injections are promising in the treatment of rotator cuff injuries, ACL reconstruction, as well as other orthopaedic injuries, but potential side effects include tendon stiffness, tendon rupture, and cardiovascular effects.
- Future work should continue to evaluate the balance between the positive effects of AAS on muscle mass and fatty degeneration and the potential negative effects on tendons, remodelling, and other organs.

References

1. Jones IA, Togashi R, Hatch GFR 3rd, Weber AE, Vangsness CT Jr. Anabolic steroids and tendons: a review of their mechanical, structural, and biologic effects. J Orthop Res. 2018;36:2830–41.
2. Li H, Guo Y, Yang Z, Roy M, Guo Q. The efficacy and safety of oxandrolone treatment for patients with severe burns: a systematic review and meta-analysis. Burns. 2016;42:717–27.
3. Hartgens F, Kuipers H. Effects of androgenic-anabolic steroids in athletes. Sports Med. 2004;34:513–54.
4. Gerber C, Meyer DC, Nuss KM, Farshad M. Anabolic steroids reduce muscle damage caused by rotator cuff tendon release in an experimental study in rabbits. J Bone Joint Surg Am. 2011;93:2189–95.
5. Gerber C, Meyer DC, Von Rechenberg B, Hoppeler H, Frigg R, Farshad M. Rotator cuff muscles lose responsiveness to anabolic steroids after tendon tear and musculotendinous retraction: an experimental study in sheep. Am J Sports Med. 2012;40:2454–61.
6. Gerber C, Meyer DC, Flück M, Benn MC, von Rechenberg B, Wieser K. Anabolic steroids reduce muscle degeneration associated with rotator cuff tendon release in sheep. Am J Sports Med. 2015;43:2393–400.
7. Noerdlinger MA, Fadale PD. The role of injectable corticosteroids in orthopedics. Orthopedics. 2001;24(4):400–5. quiz 406–7
8. Bellamy N, Campbell J, Robinson V, Gee T, Bourne R, Wells G. Intraarticular corticosteroid for treatment of osteoarthritis of the knee. Cochrane Database Syst Rev. 2006;2:CD005328.
9. Khan M, Bhandari M. Cochrane in CORR®: intra-articular corticosteroid for knee osteoarthritis. Clin Orthop Relat Res. 2018;476(7):1391–2.

10. Coombes BK, Bisset L, Vicenzino B. Efficacy and safety of corticosteroid injections and other injections for management of tendinopathy: a systematic review of randomised controlled trials. Lancet. 2010;376(9754):1751–67.

11. Sun Y, Zhang P, Liu S, Li H, Jiang J, Chen S, et al. Intra-articular steroid injection for frozen shoulder: a systematic review and meta-analysis of randomized controlled trials with trial sequential analysis. Am J Sports Med. 2017;45(9):2171–9.

12. Wernecke C, Braun HJ, Dragoo JL. The effect of intra-articular corticosteroids on articular cartilage: a systematic review. Orthop J Sports Med. 2015;3(5):2325967115581163.

13. Dragoo JL, Danial CM, Braun HJ, Pouliot MA, Kim HJ. The chondrotoxicity of single-dose corticosteroids. Knee Surg Sports Traumatol Arthrosc. 2012;20(9):1809–14.

14. Bhattacharjee S, Lee W, Lee MJ, Shi LL. Preoperative corticosteroid joint injections within 2 weeks of shoulder arthroscopies increase postoperative infection risk. J Shoulder Elb Surg. 2019;28(11):2098–102.

15. Kew ME, Cancienne JM, Christensen JE, Werner BC. The timing of corticosteroid injections after arthroscopic shoulder procedures affects postoperative infection risk. Am J Sports Med. 2019;47(4):915–21.

16. Weber AE, Trasolini NA, Mayer EN, Essilfie A, Vangsness CT Jr, Gamradt SC, et al. Injections prior to rotator cuff repair are associated with increased rotator cuff revision rates. Arthroscopy. 2019;35(3):717–24.

17. Haupt HA, Rovere GD. Anabolic steroids: a review of the literature. Am J Sports Med. 1984;12:469–84.

18. Elashoff JD, Jacknow AD, Shain SG, Braunstein GD. Effects of anabolic-androgenic steroids on muscular strength. Ann Intern Med. 1991;115:387–93.

19. Schroeder ET, Vallejo AF, Zheng L, Stewart Y, Flores C, Nakao S, Martinez C, Sattler FR. Six-week improvements in muscle mass and strength during androgen therapy in older men. J Gerontol A Biol Sci Med Sci. 2005;60:1586–92.

20. Wolf SE, Thomas SJ, Dasu MR, Ferrando AA, Chinkes DL, Wolfe RR, Herndon DN. Improved net protein balance, lean mass, and gene expression changes with oxandrolone treatment in the severely burned. Ann Surg. 2003;237:801–10. discussion 810–1

21. Jeschke MG, Finnerty CC, Suman OE, Kulp G, Mlcak RP, Herndon DN. The effect of oxandrolone on the endocrinologic, inflammatory, and hypermetabolic responses during the acute phase postburn. Ann Surg. 2007;246:351–60. discussion 360–2

22. Pope HG Jr, Kanayama G, Athey A, Ryan E, Hudson JI, Baggish A. The lifetime prevalence of anabolic-androgenic steroid use and dependence in Americans: current best estimates. Am J Addict. 2014;23:371–7.

23. Melchert RB, Welder AA. Cardiovascular effects of androgenic-anabolic steroids. Med Sci Sports Exerc. 1995;27:1252–62.

24. Clark AS, Henderson LP. Behavioral and physiological responses to anabolic-androgenic steroids. Neurosci Biobehav Rev. 2003;27:413–36.

25. O'Malley BW, Tsai MJ. Molecular pathways of steroid receptor action. Biol Reprod. 1992;46:163–7.

26. Wilson JD. Role of dihydrotestosterone in androgen action. Prostate Suppl. 1996;6:88–92.

27. van der Vies J. Pharmacokinetics of anabolic steroids. Wien Med Wochenschr. 1993;143:366–8.

28. Kicman AT. Pharmacology of anabolic steroids. Br J Pharmacol. 2008;154:502–21.

29. Kühn C. Metabolomics in animal breeding. In: Suhre K, editor. Genetics meets metabolomics: from experiment to systems biology. New York, NY: Springer New York; 2012. p. 107–23.

30. Feldkoren BI, Andersson S. Anabolic-androgenic steroid interaction with rat androgen receptor in vivo and in vitro: a comparative study. J Steroid Biochem Mol Biol. 2005;94(5):481–7.

31. Beiner JM, Jokl P, Cholewicki J, Panjabi MM. The effect of anabolic steroids and corticosteroids on healing of muscle contusion injury. Am J Sports Med. 1999;27:2–9.

32. White JP, Baltgalvis KA, Sato S, Wilson LB, Carson JA. Effect of nandrolone decanoate administration on recovery from bupivacaine-induced muscle injury. J Appl Physiol. 2009;107:1420–30.

33. Flück M, Ruoss S, Möhl CB, et al. Genomic and lipidomic actions of nandrolone on detached rotator cuff muscle in sheep. J Steroid Biochem Mol Biol. 2017;165:382–95.

34. Triantafillopoulos IK, Banes AJ, Bowman KF Jr, Maloney M, Garrett WE Jr, Karas SG. Nandrolone decanoate and load increase remodeling and strength in human supraspinatus bioartificial tendons. Am J Sports Med. 2004;32:934–43.

35. Papaspiliopoulos A, Papaparaskeva K, Papadopoulou E, Feroussis J, Papalois A, Zoubos A. The effect of local use of nandrolone decanoate on rotator cuff repair in rabbits. J Investig Surg. 2010;23:204–7.

36. Gold J, High HA, Li Y, Michelmore H, Bodsworth NJ, Finlayson R, Furner VL, Allen BJ, Oliver CJ. Safety and efficacy of nandrolone decanoate for treatment of wasting in patients with HIV infection. AIDS. 1996;10:745–52.

37. Tidermark J, Ponzer S, Carlsson P, Söderqvist A, Brismar K, Tengstrand B, Cederholm T. Effects of protein-rich supplementation and nandrolone in lean elderly women with femoral neck fractures. Clin Nutr. 2004;23:587–96.

38. Mukhopadhyay R, Sangwan SS, Gogna P, Singla R, Kundu ZS, Kamboj P, Singh A, Magu NK, Gupta NG. Anabolic steroids improve bone mineral density and quality of life in patients with osteoporotic fractures around the hip. Int J Orthop. 2014;1:164–7.

39. Sloan JP, Wing P, Dian L, Meneilly GS. A pilot study of anabolic steroids in elderly patients with hip fractures. J Am Geriatr Soc. 1992;40:1105–11.

40. Tengstrand B, Cederholm T, Söderqvist A, Tidermark J. Effects of protein-rich supplementation and nandrolone on bone tissue after a hip fracture. Clin Nutr. 2007;26:460–5.

41. Hohmann E, Tetsworth K, Hohmann S, Bryant AL. Anabolic steroids after total knee arthroplasty. A double blinded prospective pilot study. J Orthop Surg Res. 2010;5:93.

42. Tamaki T, Uchiyama S, Uchiyama Y, Akatsuka A, Roy RR, Edgerton VR. Anabolic steroids increase exercise tolerance. Am J Physiol Endocrinol Metab. 2001;280:E973–81.

43. Carson JA, Lee WJ, McClung J, Hand GA. Steroid receptor concentration in aged rat hindlimb muscle: effect of anabolic steroid administration. J Appl Physiol. 2002;93:242–50.

44. Schroeder ET, Terk M, Sattler FR. Androgen therapy improves muscle mass and strength but not muscle quality: results from two studies. Am J Physiol Endocrinol Metab. 2003;285:E16–24.

45. Inoue K, Yamasaki S, Fushiki T, Okada Y, Sugimoto E. Androgen receptor antagonist suppresses exercise-induced hypertrophy of skeletal muscle. Eur J Appl Physiol Occup Physiol. 1994;69:88–91.

46. Kadi F. Adaptation of human skeletal muscle to training and anabolic steroids. Acta Physiol Scand Suppl. 2000;646:1–52.

47. Inoue K, Yamasaki S, Fushiki T, Kano T, Moritani T, Itoh K, Sugimoto E. Rapid increase in the number of androgen receptors following electrical stimulation of the rat muscle. Eur J Appl Physiol Occup Physiol. 1993;66:134–40.

48. Sinha-Hikim I, Artaza J, Woodhouse L, Gonzalez-Cadavid N, Singh AB, Lee MI, Storer TW, Casaburi R, Shen R, Bhasin S. Testosterone-induced increase in muscle size in healthy young men is associated with muscle fiber hypertrophy. Am J Physiol Endocrinol Metab. 2002;283:E154–64.

49. Blazevich AJ, Giorgi A. Effect of testosterone administration and weight training on muscle architecture. Med Sci Sports Exerc. 2001;33:1688–93.

50. Casaburi R, Nakata J, Bistrong L, Torres E, Rambod M, Porszasz J. Effect of Megestrol acetate and testosterone on body composition and hormonal responses in COPD cachexia. Int J Chron Obstruct Pulmon Dis. 2015;3:389–97.

51. Sengelaub DR, Han Q, Liu N-K, Maczuga MA, Szalavari V, Valencia SA, Xu X-M. Protective effects of estradiol and dihydrotestosterone following spinal cord injury. J Neurotrauma. 2018;35:825–41.

52. Yarrow JF, Conover CF, Beggs LA, et al. Testosterone dose dependently prevents bone and muscle loss in rodents after spinal cord injury. J Neurotrauma. 2014;31:834–45.

53. Giangregorio L, McCartney N. Bone loss and muscle atrophy in spinal cord injury: epidemiology, fracture prediction, and rehabilitation strategies. J Spinal Cord Med. 2006;29:489–500.

54. Gladstone JN, Bishop JY, Lo IKY, Flatow EL. Fatty infiltration and atrophy of the rotator cuff do not improve after rotator cuff repair and correlate with poor functional outcome. Am J Sports Med. 2007;35:719–28.

55. Meyer DC, Wieser K, Farshad M, Gerber C. Retraction of supraspinatus muscle and tendon as predictors of success of rotator cuff repair. Am J Sports Med. 2012;40:2242–7.

56. Romani WA, Belkoff SM, Elisseeff JH. Testosterone may increase rat anterior cruciate ligament strength. Knee. 2016;23:1069–73.

57. Wu B, Lorezanza D, Badash I, Berger M, Lane C, Sum JC, Hatch GF 3rd, Schroeder ET. Perioperative testosterone supplementation increases lean mass in healthy men undergoing anterior cruciate ligament reconstruction: a randomized controlled trial. Orthop J Sports Med. 2017;5:2325967117722794.

58. Seynnes OR, Kamandulis S, Kairaitis R, Helland C, Campbell E-L, Brazaitis M, Skurvydas A, Narici MV. Effect of androgenic-anabolic steroids and heavy strength training on patellar tendon morphological and mechanical properties. J Appl Physiol. 2013;115:84–9.

59. Amory JK, Chansky HA, Chansky KL, Camuso MR, Hoey CT, Anawalt BD, Matsumoto AM, Bremner WJ. Preoperative supraphysiological testosterone in older men undergoing knee replacement surgery. J Am Geriatr Soc. 2002;50:1698–701.

60. Hedström M, Åström K, Sjöberg H, Dalén N, Sjöberg K, Brosjö E. Positive effects of anabolic steroids, vitamin D and calcium on muscle mass, bone mineral density and clinical function after a hip fracture: a randomised study of 63 women. J Bone Joint Surg Br. 2002;84:497–503

61. Laseter JT, Russell JA. Anabolic steroid-induced tendon pathology: a review of the literature. Med Sci Sports Exerc. 1991;23:1–3.

62. Karpakka JA, Pesola MK, Takala TE. The effects of anabolic steroids on collagen synthesis in rat skeletal muscle and tendon. A preliminary report. Am J Sports Med. 1992;20:262–6.

63. Miles JW, Grana WA, Egle D, Min KW, Chitwood J. The effect of anabolic steroids on the biomechanical and histological properties of rat tendon. J Bone Joint Surg Am. 1992;74:411–22.

64. Kanayama G, Hudson JI, DeLuca J, Isaacs S, Baggish A, Weiner R, Bhasin S, Pope HG Jr. Prolonged hypogonadism in males following withdrawal from anabolic-androgenic steroids: an under-recognized problem. Addiction. 2015;110:823–31.

65. Inhofe PD, Grana WA, Egle D, Min KW, Tomasek J. The effects of anabolic steroids on rat tendon. An ultrastructural, biomechanical, and biochemical analysis. Am J Sports Med. 1995;23:227–32.

Intra-articular Hyaluronic Acid Injections

8

Karan Vishwanath and Lawrence J. Bonassar

8.1 HA Chemistry and Structure

Hyaluronic acid (HA) is a high molecular weight linear non-sulfated glycosaminoglycan (GAG) composed of alternating β-D-glucuronic acid and N-acetyl-β-D-glucosamine units linked by β-1,3 and β-1,4 glycosidic bonds (Fig. 8.1). It is found in the extracellular matrix of all tissues but its concentration and molecular weight can vary greatly. In the synovial fluid (SF), the concentration of HA ranges from 1.4 to 3.6 mg/mL, and its average molecular weight is estimated to be between 1 and 5 MDa [1].

Unlike other GAG molecules, HA is produced by HA synthases—a highly specialized set of proteins located at the plasma membrane of cells [2]. At physiological pH, the large number of deprotonated carboxylate groups in the HA molecule contributes to a highly negatively charged structure, and often HA exists as an electroneutral sodium salt derivative of HA (sodium hyaluronate). Although HA is a linear polymer, high MW solutions of HA are highly viscous due to chain entanglement [3]. Long, highly entangled

K. Vishwanath
Department of Materials Science and Engineering, Cornell University, Ithaca, NY, USA

L. J. Bonassar (✉)
Meinig School of Biomedical Engineering, Cornell University, Ithaca, NY, USA

Sibley School of Mechanical and Aerospace Engineering, Cornell University, Ithaca, NY, USA
e-mail: lb244@cornell.edu

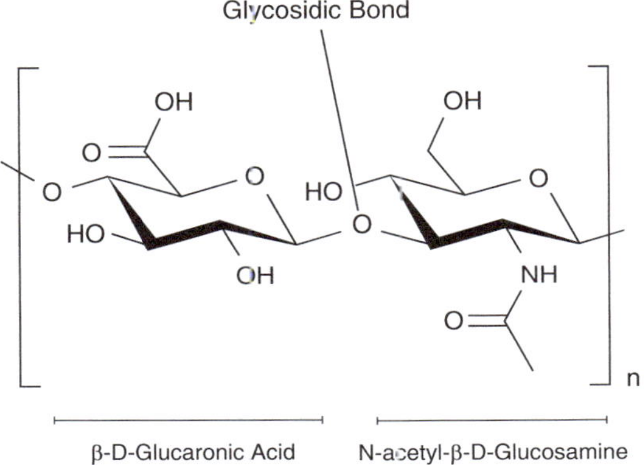

Fig. 8.1 Structural formula of hyaluronic acid

chains occupy larger volumes and thus contribute to highly viscous solutions. HA thus exhibits time-dependent deformation, often termed viscoelasticity.

The structure and configuration of HA mediate its various physiological functions: it acts as a viscoelastic support for tissue protection; it is one of the primary lubricating molecules in SF along with proteoglycan-4 (also known as lubricin or superficial zone protein); and it controls tissue hydration and water transport. In the context of joint lubrication and osteoarthritis (OA), the viscosity of SF has been strongly associated with HA, and the concentration of free HA in SF has been found to decrease in human patients with the progression of OA [4]. Previous studies investigating the lubricating ability of the SF show that there is an elevation of the coefficient of friction

© ISAKOS 2022
G. Filardo et al. (eds.), *Orthobiologics*, https://doi.org/10.1007/978-3-030-84744-9_8

in patients with OA [5]. This increase in friction was attributed to the decreased viscosity of the SF, which in turn arises from the decrease in the concentration and chain breakdown of endogenous HA in the joint.

Viscosupplementation involves the intra-articular injection of a high molecular weight HA (> 500 kDa) into the joint cavity as a means to restore the viscosity, and ultimately the lubricating ability of SF [6]. Viscosupplementation has been used clinically for more than 30 years and has been a mainstay of arthritis treatment in patients with low-grade OA. However, its efficacy and proposed mechanism of action are still controversial. Some meta-analyses have shown that HA viscosupplementation does not have a statistically significant effect over placebo saline injections with regard to restoration of lubrication or pain management in patients with OA, while others suggest that there is a benefit. Emblematic of this controversy, some clinical societies recommend the use of HA injection for mild to moderate OA, while others suggest that there is no benefit [7–9]. In spite of the controversy that surrounds the efficacy of this therapy, the size of the HA viscosupplementation industry has been valued at $4 billion in 2019, expected to grow at a rate of 9% a year [10].

8.2   Current Product Profile

There are currently eight injectable formulations of HA approved for the treatment of mild to moderate OA in the knee (Table 8.1) in the USA, but many more are available worldwide. These formulations differ depending on the source of the HA in the formulation, the molecular weight of the HA, physical properties like viscosity and lubricating ability, method of production, treatment schedule, and cost. Molecular weight and the presence of cross-links change both the mechanical properties of HA solutions as well as potential biological interactions with the carti-

Table 8.1 Currently available HA viscosupplements for OA treatment

Name	Molecular weight (kDa)	Chemical modification	Cross-linked (Y/N)
Hymovis	500–730	Sodium salt of HA	N
Hyalgan	500–720	Sodium salt of HA	N
Supartz	620–1200	Sodium salt of HA	N
ORTHOVISC	1000–2900	Sodium salt of HA	N
MONOVISC	1000–2800	Cross-linked HA derivative	Y
Euflexxa	2400–3600	Sodium salt of HA	N
Synvisc (Hylan G-F 20)	5000–6000	Cross-linked HA derivative	Y

lage extracellular matrix and with target cells such as chondrocytes and synoviocytes [11]. Subsequent sections of this chapter will review known mechanisms of mechanical and biological actions of HA solutions and how these actions are affected by chemical characteristics of formulations.

A main challenge in understanding the mechanism of action of HA injections is that there are numerous biological and mechanical modes of action that could intervene in the disease process of OA. The following sections will review the biological (Sect. 8.3) and mechanical (Sect. 8.4) roles that HA is purported to play in the treatment of OA. The presence and relative importance of the biological and mechanical roles of HA is significant, far beyond a simple academic debate. For decades, the action of HA was purported to be primarily as a mechanical lubricant, and as such, HA injections were classified by the US Food and Drug Administration (FDA) as a medical device. More recently, the FDA has indicated that the potential biological activity of HA makes it more appropriate to classify these therapies as drugs. The importance of this determination emphasizes the need to understand both the biological and mechanical actions of HA.

8.3 Biologic Effects of HA

The ubiquity of HA throughout the tissues of the joint and in SF makes the study of all of the biological actions of this molecule highly challenging. Additionally, unlike small molecules or proteins that interact with extracellular matrix and cell surface receptors, the biological activity of HA varies with molecular weight, with large versions of the molecule (>500 kDa) and smaller fragments (<10 kDa) having actions that are not only distinct but also potentially antagonistic [11, 12]. Further, large and small molecular weight versions of HA interact with common receptors (Fig. 8.2). Such receptors are also differentially expressed between target tissues in the joint such as cartilage and the synovium. This section will highlight key features of biological mechanisms of action of HA and how these may be relevant to injectable HA treatments.

Viscosupplementation with HA has been shown to suppress the in vitro activity of pro-inflammatory cytokines and chemokines by inhibiting signal transduction pathways of certain cell surface receptors, coupled with the promotion of anti-inflammatory mediators [13]. HA polymer chains have an affinity to bind to specific cell surface receptors like cluster determinant 44 (CD44), the Toll-like receptors 2 and 4 (TLR-2, TLR-4), and the intercellular adhesion molecules (ICAM-1).

Both chondrocytes and synoviocytes are potential targets for the biological activity of HA. The primary receptor of the HA ligand is CD44. CD44 is a multifunctional cell surface glycoprotein that is expressed across a wide range of cell types. Due to its abundance, it is considered to be the primary HA cellular receptor [14]. Activation of this receptor via HA binding initiates a signaling cascade associated with the activation of nuclear factor kappa-light-chain-enhancer of activated B cells (NF-kB), which is responsible for the induction of the expression of cytokines like IL-1β, IL-6, IL-8, and TNF-α. The binding affinity of HA to CD44 receptors is primarily determined by the size of the HA [15]. Larger HA chains bind more effectively to the CD44 receptor due to divalent binding, which results in a lower rate of dissociation. The expression of the CD44 receptor is responsible for maintaining homoeostasis of the articular cartilage, and the fundamental role of the receptor is

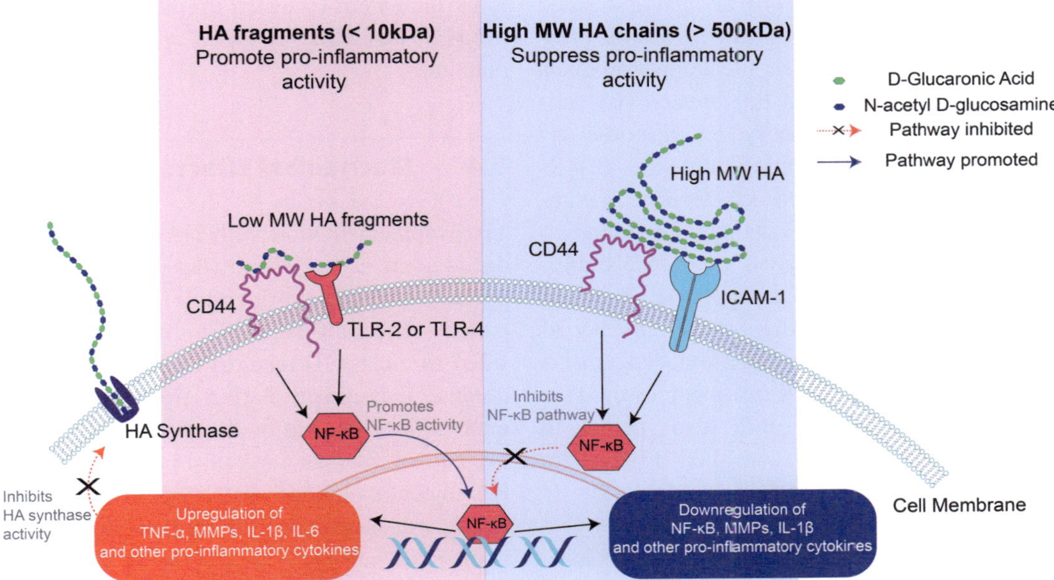

Fig. 8.2 Summary of the known biologic effects of HA in chondrocytes. Adapted from [11]

to bind to and internalize HA fragments. Various T cells and cytokines can mediate HA-CD44 binding primarily through the regulation of tumor necrosis factor-alpha (TNF-α).

To examine the role of HA-CD44 binding as a potential anti-inflammatory pathway, prior experiments conducted on CD44-deficient mice demonstrated that binding is suppressed in vivo by pro-inflammatory cytokine production and is mediated by TLR through NF-κB [16]. The effect of HA molecular weight on NF-kB activation was also studied in mouse macrophages and observed that low molecular weight HA (LMWHA) fragments activated NF-kB DNA binding, which in turn elicits a pro-inflammatory response via the release of cytokines like IL-1β, IL-6, and TNF-α as well as matrix metalloproteinases (MMPs). In contrast, high molecular weight HA (HMWHA) chains can occupy multiple CD44 receptor sites and promote anti-inflammatory effects within the cell by indirect deactivation of the NF-kB pathway. An increased proportion of divalent binding in these larger chains suppresses NF-kB production, leading to a downregulation of pro-inflammatory cytokines [17].

TLRs are responsible for signaling the cellular defense pathways against common bacteria and viruses in the immune system. TLR expression is enhanced in the presence of pathogen-associated molecular patterns (PAMPs) and also in the presence of extracellular matrix breakdown products like LMWHA fragments. Fragmented HA activates the innate immune system response via TLR-2 and TLR-4 binding and the IL-1R-associated kinase and NF-kB-dependent pathways [18]. Small fragments of HA bound to TLR-4 have also been shown to produce significant upregulation of TNF-alpha, IL-1b, IL-6, and IL-8. Campos et al. investigated the influence of HMWHA in mice with collagen-induced arthritis and showed that HMWHA treatment lowered the instances of arthritis and also decreased TNF-alpha, IL-1b, IL-16, and MMP-13 levels in these mice [19].

ICAM-1 is another cell surface receptor that preferentially binds to HA. In vitro studies have linked the upregulation of ICAM-1 expression occurs to the presence of inflamed or malignant tissue [20]. Increased expression of ICAM-1 also activates the NF-kB transcriptional system [21], which is critical to the foreign body immune response. Several studies have shown that elevated ICAM-1 levels in the presence of inflammation can be downregulated in the presence of HMWHA as HMWHA blocks the NF-kB pathway [21, 22].

In summary, HMWHA can bind to CD44, TLR-2, TLR-4, and ICAM-1 to promote anti-inflammatory effects within the cell. HA bound to CD44 has been shown to suppress the NF-kB pathway activation, which in turn lowers levels of pro-inflammatory factors like IL-8, IL-33, TNF-α, and proteinases like MMPs. It is important to note that HA is not known for having an anti-inflammatory effect within the SF. Through in vitro studies and in some small animal in vivo studies, it has been shown that HMWHA can produce some anti-inflammatory effects at the molecular level. While such studies are important and intriguing, the evidence of a direct influence of these biological phenomena to clinical outcomes in humans has not been established. Future studies on the biological effects of HA on inflamed or damaged cartilage will help determine the extent to which the anti-inflammatory properties of HMWHA play a role in clinical outcomes.

8.4 Mechanical Effects of HA

Viscosupplementation with HA was first introduced in the early 1970s by Balazs et al. with the primary goal of restoring the viscosity of SF in patients with arthritis [23]. This therapy was classified as a class III medical device by the FDA, because it is aimed at providing a mechanical intervention—the restoration of the viscosity of SF. The mechanical mechanism of intra-articular HA injections hinges on the idea that restoring SF viscosity will lower the coefficient of friction in the joint.

8.4.1 Viscosity

Viscosity can be most easily described as the resistance to flow. It is a defined ratio between the viscous stresses experienced by a material with respect to the rate of change of deformation. Most materials like fluids or polymers (including hydrogels) have an associated viscosity depending on factors like the test conditions, the chemistry of the material, the extent of chemical cross-linking, and the configuration of the test equipment.

The most commonly used methods to evaluate the viscosity of HA and HA derivatives solutions are the parallel-plate and the cone-plate configurations in a rotational rheometer (Fig. 8.3). These instruments can operate in the strain control regime, where a fixed deformation (or deformation rate) is applied and the resulting shear stress in the material is recorded, or in the stress control regime in which the applied shear stress is fixed and a deformation response is recorded.

Parallel-plate rotational rheometers have two parallel metal plates, one of which is made to rotate by an external motor at a predetermined speed while the other remains stationary. This configuration is widely used in the pharmaceutical and food and beverage industry due to its flex-ibility to accommodate a wide range of materials. The primary drawback with this configuration is the nonuniformity of shear strains that develop in the sample; greatest strains are observed at the edges of the plate, and there is no shear strain in the middle. In contrast, the cone-plate system comprises a lower flat plate on which the sample is placed and an upper rotating cone with a very shallow angle which can be lowered to be in contact with the sample. This configuration is considered to be rheologically representative as uniform strains build up in the sample, producing reliable and reproducible data. However, this technique does not allow the user to explore a wide range of shear strains since the maximum shear rate that can be applied is dependent on the gap width between the cone and plate (and this is typically fixed).

The viscosity of HA is attributed to its large size and steric interactions (charge and bulkiness) with other materials such as proteins [24]. Most commercially available HA viscosupplements exhibit shear thinning behavior, where the viscosity of the material decreases with an increase in shear rate. Notably, these viscosities can vary significantly between the different commercial HA formulations due to differences in molecular weight, cross-linking density, and composition. This complex

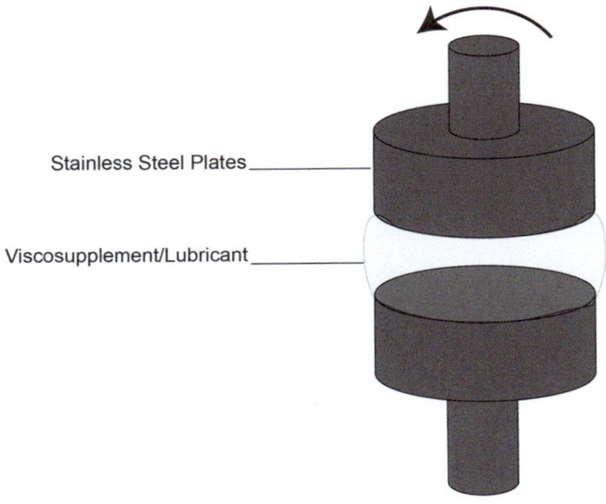
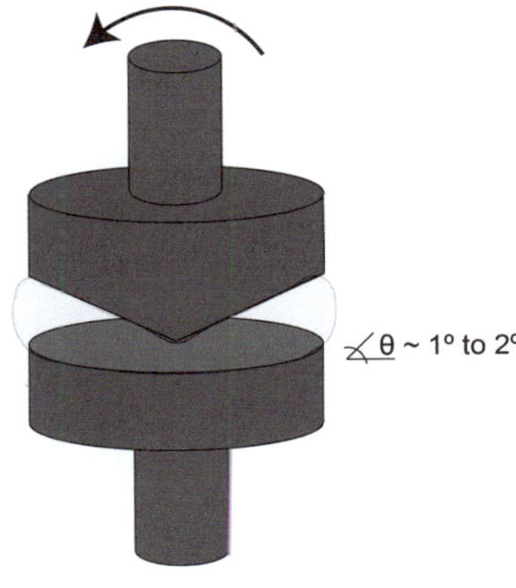

Fig. 8.3 Most commonly used rheometer configurations

behavior of HA solutions is well described by the Carreau-Yasuda model [25]:

$$\frac{\eta - \eta_\infty}{\eta_0 - \eta_\infty} = \left[1 + \left(\lambda \dot{\gamma} \right)^a \right]^{n-1/a}$$

where η is the measured viscosity at a given shear rate $\dot{\gamma}$, η_0 is the zero shear viscosity of the material at the Newtonian plateau (interpolation of data to shear rate of zero), η_∞ is the dynamic viscosity of the material at an infinite shear rate, and λ, n, and a represent fitting parameters.

Most commercial HA viscosupplements vary in molecular weight, chemical modification, and structure leading to a wide range of viscosities across shear rates. When the most common injections were evaluated in a conventional cone and plate rheometer, it was observed that the viscosupplements are shear thinning and have similar viscosities at high shear rates (Fig. 8.4). But their viscosities at low shear rates (Newtonian plateau) can span three orders of magnitude, which might have implications on the ability of the injection to restore the viscosity of SF, and can explain the wide range of clinical outcomes after viscosupplementation therapy. Since viscosity scales with the molecular weight, low viscosity HA solutions are more susceptible to be broken down or degraded by enzymes than high viscosity

HMWHA solutions, leading to a lower residence time in the joint [26]. The viscosity of an intra-articular injection is therefore an important fundamental parameter to consider while evaluating the efficacy of HA viscosupplements.

8.4.2 Lubrication

Articular cartilage is an avascular and aneural tissue responsible for providing a low friction interface during joint motion over 100 million cycles of use in a healthy knee. Friction arises from solid-solid contact between the articulating surfaces in most joints. During articulation of joints, healthy articular cartilage has been shown to exhibit one of the lowest levels of friction in nature ($\mu = 0.001$–0.02) [26].

A key component that regulates the lubricating ability of cartilage is the composition of SF present in the joint. SF is a highly viscous fluid that bathes the surfaces of articular cartilage. The viscosity of SF is attributed to the presence of large charged molecules like HA, lubricin (proteoglycan-4), surface-active phospholipids, and other proteins like albumin [27]. Previous studies have shown that the rheological properties of SF regulate the lubricating ability of joints [28, 29].

Fig. 8.4 Viscosity measurements of commercially used HA viscosupplements across a range of shear rates (low to high). Adapted from [25]

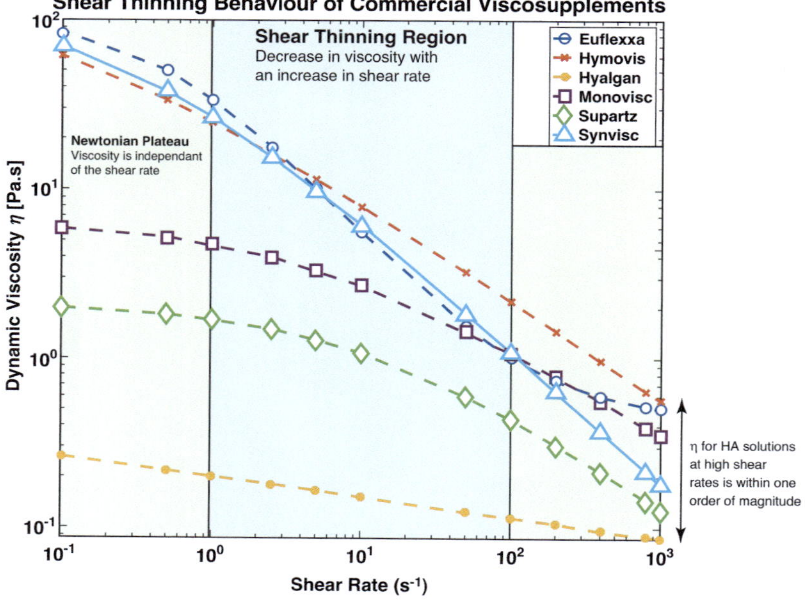

Since viscosity is an indicator of how easily two surfaces can be kept apart (fluid film separation) during motion, high viscosities can cause greater separation between contacting surfaces, thus resulting in lower rates of wear and friction during articulation. The increased viscosity from the aforementioned large molecules in the SF is therefore responsible for the low rates of wear and provides a low friction interface between the contacting surfaces in a joint. This distinct functionality of cartilage has led to decades of research focused at understanding the mechanisms that contribute to this incredibly low friction surface.

The study of a material's ability to lubricate is known as tribology. A subset of mechanical engineering, tribology offers a traditional continuum mechanics framework to study friction and also examines the surface chemistry of contacting surfaces during motion. The fundamental parameter of interest that is used to evaluate lubrication is the coefficient of friction, μ. In the last century, engineering research has been dedicated to understanding how friction changes under different operating conditions. Parameters that can influence the coefficient of friction are the relative speed between the articulating surfaces, the force or pressure that is supported by the surfaces, the viscosity of the lubricating fluid that bathes the surfaces, and the contact geometry of the system [27].

In the domain of tribology, the lubrication of incompressible materials is determined from the measurement of μ under different conditions and is characterized through distinct modes of lubrication. The systems used to measure coefficient of friction are called tribometers and can be constructed in a variety of configurations, each with their own advantages, disadvantages, and user-controlled parameters. The four most common tribometer configurations for assessment of cartilage lubrication are (i) the pin-on-disk tribometer, (ii) the pin-on-plate tribometer, (iii) the ball-on-disc tribometer, and (iv) the relative rotation configuration (Fig. 8.5).

When evaluating the lubrication of incompressible materials like journal bearings, altering the parameters in a tribometer can change the coefficient of friction by orders of magnitude—and this can be plotted on a Stribeck curve (Fig. 8.6). The changes in the friction observed are indicative of distinct modes of lubrication. Traditionally, these modes of lubrication are denoted by the boundary, mixed, and hydrodynamic modes [27].

At low speeds or viscosities and high loads, much of the force is supported by the contacting surfaces of the material. Both friction and wear are high and this mode is called the boundary mode of lubrication. Friction in this mode is dominated by the surface chemistry of the contacting surfaces and is less sensitive to the other parameters like viscosity, sliding speed, and contact pressure. Only a very thin film of lubricant, as thick as the surface roughness of the contacting surfaces, forms between the surfaces and contributes little to lowering μ. At higher sliding speeds or in the presence of high viscosity lubricants, a fluid film of considerable thickness forms between the contacting asperities, and this serves to lower the normal force being supported by the material. When impermeable materials like journal bearings are analyzed with this framework, there is a friction transition that is mapped out from the boundary mode to the region where the interfacial fluid pressurization causes the contacting surfaces to become fully separated—referred to as the hydrodynamic mode of lubrication [28].

The coefficient of friction (μ) is plotted against a dimensionless number called the Sommerfeld number (S), which is defined as a ratio of the product of the sliding speed, the viscosity of the lubricant, and the contact width of the sample to the normal force applied on the sample during the test.

While the Stribeck framework is highly applicable to impermeable metals in journal bearings, there is no scientific consensus on a similar framework that can be applied to evaluate the frictional changes in porous, viscoelastic tissues. When applied to a soft material like articular cartilage, it has been shown that there is a similar transition between the lubricating modes, but at higher speeds the system does not achieve full hydrodynamic lubrication because the lubricant can occupy the interstitial spaces in the tissue,

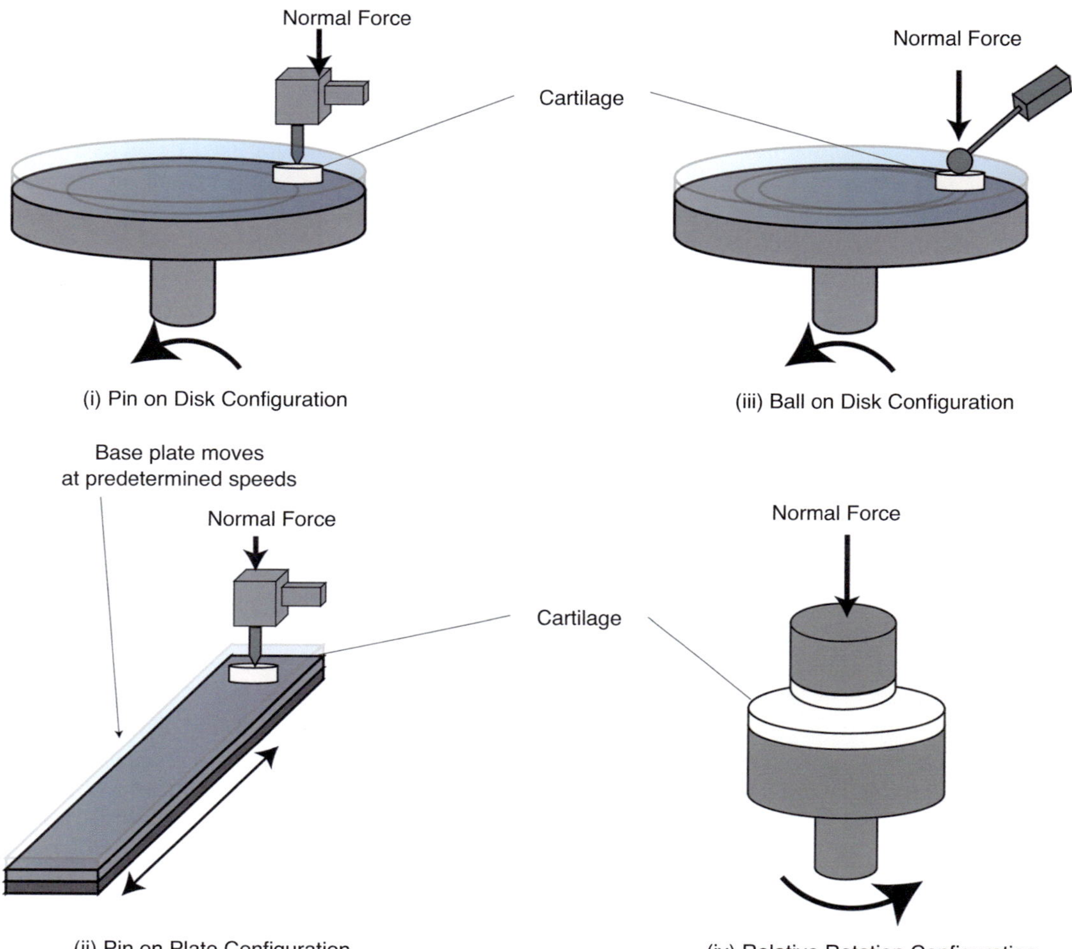

Fig. 8.5 The four commonly used tribometer configurations in cartilage tribology

preventing a full fluid film between the contacting surfaces [28]. These deviations from the classical analysis were attributed to the permeability of articular cartilage, causing fluid exchange between the bulk and the tissue, which drives this phenomenon termed the elastoviscous transition. The importance of fluid pressurization and flow at the cartilage surface during sliding has recently been confirmed by elegant experiments measuring such flows. These data demonstrate that the pressure that develops at the tissue interface during sliding drives fluid into the cartilage surface [29]. The phenomenon, termed "tribological rehydration," underscores the importance of fluid flow and pressurization at the cartilage surface in driving the lubrication of cartilage.

When analyzing commercial HA viscosupplements using this framework, a question that arises is what viscosity is appropriate to use in the analysis. Bonnevie et al. have recently shown that the coefficient of friction of commercial HA viscosupplements does not map well onto the curve

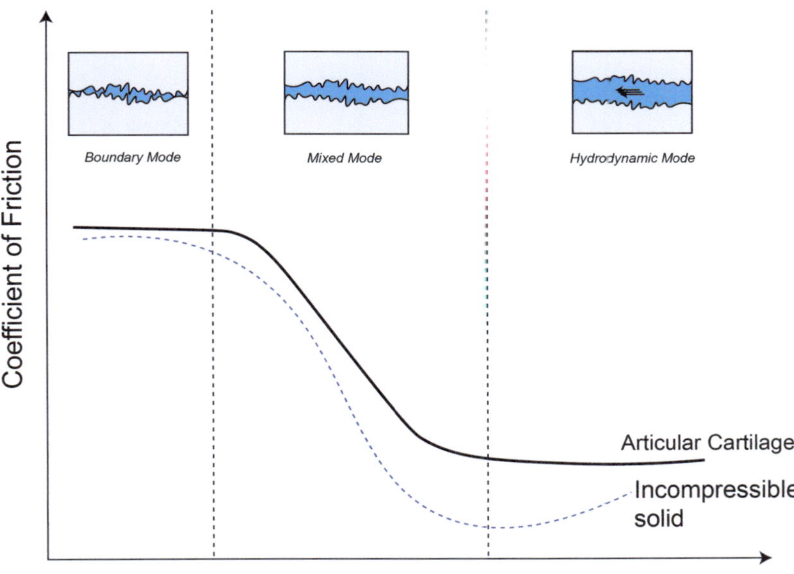

Fig. 8.6 Stribeck framework comparing transitions between modes of lubrication in an incompressible solid and articular cartilage. Adapted from [27]

when the standard zero-shear viscosity of the formulations is used [25]. The reason for this has been attributed to commercial viscosupplements having a wide range in chemical composition and rheological behavior, as can be seen in Fig. 8.4. Instead, it was shown that a numerically determined parameter called the effective viscosity (η_{eff}) of the viscosupplement was a better predictor of both the ability of the viscosupplement to lubricate effectively and the clinical improvement in pain (WOMAC score) in patients treated with the viscosupplements (Fig. 8.7). The effective viscosity correlated strongly with the measured coefficient of friction and also with the improvement in WOMAC score from baseline values in patients with OA. It is also interesting to note that the effective viscosities calculated were orders of magnitude different from the zero-shear viscosity, suggesting that only a fraction of the exogenous HA that has been injected has been localized on the surface of cartilage and contributes to enhanced viscosity and lubrication.

To further understand the mechanical effects of HA localization at the surface of cartilage, a recent study by Cook et al. showed that different molecular weight solutions of HA tested in rheometers that were functionalized with the surface layer of cartilage (Fig. 8.8) had an effective viscosity that was 20-fold times higher than when tested in conventional rheometer systems [30].

The clear implication of this work is that the interaction with the surface of a material affects the viscosity of HA solutions. In this case, the difference in effective viscosities was attributed to the formation of a viscous boundary layer between the localized HA and the surface zone proteins on the cartilage surface [31]. The effective viscosity of HA was dependent on both the molecular weight and the gap width between the functionalized surfaces. In another study by Bonnevie et al., it was hypothesized that linear HA chains freely floating in the synovial fluid physically entangle with other molecules like lubricin via its amphiphilic brush structure, enhancing HA localization at the surface (Fig. 8.9). This association leads to a higher viscosity at the surface compared to the bulk viscosity of SF away from the surface—forming a viscosity gradient away from the surface. The localization of HA therefore aims to boost the effective viscosity at the surface of the cartilage, which in turn was shown to improve the lubrication.

Both localization studies show that traditional experiments of rheology and tribology in the absence of cartilage capture the bulk or global

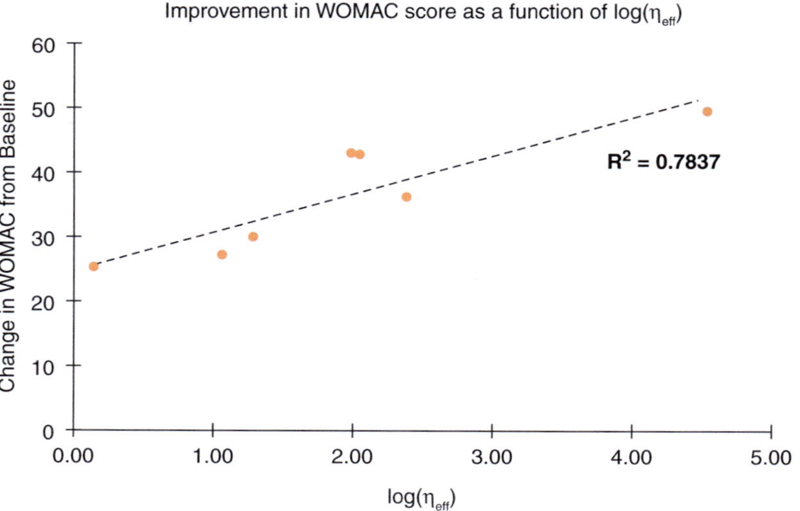

Fig. 8.7 Effective viscosity and coefficient of friction are strong predictors of clinical improvement. Adapted from [25]

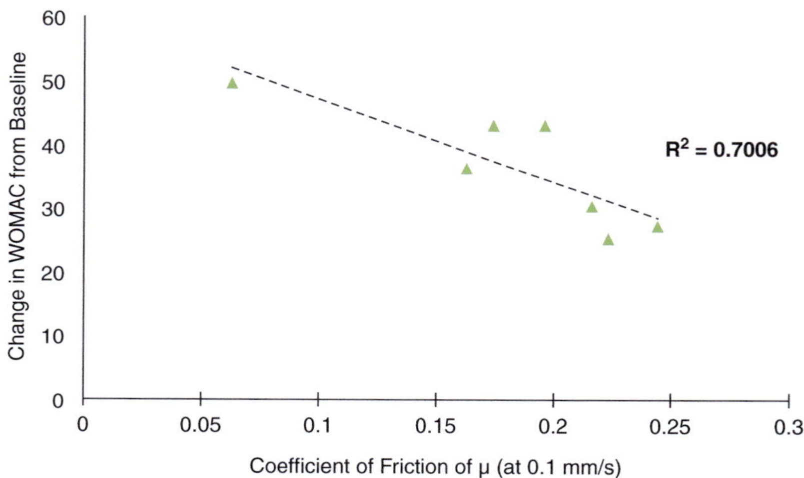

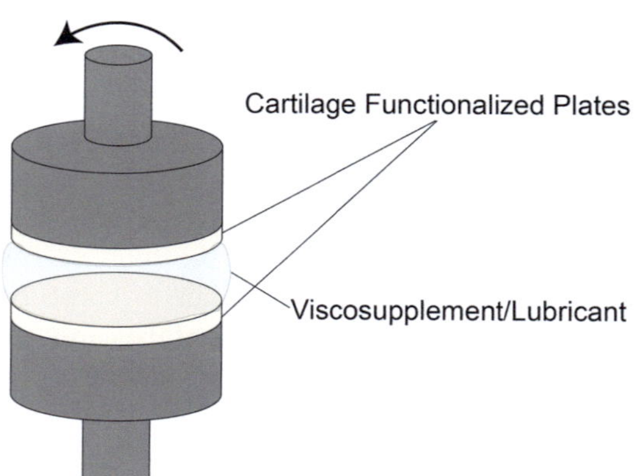

Fig. 8.8 Measurement of effective viscosity of a visco-supplement in a cartilage functionalized rheometer. Adapted from [30]

mechanical behavior of the HA formulations, but do not capture local effects on the surface of the cartilage [30]. Bound HA at the surface of the cartilage was shown to not only improve the lubricity but also increase the residence time of the viscosupplement in the joint. These studies indicate that the presence of cartilage can alter the local viscosity of HA solutions. Such data contribute to a growing body of evidence suggesting that surface interactions involving HA and cartilage can strongly affect the viscosity and friction behavior of HA viscosupplements and is an important factor to consider in developing and characterizing HA therapies. The effective viscosity represents the actual contribution of

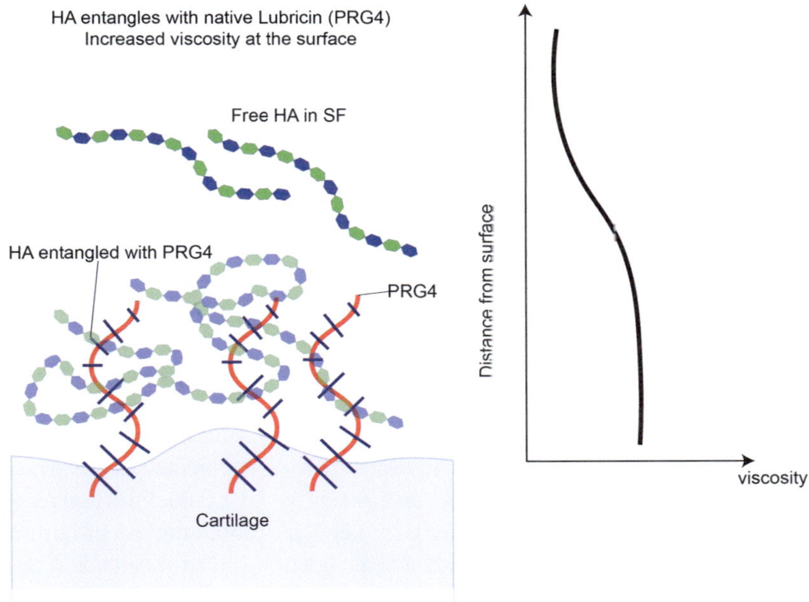

Fig. 8.9 Entanglement of HA with lubricin (PRG4) at the surface of cartilage leads to the formation of a viscous boundary layer with an elevated viscosity. Adapted from [31]

viscosity from the intra-articular injection and may be a more relevant parameter to be evaluated to predict lubrication and clinical efficacy of these solutions than bulk viscosity obtained from conventional analysis techniques.

An unresolved question in the field of HA therapeutics for OA revolves around reconciling the duration of the efficacy of these injections, which can last up to 6 months, and the reported residence times of the injected HA, which are at best a few weeks [31]. Such data are frequently noted in arguments suggesting that the effects of HA treatments must include something beyond a mechanical mechanism if the substance providing the mechanical benefit is lost from the joint, while the benefit itself endures. Notably, the evaluation of residence time involves measurements of bulk concentrations of HA in synovial fluid. The HA associated with the surface of articular cartilage (Fig. 8.9) would likely both be missed by a bulk measurement of HA in residence time studies and would be extremely effective in providing lubrication to the joint. As such, it seems likely that HA associated with the cartilage surface would have much longer residence time in the joint than that in the bulk of the SF and may still provide mechanical benefit after the bulk of HA is lost from the joint.

Recent studies of the ability of HA to augment lubrication of SF from arthritis patients shed further light on the importance of HA association with cartilage [32]. These studies identified subpopulations of patients in whose SF provided significantly less benefit in lowering friction coefficients and associated shear strains. This novel "tribological endotype" was distinguished by lower levels of lubricin, the molecule critical for localizing HA at the cartilage surface [33]. Collectively, these data underscore the necessity not only of delivering HA to a joint but also ensuring that the HA is able to effectively interact with the surface of cartilage.

8.5   Future Directions

While there are some mechanical and biological benefits to HA viscosupplementation, some big drawbacks are the cost to manufacture and purify HA formulations and the lack of consensus on the efficacy of the treatment. Equine studies involving HA injections have shown that the clearance time of HA in the joint after intra-articular injection is only 11–12 h in subjects with OA and 20 h in a normal joint, indicating that large HA chains (2–4 MDa) can depolymerize and be broken down into smaller fragments rapidly by the enzyme hyaluronidase, leading to low residence time after injection, further lowering the therapeutic effect of HA [34].

Unlike existing viscosupplements, synthetic lubricants have been shown to lubricate cartilage just as effectively as HA, can be synthesized to have a wide range of molecular weights, can be chemically modified to resist enzymatic degradation, and cost a fraction to manufacture compared to the existing HA formulations [35]. Synthetic HA-mimetic viscosupplements aim to mimic HA's linear structure and mechanical properties to be comparable to existing HA viscosupplements formulations. Wathier et al. have recently developed a biocompatible high molecular weight linear polyanion system that has been shown to lubricate cartilage as effectively as native SF, has a chondroprotective effect, and can resist enzymatic degradation in vivo [35]. There has also been some interest in the HA localization at the cartilage mediated via HA-binding peptides. Faust et al. have developed an injectable peptide-polymer composed of collagen-binding peptides (COLBP) and HA-binding peptides (HABP) which have been shown to lubricate both native and damaged cartilage effectively [36, 37]. The peptide polymers were shown to enhance the localization of both endogenous and injected HA at the surface of the cartilage, increasing the residence time and enabling a more efficient strategy to lubricate damaged cartilage.

8.6 Conclusion

Viscosupplementation with HA is used to treat mild to moderate OA and aims to restore the viscosity and lubricating ability of SF. While it has been characterized by the FDA as a class III medical device with a purely mechanical effect, HA viscosupplementation has been shown to have certain biological effects in the joint. The effect is mediated by the length of the HA in the joint. Low molecular weight fragments (less than 10 kDa) can trigger the release of pro-inflammatory cytokines, whereas high molecular weight forms (greater than 500 kDa) of HA bind to surface specific cellular receptors in chondrocytes and synoviocytes, curbing pro-inflammatory pathways and even promoting the synthesis of endogenous HA within the joint.

Despite being a mainstay of arthritis treatment for decades, the clinical efficacy and potential mechanism of action of viscosupplementation are still controversial. This has led prominent orthopedic associations like the American Academy of Orthopaedic Surgeons (AAOS), American Orthopaedic Society for Sports Medicine (AOSSM), and others to offer conflicting guidelines and recommendations to clinicians about the use of intra-articular HA injections.

The mechanical action of HA injections can be characterized by evaluating parameters such as the viscosity and the coefficient of friction. Osteoarthritic SF exhibits lower concentration of HA, which manifests in a loss in the viscosity. Supplementing SF with intra-articular injections restores the concentration of HA, increasing the viscosity and providing a low friction interface during joint articulation. Owing to differences in HA source, size, and chemical modification, the viscosity of commercial HA viscosupplements can vary widely, leading to a wide range of clinical outcomes in patients.

While HA products are typically characterized by MW and viscosity, the most important feature for cartilage protection is lubricating ability. HA's act as viscous lubricants lowering friction coefficients via a well-established Stribeck mechanism. This lubrication is influenced both by the native viscosity of HA solutions as well as their interaction with the cartilage surface and other lubricants such as lubricin. Of all the physical characteristics measured, the ability to lubricate cartilage correlates most directly with clinical outcomes.

These results indicate that experimental methods that consolidate viscosity and friction measurements of HA in the presence of cartilage may be more relevant to study the mechanical efficacy of this therapy than conventional methods. This understanding can inform the design of new and more effective HA formulations as well as synthetic HA mimics that effectively lubricate cartilage.

> **Take-Home Messages**
> - Intra-articular HA injections have been a mainstay in osteoarthritis therapy for decades, but the efficacy varies greatly between patients and by formulation.
> - Biological effects of HA depend on molecular weight, with macromolecular HA (greater than 10 kDa) exhibiting anti-inflammatory properties and HA fragments (less than 10 kDa) inducing pro-inflammatory responses.
> - The lubricating ability of HA arises from its viscosity, both of which vary greatly with composition and molecular weight.
> - Clinical outcomes of HA injections correlate highly with their ability to lubricate cartilage.
> - Interaction of HA with surfaces changes its effective viscosity and associated lubricating ability.

References

1. Fallacara A, Baldini E, Manfredini S, Vertuani S. Hyaluronic acid in the third millennium. Polymers. 2018;10(7):701.
2. Gupta R, Lall R, Srivastava A, Sinha A. Hyaluronic acid: molecular mechanisms and therapeutic trajectory. Front Vet Sci. 2019;6:192.
3. Dicker K, Gurski L, Pradhan-Bhatt S, Witt R, Farach-Carson M, Jia X. Hyaluronan: a simple polysaccharide with diverse biological functions. Acta Biomater. 2014;10(4):1558–70.
4. Nicholls M, Manjoo A, Shaw P, Niazi F, Rosen J. A comparison between rheological properties of intra-articular hyaluronic acid preparations and reported human synovial fluid. Adv Ther. 2018;35(4):523–30.
5. Bellamy N, Campbell J, Welch V, Gee T, Bourne R, Wells G. Viscosupplementation for the treatment of osteoarthritis of the knee. Cochrane Database Syst Rev. 2006;(2):CD005321.
6. Moskowitz R. Hyaluronic acid supplementation. Curr Rheumatol Rep. 2000;2(6):466–71.
7. Hochberg M, Altman R, April K, Benkhalti M, Guyatt G, McGowan J, et al. American College of Rheumatology 2012 recommendations for the use of nonpharmacologic and pharmacologic therapies in osteoarthritis of the hand, hip, and knee. Arthritis Care Res. 2012;64(4):465–74.
8. Trojian T, Concoff A, Joy S, Hatzenbuehler J, Saulsberry W, Coleman C. AMSSM scientific statement concerning viscosupplementation injections for knee osteoarthritis: importance for individual patient outcomes. Br J Sports Med. 2016;50(2):84–92.
9. Jevsevar D. Treatment of osteoarthritis of the knee: evidence-based guideline, 2nd edition. J Am Acad Orthop Surg. 2013;21(9):571–6.
10. Viscosupplementation Market Size, trends, Outlook 2020–25 [Internet]. Mordorintelligence.com. 2020 [cited 22 December 2020].
11. Ghosh P, Guidolin D. Potential mechanism of action of intra-articular hyaluronan therapy in osteoarthritis: are the effects molecular weight dependent? Semin Arthritis Rheum. 2002;32(1):10–37.
12. Altman R, Bedi A, Manjoo A, Niazi F, Shaw P, Mease P. Anti-inflammatory effects of intra-articular hyaluronic acid: a systematic review. CARTILAGE. 2018;10(1):43–52.
13. Dunn S, Kolomytkin O, Waddell D, Marino A. Hyaluronan-binding receptors: possible involvement in osteoarthritis. Mod Rheumatol. 2009;19(2):151–5.
14. Hashizume M, Koike N, Yoshida H, Suzuki M, Mihara M. High molecular weight hyaluronic acid relieved joint pain and prevented the progression of cartilage degeneration in a rabbit osteoarthritis model after onset of arthritis. Mod Rheumatol. 2010;20(5):432–8.
15. Kikuchi T, Yamada H, Shimmei M. Effect of high molecular weight hyaluronan on cartilage degeneration in a rabbit model of osteoarthritis. Osteoarthr Cartil. 1996;4(2):99–110.
16. Wang C, Lin Y, Chiang B, Lin Y, Hou S. High molecular weight hyaluronic acid down-regulates the gene expression of osteoarthritis-associated cytokines and enzymes in fibroblast-like synoviocytes from patients with early osteoarthritis. Osteoarthr Cartil. 2006;14(12):1237–47.
17. Campo G, Avenoso A, D'Ascola A, Scuruchi M, Prestipino V, Nastasi G, et al. The inhibition of hyaluronan degradation reduced pro-inflammatory cytokines in mouse synovial fibroblasts subjected to collagen-induced arthritis. J Cell Biochem. 2012;113(6):1852–67.
18. Kawana H, Karaki H, Higashi M, Miyazaki M, Hilberg F, Kitagawa M, et al. CD44 suppresses TLR-mediated inflammation. J Immunol. 2008;180(6):4235–45.
19. Campo G, Avenoso A, Nastasi G, Micali A, Prestipino V, Vaccaro M, et al. Hyaluronan reduces inflammation in experimental arthritis by modulating TLR-2 and TLR-4 cartilage expression. Biochim Biophys Acta Mol basis Dis. 2011;1812(9):1170–81.
20. McCourt P, Ek B, Forsberg N, Gustafson S. Intercellular adhesion molecule-1 is a cell surface receptor for hyaluronan. J Biol Chem. 1994;269(48):30081–4.

21. Noble P, McKee C, Cowman M, Shin H. Hyaluronan fragments activate an NF-kappa B/I-kappa B alpha autoregulatory loop in murine macrophages. J Exp Med. 1996;183(5):2373–8.
22. Yasuda T. Hyaluronan inhibits cytokine production by lipopolysaccharide-stimulated U937 macrophages through down-regulation of NF-κB via ICAM-1. Inflamm Res. 2007;56(6):246–53.
23. Balazs E, Denlinger J. Sodium hyaluronate and joint function. J Equine Vet. 1985;5(4):217–28.
24. Braithwaite G, Daley M, Toledo-Velasquez D. Rheological and molecular weight comparisons of approved hyaluronic acid products—preliminary standards for establishing class III medical device equivalence. J Biomater Sci Polym Ed. 2015;27(3): 235–46.
25. Bonnevie E, Galesso D, Secchieri C, Bonassar L. Frictional characterization of injectable hyaluronic acids is more predictive of clinical outcomes than traditional rheological or viscoelastic characterization. PLoS One. 2019;14(5):e0216702.
26. Bonnevie E, Bonassar L. A Century of Cartilage Tribology Research Is Informing Lubrication Therapies. J Biomech Eng. 2020;142(3).
27. Gleghorn J, Bonassar L. Lubrication mode analysis of articular cartilage using Stribeck surfaces. J Biomech. 2008;41(9):1910–8.
28. Ateshian G. The role of interstitial fluid pressurization in articular cartilage lubrication. J Biomech. 2009;42(9):1163–76.
29. Moore A, Burris D. Tribological rehydration of cartilage and its potential role in preserving joint health. Osteoarthr Cartil. 2017;25(1):99–107.
30. Cook S, Bonassar L. Interaction with cartilage increases the viscosity of hyaluronic acid solutions. ACS Biomater Sci Eng. 2020;6(5):2787–95.
31. Bonnevie E, Galesso D, Secchieri C, Cohen I, Bonassar L. Elastoviscous transitions of articular cartilage reveal a mechanism of synergy between lubricin and hyaluronic acid. PLoS One. 2015;10(11):e0143415.
32. Feeney E, Peal B, Inglis J, Su J, Nixon A, Bonassar L, et al. Temporal changes in synovial fluid composition and elastoviscous lubrication in the equine carpal fracture model. J Orthop Res. 2019;37(5):1071–9.
33. Irwin R, Feeney E, Galesso D, Secchieri C, Ramonda R, Cohen I, et al. Distinct tribological phenotypes of arthritic synovial fluid reveal differences in viscosupplementation efficacy. Osteoarthr Cartil. 2019;27:S165–6.
34. Brown T, Laurent U. Fraser. Turnover of hyaluronan in synovial joints: elimination of labelled hyaluronan from the knee joint of the rabbit. Exp Physiol. 1991;76(1):125–34.
35. Wathier M, Lakin B, Cooper B, Bansal P, Bendele A, Entezari V, et al. A synthetic polymeric biolubricant imparts chondroprotection in a rat meniscal tear model. Biomaterials. 2018;182:13–20.
36. Faust H, Sommerfeld S, Rathod S, Rittenbach A, Ray Banerjee S, Tsui B, et al. A hyaluronic acid binding peptide-polymer system for treating osteoarthritis. Biomaterials. 2018;183:93–101.
37. Singh A, Corvelli M, Unterman S, Wepasnick K, McDonnell P, Elisseeff J. Enhanced lubrication on tissue and biomaterial surfaces through peptide-mediated binding of hyaluronic acid. Nat Mater. 2014;13(10):988–95.

Cytokines, Chemokines, Alpha-2-Macroglobulin, Growth Factors

Claire D. Eliasberg and Scott A. Rodeo

9.1 Cytokines

9.1.1 Introduction

Cytokines are a unique subset of small proteins that play an important role in cell signaling. Cytokines can be secreted from a wide variety of cell types. They are often produced from immune cells, including leukocytes that stimulate immune responses and phagocyte activation, but they can be secreted from endothelial cells, fibroblasts, and stromal cells as well [1, 2].

Because cytokines have the ability to modulate immune responses, they have been identified as a promising target of study in the field of biologics. However, protein therapeutics are complex and present several challenges. There can be a great amount of heterogeneity in cytokines due to variations in amino acid sequence, as well as differences in glycosylation, protein folding, and protein-protein interactions [1]. Additionally, many cytokines have very short half-lives in plasma, which makes their direct administration challenging [1].

9.1.2 What Is Currently Available Clinically?

While there are over 130 known cytokines, few are available for human therapy at this time [1]. To our knowledge, there is no individual cytokine specifically approved for the treatment of orthopaedic conditions at this time (Table 9.1). However, platelet-rich plasma (PRP) preparations, while very heterogeneous in their compositions, are thought to contain several growth factors and cytokines. PRP is widely used clinically, with over 20 commercial systems currently available [2]. However, further research is necessary to better elucidate the specific cytokines that are both present in PRP preparations and that are biologically relevant in the healing response once administered. PRP is discussed in further detail in the following chapter of this text (Chap. 9). Human amniotic membranes are another commercially available technology, serving as a source of several biologically active molecules including anti-inflammatory cytokines [3]. Amniotic and placenta tissue will also be discussed in further detail in a subsequent chapter (Chap. 10).

9.1.3 What Has Shown Promise in Preclinical Studies?

Although there are no purified cytokines commercially available in the United States, several cytokines have demonstrated promise for the

C. D. Eliasberg · S. A. Rodeo (✉)
Department of Orthopaedic Surgery, Hospital for Special Surgery, New York, NY, USA
e-mail: eliasbergc@hss.edu; rodeos@hss.edu

G. Filardo et al. (eds.), *Orthobiologics*, https://doi.org/10.1007/978-3-030-84744-9_9

Table 9.1 Cytokines, chemokines, A2M, and growth factors approved for clinical use

Cytokines			
No individual cytokines currently available			
Chemokines			
No individual chemokines currently available			
Alpha-2-macroglobulin[a]			
Clinical trial name		*ClinicalTrials.gov identifier*	*Aims*
Reduction of pro-inflammatory synovial fluid biomarkers in osteoarthritis of the knee with alpha-2 macroglobulin		NCT03656575	To assess the ability of A2M to reduce the level of pro-inflammatory biomarkers in knee osteoarthritis
Injection of an autologous A2M concentrate alleviates back pain in FAC-positive patients		NCT03307876	To assess the ability of A2M to alleviate back pain in patients with low back pain from degenerative disc disease
Growth factors			
Name	*Formulations*	*Mechanism*	*Indications*
BMP-2	rh-BMP2, Infuse®	Promotes differentiation of MSCs to osteoblasts and chondrocytes	Anterior lumbar interbody spinal fusion Treatment of acute and open fractures of the tibial shaft that have been stabilized with intramedullary fixation Treatment of bone defects in oral maxillofacial procedures
BMP-7[b]	rh-BMP7, OP-1	Promotes differentiation of MSCs to osteoblasts and chondrocytes	Treatment of patients undergoing posterolateral lumbar fusion who cannot undergo autologous bone grafting and have compromising comorbidities
bFGF[c]	rh-FGF2, rh-bFGF, trafermin, Fiblast®	Stimulates angiogenesis and differentiation of MSCs, chondrocytes, myoblasts, and osteoblasts	Topical treatment of chronic wounds, pressure sores, and skin ulcers
PDGF	rhPDGF-BB, AUGMENT®, Regranex®	Recruits inflammatory cells, promotes proliferation and differentiation of MSCs, and stimulates angiogenesis	Alternative to autograft in fusion procedures of the ankle, hindfoot, and/or calcaneocuboid joints Topical treatment of lower extremity diabetic ulcers

A2M alpha-2-macroglobulin; *FAC* fibronectin-aggrecan complex; *BMP-2* bone morphogenetic protein 2; *rh-BMP2* recombinant human bone morphogenetic protein 2; *BMP-7* bone morphogenetic protein 7; *rh-BMP7* recombinant human bone morphogenetic protein 7; *bFGF* basic fibroblast growth factor; *rh-FGF2* recombinant human fibroblast growth factor 2; *rh-bFGF* recombinant human basic fibroblast growth factor; *PDGF* platelet-derived growth factor; *rhPDGF-BB* recombinant human platelet-derived growth factor BB homodimer; *MSCs* mesenchymal stromal cells
[a]Currently under investigation in clinical trials
[b]Has US FDA approval only as a humanitarian device exemption for patients with posterolateral lumbar pseudarthrosis who cannot undergo autologous bone grafting and have compromising comorbidities, but is not currently being manufactured
[c]Only approved for topical use in China and Japan—not commercially available in the United States

treatment of orthopaedic conditions in preclinical studies. Interleukin-1β (IL-1β) is a polypeptide which can be produced by inflammatory cells such as macrophages and neutrophils or by fibroblasts. It is often secreted as a response to trauma or injury and is a prototypical pro-inflammatory mediator. IL-1β has been shown to participate in the regulation of synthesis and degradation of collagen by fibroblasts and is therefore important in the process of tendon healing [4]. Koshima et al. demonstrated that IL-1β was produced in the torn rotator cuff tendon in a rabbit animal model, and Manning et al. demonstrated a dramatic upregulation of IL-1β in a canine flexor tendon transection model [5, 6]. Additionally, in vitro studies have demonstrated that IL-1β may have catabolic effects on extrinsic tendon fibroblasts [4]. Given this preclinical evidence that IL-1β likely plays an important role in the initial stages of tendon catabolism, targeted treatment strategies involving an anti-IL-1β therapeutic could hold promise in the setting of tendon repair or reconstruction procedures in orthopaedic surgery [4].

Interleukin-6 (IL-6) has been another cytokine of interest in the orthopaedic community. IL-6 is one of the several cytokines which has been shown to demonstrate pro-inflammatory actions in the tissue healing process; however, it has also been found to induce collagen production in tendon and to mediate anti-inflammatory effects as well [7]. Several studies have identified higher levels of IL-6 expression in torn rotator cuff tendons compared to non-injured tissues [8, 9]. While the results of these studies could indicate that IL-6 has a pro-inflammatory effect, a study by Lin et al. found that IL-6 knockout mice demonstrated inferior histologic and biomechanical properties in a mouse model of patellar tendinopathy [10]. This suggests that IL-6 also has an important role in the process of tendon repair. However, John et al. performed a study evaluating the treatment of cultured human tenocytes with several cytokines including IL-6 and found no significant effect [11]. Therefore, while IL-6 appears to be an important target in the tendon healing process, its specific role in the clinical context remains unclear.

Interleukin-10 (IL-10) is another cytokine which has been studied extensively in the preclinical orthopaedic literature. Unlike IL-6, IL-10 functions as an anti-inflammatory cytokine. It is secreted from immune cells including macrophages, lymphocytes, and dendritic cells but may also contribute to connective tissue cell regulation for other cell types including fibroblasts and chondrocytes [11]. Similar to the results found for IL-6, John et al. found that treatment of cultured human tenocytes with IL-10 did not have a significant effect on cytokine expression. Ackermann et al. found that IL-10 remained significantly elevated in human patients 2 weeks after undergoing Achilles tendon repair [7]. In a rat model of Achilles tendon transection and repair, Sugg et al. found that IL-10 was also upregulated post-operatively, but not until the 28-day post-operative time point, which suggests expression during the later, resolving phase of the acute postsurgical inflammatory response [12]. The preclinical work examining both IL-10 and IL-6 indicates that while these are important cytokines, their downstream effects are complex—likely involving a combination of both pro- and anti-inflammatory effects. Additionally, the changes seen in preclinical models may not align in scope or time course with what is found in humans clinically.

Recently, interleukin-17A (IL-17A) has garnered attention in the orthopaedic community for the potential treatment of tendinopathy. An IL-17A inhibitor, secukinumab, is currently commercially available under the brand name Cosentyx and produced by Novartis Pharmaceuticals Corporation (Basel, Switzerland). Millar et al. found that IL-17A-treated tenocytes obtained from human hamstring tendon exhibited increased production of pro-inflammatory cytokines, altered matrix regulation, and increased production of type III collagen, suggesting that IL-17A may have potential as a therapeutic target for tendinopathic conditions [13]. However, although secukinumab currently has FDA approval for the treatment of plaque psoriasis, psoriatic arthritis, and ankylosing spondylitis, it is not yet approved for injectable use for tendon disorders.

Finally, tumor necrosis factor alpha (TNFα) has been another cytokine of interest, but preclinical studies have demonstrated some contradictory results. While TNFα has been studied extensively in inflammatory arthropathies, it has also been shown to cause inflammation in some soft tissue disorders as well. TNFα has been localized to tenocytes in human Achilles tendon and has been measured in tendinopathic tissue [14]. The effect of TNFα has been studied in cultured equine and human tenocytes. Hosaka et al. found that treatment of equine tenocytes with TNFα resulted in increased collagen synthesis and decreased pro-matrix metalloproteinase 13 (pro-MMP-13) production [11, 15]. John et al. found that the treatment of cultured human tenocytes with TNFα significantly increased matrix metalloproteinase 1 (MMP-1) and pro-inflammatory cytokine expression [11]. Finally, Gulotta et al. demonstrated that TNFα blockade improved biomechanical properties in a rat rotator cuff repair model at early time points but that these differences were not sustained through 8 weeks post-operatively [16]. Therefore, while TNFα does seem to play a role in cytokine expression and tendon healing, further research is needed to better elucidate its role.

9.1.4 Future Directions

While several cytokines have been identified in preclinical studies as promising targets for biologic therapeutic strategies, to date none are readily available for clinical use. The field of biologics is continuing to evolve, and further research is necessary to better delineate cytokines of interest and methods for their delivery. As cytokine therapies continue to expand over the next decade, key areas of focus must include determining how to balance cytokine efficacy with their adverse effects, carefully defining the dose-response effect, limiting the heterogeneity of the formulations produced, and determining an effective method of delivery for these small proteins [1].

9.2 Chemokines

9.2.1 Introduction

Chemokines are small molecule signaling proteins which behave as regulators for leukocytes and lymphoid tissues. They have an important role in infectious, inflammatory, allergic, and autoimmune responses, as well as angiogenesis, haematopoiesis, and tumor growth [1]. Chemokines represent a subfamily of class A G protein-coupled receptors (GPCRs) and can be classified as a subgroup of cytokines. Their name derives from their ability to induce chemotaxis in target cells. These chemokine ligands act as travel signals to guide the migration of receptor-bearing cells and induce conforma-

tional changes in the receptors that trigger intracellular pathways [17].

9.2.2 What Is Currently Available Clinically?

While chemokines have demonstrated potential as drug targets for other disease processes, such as HIV and various malignancies, to our knowledge there are currently no chemokines commercially available for clinical use in musculoskeletal pathology (Table 9.1) [1].

9.2.3 What Has Shown Promise in Preclinical Studies?

Certain chemokines are thought to play a role in modulating chondrocyte metabolism. Thus, chemokines may have potential as therapeutic targets for osteoarthritis (OA). C-C motif chemokine ligand 2 (CCL2) has been shown to increase MMP3 expression in human cartilage in vitro, likely due to a catabolic effect [18]. The C-X-C motif chemokine ligand 12 and C-X-C chemokine receptor 4 (CXCL12/CXCR4) signaling pathway has also been implicated in cartilage degradation and tissue repair. Using a guinea pig OA model, Wei et al. found that signal blockade of the stromal cell-derived factor 1 (SDF-1)/CXCR4 signaling pathway may disrupt these catabolic processes and, therefore, could potentially attenuate cartilage degeneration [19]. Chen et al. found that this same SDF-1/CXCR4 pathway may promote IL-6 production in human synovial fibroblasts, which could play an important role in both OA and rheumatoid arthritis [20].

9.2.4 Future Directions

While individual chemokines are not currently readily available for clinical use in the field of orthopaedics, the most promising disease processes for chemokine therapy are likely OA and rheumatoid arthritis. Future directions for chemokine research will involve better elucidating the roles of chemokines in OA—in particular determining which are involved in the pathogenesis versus those which are merely biomarkers—in order to help develop novel drug and biologic targets [21]. Additionally, chemokines may play a role in arthritis-associated pain as part of the peripheral and central nervous system pathways [21]. Therefore, treatments involving chemokines as a potential analgesic pathway may also hold some promise.

9.3 Alpha-2-Macroglobulin

9.3.1 Introduction

Alpha-2-macroglobulin (A2M) is a large plasma protein synthesized by the liver, fibroblasts, and macrophages. It serves as a broad-spectrum MMP inhibitor, and it has been identified as a promising bioinhibitor of catabolic enzymes [22]. Because increased MMP activity has been associated with conditions involving degenerative tendinopathy, such as rotator cuff disease, Achilles tendinopathy, and patellar tendinosis, A2M has significant potential in the field of sports medicine [23]. Early research has also explored the anti-inflammatory effects of A2M in OA.

9.3.2 What Is Currently Available Clinically?

To date, there are two clinical trials underway which involve the use of A2M for orthopaedic purposes (Table 9.1). One clinical trial plans to assess the ability of A2M to reduce the level of pro-inflammatory biomarkers in knee OA and is currently actively recruiting patients [24]. Another study aims to assess the ability of A2M to alleviate back pain in patients with degenerative disc disease and has completed enrollment at this time [25]. A product to produce autologous A2M formulations is being commercially developed through the Cytonics Corporation (Jupiter, Florida, USA) [26].

9.3.3 What Has Shown Promise in Preclinical Studies?

While A2M has been less extensively studied than some other small molecule biologics, there has been some promising data in the preclinical literature. A2M has been studied in the context of post-traumatic arthritis in both mouse and rabbit animal models [27, 28]. Both studies found that there was less degeneration in the joints treated with A2M compared to controls, suggesting that A2M may attenuate cartilage damage in a post-traumatic setting [27, 28]. Similarly, Zhang et al. developed an A2M variant and also demonstrated that it may have chondroprotective effects after ACL transection in a rat knee model [22]. Bedi et al. examined the effects of A2M in tendon-to-bone healing and found that local A2M delivery was associated with some distinct histologic differences at the healing tendon-bone interface following rotator cuff repair in a rat model [23].

9.3.4 Future Directions

Following up on the results of the current clinical trials investigating the effects of A2M in knee OA and in back pain following degenerative disc disease will be very informative and guide the way for future studies. Additional clinical trials to assess the safety and efficacy of A2M will also be necessary if these trials are effective. Finally, engineered, recombinant A2M may be promising for future commercial use [22].

9.4 Growth Factors

9.4.1 Introduction

Growth factors are a broader class of signaling molecules which bind to specific target cell surface receptors to stimulate various cell functions, including cell growth, chemotaxis, and matrix synthesis. While the terms "growth factor" and "cytokine" are sometimes used interchangeably, there are some distinct differences between the two [2]. The category "growth factor" encompasses both proteins and steroid hormones which lead to proliferative effects, whereas "cytokines" are exclusively peptides. Conversely, cytokines may have a variety of downstream effects which are not just limited to cell growth.

Growth factors participate in the physiologic reaction to injury to promote the body's innate regenerative healing response [2]. Due to their ability to promote regeneration of soft tissues, growth factors have been identified as a potential source of biologic augmentation in musculoskeletal medicine. However, there are several issues with growth factors that limit their clinical use at this time. Because many growth factors are pleiotropic, they have more than one downstream effect and can simultaneously act on a wide array of cell types [29]. Additionally, many growth factors function only in conjunction with other proteins or signaling molecules. Thus, using growth factors in vitro or in vivo in isolation may not produce the desired effect [29].

9.4.2 What Is Currently Available Clinically?

To date, there are only a few isolated growth factors that are commercially available. Bone morphogenetic protein-2 (BMP-2) is a recombinant human BMP-2, which is approved as the product Infuse® (Medtronic, Minneapolis). It was initially approved for use in anterior lumbar interbody spinal fusion procedures in patients with lumbar degenerative disc disease for patients who may be at increased risk for developing nonunions [30]. Subsequently, rhBMP-2 was approved for the treatment of acute and open fractures of the tibial shaft that have been stabilized with intramedullary nail fixation after appropriate wound management and for the treatment of bone defects in oral maxillofacial procedures [31]. Bone morphogenetic protein-7 (BMP-7) is another recombinant human BMP, which was initially approved as osteogenic protein-1 (OP-1) Putty or OP-1 Implant. It was approved for posterolateral lumbar spinal fusion as an alternative or an adjunct to autograft and for tibial nonunions of at least 9 months [32]. However, despite these

relatively narrow indications and approvals for rhBMP-2 and rhBMP-7, it was estimated that up to 85% of their usage is off-label as of 2014 [1]. Additionally, after follow-up trials failed to demonstrate the non-inferiority of rhBMP-7 to iliac crest bone graft for spine fusion and after the OP-1 assets were sold, the sale of rhBMP-7 products was discontinued in 2014 [31].

Another growth factor, basic fibroblast growth factor (bFGF), has been shown to improve lower extremity wound healing [33]. bFGF, also known as FGF-2, has potential for the treatment of pressure ulcers, second-degree burns, and diabetic foot ulcers. To date, it has been approved for topical use in both China and Japan, but not yet in the United States [34]. Finally, platelet-derived growth factor (PDGF) was recently approved and is commercially available as a recombinant human PDGF BB homodimer (rhPDGF-BB) (AUGMENT Bone Graft®, Wright Medical, Memphis). It can be used as an alternative to autograft in fusion procedures of the ankle, hindfoot, and/or calcaneocuboid joints. Another formulation (Regranex® [becaplermin], Smith & Nephew, London) is available as a topical gel for patients with lower extremity diabetic ulcers (Table 9.1) [1].

Despite the significant interest in growth factors for clinical use, several adverse effects have been cited in the literature. Adverse events following the use of rhBMP-2 in spinal fusion procedures have included vertebral osteolysis, heterotopic bone formation, graft migration and subsidence, hematoma, infections, and development of antibodies to BMP-2 [1, 35, 36]. rhBMP-7 use resulted in some mild to moderate adverse events such as fever, oedema, hematoma, and low levels of anti-BMP-7 antibody formation following tibial nonunion treatment [32] and transient brachialgia and dysphagia in few patients undergoing anterior cervical spine fusion [37]. Documented adverse effects associated with becaplermin use have been rare; however, becaplermin does carry a theoretical risk of increasing cancer rates, although this has not been demonstrated in the literature [1]. It is important for physicians to be aware of the potential adverse events which may be associated with the use of these biologics in order to provide patient counseling and to weigh the risks and benefits of their use.

9.4.3 What Has Shown Promise in Preclinical Studies?

Several other growth factors have shown promise in preclinical studies. Transforming growth factor-beta1 (TGF-β1) is an important mediator of extracellular matrix degradation and repair and has been shown in several studies to play an important role in ligament healing [38–40]. Kondo et al. demonstrated improvement in rabbit ACL biomechanical and histologic properties after administration of TGF-β1 in a partial ACL injury model [39]. Xie et al. found that TGF-β1-induced injured human ACL fibroblasts express higher levels of MMPs than injured human MCL fibroblasts [38]. Spindler et al. conducted an in vitro study using sheep ACL and patellar tendon explants and found that TGF-β1 stimulates ACL cells, suggesting the potential to promote the initial healing of the ACL; however, TGF-β1 did not have the same effect on the patellar tendon, suggesting that growth factors may elicit different responses in various soft tissues.

Platelet-derived growth factor (PDGF) has also shown promise in preclinical orthopaedic studies for soft tissue healing. Kovacevic et al. demonstrated that rhPDGF-BB augmentation led to increased angiogenesis and cellular proliferation in early stages following rat rotator cuff repair; however, PDGF had detrimental effects on biomechanical properties at 4 weeks post-operatively [41]. PDGF has also been shown to stimulate fibroblast activity, which may be advantageous in ACL graft ligamentization and tunnel integration [42].

Vascular endothelial growth factor (VEGF) is an interesting growth factor which has garnered interest due to its ability to promote angiogenesis—and therefore promote healing—however, overexpression of VEGF has proven to be detrimental to soft tissue healing in several studies. Yoshikawa et al. demonstrated that in a sheep model of ACL reconstruction, VEGF-soaked tendons demonstrated increased angiogenesis at the time of harvest but that these grafts also had increased laxity and decreased stiffness compared to controls at 12 weeks post-operatively [43]. Additionally, Takayama et al. found that blocking VEGF had detrimental effects on graft maturation and biomechanical properties in a rat ACL reconstruction model but that overexpression may also have detrimental effects [44].

Two other growth factors of interest include insulin-like growth factor-1 (IGF-1) and epidermal growth factor (EGF). Lyras et al. demonstrated that IGF-1 overexpression may improve tendon healing in a rabbit model of patellar tendon defects [45]. EGF was found to have a stimulatory effect on tenoblast migration in an in vitro study of chicken long digital flexor tendons [46]. However, in the same study, IGF was found to be stimulatory at low concentrations but inhibitory at high concentrations, further highlighting the complex nature of growth factor physiology at various concentrations, locations, and across different tissue types and animal models. Additionally, a recent study by Ikeda et al. found that the introduction of the IGF-1 gene into human mesenchymal progenitor cells cultured in chondrogenic conditions led to improved chondrogenic differentiation capacity without stimulating a hypertrophic or osteogenic phenotype [47].

9.4.4 Future Directions

Growth factors offer a promising pathway for the study of biologics in orthopaedic surgery, but there are several issues which need to be optimized prior to more routine use of isolated growth factors. First, the safety of growth factors in general needs to be verified. BMPs have been a good example of growth factors which have pleiotropic effects, and controversy arose when several industry-sponsored publications seemed to underrepresent the incidence of complications and adverse events [48]. This will lead to increased scrutiny as future growth factors are examined. Other issues will include the logistics of delivering growth factors, as localized (non-systemic) and controlled-release growth factor administration could both help to minimize unwanted effects [29]. Finally, ultimately, use of growth factors in combination with cell-based therapies is likely necessary to provide the best possible outcomes for patients. Future research is needed to determine the safety and synergistic effects of the combination of growth factors and cell therapy as well as the optimal combinations of biologics for each orthopaedic condition.

9.5 Conclusions

Cytokines, chemokines, A2M, and growth factors constitute a unique set of small molecules which have promising potential in the field of biologics. To date, only a limited number of growth factors are available for clinical use, but preclinical studies have identified several cytokines, chemokines, and novel growth factors which may be utilized for the treatment of bone, cartilage, tendon, and ligament injuries in the future. A2M has also shown promise in preclinical studies for the treatment of OA and tendon-to-bone healing, and it is currently being studied in clinical trials. Further research is needed in the preclinical area to clarify the mechanisms by which these small molecules function and to better elucidate their effects. Additionally, the efficacy and safety of these potential treatments need to be verified in clinical trials. However, given the promising preclinical data published to date and the increasing scientific and popular interest in the field of biologics, this research will continue to be pertinent to the orthopaedic community.

Take-Home Messages
- Cytokines, chemokines, alpha-2-macroglobulin, and growth factors have been identified as promising targets to augment bone and soft tissue healing.
- Specific cytokines and chemokines have shown promise in preclinical studies, but to date, none are available for clinical use.
- Alpha-2-macroglobulin is a broad-spectrum matrix metalloproteinase inhibitor which may have potential in the treatment of arthritic conditions and degenerative tendinopathies.
- Growth factors have been studied extensively with regard to bone healing, but several preclinical studies have identified growth factors which may promote soft tissue regeneration as well.
- Future directions for small molecule biologic therapies include further elucidating their roles in tendon and cartilage pathogenesis, limiting formulation heterogeneity, and determining effective delivery methods.

References

1. Baldo BA. Side effects of cytokines approved for therapy. Drug Saf. 2014;37(11):921–43.
2. Murray IR, Safran MR, LaPrade RF. Biologics in orthopaedics. Bone Jt 360. 2018;7(6):2–8.
3. Riboh JC, Saltzman BM, Yanke AB, Cole BJ. Human amniotic membrane-derived products in sports medicine: basic science, early results, and potential clinical applications. Am J Sports Med. 2016;44(9):2425–34.
4. Tohyama H, Yasuda K, Uchida H, Nishihira J. The responses of extrinsic fibroblasts infiltrating the devitalised patellar tendon to IL-1β are different from those of normal tendon fibroblasts. J Bone Jt Surg Ser B. 2007;89(9):1261–7.
5. Koshima H, Kondo S, Mishima S, Choi H-R, Shimpo H, Sakai T, et al. Expression of interleukin-1beta, cyclooxygenase-2, and prostaglandin E2 in a rotator cuff tear in rabbits. J Orthop Res. 2007;25(1):92–7.
6. Manning CN, Havlioglu N, Knutsen E, Sakiyama-Elbert SE, Silva MJ, Thomopoulos S, et al. The early inflammatory response after flexor tendon healing: a gene expression and histological analysis. J Orthop Res. 2014;32(5):645–52.
7. Ackermann PW, Domeij-Arverud E, Leclerc P, Amoudrouz P, Nader GA. Anti-inflammatory cytokine profile in early human tendon repair. Knee Surgery Sport Traumatol Arthrosc. 2013;21(8):1801–6.
8. Millar NL, Wei AQ, Molloy TJ, Bonar F, Murrell GAC. Cytokines and apoptosis in supraspinatus tendinopathy. J Bone Jt Surg Ser B. 2009;91(3):417–24.
9. Nakama K, Gotoh M, Yamada T, Mitsui Y, Yasukawa H, Imaizumi T, et al. Interleukin-6-induced activation of signal transducer and activator of transcription-3

in ruptured rotator cuff tendon. J Int Med Res. 2006;34(6):624–31.

10. Lin TW, Cardenas L, Glaser DL, Soslowsky LJ. Tendon healing in interleukin-4 and interleukin-6 knockout mice. J Biomech. 2006;39(1):61–9.

11. John T, Lodka D, Kohl B, Ertel W, Jammrath J, Conrad C, et al. Effect of pro-inflammatory and immunoregulatory cytokines on human tenocytes. J Orthop Res. 2010;28(8):1071–7.

12. Sugg KB, Lubardic J, Gumucio JP, Mendias CL. Changes in macrophage phenotype and induction of epithelial-to-mesenchymal transition genes following acute Achilles tenotomy and repair. J Orthop Res. 2014;32(7):944–51.

13. Millar NL, Akbar M, Campbell AL, Reilly JH, Kerr SC, McLean M, et al. IL-17A mediates inflammatory and tissue remodelling events in early human tendinopathy. Sci Rep. 2016;6:27149.

14. Gaida JE, Bagge J, Purdam C, Cook J, Alfredson H, Forsgren S. Evidence of the TNF-α system in the human Achilles tendon: expression of TNF-α and TNF receptor at both protein and mRNA levels in the tenocytes. Cells Tissues Organs. 2012;196(4):339–52.

15. Hosaka YZ, Uratsuji T, Ueda H, Uehara M, Takehana K. Comparative study of the properties of tendinocytes derived from three different sites in the equine superficial digital flexor tendon. Biomed Res. 2010;31(1):35–44.

16. Gulotta LV, Kovacevic D, Cordasco F, Rodeo SA. Evaluation of tumor necrosis factor α blockade on early tendon-to-bone healing in a rat rotator cuff repair model. Arthrosc J Arthrosc Relat Surg. 2011;27(10):1351–7.

17. O'Hayre M, Salanga CL, Handel TM, Hamel DJ. Emerging concepts and approaches for chemokine-receptor drug discovery. Expert Opin Drug Discovery. 2010;5(11):1109–22.

18. Borzì RM, Mazzetti I, Cattini L, Uguccioni M, Baggiolini M, Facchini A. Human chondrocytes express functional chemokine receptors and release matrix-degrading enzymes in response to C-X-C and C-C chemokines. Arthritis Rheum. 2000;43(8):1734–41.

19. Wei F, Moore DC, Wei L, Li Y, Zhang G, Wei X, et al. Attenuation of osteoarthritis via blockade of the SDF-1/CXCR4 signaling pathway. Arthritis Res Ther. 2012;14(4):R177.

20. Te Chen H, Tsou HK, Hsu CJ, Tsai CH, Kao CH, Fong YC, et al. Stromal cell-derived factor-1/CXCR4 promotes IL-6 production in human synovial fibroblasts. J Cell Biochem. 2011;112(4):1219–27.

21. Scanzello CR. Chemokines and inflammation in osteoarthritis: insights from patients and animal models. J Orthop Res. 2017;35:735–9.

22. Zhang Y, Wei X, Browning S, Scuderi G, Hanna LS, Wei L. Targeted designed variants of alpha-2-macroglobulin (A2M) attenuate cartilage degeneration in a rat model of osteoarthritis induced by anterior cruciate ligament transection. Arthritis Res Ther. 2017;19(1):1–11.

23. Bedi A, Kovacevic D, Hettrich C, Gulotta LV, Ehteshami JR, Warren RF, et al. The effect of matrix metalloproteinase inhibition on tendon-to-bone healing in a rotator cuff repair model. J Shoulder Elb Surg. 2010;19(3):384–91.

24. Reduction of Pro-Inflammatory Synovial Fluid Biomarkers in Osteoarthritis of the Knee With Alpha-2 Macroglobulin. National Library of Medicine (US). 2018.

25. Injection of an Autologous A2M Concentrate Alleviates Back Pain in FAC-positive Patients. National Library of Medicine (US). 2017.

26. Cytonics.

27. Demirag B, Sarisozen B, Durak K, Faruk ÖF, Ozturk C. The effect of alpha-2 macroglobulin on the healing of ruptured anterior cruciate ligaments in rabbits. Connect Tissue Res. 2004;45(1):23–7.

28. Li S, Xiang C, Wei X, Sun X, Li R, Li P, et al. Early supplemental α2-macroglobulin attenuates cartilage and bone damage by inhibiting inflammation in collagen II-induced arthritis model. Int J Rheum Dis. 2019;22(4):654–65.

29. LaPrade RF, Geeslin AG, Murray IR, Musahl V, Zlotnicki JP, Petrigliano F, et al. Biologic treatments for sports injuries II think tank—current concepts, future research, and barriers to advancement, part 1. Am J Sports Med. 2016;44(12):3270–83.

30. Govender S, Csimma C, Genant HK, Valentin-Opran A, Amit Y, Arbel R, et al. Recombinant human bone morphogenetic protein-2 for treatment of open tibial fractures a prospective, controlled, randomized study of four hundred and fifty patients. J Bone Jt Surg Ser A. 2002;84(12):2123–34.

31. El Bialy I, Jiskoot W, Reza Nejadnik M. Formulation, delivery and stability of bone morphogenetic proteins for effective bone regeneration. Pharm Res. 2017;34(6):1152–70.

32. Friedlaender GE, Perry CR, Cole JD, Cook SD, Cierny G, Muschler GF, et al. Osteogenic protein-1 (bone morphogenetic protein-7) in the treatment of tibial nonunions. J Bone Joint Surg Am. 2001;83-A Suppl 1(Pt 2):S151–8.

33. Akita S, Akino K, Tanaka K, Anraku K, Hirano A. A basic fibroblast growth factor improves lower extremity wound healing with a porcine-derived skin substitute. J Trauma Inj Infect Crit Care. 2008;64(3):809–15.

34. Hui Q, Jin Z, Li X, Liu C, Wang X. FGF family: from drug development to clinical application. Int J Mol Sci. 2018;19(7):1875.

35. Cahill KS, Chi JH, Day A, Claus EB. Prevalence, complications, and hospital charges associated with use of bone-morphogenetic proteins in spinal fusion procedures. JAMA J Am Med Assoc. 2009;302(1):58–66.

36. Woo EJ. Recombinant human bone morphogenetic protein-2: adverse events reported to the manufacturer and user facility device experience database. Spine J. 2012;12(10):894–9.

37. Leach J, Bittar RG. BMP-7 (OP-1®) safety in anterior cervical fusion surgery. J Clin Neurosci. 2009;16(11):1417–20.

38. Xie J, Wang C, Huang D, Zhang Y, Xu J, Kolesnikov SS, et al. TGF-beta1 induces the different expressions of lysyl oxidases and matrix metalloproteinases in anterior cruciate ligament and medial collateral ligament fibroblasts after mechanical injury. J Biomech. 2013;46(5):890–8.

39. Kondo E, Yasuda K, Yamanaka M, Minami A, Tohyama H. Effects of administration of exogenous growth factors on biomechanical properties of the elongation-type anterior cruciate ligament injury with partial laceration. Am J Sports Med. 2005;33(2):188–96.

40. Spindler KP, Imro AK, Mayes CE, Davidson JM. Patellar tendon and anterior cruciate ligament have different mitogenic responses to platelet-derived growth factor and transforming growth factor? J Orthop Res. 1996;14(4):542–6.

41. Kovacevic D, Gulotta LV, Ying L, Ehteshami JR, Deng XH, Rodeo SA. rhPDGF-BB promotes early healing in a rat rotator cuff repair model. Clin Orthop Relat Res. 2015;473(5):1644–54.

42. Molloy T, Wang Y, Murrell GAC. The roles of growth factors in tendon and ligament healing. Sports Med. 2003;33(5):381–94.

43. Yoshikawa T, Tohyama H, Katsura T, Kondo E, Kotani Y, Matsumoto H, et al. Effects of local administration of vascular endothelial growth factor on mechanical characteristics of the semitendinosus tendon graft after anterior cruciate ligament reconstruction in sheep. Am J Sports Med. 2006;34(12):1918–25.

44. Takayama K, Kawakami Y, Mifune Y, Matsumoto T, Tang Y, Cummins JH, et al. The effect of blocking angiogenesis on anterior cruciate ligament healing following stem cell transplantation. Biomaterials. 2015;60:9–19.

45. Lyras DN, Kazakos K, Agrogiannis G, Verettas D, Kokka A, Kiziridis G, et al. Experimental study of tendon healing early phase: is IGF-1 expression influenced by platelet rich plasma gel? Orthop Traumatol Surg Res. 2010;96(4):381–7.

46. Jann HW, Stein LE, Slater DA. In vitro effects of epidermal growth factor or insulin-like growth factor on tenoblast migration on absorbable suture material. Vet Surg. 1999;28(4):268–78.

47. Ikeda Y, Sakaue M, Chijimatsu R, Hart DA, Otsubo H, Shimomura K, et al. IGF-1 gene transfer to human synovial MSCs promotes their chondrogenic differentiation potential without induction of the hypertrophic phenotype. Stem Cells Int. 2017;2017:5804147.

48. Carragee EJ, Mitsunaga KA, Hurwitz EL, Scuderi GJ. Retrograde ejaculation after anterior lumbar interbody fusion using rhBMP-2: a cohort controlled study. Spine J. 2011;11(6):511–6.

Platelet-Rich Plasma: Processing and Composition

10

Spencer M. Stein and Bert R. Mandelbaum

10.1 Introduction

The use of injectable biologic therapy in treating musculoskeletal ailments holds great potential for improving treatment options in sports medicine. The exciting benefits of orthobiologics are largely due to their promising ability to improve healing and recovery times in a minimally invasive manner.

One commonly studied and promising orthobiologic is platelet-rich plasma (PRP), a prepared autologous plasma that is rich in platelets. These platelets hold multiple growth factors and mediators, which can be greatly beneficial to healing or disrupted human tissue. The ability of these growth factors to modify the inflammatory response, augment the natural healing process, and possibly affect cell proliferation and differentiation provides promise in treating musculoskeletal disease [1, 2].

Normal human serum contains 150,000–450,000 platelets per microliter. While the original definition of PRP was "a platelet count above baseline," it is thought that PRP concentrations of at least 1,000,000 platelets per microliter, a three to six times increase of baseline, are required to improve the healing response in the musculoskel-

etal system [3, 4]. In fact, a systematic review of PRP studies showed the mean final platelet count in PRP preparations is 1,473,000 platelets per microliter [5]. However, the ideal platelet count has yet to be determined and likely depends on the clinical scenario.

To date, PRP as an injectable form (Fig. 10.1) has been used and studied in a variety of musculoskeletal disorders. This includes lateral epicondylitis, patella tendinopathy, rotator cuff tendinopathy, ankle sprains and plantar fasciitis. It has also been investigated in the treatment of osteoarthritis of the knee and hip and as an augment to surgical reconstruction of the anterior cruciate ligament (ACL) and in rotator cuff repair [6, 7].

The optimal composition of PRP is yet to be determined and the ideal preparation may depend on the clinical indication. While multiple PRP preparation systems are FDA approved, there are no clear guidelines regulating the required composition of PRP. In fact, PRP can vary greatly in platelet, leukocyte, and growth factor concentration depending on patient factors and protocol used [5, 8–10].

Variables in preparation of PRP include the amount of blood obtained, the process of concentration, the use of anticoagulant, and the use of activation substance. The final product can differ in platelet cell count, leukocyte count, and composition of growth factors, and while different PRP processing systems exist, most published reports do not define the composition of their

S. M. Stein (✉)
NYU Langone Health, New York, NY, USA

B. R. Mandelbaum
Cedars-Sinai Kerlan-Jobe Institute,
Santa Monica, CA, USA

© ISAKOS 2022
G. Filardo et al. (eds.), *Orthobiologics*, https://doi.org/10.1007/978-3-030-84744-9_10

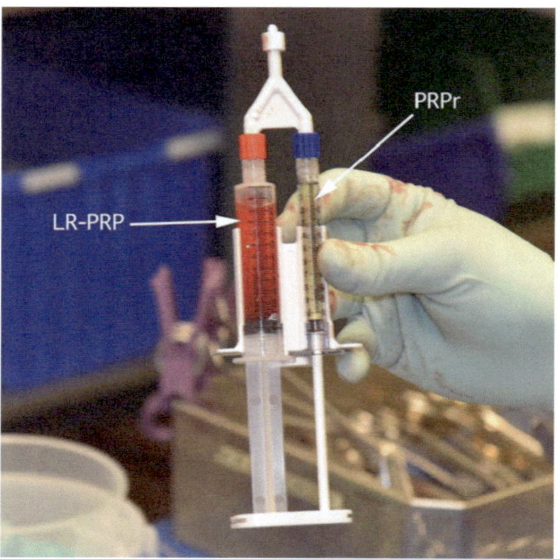

Fig. 10.1 Photograph of a double-syringe system for injection. The device provides a homologous mix of two solutions—LR-PRP (left) and PRP releasate (PRPr) (right). (Reprinted with permission from Hussain ZB, Chahla J, LaPrade RF, Mandelbaum BR. Orthobiologics: Today and Tomorrow. In: Farr J, Gomoll AH, editors. Cartilage Restoration: Practical Clinical Applications. Cham: Springer International Publishing; 2018. p. 131–42)

PRP product, making direct comparisons difficult [5, 8]. Here we will review the variables in processing and formulations, as well as available outcomes of PRP.

10.2 Processing

10.2.1 Whole Blood

PRP is a product of autologous whole blood that can be rapidly prepared in a one stage setting. Preparation of PRP begins with whole blood which can be obtained by peripheral venipuncture. There is variability in the quantity of whole blood required based on the preparation protocols; however, the reported mean and median volume required are 44 mL and 52 mL, respectively (range: 11–54 mL) (Table 10.1) [11]. Additionally, the composition of whole blood obtained can have an effect on the final composi-

tion and efficacy of the PRP end product [12]. Therefore, practitioners may consider screening patients for cytopenias, especially thrombocytopenia when preparing for PRP processing.

10.2.2 Anticoagulant

Once obtained, whole blood is typically mixed with an anticoagulant to prevent premature clotting and to facilitate processing. Some of the more common anticoagulants used are acid-citrate-dextrose, sodium citrate, and ethylene diamine tetra-acetic acid [5, 11]. It appears that these anticoagulants are appropriate during PRP preparation, although data is limited with regard to anticoagulants and their effect on the final PRP end product or on its clinical outcomes. However, acid-citrate-dextrose and citrate-theophylline-adenosine-dipyridamole have been shown to be superior to heparin in maintaining platelet structure, releasing transforming growth factor beta and proliferating human marrow stromal cells in vitro [13].

10.2.3 Isolation and Concentration Method

Isolation method is probably the most important step in PRP processing. Mechanical forces not only concentrate platelets, but can activate platelets and therefore cause the release of growth factor-rich alpha granules [14]. There is considerable variability in centrifugation protocols: it can be one- or two-step; and the speed of centrifugation, type of collection tube system, and processing can vary. Typically, however, the first spin aims to separate out red blood cells and platelet-poor plasma from the "buffy coat," a layer of leukocytes and platelets (Fig. 10.2). The second spin can separate out leukocytes if desired [8].

According to Chahla et al., who reviewed published PRP preparation protocols, the median rate for the first spin was 1500 revolutions per minute (rpm) (range: 120–5800) with

Table 10.1 Characteristics of PRP preparations from different commercially available systems

System	Company	Blood volume required, mL	Concentrated volume produced, mL	Processing time, min	Increase in [platelets], times baseline	Platelet capture efficiency, % yield
Leukocyte-rich PRP						
Angel	Arthrex (Florida, USA)	52	1–20	17	10	56–75%
GenesisCS	EmCyte (Florida, USA)	54	6	10	4–7	61% +/− 12%
GPS III	Biomet (Zimmer Biomet, Indiana, USA)	54	6	15	3–10	70% +/− 30%
Magellan	Isto Biologics/ Arteriocyte (now known as Isto Biologics, Massachusetts, USA)	52	3.5–7	17	3–15	86% +/− 41%
SmartPreP 2	Harvest (now known as Terumo BCT, Colorado, USA)	54	7	14	5–9	94% +/− 12%
Leukocyte-poor PRP						
Autologous Conditioned Plasma (ACP)	Arthrex	11	4	5	1.3	48% +/− 7%
Cascade	MTF (New Jersey, USA)	18	7.5	6	1.6	68% +/− 4%
Clear PRP	Harvest	54	6.5	18	3–6	62% +/− 5%
Pure PRP	EmCyte	50	6.5	8.5	4–7	76% +/− 4%

Abbreviations: *PPP* platelet-poor plasma; *PRP* platelet-rich plasma
Plus minus sign signifies reported variance of platelet capture efficiency
Reprinted with permission from Le ADK, Enweze L, DeBaun MR, Dragoo JL. Platelet-Rich Plasma. Clin Sports Med. 2019;38(1):17–44

a median spin time of 14 min (range: 3–15 min), and the median rate of the second spin was 3300 rpm (range: 200–4500) for a median time of 10 min (range: 2–25 min) [5]. Sabarish et al. [15] evaluated the effect of three published spin rates and times on final platelet concentration and found an initial centrifugation of 1000 rpm for 4 min followed by 900 rpm for 9 min, as compared to protocols with faster rpm for longer times, produced the highest concentration of platelets. However, the ideal centrifugation time and spin rate are debatable. A systematic review that aimed to report the ideal PRP preparation protocol showed that a larger initial blood volume using a higher spin force was correlated with increased platelet count, but a longer spin time did not correlate with higher platelet count [8]. Additionally, the ultimate force on cells is dependent on not only centrifugation speed (rpm) but also on rotor diameter, making it difficult to directly compare centrifugation protocols reported in rpm across different systems [12].

Apheresis is a similar PRP processing technique that uses centrifugation to separate whole blood products into components. However, the process of apheresis allows the non-platelet blood products to be returned to the donor. Apheresis has been shown to be an effective method, with a similar ability to concentrate platelets, maintain platelet viability, and produce a product rich in chondrogenic growth factors [16–19].

While the available literature has helped demonstrate the most effective way to increase platelet concentration, the ideal platelet count has not yet been determined. In fact, more is not necessarily better. In a rat model that assessed intestinal healing, PRP platelet counts greater than

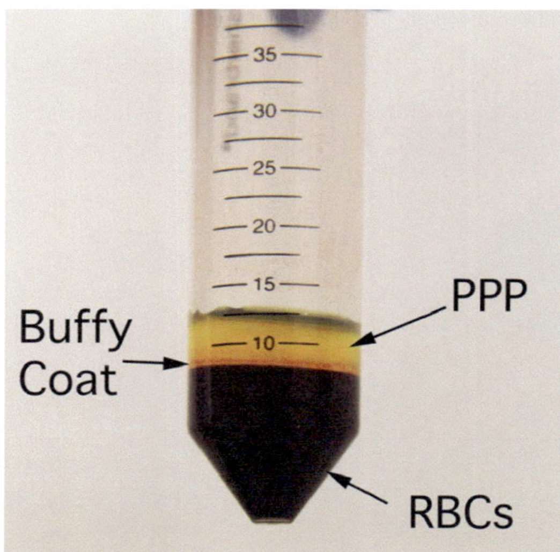

Fig. 10.2 Photograph of three distinct layers of cellular material after the first centrifugation. The top layer is platelet-poor plasma (PPP), beneath this layer is the buffy coat where most platelets lie, and at the bottom are the red blood cells (RBCs). (Reprinted with permission from Hussain ZB, Chahla J, LaPrade RF, Mandelbaum BR. Orthobiologics: Today and Tomorrow. In: Farr J, Gomoll AH, editors. Cartilage Restoration: Practical Clinical Applications. Cham: Springer International Publishing; 2018. p. 131–42)

5,000,000 per microliter had a detrimental effect on healing and produced cell death [20]. The most desirable platelet concentration for healing likely depends on the tissue being treated. For example, Fleming et al. [21] evaluated the effect of PRP supplementation on graft healing following ACL reconstruction in minipigs using either 1×, 3×, or 5× PRP concentrations. Only the 1× platelet concentration improved healing over traditional ACL reconstruction. Similarly, Yoshida et al. [22] found that porcine ACL fibroblasts suspended in 1× PRP concentration had improved type I and type III collagen gene expression, apoptosis prevention, and cell metabolism as compared to 5× PRP concentrations. Weibrich et al. [23] reported an intermediate concentration (2–6×) resulted in optimal peri-implant bone regeneration in rabbits. Taken together, this basic science data suggests that the optimal PRP preparation is dependent on the tissue and pathology being treated.

10.2.4 Activation

The key to the healing potential of PRP is thought to be the multiple growth factors held in the platelet's alpha granules. Known growth factors harnessed by platelets include platelet-derived growth factor (PDGF), beta transforming growth factor (TGF-b1 and TGF-b2), vascular endothelial growth factor (VEGF), basic fibroblastic growth factor (bFGF), epidemic growth factor (EGF), and insulin-like growth factor (IGF-1, IGF-2, IGF-3) among others [24]. As discussed, activation of PRP begins with processing through mechanical forces; however activating agents can be added to enhance the release of growth factors.

The most commonly used activating agents in clinical practice are calcium chloride, thrombin, and calcium chloride/thrombin combination. It is known that these activators increase the release of growth factors from PRP. However, each activating substance results in different amounts of growth factor release and can alter the mechanical properties of the solution.

Activating PRP does initiate the clotting cascade. PRP activated with calcium chloride, thrombin, or a mixture can result in clot detected at 15 min and can persist up to 24 h. On the other hand, an in vitro model using collagen type I as an activator designed to mimic PRP injection in resting form demonstrated no clot formation. With respect to PDGF, TGF-b, and VEGF, it appears that a 10% calcium chloride with thrombin is most effective while just collagen type 1 is least effective at growth factor release [25]. Calcium chloride with thrombin can still be used as an injectable; however the practitioner should note the clot formation time. Clot formation can be advantageous as it helps prevent PRP from washing away. Practically, the usefulness of clot formation likely depends on the therapeutic application. For example, an intra-articular PRP injection for osteoarthritis is less likely to benefit from clot formation than an intra-tendinous or muscular application of PRP where stabilization of the substance in the desired target is advantageous. Despite this in vitro evidence, there is no clinical consensus on which agent is optimal or if activation is necessary at all [8].

There is considerable variability in PRP preparation protocols. Despite the basic formula of whole blood collection, anticoagulation, centrifugation, concentration, and option activation factor usage, there is a lack of consensus on what parameters are ideal. What's more, many published studies have limited or no description of their process and PRP outcome. Chahla et al. [5] noted only 10% of studies reviewed provided comprehensive reporting with a description of the preparation protocol and only 16% of studies provided metrics on the final PRP product. This led authors to call for standardization in reporting for PRP preparation protocols and the PRP end product. The minimum parameters for PRP product should include at least platelet concentration, factor increase from whole blood, leukocyte concentration, leukocyte differential, and presence of growth factors. The minimum description of PRP preparation protocol should increase at least baseline whole blood platelet count, centrifugal force, centrifuge time, and anticoagulant used [8].

The need for standardization of reporting has been echoed in a recent work by Kon et al. [26]. They emphasized the need for a standardization of reporting in order to appropriately analyze the effectiveness and establish indications for PRP, and they have aptly proposed a coding system to describe components (platelets, red blood cells, and white blood cells) and the activation of PRP.

10.3 Product Composition

10.3.1 Growth Factors

Platelets play an important role in coordinating the response to injury, especially during the early and inflammatory phases. Multiple growth factors, which are released by degranulation of platelet alpha granules, are key to the healing potential of PRP. PDGF, TGF-b, VEGF, and FGF in PRP are known to play critical roles in cellular activities such as angiogenesis, stem cell trafficking, proliferation, and differentiation [25, 27, 28].

Each growth factor plays a different role in the healing potential of PRP. PDGF is a mediator of the proliferation of fibroblast and smooth muscle cells and therefore is important in angiogenesis, formation of fibrous tissue, and the reepithelization phases of wound healing [29]. TGF-b stimulates the proliferation and differentiation of mesenchymal stem cells and plays a role in the chemotaxis of endothelial cells [29, 30]. VEGF is a key mitogen for endothelial cells and is a strong regulator of angiogenesis and vascular permeability [31]. FGF is a mediator of proliferation and differentiation of a wide variety of cells and tissues and can mediate the formation of angiogenesis during wound healing. There is also evidence that FGF and PDGF can act synergistically in improving healing potential [32].

Cytokines, which are also present in PRP, can play a pivotal role in the inflammatory properties of PRP. For example, interleukin (IL)-1beta and matrix metalloproteinases (MMPs) are catabolic cytokines present in PRP that are known to play a role in inflammation and matrix degeneration, respectively [33, 34]. When considering the inflammatory prosperities in PRP, it is important to note that the first phase of tendon healing is inflammation, which ultimately may explain why "pro-inflammatory" PRP preparations may be more desirable when treating tendinopathies [35].

The key to growth factor and cytokine effects in PRP is multifactorial and probably related not only to the concentration of growth factors but also to their temporal nature. Oh et al. [36] showed in vitro that PDGF and VEGF release was constant and sustained over 7 days while FGF and TGF release occurred quickly and decreased over time.

Knowledge of growth factors in samples is undoubtedly essential for appropriate clinical treatment and for improving our understanding of PRP's mechanism. Unfortunately, most literature to date has not reported quantitative growth factor statistics [5]. The importance of reporting on growth factors within PRP cannot be overstated as small variations in growth factor concentrations can exert very different effects on tissue. For example, TGF-beta is known to play a role in the resolution of inflammation and promotion of tissue repair. However, excessive TGF-beta1 has been associated with inflammation and fibrotic events [28]. Additionally, there is a considerable amount of variation from individual to individual in growth factor concentration with differing reports of correlations between growth factor concentration, age, and platelet count [9, 10].

Ultimately, more research will be required to determine how to prepare the ideal PRP with respect to growth factor concentration. Differing reports of the clinical efficacy of PRP reported may be related to the variation in PRP's growth factor concentration. This variation in growth factor composition is possibly multifactorial and related to patient as well as processing factors. With ongoing research and a call for PRP reporting standards, advances can be made toward understanding how to harness the great healing potential of the growth factors present in PRP.

10.3.2 Leukocytes

White blood cell concentration is an important component in the final composition of PRP. In a systematic review of peer-reviewed published data on PRP preparations from commercially available systems, Fadadu et al. [8] noted a wide range of leukocyte concentrations in PRP products. The average leukocyte cell concentration count was $41.66 +/- 95.16 \times 10^3$ cells per microliter. Interestingly, there was no correlation between PRP platelet concentration and leukocyte concentration. However, this study did not differentiate between leukocyte-rich PRP (LR-PRP) and leukocyte-poor PRP (LP-PRP).

Based on preparation protocols and final leukocyte concentrations, PRP can be classified as LR-PRP or LP-PRP (Fig. 10.3) [11, 37, 38]. While there is some variability in specific commercially available protocols, the initial spin during centrifugation separates the "buffy coat," which is platelet-rich plasma with leukocytes, from red blood cells and platelet-poor plasma. A second spin can reduce the leukocyte concentra-

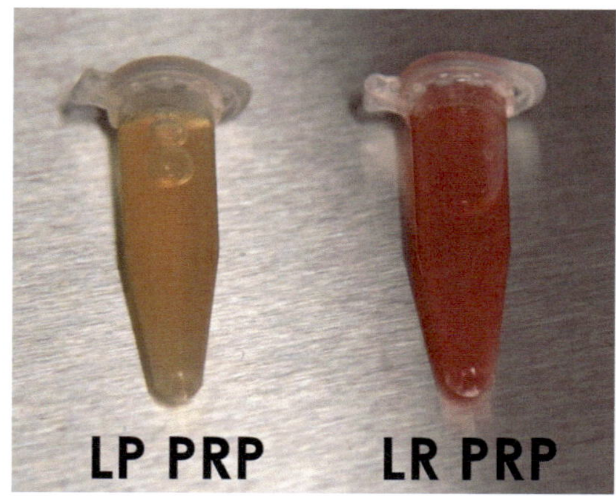

Fig. 10.3 Photograph illustrating the external appearance of leukocyte-poor (LP-PRP) (left) and leukocyte-rich (LR-PRP) (right) platelet-rich plasma. LR-PRP appears better suited for intraarticular applications such as cartilage restoration, while LR-PRP seems to be superior for muscle or tendinous applications. (Reprinted with permission from Hussain ZB, Chahla J, LaPrade RF, Mandelbaum BR. Orthobiologics: Today and Tomorrow. In: Farr J, Gomoll AH, editors. Cartilage Restoration: Practical Clinical Applications. Cham: Springer International Publishing; 2018. p. 131–42)

tion. In recent publications, Le et al. conveniently classified commercially available processing systems as LR-PRP or LP-PRP [6, 11].

In general, there is a debate regarding the ideal composition of PRP. This could not be truer than in the case of leukocyte concentration. Initially, some groups argued that the presence of leukocytes could be detrimental to the healing process due to an increased inflammatory response, while others promoted LR-PRP as richer in growth factors, anti-pain mediators, and anti-infectious potential [35, 39–41].

Adding depth to the issue is that not all leukocytes are the same. Lymphocytes, monocytes, and granulocytes behave differently, and the quantity and in which state these cells are in (centrifugation can activate, stimulate, or destroy leukocytes) likely contribute to their effect on target tissues [37].

The leukocytes present in LR-PRP release catabolic and pro-inflammatory mediators such as MMP-1 and MMP-13 and cytokines such as IL-1 beta, IL-6, and tumor necrosis factor (TNF) alpha. This has led to a surge in interest into the role leukocytes play on the clinical effects of PRP. Some authors have suggested that leukocytes in PRP can have a potential detrimental effect due to catabolic activity [42–44]. However, others have suggested that LR-PRP could be superior in scenarios where tendon or muscle healing is required. For example, Ziegler et al. [45] analyzed human PRP for inflammatory mediators and found LR-PRP had higher levels of not only PDGF, TGFbeta, and VEGF but also higher concentrations of interleukin-1 receptor antagonist (IL-1Ra), an important inhibitor of the pro-inflammatory molecule IL-1 beta. The authors concluded that LR-PRP may be preferable in cases where increased vascularity and healing are desired for injured tissue such as muscles and tendons.

Dragoo et al. [44] performed a comparative study on the effect of a commercially available LR-PRP and a commercially available LP-PRP system on the patella tendons of New Zealand White rabbits. They noted that, after 5 days, tendons treated with LR-PRP group had significantly greater overall tendon scores, vascularity, and fiber structure, which they accredited to the greater inflammatory response after LR-PRP injection. They did accurately note, however, that the LR-PRP system had significantly more platelets and a lower platelet/WBC ratio, which could have confounded their results. Additionally, the difference between cellularity, fiber orientation, and vascularity dissipated 14 days after injection. Still, this work highlighted the important difference between LR-PRP and LP-PRP in an in vivo model.

Based on a growing amount of literature supporting LR-PRP in treating tendinopathies, Dragoo et al. [46] compared the effectiveness of LR-PRP on human patellar tendinopathy vs. dry needling in a randomized controlled trial and found significant improvement in the LR-PRP group in patient outcome scores at 12 weeks. Another randomized controlled trial that compared LR-PRP, LP-PRP, and saline injections for patellar tendinopathy found no difference at any timepoint up to a year between groups [47]. That study, though, was underpowered and had leukocyte and platelet counts measured for only a subset of injections. For those injections that were measured, there was a notable variability in leu-

kocyte fold increases from baseline in the LR-PRP group, leaving one to question whether there was appropriate leukocyte enrichment in all of the LR-PRP injections [48, 49].

One of the most widely cited reports of LR-PRP in tendinopathies was conducted by Mishra et al. [50]. In a double-blind, prospective randomized controlled trial, the authors evaluated 230 patients with chronic lateral epicondylitis and reported that patients treated with LR-PRP had significantly and clinically meaningful improvement at 24 weeks' follow-up as compared to the control group.

The importance of understanding and reporting the differences in leukocyte concentration is highlighted by the recent randomized controlled trial by Linnanmäki et al. [51] who concluded there is no benefit to PRP in the treatment of lateral epicondylitis. However, that study utilized a LP-PRP system, which, based on the available literature, is not the ideal formulation for this pathology.

There has also been a substantial amount of research into the use of PRP in osteoarthritis (OA) of the knee. Multiple studies have shown that PRP is effective in reducing pain and improving patient-reported outcomes especially in those with mild to moderate OA [52–58]. There is a growing amount of evidence suggesting specifically LP-PRP's effectiveness in treating knee OA. For example, Duif et al. [59] showed superior patient recovery after PRP injection versus control in patients undergoing knee arthroscopy for degenerative cartilage of meniscal pathology. In a meta-analysis of six randomized controlled trials and three prospective comparative studies totaling 1055 patients, Riboh et al. [60] compared LP-PRP and LR-PRP in the treatment of knee OA and found injections of LP-PRP, but not LR-PRP, resulted in significantly improved Western Ontario and McMaster University Osteoarthritis Index scores as compared with hyaluronic acid or placebo. At the same time, the results of LR-PRP have shown minimal effect in randomized control trials versus viscosupplementation control in treating knee OA [61, 62].

It should be noted, however, that meta-analyses comparing PRP formulations have inherent confounders, such as differences in PRP composition. Although evidence has suggested LP-PRP may be better suited to the treatment of OA, a comparative trial found no differences between leukocyte-rich and leukocyte-poor preparations at 1 year, and reported increased postinjection pain and swelling associated with leukocyte-rich preparations [63]. Moreover, other literature analyses highlight the lack of direct comparative studies and conclusive evidence to recommend one preparation over another [64]. Still, considering this limitation of the available evidence, LP-PRP holds promise for the treatment for mild to moderate OA likely due to its anti-inflammatory nature and its notable lack of pro-inflammatory cytokines.

Autologous conditioned serum (ACS) uses a slight variation in PRP processing. By using medical-grade glass beads in syringes during blood collection, it induces the production of IL1-Ra by leukocytes. IL1-Ra is a known powerful inhibitor of the IL1 and therefore holds potential as an anti-inflammatory mediator. Baltzer et al. [65] evaluated autologous conditioned serum in a randomized controlled trial for patients with knee OA. They noted superior patient-reported outcomes in the autologous conditioned serum group as compared to the hyaluronic acid and saline groups. However, other studies have demonstrated a lack of clinical benefit at longer-term follow-up (greater than 1 year) or in patients with end-stage OA [66, 67].

The potential benefits of leukocytes in PRP preparations are controversial and continue to be debated. Their presence can potentially alter the pro- or anti-inflammatory properties of the final product. Ideally, continued research will help the treating clinician tailor PRP preparation protocols to the specific pathology being treated.

10.4 Conclusion

With a wide variety of applications, PRP is a complex but very exciting tool in the clinician's toolbox. The high concentration of platelets along with their powerful growth factors and cytokines holds immense potential in the minimally invasive injectable treatment of musculoskeletal ailments.

There has been a variety of clinical results making the use of PRPs somewhat debatable. The complex composition, multiple different methods of preparation, and lack of standardized criteria for PRP are probably responsible for the variety of treatment efficacy outcomes. Multiple authors have called for standardization in reporting with respect to preparation parameters and end product. Doing so will help further delineate the most appropriate composition and uses for PRP. It's likely that in the near future, PRP preparations will be customized based on the patient and the pathology being treated. As our knowledge of PRP and its components continues, there is promise in our ultimate quest of improving the quality of life of our patients in the most efficient and effective manner.

Take-Home Messages
- Platelet-rich plasma (PRP) is an autologous concentrated platelet preparation; however, the end product can vary based on patient-specific factors and sample processing.
- Processing methods differ based on activation, anticoagulation, and the isolation/concentration method.
- The ideal PRP concentration of platelets, growth factors, and leukocyte concentration is yet to be determine but will likely depend on the pathology being treated.
- There has been a call for reporting standards in future research, which may help identify the ideal composition.

References

1. Foster TE, Puskas BL, Mandelbaum BR, Gerhardt MB, Rodeo SA. Platelet-rich plasma: from basic science to clinical applications. Am J Sports Med. 2009;37(11):2259–72.

2. LaPrade RF, Geeslin AG, Murray IR, Musahl V, Zlotnicki JP, Petrigliano F, et al. Biologic treatments for sports injuries II think tank-current concepts, future research, and barriers to advancement, part 1: biologics overview, ligament injury, tendinopathy. Am J Sports Med. 2016;44(12):3270–83.

3. Marx RE. Platelet-rich plasma (PRP): what is PRP and what is not PRP? Implant Dent. 2001;10(4):225–8.

4. Dhillon RS, Schwarz EM, Maloney MD. Platelet-rich plasma therapy—future or trend? Arthritis Res Ther. 2012;14(4):219.

5. Chahla J, Cinque ME, Piuzzi NS, Mannava S, Geeslin AG, Murray IR, et al. A call for standardization in platelet-rich plasma preparation protocols and composition reporting: a systematic review of the clinical orthopaedic literature. J Bone Joint Surg Am. 2017;99(20):1769–79.

6. Le ADK, Enweze L, DeBaun MR, Dragoo JL. Current clinical recommendations for use of platelet-rich plasma. Curr Rev Musculoskelet Med. 2018;11(4):624–34.

7. Hussain ZB, Chahla J, LaPrade RF, Mandelbaum BR. Orthobiologics: today and tomorrow. In: Farr J, Gomoll AH, editors. Cartilage restoration: practical clinical applications. Cham: Springer International Publishing; 2018. p. 131–42.

8. Fadadu PP, Mazzola AJ, Hunter CW, Davis TT. Review of concentration yields in commercially available platelet-rich plasma (PRP) systems: a call for PRP standardization. Reg Anesth Pain Med. 2019.

9. Weibrich G, Kleis WK, Hafner G, Hitzler WE. Growth factor levels in platelet-rich plasma and correlations with donor age, sex, and platelet count. J Craniomaxillofac Surg. 2002;30(2):97–102.

10. Taniguchi Y, Yoshioka T, Sugaya H, Gosho M, Aoto K, Kanamori A, et al. Growth factor levels in leukocyte-poor platelet-rich plasma and correlations with donor age, gender, and platelets in the Japanese population. J Exp Orthop. 2019;6(1):4.

11. Le ADK, Enweze L, DeBaun MR, Dragoo JL. Platelet-rich plasma. Clin Sports Med. 2019;38(1):17–44.

12. Andrade MG, de Freitas Brandão CJ, Sá CN, de Bittencourt TC, Sadigursky M. Evaluation of factors that can modify platelet-rich plasma properties. Oral Surg Oral Med Oral Pathol Oral Radiol Endod. 2008;105(1):e5–e12.

13. Lei H, Gui L, Xiao R. The effect of anticoagulants on the quality and biological efficacy of platelet-rich plasma. Clin Biochem. 2009;42(13–14):1452–60.

14. Dugrillon A, Eichler H, Kern S, Klüter H. Autologous concentrated platelet-rich plasma (cPRP) for local application in bone regeneration. Int J Oral Maxillofac Surg. 2002;31(6):615–9.

15. Sabarish R, Lavu V, Rao SR. A comparison of platelet count and enrichment percentages in the Platelet Rich Plasma (PRP) obtained following preparation by three different methods. J Clin Diagn Res. 2015;9(2):Zc10-2.

16. Krüger JP, Freymannx U, Vetterlein S, Neumann K, Endres M, Kaps C. Bioactive factors in platelet-rich plasma obtained by apheresis. Transfus Med Hemother. 2013;40(6):432–40.

17. de Vries RA, de Bruin M, Marx JJ, Hart HC, Van de Wiel A. Viability of platelets collected by apheresis versus the platelet-rich plasma technique: a direct comparison. Transfus Sci. 1993;14(4):391–8.

18. O'Neill EM, Zalewski WM, Eaton LJ, Popovsky MA, Pivacek LE, Ragno G, et al. Autologous platelet-rich plasma isolated using the Haemonetics Cell Saver 5 and Haemonetics MCS+ for the preparation of platelet gel. Vox Sang. 2001;81(3):172–5.

19. van der Meer PF. Platelet concentrates, from whole blood or collected by apheresis? Transfus Apher Sci. 2013;48(2):129–31.

20. Yamaguchi R, Terashima H, Yoneyama S, Tadano S, Ohkohchi N. Effects of platelet-rich plasma on intestinal anastomotic healing in rats: PRP concentration is a key factor. J Surg Res. 2012;173(2):258–66.

21. Fleming BC, Proffen BL, Vavken P, Shalvoy MR, Machan JT, Murray MM. Increased platelet concentration does not improve functional graft healing in bio-enhanced ACL reconstruction. Knee Surg Sports Traumatol Arthrosc. 2015;23(4):1161–70.

22. Yoshida R, Cheng M, Murray MM. Increasing platelet concentration in platelet-rich plasma inhibits anterior cruciate ligament cell function in three-dimensional culture. J Orthop Res. 2014;32(2):291–5.

23. Weibrich G, Hansen T, Kleis W, Buch R, Hitzler WE. Effect of platelet concentration in platelet-rich plasma on peri-implant bone regeneration. Bone. 2004;34(4):665–71.

24. Lana JFbSD, Santana MHA, Belangero WD, Luzo ACM. Platelet-rich plasma: regenerative medicine: sports medicine, orthopedic, and recovery of musculoskeletal injuries. Heidelberg: Springer; 2014. xvii, 360 p.

25. Cavallo C, Roffi A, Grigolo B, Mariani E, Pratelli L, Merli G, et al. Platelet-rich plasma: the choice of activation method affects the release of bioactive molecules. Biomed Res Int. 2016;2016:6591717.

26. Kon E, Di Matteo B, Delgado D, Cole BJ, Dorotei A, Dragoo JL, et al. Platelet-rich plasma for the treatment of knee osteoarthritis: an expert opinion and proposal for a novel classification and coding system. Expert Opin Biol Ther. 2020;20(12):1447–60.

27. Sánchez-González DJ, Méndez-Bolaina E, Trejo-Bahena NI. Platelet-rich plasma peptides: key for regeneration. Int J Pept. 2012;2012:532519.

28. Tschon M, Fini M, Giardino R, Filardo G, Dallari D, Torricelli P, et al. Lights and shadows concerning platelet products for musculoskeletal regeneration. Front Biosci (Elite Ed). 2011;3:96–107.

29. Hosgood G. Wound healing. The role of platelet-derived growth factor and transforming growth factor beta. Vet Surg. 1993;22(6):490–5.

30. Civinini R, Nistri L, Martini C, Redl B, Ristori G, Innocenti M. Growth factors in the treatment of early osteoarthritis. Clin Cases Miner Bone Metab. 2013;10(1):26–9.

31. Hoeben A, Landuyt B, Highley MS, Wildiers H, Van Oosterom AT, De Bruijn EA. Vascular endothelial growth factor and angiogenesis. Pharmacol Rev. 2004;56(4):549–80.

32. Cao R, Bråkenhielm E, Pawliuk R, Wariaro D, Post MJ, Wahlberg E, et al. Angiogenic synergism, vascular stability and improvement of hind-limb ischemia by a combination of PDGF-BB and FGF-2. Nat Med. 2003;9(5):604–13.

33. Sundman EA, Cole BJ, Fortier LA. Growth factor and catabolic cytokine concentrations are influenced by the cellular composition of platelet-rich plasma. Am J Sports Med. 2011;39(10):2135–40.

34. Thampatty BP, Li H, Im HJ, Wang JH. EP4 receptor regulates collagen type-I, MMP-1, and MMP-3 gene expression in human tendon fibroblasts in response to IL-1 beta treatment. Gene. 2007;386(1–2):154–61.

35. Braun HJ, Kim HJ, Chu CR, Dragoo JL. The effect of platelet-rich plasma formulations and blood products on human synoviocytes: implications for intra-articular injury and therapy. Am J Sports Med. 2014;42(5):1204–10.

36. Oh JH, Kim W, Park KU, Roh YH. Comparison of the cellular composition and cytokine-release kinetics of various platelet-rich plasma preparations. Am J Sports Med. 2015;43(12):3062–70.

37. Dohan Ehrenfest DM, Andia I, Zumstein MA, Zhang CQ, Pinto NR, Bielecki T. Classification of platelet concentrates (Platelet-Rich Plasma-PRP, Platelet-Rich Fibrin-PRF) for topical and infiltrative use in orthopedic and sports medicine: current consensus, clinical implications and perspectives. Muscles Ligaments Tendons J. 2014;4(1):3–9.

38. Everts PA, Hoffmann J, Weibrich G, Mahoney CB, Schönberger JP, van Zundert A, et al. Differences in platelet growth factor release and leucocyte kinetics during autologous platelet gel formation. Transfus Med. 2006;16(5):363–8.

39. Anitua E, Sánchez M, Orive G, Andía I. The potential impact of the preparation rich in growth factors (PRGF) in different medical fields. Biomaterials. 2007;28(31):4551–60.

40. Bielecki T, Dohan Ehrenfest DM, Everts PA, Wiczkowski A. The role of leukocytes from L-PRP/L-PRF in wound healing and immune defense: new perspectives. Curr Pharm Biotechnol. 2012;13(7):1153–62.

41. Moojen DJ, Everts PA, Schure RM, Overdevest EP, van Zundert A, Knape JT, et al. Antimicrobial activity of platelet-leukocyte gel against Staphylococcus aureus. J Orthop Res. 2008;26(3):404–10.

42. Zhou Y, Zhang J, Wu H, Hogan MV, Wang JH. The differential effects of leukocyte-containing and pure platelet-rich plasma (PRP) on tendon stem/progenitor cells—implications of PRP application for the clinical treatment of tendon injuries. Stem Cell Res Ther. 2015;6:173.

43. Lana JF, Huber SC, Purita J, Tambeli CH, Santos GS, Paulus C, et al. Leukocyte-rich PRP versus leukocyte-poor PRP. J Clin Orthop Trauma. 2019;10(Suppl 1):S7–S12.

44. Dragoo JL, Braun HJ, Durham JL, Ridley BA, Odegaard JI, Luong R, et al. Comparison of the acute inflammatory response of two commercial platelet-rich plasma systems in healthy rabbit tendons. Am J Sports Med. 2012;40(6):1274–81.

45. Ziegler CG, Van Sloun R, Gonzalez S, Whitney KE, DePhillipo NN, Kennedy MI, et al. Characterization of growth factors, cytokines, and chemokines in bone marrow concentrate and platelet-rich plasma: a prospective analysis. Am J Sports Med. 2019;47(9):2174–87.

46. Dragoo JL, Wasterlain AS, Braun HJ, Nead KT. Platelet-rich plasma as a treatment for patellar tendinopathy: a double-blind, randomized controlled trial. Am J Sports Med. 2014;42(3):610–8.

47. Scott A, LaPrade RF, Harmon KG, Filardo G, Kon E, Della Villa S, et al. Platelet-rich plasma for patellar tendinopathy: a randomized controlled trial of Leukocyte-Rich PRP or Leukocyte-Poor PRP versus saline. Am J Sports Med. 2019;47(7):1654–61.

48. Baria MR, Vasileff WK, Kaeding C. Platelet-rich plasma for patellar tendinopathy: letter to the editor. Am J Sports Med. 2020;48(2):NP21–NP2.

49. Scott A, Harmon KG. Platelet-rich plasma for patellar tendinopathy: response. Am J Sports Med. 2020;48(2):NP22.

50. Mishra AK, Skrepnik NV, Edwards SG, Jones GL, Sampson S, Vermillion DA, et al. Efficacy of platelet-rich plasma for chronic tennis elbow: a double-blind, prospective, multicenter, randomized controlled trial of 230 patients. Am J Sports Med. 2014;42(2):463–71.

51. Linnanmäki L, Kanto K, Karjalainen T, Leppänen OV, Lehtinen J. Platelet-rich plasma or autologous blood do not reduce pain or improve function in patients with lateral epicondylitis: a randomized controlled trial. Clin Orthop Relat Res. 2020;478(8):1892–900.

52. Duymus TM, Mutlu S, Dernek B, Komur B, Aydogmus S, Kesiktas FN. Choice of intra-articular injection in treatment of knee osteoarthritis: platelet-rich plasma, hyaluronic acid or ozone options. Knee Surg Sports Traumatol Arthrosc. 2017;25(2):485–92.

53. Görmeli G, Görmeli CA, Ataoglu B, Çolak C, Aslantürk O, Ertem K. Multiple PRP injections are more effective than single injections and hyaluronic acid in knees with early osteoarthritis: a randomized, double-blind, placebo-controlled trial. Knee Surg Sports Traumatol Arthrosc. 2017;25(3):958–65.

54. Lana JF, Weglein A, Sampson SE, Vicente EF, Huber SC, Souza CV, et al. Randomized controlled trial comparing hyaluronic acid, platelet-rich plasma and the combination of both in the treatment of mild and moderate osteoarthritis of the knee. J Stem Cells Regen Med. 2016;12(2):69–78.

55. Montañez-Heredia E, Irízar S, Huertas PJ, Otero E, Del Valle M, Prat I, et al. Intra-articular injections of platelet-rich plasma versus hyaluronic acid in the treatment of osteoarthritic knee pain: a randomized

clinical trial in the context of the Spanish National Health Care System. Int J Mol Sci. 2016;17(7).

56. Paterson KL, Nicholls M, Bennell KL, Bates D. Intra-articular injection of photo-activated platelet-rich plasma in patients with knee osteoarthritis: a double-blind, randomized controlled pilot study. BMC Musculoskelet Disord. 2016;17:67.

57. Raeissadat SA, Rayegani SM, Hassanabadi H, Fathi M, Ghorbani E, Babaee M, et al. Knee osteoarthritis injection choices: Platelet-Rich Plasma (PRP) versus hyaluronic acid (A one-year randomized clinical trial). Clin Med Insights Arthritis Musculoskelet Disord. 2015;8:1–8.

58. Patel S, Dhillon MS, Aggarwal S, Marwaha N, Jain A. Treatment with platelet-rich plasma is more effective than placebo for knee osteoarthritis: a prospective, double-blind, randomized trial. Am J Sports Med. 2013;41(2):356–64.

59. Duif C, Vogel T, Topcuoglu F, Spyrou G, von Schulze PC, Lahner M. Does intraoperative application of leukocyte-poor platelet-rich plasma during arthroscopy for knee degeneration affect postoperative pain, function and quality of life? A 12-month randomized controlled double-blind trial. Arch Orthop Trauma Surg. 2015;135(7):971–7.

60. Riboh JC, Saltzman BM, Yanke AB, Fortier L, Cole BJ. Effect of leukocyte concentration on the efficacy of platelet-rich plasma in the treatment of knee osteoarthritis. Am J Sports Med. 2016;44(3):792–800.

61. Di Martino A, Di Matteo B, Papio T, Tentoni F, Selleri F, Cenacchi A, et al. Platelet-rich plasma versus hyaluronic acid injections for the treatment of knee osteoarthritis: results at 5 years of a double-blind, randomized controlled trial. Am J Sports Med. 2019;47(2):347–54.

62. Filardo G, Di Matteo B, Di Martino A, Merli ML, Cenacchi A, Fornasari P, et al. Platelet-rich plasma intra-articular knee injections show no superiority versus viscosupplementation: a randomized controlled trial. Am J Sports Med. 2015;43(7):1575–82.

63. Filardo G, Kon E, Pereira Ruiz MT, Vaccaro F, Guitaldi R, Di Martino A, et al. Platelet-rich plasma intra-articular injections for cartilage degeneration and osteoarthritis: single- versus double-spinning approach. Knee Surg Sports Traumatol Arthrosc. 2012;20(10):2082–91.

64. Filardo G, Previtali D, Napoli F, Candrian C, Zaffagnini S, Grassi A. PRP injections for the treatment of knee osteoarthritis: a meta-analysis of randomized controlled trials. Cartilage 2020:1947603520931170.

65. Baltzer AW, Moser C, Jansen SA, Krauspe R. Autologous conditioned serum (Orthokine) is an effective treatment for knee osteoarthritis. Osteoarthr Cartil. 2009;17(2):152–60.

66. Zarringam D, Bekkers JEJ, Saris DBF. Long-term effect of injection treatment for osteoarthritis in the knee by orthokin autologous conditioned serum. Cartilage. 2018;9(2):140–5.

67. Rutgers M, Creemers LB, Auw Yang KG, Raijmakers NJ, Dhert WJ, Saris DB. Osteoarthritis treatment using autologous conditioned serum after placebo. Acta Orthop. 2015;86(1):114–8.

Placental Tissue Extracts

11

Bogdan A. Matache, Eric J. Strauss, and Jack Farr

11.1 Introduction

The investigational use of tissue-derived non-culture-expanded cells is a rapidly expanding field of orthopedic interest due to the theoretical advantages these offer over other biologic strategies. Unlike other biologics, such as platelet-rich plasma (PRP) and hyaluronic acid (HA), tissue resident cells, including potential stem and progenitor populations, are available as an inexpensive autogenous resource [1, 2]. Cells can be extracted from autologous or allogeneic sources. In addition to many mature tissue resident cells, these cellular isolates may contain a small fraction of cells that are capable of proliferation. This heterogeneous population of native tissue-specific stem and progenitor cells in connective tissues is referred to as connective tissue progenitors (CTPs).

The concentration, prevalence, yields, and biological potential of native CTPs in any tissue are a function of age (possibly age-related senescence) [3], donor-site location [4], harvest and processing technique, and location-dependent variability in tissue health [5]. This variability has led to exploration of alternative tissue sources, including allograft tissues. There has also been great interest in placing CTP populations into culture and expanding the diversity of clones that are present in the competitive environment of in vitro tissue culture to generate large numbers of culture-expanded adherent cells that have been referred to as mesenchymal stromal cells (MSC). However, this chapter does not focus on culture-expanded cells.

Allogeneic stem cells for orthopedic use can be obtained from the same tissues as autologous ones, including bone marrow, adipose, synovial tissue, and periosteum. More recently, research has explored the use of placental tissue as a cell source. Promising outcomes have been reported using placental-derived cells to treat burn, cardiac, and hepatic patients, among others [6–8].

11.2 Clinically Relevant Anatomy

The placenta at full-term is comprised of an internal cavity, contained by an internal and external lining, and the umbilical cord. The internal, or fetal, lining of the placenta contributes to the containment and production of amniotic fluid, which nourishes the developing fetus, termed the amniotic membrane. The outer layer of the amniotic membrane is of clinical importance since it repre-

B. A. Matache · E. J. Strauss
Division of Sports Medicine, Department of
Orthopedic Surgery, NYU Langone Health,
New York, NY, USA
e-mail: Bogdan.matache@nyulangone.org;
eric.strauss@nyulangone.org

J. Farr (✉)
Knee Preservation and Cartilage Restoration Center,
OrthoIndy Hospital, Greenwood, IN, USA
e-mail: jfarr@orthoindy.com

© ISAKOS 2022
G. Filardo et al. (eds.), *Orthobiologics*, https://doi.org/10.1007/978-3-030-84744-9_11

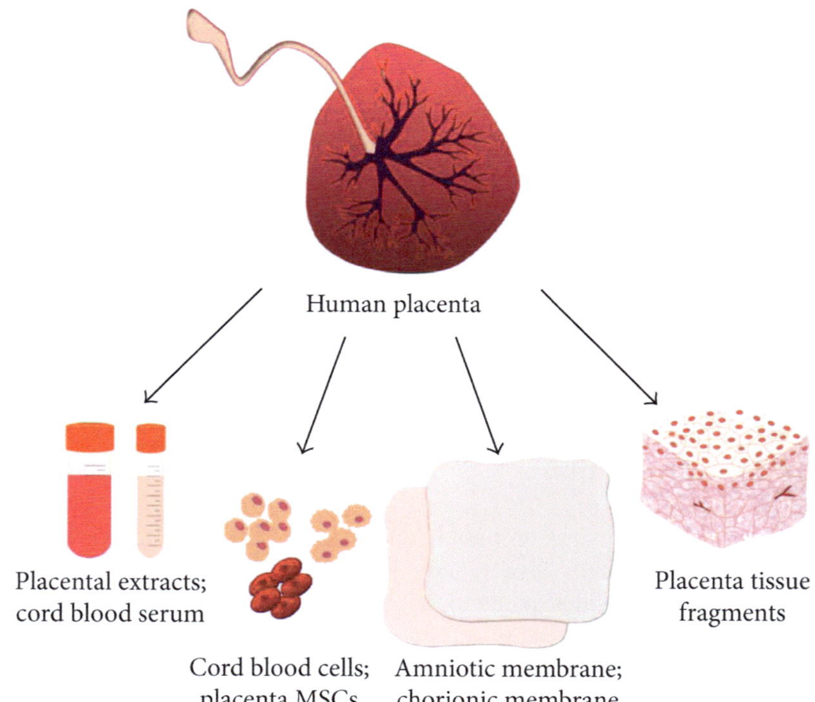

Fig. 11.1 Commonly utilized cell sources from placental tissues. (Copyright © 2018 Olena Pogozhykh et al., licensed under CC BY 4.0. Reprinted from Pogozhykh O, Prokopyuk V, Figueiredo C, Denys Pogozhykh D. Placenta and placental derivatives in regenerative therapies: experimental studies, history, and prospects. Stem Cells International. 2018:4837930)

sents the mesenchymal layer, which contains CTPs that can differentiate along a fibroblastic lineage. The term AM-MSC has been used to describe a presumed stem and progenitor cell population in the amniotic membrane [9]; however, these heterogeneous native cells and its heterogeneous population of native stem and progenitor cells do not meet the International Society for Cell and Gene Therapy (ISCT) criteria for MSCs. They are not a culture-expanded, plastic adherent population. Therefore, the standards in this book dictate the designation of native amniotic progenitors as AM-CTPs or CTP-AMs.

The external, or maternal, lining of the placenta comprises the chorionic membrane [10], where the native stem and progenitor population can be defined as CM-CTPs or CTP-CMs. The amniotic fluid contained within the amniotic membrane also contains a cell population that is capable of proliferation. These have been termed amniotic fluid (AF)-MSCs [9], but respecting the ISCT definition of MSC we refer to them here as CTP-AFs.

Not surprisingly, culture-expanded populations that may meet the criteria of MSCs can be derived by placing a mixed CTP population into culture, followed by a period of competitive expansion. Cell populations from the chorion are reported to exhibit highest proliferation and multi-lineage differentiation potential out of all placental stem cells (Fig. 11.1) [10, 11]. These expanded populations will have a secretome that includes a variety of soluble factors and extracellular vesicles that may contribute to paracrine signaling [12].

Delivery of exosomes rather than cells has recently drawn interest due to the theoretical advantages, namely, a reduced concern of tumorigenesis and an ability to cross the blood-brain barrier due to their relatively smaller size [12]. Culture-expanded MSC populations derived from CTPs isolated from amniotic fluid have been reported to produce a significantly higher total yield than MSC populations generated from other tissue sources, such as bone marrow [12]. However, these products are not yet FDA-approved. Further research into their safety, efficacy, and clinically applicable delivery is required. Exosomes are discussed in detail in Chap. 11.

The umbilical cord is involved in the exchange of maternal nutrients and fetal waste by providing a safe conduit for this process to occur. Although it forms part of the placenta, umbilical tissue is considered extraembryonic. Placental tissues and other extraembryonic tissues are highly metabolic tissues undergoing rapid turnover and remodeling. These kinetics makes these tissues, like embryonic tissues, a rich potential source for native CTPs. Colony founding cells are found in particular abundance in the perivascular tissues and epithelium [13], Wharton's jelly, and also umbilical cord blood (UCB). Wharton's jelly has been shown to possess up to two orders of magnitude greater concentration of colony founding cells when compared to other placental sources [14]. MSC populations that are generated from UCB are reported to be similar in composition to those derived from the bone marrow [15]. However, there is persistent controversy surrounding the clinical utility of umbilical (U)- and UCB-derived MSCs due to concerns about their ability to differentiate [16–19].

11.3 Evidence in Sports Medicine

11.3.1 Cartilage Injury and Osteoarthritis

OA is a progressive, degenerative disease involving the chronic degradation of articular cartilage. In the early stages of the disease, there is a release of inflammatory mediators such as interleukin (IL)-1ß, tumor necrosis factor (TNF)-α, and IL-6 by cartilage, bone, and synovium (Fig. 11.2) [20,

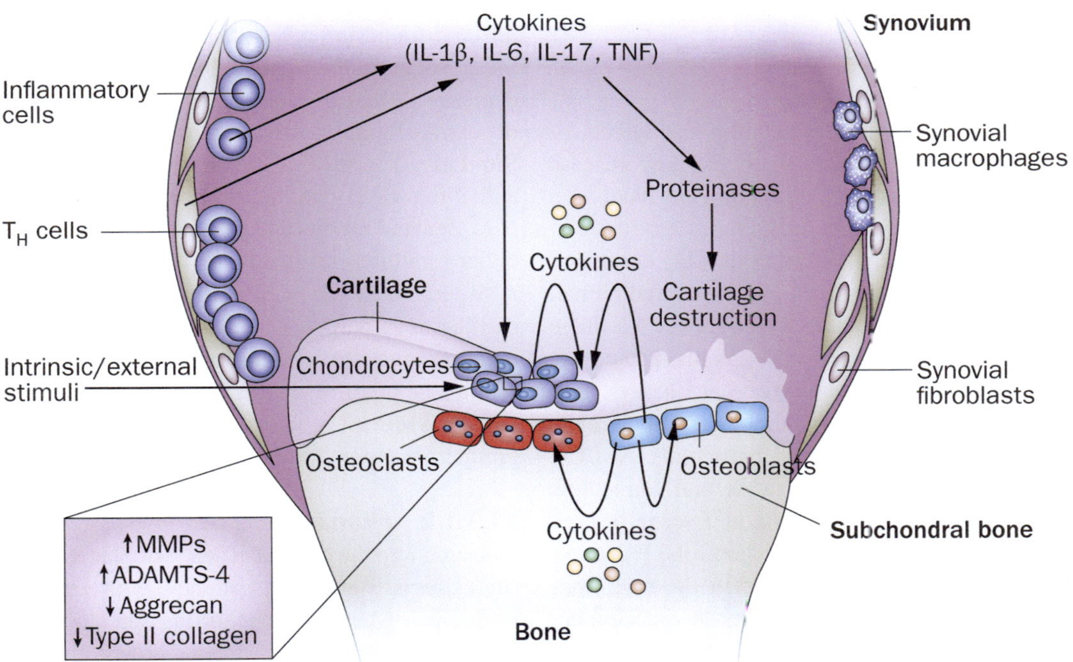

Fig. 11.2 The role of pro-inflammatory cytokines in the pathophysiology of OA. The levels of pro-inflammatory cytokines, including IL-1β, TNF, and IL-6, are elevated in OA. These cytokines contribute to the pathogenesis of OA through several mechanisms including downregulation of anabolic events and upregulation of catabolic and inflammatory responses; effects that result in structural damage to the OA joint. Abbreviations: *ADAMTS* a disintegrin- like and metalloproteinase with thrombospondin type 1 motifs; *IL* interleukin; *MMP* matrix metalloproteinase; *OA* osteoarthritis; *TNF* tumor necrosis factor. (Copyright © 2011. Reprinted with permission by Springer Nature from Kapoor M, Martel-Pelletier J, Lajeunesse D, et al. Role of pro-inflammatory cytokines in the pathophysiology of osteoarthritis. Nat Rev. Rheumatol. 2011;7:33–42)

21]. These pro-inflammatory cytokines are thought to be involved in the progression of cartilage degeneration and the resultant pain response and joint effusions observed in affected patients. It follows that much scientific interest has been directed at developing pharmacologic and injectable interventions that halt the inflammatory cascade and potentially diminish the rate of disease progression. However, despite these efforts, no disease-modifying agent has yet been identified to arrest or potentially reverse the effects of inflammation-mediated cartilage degradation in OA. Placental tissue extracts have been studied and used in wound therapy for many years due to the ability to suppress an excessive host tissue inflammatory response [22, 23]. Given the inflammatory component in the pathogenesis, symptoms, and progression of OA, the reported anti-inflammatory effects of culture-expanded MSC populations have been a source of recent scientific interest.

11.3.1.1 Basic Science

Small animal models examining the effects of placental tissue extracts in modulating disease progression of cartilage injury and OA have been performed in the knees of rats and rabbits [24–30]. In a rat model, Willet et al. [24] compared the intra-articular injection of dehydrated human AM/CM to saline and found fewer cartilage lesions, less articular erosion, and less proteoglycan loss in the treatment group at 21 days. In a similar study, Raines et al. [25] compared the intra-articular injection of particulate AM/UC matrix (2.5 vs. 5 mg) to saline and found less articular cartilage destruction at 1 week in both AM/UC groups compared to controls. However, this effect was sustained only in the high-dose group at 4 weeks, suggesting a dose-dependent effect of particulate AM/UC matrix. In a rat pain model of OA, Kimmerling et al. [27] tested an amniotic suspension allograft (ASA) (25 μL vs. 50 μL) against triamcinolone and saline and found that both intervention groups significantly reduced pain, as assessed by the animal's aversion to weight-bear, and swelling compared to the saline control group. Interestingly, synovial fluid IL-10 levels were significantly increased

after treatment in the ASA group, suggesting an unexpected anti-inflammatory response in these animals.

Recently, Reece et al. [28] showed that particle size and factor elution play a role in the clinical effectiveness of micronized dehydrated AM/CM, with smaller particles demonstrating a reduced therapeutic efficacy in an OA model. Also, Wang et al. [29] observed a superior effect when culture-expanded AM-derived MSCs were combined with HA in intra-articular injections in rats with articular cartilage injuries as compared to AM-MSCs and HA alone. This suggests that factors other than tissue source and dose impact the clinical efficacy of placental tissue extracts.

Rabbit models have also been used to test the effects of placental tissue extracts on the progression of OA [26, 30]. Marino-Martínez et al. [26] injected pulverized AM into the right knees of New Zealand rabbits with OA and saline in the left and found improved histological features of the cartilage in the intervention limbs, as well as decreased disease progression and delayed loss of extracellular matrix compared to controls. Similarly, You et al. [30] created 3.5-mm-wide and 3-mm-deep patellar chondral defects in 20 New Zealand rabbits and filled these defects with either culture-expanded AM-derived MSCs, AM-derived MSCs with cartilaginous particles, cartilaginous particles alone, or fibrin glue. The authors demonstrated a histologically and macroscopically superior quality of regenerated tissue in the AM-MSCs with cartilaginous particles group as compared to the other cohorts.

11.3.1.2 Human Studies

Research into the clinical efficacy of placental tissue extracts is still in its infancy, with very few studies published to date reporting outcomes [31–33]. Vines et al. [33] performed a prospective feasibility study for a larger trial and tested a single intra-articular injection of ASA in six patients with Kellgren-Lawrence (K-L) grades 3/4 knee OA and followed them for 1 year. The authors found an overall trend of improvement in the Knee Injury and Osteoarthritis Outcome Score (KOOS), International Knee Documentation Committee (IKDC), and Single Assessment

Numeric Evaluation scores, with no significant adverse events.

In large retrospective case series of 82 patients (100 knees) with knee OA treated with intra-articular injections of micronized, dehydrated acellular AM/CM, Alden et al. [31] found a 32%, 56%, and 65% increase in the KOOS score at 6 weeks, 3 months, and 6 months post-injection as compared to baseline. Furthermore, quality of life, sports/recreation, and pain scores improved by 111%, 118%, and 67% at 6 months.

In a recent multicenter single-blinded clinical trial, Farr et al. [32] randomized 200 patients with symptomatic [7-day visual analogue pain (VAS) score $\geq 4/10$] knee OA (K-L grades 2/3) to receive a single intra-articular injection of either ASA, HA, or saline. Patients reporting unacceptable pain at 3 months were considered treatment failures; the percentage of failures in each group were 13.2%, 68.8%, and 75% in the ASA, HA, and saline groups, respectively. Patients who received the ASA demonstrated significantly greater improvements from baseline in terms of the VAS, KOOS pain, and KOOS-activities of daily living scores as compared to the HA group at 3 months and both the HA and saline groups at 6 months.

At this stage, the basic science and early clinical evidence is positive and supports the continued investigation into the efficacy of placental tissue extracts in the treatment of OA. Specifically, research aimed at identifying the optimal method of preparation, tissue sourcing, and particle sizing are needed for standardization and ease of comparison between studies to improve the overall quality of available evidence.

11.3.2 Tendon Injury

Basic science studies have shown that particulated amniotic membrane and amniotic fluid processed in way that preserves native cell viability can promote increased cell density and matrix deposition, improve cell migration, and down-regulate certain pro-inflammatory cytokines in tenocytes [34]. These and other findings have led to the study of cellular placental tissue extracts in small and large animal models in the context of Achilles tendon injury [35–38], patellar tendinopathy [39], and rotator cuff tears [40, 41].

11.3.2.1 Basic Science

de Girolamo et al. [36] induced Achilles tendinopathy in the right limbs of rats using intratendinous injections of collagenase type I, then subsequently injected the tendons with either ASA, saline, or nothing 7 days following disease induction. The authors found significantly improved fiber organization, cell density, and less fatty deposition in the intervention group at 14 days as compared to controls. However, no significant differences between groups were found at the 28-day timepoint. Similarly, Coban et al. [35] compared primary repair augmented with AM and AF-derived culture-expanded MSCs to primary repair with/without augmentation with AF-MSCs in a rat model of 72 Achilles tendon ruptures and found no significant differences in terms of histological grade between groups at any timepoint. Liu et al. [37] compared human AM to porcine small intestinal mucosa in a rabbit model of Achilles rupture and found no significant differences in filamentous adhesion, cross-sectional areas of the rupture sites, levels of pro-inflammatory cytokines, and collagen type I expression between the two groups. Further, the authors found lower ultimate stress and Young's modulus and poorer vascularity in the AM cohort. In a sheep model of partial Achilles tendon rupture, Barboni et al. [38] showed that AM-derived culture-expanded MSCs embedded in fibrin glue accelerated tendon healing as compared to fibrin glue alone and performed better than controls during biomechanical testing.

In a collagenase-induced model of patellar tendinopathy, Ma et al. [39] compared CM-derived culture-expanded MSCs to saline in 60 male Sprague-Dawley rats and found a significantly higher load-to-failure and higher levels of pro-inflammatory markers IL-1ß and IL-6 in the MSC-treated tendons than saline controls at 2 weeks. However, there was no significant difference between the two groups in terms of load-to-failure at 4 weeks and stiffness at either timepoint.

Two large animal models have tested the modulatory effects of placental tissue extracts on rotator cuff tendon healing [38, 40, 41]. In a rabbit model of full-thickness subscapularis tears, Park et al. [41] showed that 7/10 tears were partially healed 4 weeks after ultrasound-guided injection with UCB-derived culture-expanded MSCs, while 3/10 remained completely torn. This was compared to HA and saline controls, which had no evidence of tendon healing in any of the 10 surgically created tears. Recently, Smith et al. [40] tested different biological scaffolds for augmentation of healing of 50% partial-thickness articular-sided supraspinatus rotator cuff tears in a large animal model using 16 purpose-bred dogs. They compared augmentation with either an AM, decellularized human dermal allograft, or bovine collagen patch to debridement alone. At 3 months, all treatment groups exhibited significantly higher VAS scores and less comfortable shoulder range of motion (ROM) as compared to controls, with no intergroup differences. However, at 6 months, the AM and dermal allograft groups had significantly lower VAS scores and more comfortable shoulder ROM as compared to the bovine collagen and debridement groups. In terms of MRI appearance, AM- and dermal allograft-treated groups demonstrated similar findings, with tendon fibers approaching normal appearance at 6 months.

These studies suggest that there may be a role for the use of placental tissue extracts in managing certain tendon injuries. Specifically, there may be a potential early beneficial effect of its use in the treatment of patellar and Achilles tendinopathy. However, the effects appear to be generally short-lived.

11.3.2.2 Human Studies

There have not been any human studies to date examining the effects of placental tissue extracts on tendon injuries in the orthopedic sports medicine literature.

11.3.3 Ligament Injury

The anterior cruciate ligament (ACL) is commonly injured in athletes, especially those participating in cutting and jumping sports [42].

ACL reconstruction has been shown to restore knee stability, increase the rate of return-to-play, and reduce societal costs as compared to rehabilitation alone [43–45]. Despite improved techniques of performing ACL reconstruction, failure rates remain relatively high. Given this, there is an ongoing interest in the use of biologics to augment the healing response and reduce the re-rupture rate after ACL reconstruction. Recently, this has resulted in the experimental application of placental tissue extracts in the setting of ACL surgery.

11.3.3.1 Basic Science

Li et al. [46] studied whether AM-derived culture-expanded MSCs would differentiate into ACL fibroblasts after induction with basic fibroblast growth factor-2 and transforming growth factor-ß1 in a human coculture in vitro model. The authors found that the monolayer and Transwell coculture systems resulted in MSC expression of ligament-related genes and proteins (collagen types I and III). Furthermore, they observed a time-dependent effect in the MSCs, with increased bioactivity over time. They surmised that this effect may have occurred because of the ACL fibroblasts producing MSC-regulating cytokines.

Jang et al. [47] investigated the transplantation of UCB-MSCs after semitendinosus tendon autograft ACL reconstruction in a rabbit model. An MSC-laden fibrin glue was injected at the tunnel bone-tendon interface after reconstruction in the intervention cohort, while controls received ACL reconstruction alone. The authors found significantly less tunnel enlargement and higher histologic scores at the bone-tendon interface in the intervention group, with fibrocartilaginous healing resembling a normal ACL enthesis, as compared to controls. Furthermore, there was no evidence of graft rejection in any of the treated animals.

11.3.3.2 Human Studies

Only two human applications of placental tissue extracts have been used in the context of ACL reconstruction [48, 49]. Woodall et al. [48] described a technique of augmenting a doubled semitendinosus allograft with an AM-containing matrix during ACL reconstruction, performed using suspensory fixation on the femur and screw-in-sheath fixation on the tibia. Subsequently, Lavender et al. [49] described a similar technique, but injected bone marrow aspirate mixed with allograft bone putty into the tunnels prior to the graft passage, and bone marrow aspirate between the AM matrix and the semitendinosus allograft after graft fixation. Both studies were technique articles and lacked a clinical follow-up component.

11.3.4 Plantar Fasciitis

Plantar fasciitis is the most common cause of plantar heel pain, affecting up to 10% of the general population over a lifetime [50]. It is most prevalent in runners, as well as in people with pes planus and reduced ankle ROM [51, 52]. In recalcitrant cases, plantar fasciitis can be an extremely debilitating condition with few proven therapeutic solutions. As such, several studies have tested the effects of injection with biologics on the natural history of this condition, including placental tissue extracts.

11.3.4.1 Basic Science
There have not been any basic science studies to date examining the effects of placental tissue extracts on plantar fasciitis.

11.3.4.2 Human Studies
Three randomized controlled trials have been published to date studying the use of acellular placental tissue extracts in chronic, recalcitrant plantar fasciitis [53–55]. In a feasibility study for a larger trial, Hanselman et al. [53] randomized 23 patients with chronic plantar fasciitis to two injection groups: cryopreserved human amniotic membrane (c-hAM) and corticosteroid. Three participants from each group received a second injection 6 weeks after the initial intervention. The authors found significant improvements in the Foot Health Status Questionnaire score at 18 weeks, but no differences in the VAS score or verbal percentage improvement at any follow-up timepoint. In a larger study, Zelen et al. [54] ran-

domized 45 patients with chronic plantar fasciitis to three injection groups: 1.25 cc micronized dehydrated human amniotic/chorionic membrane (mDHACM) with Marcaine, 0.5 cc mDHACM with Marcaine, and Marcaine-only. As compared to controls, the two intervention groups demonstrated significantly greater improvements in the American Orthopaedic Foot and Ankle Society Hindfoot score at all follow-ups, with no differences observed between the two mDHACM groups.

Recently, Cazzell et al. [55] performed a trial examining the 3-month safety and efficacy of acellular micronized dehydrated human amnion/chorion membrane (dHACM) injection in the treatment of chronic plantar fasciitis. The authors randomized 145 patients, with patients receiving a single injection of either dHACM or saline. There were no significant differences in patient demographics between the groups, aside from a slightly older age population in the control group (53 vs. 49 years). At 3 months, the intervention group demonstrated a significantly greater reduction in the VAS and Foot Function Index-Revised (FFI-R) scores from baseline as compared to controls (VAS: 76% vs. 45%, $p < 0.0001$; FFI-R: 60% vs. 40%, $p = 0.0004$). However, given that both groups experienced significant improvements in functional scores from baseline and that follow-up was short, further research is required to validate these results.

11.4 Conclusion

Within the relatively short history of the use of placental tissue-derived cells or extracellular matrix extracts in orthopedic sports medicine, the techniques used to extract, preserve, and apply these cells have varied widely, but also improved considerably, allowing preparation of therapeutic products with definable composition. Optimal sourcing, preparation, and clinical delivery of processed tissue matrix, tissue-derived cells, culture-expanded cells, or the secretory products of culture-expanded cells are actively being investigated as biological strategies to augment the healing response in a variety of orthopedic conditions. Further research into this field will help refine the clinical indications for the use and rigorous standards for the processing, quantitative characterization, and reproducible manufacturing of placental tissue-derived products for orthopedics and sports medicine applications.

> **Take-Home Messages**
> - Placental tissue extracts used for orthopedic conditions have advanced rapidly in recent years, after promising results were reported for treatment of corneal lesion, extremity ulcers, burn, cardiac, and hepatic patients.
> - Placental tissues advancements include improved techniques of extraction, preparation, storage, and delivery of human amniotic- and placenta-derived cells.
> - The biologic response may vary when comparing placental tissue and its component layers of chorion and amnion, whether amniotic fluid cells are included, and whether they are alive or dead.
> - Placental tissue extract provides the theoretical advantages of reduced immunogenicity, greater potency of younger progenitor populations, and absence of donor-site morbidity compared to stem cells obtained from autologous sources.
> - The difference in extracellular matrix composition and cellular composition may offer biological properties that other biologic injectable therapies (e.g., platelet-rich plasma and hyaluronic acid) do not provide.
> - The optimal preparation, storage, and injection dosing/timing remain under investigation.

References

1. Caplan AI, Correa D. PDGF in bone formation and regeneration: new insights into a novel mechanism involving MSCs. J Orthop. 2011;29(12):1795–803. Epub 2011/05/28.
2. Liang X, Ding Y, Zhang Y, Tse HF, Lian Q. Paracrine mechanisms of mesenchymal stem cell-based therapy: current status and perspectives. Cell Transplant. 2014;23(9):1045–59. Epub 2013/05/17.
3. Ganguly P, El-Jawhari JJ, Giannoudis PV, Burska AN, Ponchel F, Jones EA. Age-related changes in bone marrow mesenchymal stromal cells: a potential impact on osteoporosis and osteoarthritis development. Cell Transplant. 2017;26(9):1520–9. Epub 2017/11/09.
4. Nam TW, Oh HM, Lee JE, Kim JH, Hwang JM, Park E, et al. An unusual complication of sacral nerve root injury following bone marrow harvesting: a case report. BMC Cancer. 2019;19(1):347. Epub 2019/04/13.
5. Hyer CF, Berlet GC, Bussewitz BW, Hankins T, Ziegler HL, Philbin TM. Quantitative assessment of the yield of osteoblastic connective tissue progenitors in bone marrow aspirate from the iliac crest, tibia, and calcaneus. J Bone Joint Surg Am. 2013;95(14):1312–6. Epub 2013/07/19.
6. Parolini O, Alviano F, Bagnara GP, Bilic G, Buhring HJ, Evangelista M, et al. Concise review: isolation and characterization of cells from human term placenta: outcome of the first international workshop on placenta derived stem cells. Stem Cells. 2008;26(2):300–11. Epub 2007/11/03.
7. Trounson A, McDonald C. Stem cell therapies in clinical trials: progress and challenges. Cell Stem Cell. 2015;17(1):11–22. Epub 2015/07/04.
8. Malek A, Bersinger NA. Human placental stem cells: biomedical potential and clinical relevance. J Stem Cells. 2011;6(2):75–92. Epub 2011/01/01.
9. Roubelakis MG, Trohatou O, Anagnou NP. Amniotic fluid and amniotic membrane stem cells: marker discovery. Stem Cells Int. 2012;2012:107836. Epub 2012/06/16.
10. Choi YS, Park YB, Ha CW, Kim JA, Heo JC, Han WJ, et al. Different characteristics of mesenchymal stem cells isolated from different layers of full term placenta. PLoS One. 2017;12(2):e0172642. Epub 2017/02/23.
11. Bailo M, Soncini M, Vertua E, Signoroni PB, Sanzone S, Lombardi G, et al. Engraftment potential of human amnion and chorion cells derived from term placenta. Transplantation. 2004;78(10):1439–48. Epub 2004/12/16.
12. Tracy SA, Ahmed A, Tigges JC, Ericsson M, Pal AK, Zurakowski D, et al. A comparison of clinically relevant sources of mesenchymal stem cell-derived exosomes: bone marrow and amniotic fluid. J Pediatr Surg. 2019;54(1):86–90. Epub 2018/10/27.
13. Sarugaser R, Lickorish D, Baksh D, Hosseini MM, Davies JE. Human umbilical cord perivascular (HUCPV) cells: a source of mesenchymal progenitors. Stem Cells. 2005;23(2):220–9. Epub 2005/01/27.
14. Vangsness CT Jr, Sternberg H, Harris L. Umbilical cord tissue offers the greatest number of harvestable mesenchymal stem cells for research and clinical application: a literature review of differ-

ent harvest sites. Arthroscopy. 2015;31(9):1836–43. Epub 2015/09/12.

15. Lee OK, Kuo TK, Chen WM, Lee KD, Hsieh SL, Chen TH. Isolation of multipotent mesenchymal stem cells from umbilical cord blood. Blood. 2004;103(5):1669–75. Epub 2003/10/25.

16. De la Fuente A, Mateos J, Lesende-Rodriguez I, Calamia V, Fuentes-Boquete I, de Toro FJ, et al. Proteome analysis during chondrocyte differentiation in a new chondrogenesis model using human umbilical cord stroma mesenchymal stem cells. Mol Cell Proteomics. 2012;11(2):M111.010496. Epub 2011/10/20.

17. Majore I, Moretti P, Stahl F, Hass R, Kasper C. Growth and differentiation properties of mesenchymal stromal cell populations derived from whole human umbilical cord. Stem Cell Rev Rep. 2011;7(1):17–31. Epub 2010/07/03.

18. Marmotti A, Mattia S, Bruzzone M, Buttiglieri S, Risso A, Bonasia DE, et al. Minced umbilical cord fragments as a source of cells for orthopaedic tissue engineering: an in vitro study. Stem Cells Int. 2012;2012:326813. Epub 2012/05/03.

19. Prasanna SJ, Gopalakrishnan D, Shankar SR, Vasandan AB. Pro-inflammatory cytokines, IFNgamma and TNFalpha, influence immune properties of human bone marrow and Wharton jelly mesenchymal stem cells differentially. PLoS One. 2010;5(2):e9016. Epub 2010/02/04.

20. Berenbaum F. Osteoarthritis as an inflammatory disease (osteoarthritis is not osteoarthrosis!). Osteoarthr Cartil. 2013;21(1):16–21. Epub 2012/12/01.

21. Kapoor M, Martel-Pelletier J, Lajeunesse D, Pelletier JP, Fahmi H. Role of proinflammatory cytokines in the pathophysiology of osteoarthritis. Nat Rev Rheumatol. 2011;7(1):33–42. Epub 2010/12/02.

22. Liu J, Sheha H, Fu Y, Liang L, Tseng SC. Update on amniotic membrane transplantation. Expert Rev Ophthalmol. 2010;5(5):645–61. Epub 2011/03/26.

23. Adly OA, Moghazy AM, Abbas AH, Ellabban AM, Ali OS, Mohamed BA. Assessment of amniotic and polyurethane membrane dressings in the treatment of burns. Burns. 2010;36(5):703–10. Epub 2009/12/17.

24. Willett NJ, Thote T, Lin AS, Moran S, Raji Y, Sridaran S, et al. Intra-articular injection of micronized dehydrated human amnion/chorion membrane attenuates osteoarthritis development. Arthritis Res Ther. 2014;16(1):R47. Epub 2014/02/07.

25. Raines AL, Shih MS, Chua L, Su CW, Tseng SC, O'Connell J. Efficacy of particulate amniotic membrane and umbilical cord tissues in attenuating cartilage destruction in an osteoarthritis model. Tissue Eng Part A. 2017;23(1–2):12–9. Epub 2016/10/07.

26. Marino-Martinez IA, Martinez-Castro AG, Pena-Martinez VM, Acosta-Olivo CA, Vilchez-Cavazos F, Guzman-Lopez A, et al. Human amniotic membrane intra-articular injection prevents cartilage damage in an osteoarthritis model. Exp Ther Med. 2019;17(1):11–6. Epub 2019/01/18.

27. Kimmerling KA, Gomoll AH, Farr J, Mowry KC. Amniotic suspension allograft modulates inflammation in a rat pain model of osteoarthritis. J Orthop. 2019;38(5):1141–9.

28. Reece DS, Burnsed OA, Parchinski K, Marr EE, White RM, Salazar-Noratto GE, et al. Reduced size profile of amniotic membrane particles decreases osteoarthritis therapeutic efficacy. Tissue Eng Part A. 2020;26(1–2):28–37. Epub 2019/07/05.

29. Wang AT, Zhang QF, Wang NX, Yu CY, Liu RM, Luo Y, et al. Cocktail of hyaluronic acid and human amniotic mesenchymal cells effectively repairs cartilage injuries in sodium iodoacetate-induced osteoarthritis rats. Front Bioeng Biotechnol. 2020;8:87. Epub 2020/03/27.

30. You Q, Liu Z, Zhang J, Shen M, Li Y, Jin Y, et al. Human amniotic mesenchymal stem cell sheets encapsulating cartilage particles facilitate repair of rabbit osteochondral defects. Am J Sports Med. 2020;48(3):599–611. Epub 2020/01/16.

31. Alden KJ, Harris S, Hubbs B, Kot K, Istwan NB, Mason D. Micronized dehydrated human amnion chorion membrane injection in the treatment of knee osteoarthritis-a large retrospective case series. J Knee Surg. 2021;34(8):841–5. Epub 2019/11/30.

32. Farr J, Gomoll AH, Yanke AB, Strauss EJ, Mowry KC. A randomized controlled single-blind study demonstrating superiority of amniotic suspension allograft injection over hyaluronic acid and saline control for modification of knee osteoarthritis symptoms. J Knee Surg. 2019;32(11):1143–54. Epub 2019/09/19.

33. Vines JB, Aliprantis AO, Gomoll AH, Farr J. Cryopreserved amniotic suspension for the treatment of knee osteoarthritis. J Knee Surg. 2016;29(6):443–50. Epub 2015/12/20.

34. Kimmerling KA, McQuilling JP, Staples MC, Mowry KC. Tenocyte cell density, migration, and extracellular matrix deposition with amniotic suspension allograft. J Orthop. 2019;37(2):412–20. Epub 2018/11/01.

35. Coban I, Satoglu IS, Gultekin A, Tuna B, Tatari H, Fidan M. Effects of human amniotic fluid and membrane in the treatment of Achilles tendon ruptures in locally corticosteroid-induced Achilles tendinosis: an experimental study on rats. Foot Ankle Surg. 2009;15(1):22–7. Epub 2009/02/17.

36. de Girolamo L, Morlin Ambra LF, Perucca Orfei C, McQuilling JP, Kimmerling KA, Mowry KC, et al. Treatment with human amniotic suspension allograft improves tendon healing in a rat model of collagenase-induced tendinopathy. Cells. 2019;8(11). Epub 2019/11/14.

37. Liu Y, Peng Y, Fang Y, Yao M, Redmond RW, Ni T. No midterm advantages in the middle term using small intestinal submucosa and human amniotic membrane in Achilles tendon transverse tenotomy. J Orthop Surg Res. 2016;11(1):125. Epub 2016/11/25.

38. Barboni B, Russo V, Curini V, Mauro A, Martelli A, Muttini A, et al. Achilles tendon regeneration can be improved by amniotic epithelial cell allotransplanta-

tion. Cell Transplant. 2012;21(11):2377–95. Epub 2012/04/18.

39. Ma R, Schar M, Chen T, Wang H, Wada S, Ju X, et al. Use of human placenta-derived cells in a preclinical model of tendon injury. J Bone Joint Surg Am. 2019;101(13):e61. Epub 2019/07/06.

40. Smith MJ, Bozynski CC, Kuroki K, Cook CR, Stoker AM, Cook JL. Comparison of biologic scaffolds for augmentation of partial rotator cuff tears in a canine model. J Shoulder Elb Surg. 2020;29(8):1573–83. Epub 2020/03/15.

41. Park GY, Kwon DR, Lee SC. Regeneration of full-thickness rotator cuff tendon tear after ultrasound-guided injection with umbilical cord blood-derived mesenchymal stem cells in a rabbit model. Stem Cells Transl Med. 2015;4(11):1344–51. Epub 2015/09/16.

42. Spindler KP, Huston LJ, Chagin KM, Kattan MW, Reinke EK, Amendola A, et al. Ten-year outcomes and risk factors after anterior cruciate ligament reconstruction: a MOON longitudinal prospective cohort study. Am J Sports Med. 2018;46(4):815–25. Epub 2018/03/16.

43. Brophy RH, Schmitz L, Wright RW, Dunn WR, Parker RD, Andrish JT, et al. Return to play and future ACL injury risk after ACL reconstruction in soccer athletes from the Multicenter Orthopaedic Outcomes Network (MOON) group. Am J Sports Med. 2012;40(11):2517–22. Epub 2012/09/25.

44. Mather RC 3rd, Koenig L, Kocher MS, Dall TM, Gallo P, Scott DJ, et al. Societal and economic impact of anterior cruciate ligament tears. J Bone Joint Surg Am. 2013;95(19):1751–9. Epub 2013/10/04.

45. Filbay SR, Culvenor AG, Ackerman IN, Russell TG, Crossley KM. Quality of life in anterior cruciate ligament-deficient individuals: a systematic review and meta-analysis. Br J Sports Med. 2015;49(16):1033–41. Epub 2015/08/01.

46. Li Y, Liu Z, Jin Y, Zhu X, Wang S, Yang J, et al. Differentiation of human amniotic mesenchymal stem cells into human anterior cruciate ligament fibroblast cells by in vitro coculture. Biomed Res Int. 2017;2017:7360354. Epub 2017/11/01.

47. Jang KM, Lim HC, Jung WY, Moon SW, Wang JH. Efficacy and safety of human umbilical cord blood-derived mesenchymal stem cells in anterior cruciate ligament reconstruction of a rabbit model: new strategy to enhance tendon graft healing. Arthroscopy. 2015;31(8):1530–9. Epub 2015/04/18.

48. Woodall BM, Elena N, Gamboa JT, Shin EC, Pathare N, McGahan PJ, et al. Anterior cruciate ligament reconstruction with amnion biological augmentation. Arthrosc Tech. 2018;7(4):e355–e60. Epub 2018/06/06.

49. Lavender C, Bishop C. The fertilized anterior cruciate ligament: an all-inside anterior cruciate ligament reconstruction augmented with amnion, bone marrow concentrate, and a suture tape. Arthrosc Tech. 2019;8(6):e555–e9. Epub 2019/07/25.

50. Crawford F, Thomson C. Interventions for treating plantar heel pain. Cochrane Database Syst Rev. 2003(3):Cd000416. Epub 2003/08/15.

51. Buchbinder R. Clinical practice. Plantar fasciitis. N Engl J Med. 2004;350(21):2159–66. Epub 2004/05/21.

52. Riddle DL, Pulisic M, Pidcoe P, Johnson RE. Risk factors for plantar fasciitis: a matched case-control study. J Bone Joint Surg Am. 2003;85(5):872–7. Epub 2003/05/03.

53. Zelen CM, Poka A, Andrews J. Prospective, randomized, blinded, comparative study of injectable micronized dehydrated amniotic/chorionic membrane allograft for plantar fasciitis—a feasibility study. Foot Ankle Int. 2013;34(10):1332–9. Epub 2013/08/16.

54. Hanselman AE, Tidwell JE, Santrock RD. Cryopreserved human amniotic membrane injection for plantar fasciitis: a randomized, controlled, double-blind pilot study. Foot Ankle Int. 2015;36(2):151–8. Epub 2014/09/25.

55. Cazzell S, Stewart J, Agnew PS, Senatore J, Walters J, Murdoch D, et al. Randomized controlled trial of micronized Dehydrated Human Amnion/Chorion Membrane (dHACM) injection compared to placebo for the treatment of plantar fasciitis. Foot Ankle Int. 2018;39(10):1151–61. Epub 2018/07/31.

Secretome, Extracellular Vesicles, Exosomes

12

Florien Jenner and Iris Ribitsch

12.1 Introduction

Culture-expanded mesenchymal stromal cells (MSCs) have demonstrated their therapeutic potential in numerous preclinical models and clinical trials of inflammatory and degenerative pathologies, including osteochondral lesions, bone defects, and tendon injuries [1–11].

The therapeutic administration of MSCs was initially predicated on the premise that they regenerate tissue via engraftment and differentiation. However, the engraftment and survival of MSCs in the target tissue is negligible and mounting evidence suggests that MSCs exert their therapeutic effect predominantly by secreting a plethora of bioactive factors, collectively termed the secretome, to induce and support endogenous regeneration [4, 12–23]. This new paradigm of paracrine-mediated beneficial effects is supported by many studies achieving therapeutic efficacy with MSC treatments without relevant cell engraftment [13, 15, 17, 19, 24–40].

The secretome mirrors the ability of the donor cells to condition and program the surrounding microenvironment and has shown equivalent therapeutic potential to its donor cells in stroke, graft-versus-host disease, tendinopathy, osseous defects, and osteoarthritis (OA) in vitro and in vivo, thus paving the way for the development of cell-free therapies [15, 19, 25–37, 39, 40]. The therapeutic application of secreted molecules instead of living, replicating cells provides several advantages, including reduced safety concerns, easier standardization, process optimization, storage and clinical upscaling, and their potential as a ready-to-go, off-the-shelf biological therapeutic agent [6, 40–49]. Accordingly, the secretome has become a subject of intensive research to identify the mechanism of action and key bioactive molecules for regenerative medicine applications.

12.2 Composition of the Secretome and Its Extracellular Vesicle Fraction

The secretome is a composite product secreted by cells in vitro (conditioned medium, MSC-CM) and in vivo (in the extracellular milieu), which consists of soluble and extracellular vesicle (EV) bound proteins, lipids, and nucleic acids [4, 12, 14, 16, 18–23, 25, 50–52]. These bioactive molecules are important mediators of intercellular communication and orchestrate the regenerative processes in damaged tissues by modulating the immune response, regulating cell survival, renewal, and differentiation, inhibiting fibrosis, stimulating endogenous cell recruitment and proliferation, and promoting vascularization and matrix production [15, 39, 53–59].

F. Jenner (✉) · I. Ribitsch
Veterinary Tissue Engineering and Regenerative Medicine Laboratory, Equine Surgery Unit, University of Veterinary Medicine Vienna, Vienna, Austria
e-mail: Florien.Jenner@vetmeduni.ac.at

G. Filardo et al. (eds.), *Orthobiologics*, https://doi.org/10.1007/978-3-030-84744-9_12

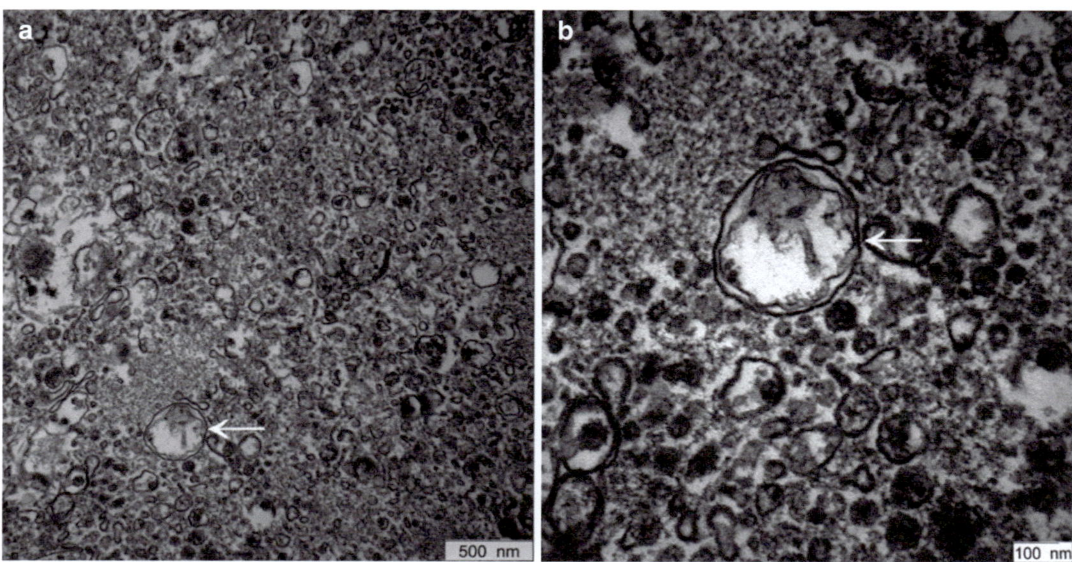

Fig. 12.1 Electron microscopy images showing EVs derived from equine MSCs by ultracentrifugation in 30,000× (**a**) resp. 85,000× (**b**) magnification. The white arrows indicate the same EV in both images

EVs (Fig. 12.1) are cell-derived membranous structures surrounded by a phospholipidic bilayer, which serve as intercellular messengers via receptor-mediated interaction and by transferring bioactive lipid, protein, and nucleic acid cargo to recipient cells to elicit regenerative processes and homeostasis [23, 39, 56, 60–62]. EVs harbour a specific subset of bioactive molecules, rather than random cellular factors, and contain both a common set of components, involved in vesicle structure, biogenesis, and trafficking, and specific subsets with cell type-associated pathophysiological functions [63]. The heterogeneous population of EVs can be further defined by their physical characteristics, such as size or density, their biochemical composition, the cells of origin, and culture conditions [56]. To avoid inconsistencies in the nomenclature, the traditional categorization into three subtypes exosomes (30–150 nm diameter, biogenesis: fusion of multivesicular endosomes with the plasma membrane), plasma membrane-derived microvesicles (100–1000 nm, biogenesis: outward budding of the plasma membrane), and apoptotic bodies (1–5 μm diameter, biogenesis: disassembly of apoptotic cells) has been discontinued as these terms are historically burdened by manifold, contradictory definitions, and inaccurate expectations of unique biogenesis [56].

Multi-omics compositional analyses of secretome and its fractions have identified a variety of growth factors, cytokines, chemokines, extracellular matrix proteins, and remodelling enzymes, including transforming growth factor-β (TGF-β) vascular endothelial growth factor (VEGF), basic fibroblast growth factor (bFGF), insulin-like growth factor (IGF), bone morphogenetic protein-2 (BMP-2), interleukin-1 receptor antagonist (IL-1Ra), interleukins-6 (IL-6) and 10 (IL-10), monocyte chemoattractant proteins-1 (MCP-1), and matrix metalloproteinases-1 (MMP-1) and 3 (MMP-3) [19, 59, 64–68]. In addition, total lipid analysis by mass spectrometry identified nearly two thousand lipids in MSC-CM [69]. The bioactive lipids secreted by culture-expanded MSC populations, such as leukotrienes, arachidonic acid, prostaglandins, diacylglycerols, lysophosphatidylcholine, phosphatidylserine, and sphingosine-1-phosphate, have been shown to contribute to diverse biological effects including intercellular communication, chemoattraction, angiogenesis, immune modulation, cell proliferation, and migration [21, 52, 69–71]. Furthermore, specialized pro-resolving lipid mediators, including resolvins, protectins, and maresins, stimulate inflammation resolution by modulating proinflammatory mechanisms and promoting tissue repair and homeostasis [34–36, 52, 63, 69–73]. In addition to proteins and lipids, EVs also contain a broad variety of nucleic acids, predominantly miRNAs, with each miRNA possibly affecting hundreds of different mRNA targets in the surrounding microenvironment, and therefore being able to regulate gene transcription and the functions of recipient cells and to affect numerous signalling pathways [50, 51, 72–74]. Comprehensive characterization of the miRNA cargo of EVs using high-throughput RNA sequencing revealed that EVs are selectively enriched for distinct classes of nucleic acids, including RNAs encoding TGF-β signalling and transcription factors, which are involved in angiogenesis, cellular transport, apoptosis, and extracellular matrix turnover. In contrast, calcium signalling, mitochondrial, and cytoskeleton genes are selectively depleted from EVs compared to the donor MSCs [72]. Furthermore, EVs contain many well-characterized anti-inflammatory miRNAs, including the miR-let7 family, which plays an important role in macrophage polarization towards the anti-inflammatory M2 phenotype and in reducing fibrosis and miRNAs targeting other downstream molecules in Toll-like receptor signalling including nuclear factor kappa-light-chain-enhancer of activated B cells (NF-κB), IL-6, and tumor necrosis factor α (TNF-α) [72, 73].

The composition of the secretome (and its fractions) and hence its therapeutic effects differ depending on the donor cell, its somatic function, developmental and anatomical origin, extent of differentiation, and the microenvironment surrounding the cells [54, 55, 58, 59, 75–79]. Correspondingly, the therapeutic potential of MSC-CM and EVs can be optimized for a given therapeutic application by careful selection of the donor cell, culture conditions, and cellular preconditioning, such as 3D culture, inflammatory stimulation, or hypoxia [52, 54, 55, 58, 59, 61, 75–81]. Conversely, EVs derived from IL-1β-pre-treated MSCs have been shown to induce OA changes in chondrocyte culture [82], and bone regenerative effect of MSC-derived EVs is impaired in type 1 diabetes [83], emphasizing the

limitations of autologous transplantation and the requirement for careful selection of the secretome/EVs donor cell and standardized culture conditions.

Thus far, the principal therapeutic fractions or bioactive factors underpinning the regenerative and immunomodulatory capabilities of MSC-CM in tissue repair have not been identified. To date very few studies have compared the therapeutic potential of MSC-CM to the corresponding EVs, and the difficulties in achieving equal dosage make these comparisons challenging [61]. However, recent research implies that the EVs and soluble fractions of the MSC secretome each have a distinct spectrum of bioactive factors and act in a synergistic manner to promote regeneration by simultaneously regulating multiple signalling pathways [21, 54, 73, 84, 85]. For example, in a muscle injury model, only the total secretomes but not the EVs fractions were able to reduce the number of senescent cells, whereas the anti-inflammatory effects of the secretome were mainly mediated by the EVs fraction [73]. Accordingly, complete MSC-CM has superior therapeutic potential compared to its EVs-, protein-, or lipid-enriched fractions alone [21, 54, 73, 84, 86].

12.2.1 Secretome: Preclinical and Clinical Evidence

Evaluating the therapeutic efficacy of secretome for cartilage regeneration in vitro, MSC-CM was shown to interfere with the NF-κB pathway and downregulate inflammation-associated and free-radical-related genes, such as TNF-α, IL-1β, and iNOS, to mediate anti-inflammatory, cytoprotective, and anti-catabolic effects, and to influence matrix turnover in OA-derived chondrocytes, synovium, and cartilage explants [62, 87–89]. In addition, MSC-CM enhanced production of immunosuppressive IL-10 in IL-1β-activated OA chondrocytes [90]. Furthermore, MSC-derived paracrine factors were shown to enhance chondrocyte proliferation, matrix synthesis, and phenotype maintenance [68, 91, 92]. MSC-CM was also able to counter the premature senescence of

OA chondrocytes and protect against degenerative changes by reducing p53 acetylation, downregulating senescence markers, and decreasing oxidative stress [93].

In preclinical in vivo studies, an intra-articular injection of the MSC secretome in a murine OA model provided early pain reduction and limited the development of cartilage damage analogous to MSC therapy [94]. Similarly, intra-articular injection of murine MSC-CM was shown to be effective in reducing disease severity, cartilage damage, knee-joint swelling, and histopathological changes in an antigen-induced model of inflammatory arthritis by suppressing TNFα induction, decreasing aggrecan cleavage, enhancing Treg function, and adjusting the Treg:Th17 ratio [95].

Evaluating the effects of MSC-CM on bone regeneration in vitro, MSC-CM enhanced the migration, proliferation, and expression of alkaline phosphatase and osteogenic differentiation of MSCs including the expression of osteogenic marker genes, such as osteocalcin and Runx2 [96, 97].

In an in vivo model of surgically induced bone lesions, MSC-CM induced bone regeneration analogous to MSC therapy with equivalent quantity and quality of newly formed bone, repaired area, bone density, arrangement of collagen fibres, maturation, and inorganic matrix calcification [19]. Similarly, MSC-CM promoted the formation of the new bone callus in a murine distraction osteogenesis gap by recruiting endogenous murine bone marrow MSCs and promoting their osteogenic differentiation via MCP-1/-3 and IL-3/-6 signalling [98]. The recruitment of endogenous cells by MSC-CM was also confirmed in a rat calvarial bone defect model, which showed increased new bone formation and migration of endogenous stem cells to the bone lesion following MSC-CM therapy [97]. Finally, the local application of MSC-CM in a rat distraction osteogenesis model significantly improved bone consolidation on μCT images, mechanical strength, and histological and immunohistochemistry parameters [96].

In studies evaluating the therapeutic potential of MSC-CM on muscle regeneration, secretome was shown to promote stem cell proliferation, migration, and skeletal muscle differentiation and to provide protection against cellular senescence in vitro [73, 86]. In vivo, in a murine muscle injury model, systemic administration of MSC-CM promoted muscle regeneration and yielded significantly larger newly formed fibres, higher number of capillaries/fibre, more committed muscle progenitors, and a decreased number of infiltrating macrophages [73, 86].

In a next step towards clinical translation, a few clinical trials using MSC-CM for bone regeneration established the safety and feasibility of secretome administration [57, 99–102]. Furthermore, the administration of MSC-CM to treat alveolar bone atrophy enhanced vascularization and early bone formation and reduced inflammation and infiltration of inflammatory cells, yielding improved bone quality in a shortened treatment time [100, 101]. Similarly, application of allograft MSC-CM has improved wound healing after fractional carbon dioxide laser resurfacing by reducing transient adverse effects such as erythema, hyperpigmentation, and transepidermal water loss [99].

12.2.2 Extracellular Vesicles: Preclinical and Clinical Evidence

In recent years, EVs have come into focus in regenerative medicine due to their pleiotropic trophic and immunological functions and their ability to induce functional and phenotypic changes in recipient cells by transfer of bioactive molecules [21, 30, 31, 44, 46, 51, 103, 104]. They have been established as a major paracrine effector of the therapeutic responses obtained with MSC transplantation and as a potential cell-free therapy for musculoskeletal regeneration in various orthopaedic conditions, including cartilage lesions and OA, critical-size bone defect healing, and muscle and tendon injuries [11, 13, 24, 38, 50–53, 58, 63, 64, 69–74, 77, 83, 89, 90, 102, 105–130].

Several studies have established the ability of EVs to treat cartilage injuries and OA. EVs have been shown to penetrate into articular cartilage, where they are internalized by chondrocytes and promote cartilage regeneration, reduce the inflammatory response in OA chondrocytes, and stimulate extracellular matrix production [7, 131, 132]. EVs increased the viability, proliferation, and migration of OA chondrocytes or chondrocyte progenitor cells in vitro with some studies showing a dose-dependent effect [108, 119, 121, 123, 132, 133]. Furthermore, EV treatment increased the expression of chondrogenic genes SRY-box 9 (SOX9) and Wnt-7A, collagen type II and aggrecan, and anti-inflammatory IL-10. In contrast, EVs downregulated the production of proinflammatory mediators IL-1α, IL-1β, IL-6, IL-8, IL-17, TNF-α, and cyclooxygenase-2, catabolic markers matrix metalloproteinase 13 and ADAMTS5, and hypertrophy markers RUNX2, collagen type-X, alkaline phosphatase, and osteocalcin [7, 74, 85, 108, 110, 111, 119, 123, 132, 134–136]. The latter was in part achieved by inhibition of NF-κB and activator protein-1 and the downregulation of inducible nitric oxide synthase [110]. EVs were also shown to protect chondrocytes from apoptosis and to downregulate senescence-associated β-galactosidase activity in osteoarthritic osteoblasts [110, 132]. EVs are also incorporated into synoviocytes, where they reduced the expression of proinflammatory cytokines and chemokines in a chronic model of osteoarthritic synoviocytes [111].

In vivo, intra-articularly injected EVs protected from joint damage and attenuated disease scores in chemically induced rodent OA models, a murine destabilized meniscus model, and an antigen-induced synovitis porcine model [35, 36, 108, 118, 132]. In a rat osteochondral and a lapine chondral defect model, EV treatment induced chondrocyte proliferation, matrix deposition, and cartilage defect repair yielding hyaline cartilage with good surface integrity, complete bonding to adjacent cartilage, subchondral bone restoration, and increased M2 macrophage polarization with a concomitant decrease in inflammatory cytokines [34, 120, 123, 133]. Finally, EV treatment reduced pain and corresponding lameness associated with murine OA [108, 120].

In addition, MSC-derived EVs also show great potential in treating osteoporosis and bone defects. In vitro, EVs were able to promote the proliferation and osteogenic differentiation, including upregulation of osteogenic genes in primary osteoblastic cells and in MSCs from ovariectomized rats [117, 121, 122] Furthermore, EVs antagonized hypoxia and serum deprivation-induced osteocyte apoptosis and osteocyte-mediated osteoclastogenesis, confirming the therapeutic potential of EVs in age-related bone disease [113].

In vivo studies showed that EVs significantly stimulated bone regeneration and angiogenesis in critical-sized calvarial defects in rats with and without ovariectomy and accelerated bone healing in a murine femur fracture model [83, 116, 121, 122].

The intravenous injection of EVs attenuated bone loss in old mice by stimulating bone formation and inhibiting bone resorption, yielding increased trabecular and cortical bone mass, enhanced osteoblast formation, and reduced osteoclast formation compared to the control mice [114].

EVs also promoted myogenesis and angiogenesis in vitro and muscle regeneration with a modulated inflammatory response in an in vivo model of muscle injury [73, 86, 107]. Similarly, in a mouse Achilles tendon rupture model, EVs supported healing by increasing the number of endothelial cells and decreasing the M1/M2 ratio [105]. The anti-inflammatory and proregenerative effect of EV treatment in tendon healing was also confirmed by a reduced rate of post-repair tendon gap formation and rupture, increased collagen formation at the injury site, and a decrease in proinflammatory genes Il1b and Ifng via modulation of the macrophage inflammatory response [106].

12.3 Conclusions

MSC-derived secretome and its EV fraction have shown their therapeutic potential in the treatment of numerous orthopaedic indications, paving the way for cell-free, off-the-shelf regenerative medicine applications. The absence of replicating (allogeneic) cells in secretome-based therapies improves patient safety and facilitates product standardization, quality control, and cost-efficient production and storage. In addition, donor cells, culture conditions, and preconditioning strategies can be optimized to tailor the secretome/EVs therapy for each indication specifically and provide off-the-shelf therapies for immediate application in acute conditions. Furthermore, secretome-based biological medicines may be evaluated for safety, dosage, potency, and efficacy analogous to conventional pharmaceutical agents which facilitates their safe translation into clinical practice.

However, much still needs to be done to facilitate successful clinical translation. Rational therapy design will rely on a comprehensive understanding of the common and donor cell-specific molecular secretome/EV components and their in vivo functionality and the identification of optimal secretome donor cells, preconditioning, and culture methods for each indication. This may include the use of transgenic donor cells that are genetically modified to overexpress certain proteins and miRNAs or strategies to exogenously enrich EVs with specific proteins and nucleic acids via, e.g., electroporation, freeze-thaw cycles, saponin-mediated loading, or hypotonic dialysis, to enhance their functionality and therapeutic potential. Also, GMP-compliant protocols for the preparation and storage of the MSC secretome and quality control parameters need to be developed to ensure the safety and efficacy of MSC-CM and its fractions. Additional studies focusing on the optimal dosage and administration frequency are also essential to facilitate the successful translation of secretome/EV-based treatments into the clinical setting.

Take-Home Messages
- Mesenchymal stromal cells (MSCs) exert their therapeutic effect predominantly by secreting a plethora of bioactive factors, collectively termed secretome, to induce and support endogenous regeneration.
- The secretome is a composite product secreted by cells in vitro (conditioned medium) and in vivo, which consists of soluble and extracellular vesicle (EV) bound proteins, lipids, and nucleic acids. These bioactive molecules are important mediators of intercellular communication and orchestrate the regenerative processes in damaged tissues by modulating the immune response, regulating cell survival, renewal, and differentiation, inhibiting fibrosis, stimulating endogenous cell recruitment and proliferation, and promoting vascularization and matrix production.
- EVs are cell-derived membranous structures surrounded by a phospholipidic bilayer, which serve as intercellular messengers via receptor-mediated interaction and by transferring bioactive lipid, protein, and nucleic acid cargo to recipient cells to elicit regenerative processes and homeostasis. The heterogeneous population of EVs can be defined by their physical characteristics (e.g., size, density, morphology), their biochemical composition, the cells of origin, and culture conditions.
- The secretome has proven equivalent therapeutic potential to its donor cells. The systemic and local administration of conditioned medium and/or its EV fraction has yielded promising results in a wide variety of currently intractable musculoskeletal problems including cartilage defects, osteoarthritis, bone defects, osteoporosis, muscle, and tendon injuries.
- The composition of the secretome (and its fractions) and hence its therapeutic effects differ depending on the donor cell, its somatic function, developmental and anatomical origin, extent of differentiation, and the microenvironment surrounding the cells. Correspondingly, the therapeutic potential of MSC-CM and EVs could be optimized for a given therapeutic application by careful selection of the donor cells, culture condi-

tions, and cellular preconditioning, such as 3D culture, inflammatory stimulation, or hypoxia.

- The therapeutic application of cell-free preparations comprising secretome derivates (e.g., conditioned medium, EV fraction) shows excellent promise as a new treatment approach for orthopaedic disorders, as it circumvents safety concerns associated with transplantation of living, replicating cells, facilitates standardization, quality control, and clinical upscaling, and can be produced as a ready-to-go, off-the-shelf biological therapeutic agent.

References

1. Caplan AI. Mesenchymal stem cells. J Orthop Res. 1991;9(5):641–50.
2. Awad HA, Butler DL, Boivin GP, Smith FN, Malaviya P, Huibregtse B, et al. Autologous mesenchymal stem cell-mediated repair of tendon. Tissue Eng. 1999;5(3):267–77.
3. Murphy JM, Fink DJ, Hunziker EB, Barry FP. Stem cell therapy in a caprine model of osteoarthritis. Arthritis Rheum. 2003;48(12):3464–74.
4. Murphy MB, Moncivais K, Caplan AI. Mesenchymal stem cells: environmentally responsive therapeutics for regenerative medicine. Exp Mol Med. 2013;45(11):e54.
5. Barry F, Murphy M. Mesenchymal stem cells in joint disease and repair. Nat Rev Rheumatol. 2013;9(10):584–94.
6. Harrell CR, Markovic BS, Fellabaum C, Arsenijevic A, Volarevic V. Mesenchymal stem cell-based therapy of osteoarthritis: current knowledge and future perspectives. Biomed Pharmacother. 2019;109:2318–26.
7. Vonk LA, van Dooremalen SFJ, Liv N, Klumperman J, Coffer PJ, Saris DBF, et al. Mesenchymal stromal/stem cell-derived extracellular vesicles promote human cartilage regeneration in vitro. Theranostics. 2018;8(4):906–20.
8. Zhang W, Ouyang H, Dass CR, Xu J. Current research on pharmacologic and regenerative therapies for osteoarthritis. Bone Res. 2016;4(1):15040.
9. Abou-El-Enein M, Elsanhoury A, Reinke P. Overcoming challenges facing advanced therapies in the EU market. Cell Stem Cell. 2016;19(3):293–7.
10. Kang Y, Zheng L. Rejuvenation: an integrated approach to regenerative medicine. Regen Med Res. 2013;1(1):7.
11. Jo CH, Lee YG, Shin WH, Kim H, Chai JW, Jeong EC, et al. Intra-articular injection of mesenchymal stem cells for the treatment of osteoarthritis of the knee: a proof-of-concept clinical trial. Stem Cells. 2014;32(5):1254–66.
12. Gnecchi M, He H, Liang OD, Melo LG, Morello F, Mu H, et al. Paracrine action accounts for marked protection of ischemic heart by Akt-modified mesenchymal stem cells. Nat Med. 2005;11(4):367–8.
13. Tögel F, Hu Z, Weiss K, Isaac J, Lange C, Westenfelder C. Administered mesenchymal stem cells protect against ischemic acute renal failure through differentiation-independent mechanisms. Am J Physiol-Renal. 2005;289(1):F31–42.
14. Caplan AI, Dennis JE. Mesenchymal stem cells as trophic mediators. J Cell Biochem. 2006;98(5):1076–84.
15. da Silva Meirelles L, Fontes AM, Covas DT, Caplan AI. Mechanisms involved in the therapeutic properties of mesenchymal stem cells. Cytokine Growth Factor Rev. 2009;20(5–6):419–27.
16. Caplan AI, Correa D. The MSC: an injury drugstore. Cell Stem Cell. 2011;9(1):11–5.
17. von Bahr L, Batsis I, Moll G, Hägg M, Szakos A, Sundberg B, et al. Analysis of tissues following mesenchymal stromal cell therapy in humans indicates limited long-term engraftment and no ectopic tissue formation. Stem Cells. 2012;30(7):1575–8.
18. Madrigal M, Rao KS, Riordan NH. A review of therapeutic effects of mesenchymal stem cell secretions and induction of secretory modification by different culture methods. J Transl Med. 2014;12(1):1.
19. Linero I, Chaparro O. Paracrine effect of mesenchymal stem cells derived from human adipose tissue in bone regeneration. PLoS One. 2014;9(9):e107001.
20. Gnecchi M, Danieli P, Malpasso G, Ciuffreda MC. Paracrine mechanisms of mesenchymal stem cells in tissue repair. Methods Mol Biol. 2016;1416(Suppl 1):123–46.
21. Beer L, Zimmermann M, Mitterbauer A, Ellinger A, Gruber F, Narzt M-S, et al. Analysis of the secretome of apoptotic peripheral blood mononuclear cells: impact of released proteins and exosomes for tissue regeneration. Apoptosis. 2015;21:1–18.
22. Shim G, Lee S, Han J, Kim G, Jin H, Miao W, et al. Pharmacokinetics and in vivo fate of intra-articularly transplanted human bone marrow-derived clonal mesenchymal stem cells. Stem Cells Dev. 2015;24(9):1124–32.
23. Mancuso P, Raman S, Glynn A, Barry F, Murphy JM. Mesenchymal stem cell therapy for osteoarthritis: the critical role of the cell secretome. Front Bioeng Biotechnol. 2019;7:1315.
24. Crisostomo PR, Markel TA, Wang Y, Meldrum DR. Surgically relevant aspects of stem cell paracrine effects. Surgery. 2008;143(5):577–81.

25. Lai RC, Arslan F, Lee MM, Sze NSK, Choo A, Chen TS, et al. Exosome secreted by MSC reduces myocardial ischemia/reperfusion injury. Stem Cell Res. 2010;4(3):214–22.

26. Vonk LA, de Windt TS, Kragten AHM, Beekhuizen M, Mastbergen SC, Dhert WJA, et al. Enhanced cell-induced articular cartilage regeneration by chondrons; the influence of joint damage and harvest site. Osteoarthr Cartil. 2014;22(11):1910–7.

27. Lener T, Gimona M, Aigner L, Börger V, Buzas E, Camussi G, et al. Applying extracellular vesicles based therapeutics in clinical trials—an ISEV position paper. J Extracell Vesicles. 2015;4(1):30087.

28. Doeppner TR, Herz J, Görgens A, Schlechter J, Ludwig A-K, Radtke S, et al. Extracellular vesicles improve post-stroke neuroregeneration and prevent postischemic immunosuppression. Stem Cells Transl Med. 2015;4(10):1131–43.

29. Bach FC, de Vries S, Krouwels A, Creemers LB, Ito K, Meij BP, et al. The species-specific regenerative effects of notochordal cell-conditioned medium on chondrocyte-like cells derived from degenerated human intervertebral discs. Eur Cell Mater. 2015;30:132–47.

30. Sevivas N, Teixeira FG, Portugal R, Araújo L, Carriço LF, Ferreira N, et al. Mesenchymal stem cell secretome: a potential tool for the prevention of muscle degenerative changes associated with chronic rotator cuff tears. Am J Sports Med. 2016;45(1):179–88.

31. Teixeira FG, Panchalingam KM, Assunção-Silva R, Serra SC, Mendes-Pinheiro B, Patrício P, et al. Modulation of the mesenchymal stem cell secretome using computer-controlled bioreactors: impact on neuronal cell proliferation, survival and differentiation. Sci Rep. 2016;6:27791.

32. Lombardi F, Palumbo P, Augello FR, Cifone MG, Cinque B, Giuliani M. Secretome of Adipose Tissue-Derived Stem Cells (ASCs) as a novel trend in chronic non-healing wounds: an overview of experimental in vitro and in vivo studies and methodological variables. IJMS. 2019;20(15):3721.

33. Kim MJ, Kim ZH, Kim S-M, Choi Y-S. Conditioned medium derived from umbilical cord mesenchymal stem cells regenerates atrophied muscles. Tissue Cell. 2016;48(5):533–43.

34. Zhang S, Chu WC, Lai RC, Lim SK, Hui JHP, Toh WS. Exosomes derived from human embryonic mesenchymal stem cells promote osteochondral regeneration. Osteoarthr Cartil. 2016;24(12):2135–40.

35. Wang Y, Yu D, Liu Z, Zhou F, Dai J, Wu B, et al. Exosomes from embryonic mesenchymal stem cells alleviate osteoarthritis through balancing synthesis and degradation of cartilage extracellular matrix. Stem Cell Res Ther. 2017;8(1):376.

36. Zhu Y, Wang Y, Zhao B, Niu X, Hu B, Li Q, et al. Comparison of exosomes secreted by induced pluripotent stem cell-derived mesenchymal stem cells and synovial membrane-derived mesenchymal stem cells for the treatment of osteoarthritis. Stem Cell Res Ther. 2017;8(1):376.

37. Tao S-C, Yuan T, Zhang Y-L, Yin W-J, Guo S-C, Zhang C-Q. Exosomes derived from miR-140-5p-overexpressing human synovial mesenchymal stem cells enhance cartilage tissue regeneration and prevent osteoarthritis of the knee in a rat model. Theranostics. 2017;7(1):180–95.

38. Park YB, Ha CW, Kim JA, Han WJ, Rhim JH, Lee HJ, et al. Single-stage cell-based cartilage repair in a rabbit model: cell tracking and in vivo chondrogenesis of human umbilical cord blood-derived mesenchymal stem cells and hyaluronic acid hydrogel composite. Osteoarthr Cartil. 2017;25(4):570–80.

39. Muhammad SA, Nordin N, Mehat MZ, Fakurazi S. Comparative efficacy of stem cells and secretome in articular cartilage regeneration: a systematic review and meta-analysis. Cell Tissue Res. 2018;5(4):e37976.

40. D'Arrigo D, Roffi A, Cucchiarini M, Moretti M, Candrian C, Filardo G. Secretome and extracellular vesicles as new biological therapies for knee osteoarthritis: a systematic review. J Clin Med. 2019;8(11):1867.

41. Murphy JM, Dixon K, Beck S, Fabian D, Feldman A, Barry F. Reduced chondrogenic and adipogenic activity of mesenchymal stem cells from patients with advanced osteoarthritis. Arthritis &. Rheumatology. 2002;46(3):704–13.

42. Somoza RA, Welter JF, Correa D, Caplan AI. Chondrogenic differentiation of mesenchymal stem cells: challenges and unfulfilled expectations. Tissue Eng Part B Rev. 2014;20(6):596–608.

43. Toh WS, Brittberg M, Farr J, Foldager CB, Gomoll AH, Hui JHP, et al. Cellular senescence in aging and osteoarthritis. Acta Orthop. 2016;87(sup363):6–14.

44. Konala VBR, Mamidi MK, Bhonde R, Das AK, Pochampally R, Pal R. The current landscape of the mesenchymal stromal cell secretome: a new paradigm for cell-free regeneration. J Cytotherapy. 2016;18(1):13–24.

45. Reiner AT, Witwer KW, van Balkom BWM, de Beer J, Brodie C, Corteling RL, et al. Concise review: developing best-practice models for the therapeutic use of extracellular vesicles. Stem Cells Transl Med. 2017;6(8):1730–9.

46. Toh WS, Lai RC, Hui JHP, Lim S-K. MSC exosome as a cell-free MSC therapy for cartilage regeneration: implications for osteoarthritis treatment. Semin Cell Dev Biol. 2017;67:56–64.

47. Vizoso F, Eiro N, Cid S, Schneider J, Perez-Fernandez R. Mesenchymal stem cell secretome: toward cell-free therapeutic strategies in regenerative medicine. IJMS. 2017;18(9):1852.

48. Iijima H, Isho T, Kuroki H, Takahashi M, Aoyama T. Effectiveness of mesenchymal stem cells for treating patients with knee osteoarthritis: a meta-analysis toward the establishment of effective regenerative rehabilitation. NPJ Regen Med. 2018;3(1):15.

49. Li JJ, Hosseini-Beheshti E, Grau GE, Zreiqat H, Little CB. Stem cell-derived extracellular

vesicles for treating joint injury and osteoarthritis. Nanomaterials (Basel). 2019;9(2):261.

50. Krek A, Grün D, Poy MN, Wolf R, Rosenberg L, Epstein EJ, et al. Combinatorial microRNA target predictions. Nat Genet. 2005;37(5):495–500.

51. Valadi H, Ekstrom K, Bossios A, Sjöstrand M, Lee JJ, Lötvall JO. Exosome-mediated transfer of mRNAs and microRNAs is a novel mechanism of genetic exchange between cells. Nat Cell Biol. 2007;9(6):654–9.

52. Maacha S, Sidahmed H, Jacob S, Gentilcore G, Calzone R, Grivel J-C, et al. Paracrine mechanisms of mesenchymal stromal cells in angiogenesis. Stem Cells Int. 2020;2020:1–12.

53. Haynesworth SE, Baber MA, Caplan A. Cytokine expression by human marrow-derived mesenchymal progenitor cells in vitro: effects of dexamethasone and IL-1 alpha. J Cell Physiol. 1996;166(3):585–92.

54. Lavoie JR, Rosu-Myles M. Uncovering the secretes of mesenchymal stem cells. Biochimie. 2013;95(12):1–10.

55. Severino V, Alessio N, Farina A, Sandomenico A, Cipollaro M, Peluso G, et al. Insulin-like growth factor binding proteins 4 and 7 released by senescent cells promote premature senescence in mesenchymal stem cells. Cell Death Dis. 2013;4:e911.

56. Théry C, Witwer KW, Aikawa E, Alcaraz MJ, Anderson JD, Andriantsitohaina R, et al. Minimal information for studies of extracellular vesicles 2018 (MISEV2018): a position statement of the International Society for Extracellular Vesicles and update of the MISEV2014 guidelines. J Extracell Vesicles. 2018;7(1):1535750.

57. Kumar P, Kandoi S, Misra R, Vijayalakshmi S, Rajagopal K, Verma RS. The mesenchymal stem cell secretome: a new paradigm towards cell-free therapeutic mode in regenerative medicine. Cytokine Growth Factor Rev. 2019;46:1–9.

58. Ferreira JR, Teixeira GQ, Santos SG, Barbosa MA, Almeida-Porada G, Gonçalves RM. Mesenchymal stromal cell secretome: influencing therapeutic potential by cellular pre-conditioning. Front Immunol. 2018;9:2.

59. Islam A, Urbarova I, Bruun JA, Martinez-Zubiaurre I. Large-scale secretome analyses unveil the superior immunosuppressive phenotype of umbilical cord stromal cells as compared to other adult mesenchymal stem cells. Eur Cell Mater. 2019;37:153–74.

60. Munshi A, Mehic J, Creskey M, Gobin J, Gao J, Rigg E, et al. A comprehensive proteomics profiling identifies NRP1 as a novel identity marker of human bone marrow mesenchymal stromal cell-derived small extracellular vesicles. Stem Cell Res Ther. 2019;10(1):1–18.

61. Piazza N, Dehghani M, Gaborski TR, Wuertz-Kozak K. Therapeutic potential of extracellular vesicles in degenerative diseases of the intervertebral disc. Front Bioeng Biotechnol. 2020;8:311.

62. Palamà MEF, Shaw GM, Carluccio S, Reverberi D, Sercia L, Persano L, et al. The secretome derived from mesenchymal stromal cells cultured in a xeno-free medium promotes human cartilage recovery in vitro. Front Bioeng Biotechnol. 2020;8:302.

63. Choi D-S, Kim D-K, Kim Y-K, Gho YS. Proteomics, transcriptomics and lipidomics of exosomes and ectosomes. Proteomics. 2013;13(10–11):1554–71.

64. Skalnikova H, MOTLÍK J, Gadher SJ, Kovarova H. Mapping of the secretome of primary isolates of mammalian cells, stem cells and derived cell lines. Proteomics. 2011;11(4):691–708.

65. Skalnikova HK. Proteomic techniques for characterisation of mesenchymal stem cell secretome. Biochimie. 2013;95(12):2196–211.

66. Harrell C, Fellabaum C, Jovicic N, Djonov V, Arsenijevic N, Volarevic V. Molecular mechanisms responsible for therapeutic potential of mesenchymal stem cell-derived secretome. Cell. 2019;8(5):467.

67. Baberg F, Geyh S, Waldera-Lupa D, Stefanski A, Zilkens C, Haas R, et al. Secretome analysis of human bone marrow derived mesenchymal stromal cells. Biochim Biophys Acta Proteins Proteom. 2019;1867(4):434–41.

68. Parate D, Kadir ND, Celik C, Lee EH, Hui JHP, Franco-Obregón A, et al. Pulsed electromagnetic fields potentiate the paracrine function of mesenchymal stem cells for cartilage regeneration. Stem Cell Res Ther. 2020;11(1):S11.

69. Lakatos K, Kalomoiris S, Merkely B, Nolta JA, Fierro FA. Mesenchymal stem cells respond to hypoxia by increasing diacylglycerols. J Cell Biochem. 2016;117(2):300–7.

70. Lauber K, Bohn E, Kröber SM, Xiao Y-J, Blumenthal SG, Lindemann RK, et al. Apoptotic cells induce migration of phagocytes via caspase-3-mediated release of a lipid attraction signal. Cell Transplant. 2003;113(6):717–30.

71. Romano M, Patruno S, Pomilio A, Recchiuti A. Proresolving lipid mediators and receptors in stem cell biology: concise review. Stem Cells Transl Med. 2019;8(10):992–8.

72. Eirin A, Riester SM, Zhu X-Y, Tang H, Evans JM, O'Brien D, et al. MicroRNA and mRNA cargo of extracellular vesicles from porcine adipose tissue-derived mesenchymal stem cells. Gene. 2014;551(1):55–64.

73. Mitchell R, Mellows B, Sheard J, Antonioli M, Kretz O, Chambers D. et al. Secretome of adipose-derived mesenchymal stem cells promotes skeletal muscle regeneration through synergistic action of extracellular vesicle cargo and soluble proteins. Stem Cell Res Ther. 2019;10(1):116.

74. Sun H, Hu S, Zhang Z, Lun J, Liao W, Zhang Z. Expression of exosomal microRNAs during chondrogenic differentiation of human bone mesenchymal stem cells. J Cell Biochem. 2019;120(1):171–81.

75. Baraniak PR, Mcdevitt TC. Stem cell paracrine actions and tissue regeneration. Regen Med. 2010;5(1):121–43.

76. Polacek M, Bruun J-A, Elvenes J, Figenschau Y, Martinez I. The secretory profiles of cultured human

articular chondrocytes and mesenchymal stem cells: implications for autologous cell transplantation strategies. Cell Transplant. 2011;20(9):1381–93.

77. Amable PR, Teixeira MVT, Carias RBV, Granjeiro JM, Borojevic R. Protein synthesis and secretion in human mesenchymal cells derived from bone marrow, adipose tissue and Wharton's jelly. Stem Cell Res Ther. 2014;5(2):53.

78. Dabrowski FA, Burdzinska A, Kulesza A, Sladowska A, Zolocinska A, Gala K, et al. Comparison of the paracrine activity of mesenchymal stem cells derived from human umbilical cord, amniotic membrane and adipose tissue. J Obstet Gynaecol Res. 2017;43(11):1758–68.

79. Pires AO, Mendes-Pinheiro B, Teixeira FG, Anjo SI, Ribeiro-Samy S, Gomes ED, et al. Unveiling the differences of secretome of human bone marrow mesenchymal stem cells, adipose tissue-derived stem cells, and human umbilical cord perivascular cells: a proteomic analysis. Stem Cells Dev. 2016;25(14):1073–83.

80. Park K-S, Bandeira E, Shelke GV, Lässer C, Lötvall J. Enhancement of therapeutic potential of mesenchymal stem cell-derived extracellular vesicles. Stem Cell Res Ther. 2019;10(1):1–15.

81. Wu X, Wang Y, Xiao Y, Crawford R, Mao X, Prasadam I. Extracellular vesicles: potential role in osteoarthritis regenerative medicine. J Orthop Translat. 2020;21:73–80.

82. Kato T, Miyaki S, Ishitobi H, Nakamura Y, Nakasa T, Lotz MK, et al. Exosomes from IL-1β stimulated synovial fibroblasts induce osteoarthritic changes in articular chondrocytes. Arthritis Res Ther. 2014;16(4):R163.

83. Zhu Y, Jia Y, Wang Y, Xu J, Chai Y. Impaired bone regenerative effect of exosomes derived from bone marrow mesenchymal stem cells in type 1 diabetes. Stem Cells Transl Med. 2019;8(6):593–605.

84. Wagner T, Traxler D, Simader E, Beer L, Narzt M-S, Gruber F, et al. Different pro-angiogenic potential of γ-irradiated PBMC-derived secretome and its subfractions. Sci Rep. 2018;8(1):18016.

85. Niada S, Giannasi C, Gualerzi A, Banfi G, Brini AT. Differential proteomic analysis predicts appropriate applications for the secretome of adipose-derived mesenchymal stem/stromal cells and dermal fibroblasts. Stem Cells Int. 2018;2018(2):1–11.

86. Mellows B, Mitchell R, Antonioli M, Kretz O, Chambers D, Zeuner M-T, et al. Protein and molecular characterization of a clinically compliant amniotic fluid stem cell-derived extracellular vesicle fraction capable of accelerating muscle regeneration through enhancement of angiogenesis. Stem Cells Dev. 2017;26(18):1316–33.

87. van Buul GM, Villafuertes E, Bos PK, Waarsing JH, Kops N, Narcisi R, et al. Mesenchymal stem cells secrete factors that inhibit inflammatory processes in short-term osteoarthritic synovium and cartilage explant culture. Osteoarthr Cartil. 2012;20(10):1186–96.

88. Platas J, Guillén MI, del Caz MDP, Gomar F, Mirabet V, Alcaraz MJ. Conditioned media from adipose-tissue-derived mesenchymal stem cells downregulate degradative mediators induced by interleukin-1β in osteoarthritic chondrocytes. Mediat Inflamm. 2013;2013(1–2):1–10.

89. Chen Y-C, Chang Y-W, Tan KP, Shen Y-S, Wang Y-H, Chang C-H. Can mesenchymal stem cells and their conditioned medium assist inflammatory chondrocytes recovery? Vinci MC, editor. PLoS One. 2018;13(11):e0205563.

90. Tofiño-Vian M, Guillén MI, del Caz MDP, Silvestre A, Alcaraz MJ. Microvesicles from human adipose tissue-derived mesenchymal stem cells as a new protective strategy in osteoarthritic chondrocytes. Cell Physiol Biochem. 2018;47(1):11–25.

91. Wu L, Leijten JCH, Georgi N, Post JN, van Blitterswijk CA, Karperien M. Trophic effects of mesenchymal stem cells increase chondrocyte proliferation and matrix formation. Tissue Eng Part A. 2011;17(9–10):1425–36.

92. Levorson EJ, Santoro M, Kurtis Kasper F, Mikos AG. Direct and indirect co-culture of chondrocytes and mesenchymal stem cells for the generation of polymer/extracellular matrix hybrid constructs. Acta Biomater. 2014;10(5):1824–35.

93. Platas J, Guillén M, Pérez del Caz M, Gomar F, Castejón M, Mirabet V, et al. Paracrine effects of human adipose-derived mesenchymal stem cells in inflammatory stress-induced senescence features of osteoarthritic chondrocytes. Aging. 2016;8(8):1703–17.

94. Khatab S, van Osch GJ, Kops N, Bastiaansen-Jenniskens YM, Bos PK, Verhaar JA, et al. Mesenchymal stem cell secretome reduces pain and prevents cartilage damage in a murine osteoarthritis model. Eur Cell Mater. 2018;36:218–30.

95. Kay AG, Long G, Tyler G, Stefan A, Broadfoot SJ, Piccinini AM, et al. Mesenchymal stem cell-conditioned medium reduces disease severity and immune responses in inflammatory arthritis. Sci Rep. 2017;7(1):S2.

96. Xu J, Wang B, Sun Y, Wu T, Liu Y, Zhang J, et al. Human fetal mesenchymal stem cell secretome enhances bone consolidation in distraction osteogenesis. Stem Cell Res Ther. 2016;7(1):849.

97. Osugi M, Katagiri W, Yoshimi R, Inukai T, Hibi H, Ueda M. Conditioned media from mesenchymal stem cells enhanced bone regeneration in rat calvarial bone defects. Tissue Eng Part A. 2012;18(13–14):1479–89.

98. Ando Y, Matsubara K, Ishikawa J, Fujio M, Shohara R, Hibi H, et al. Stem cell-conditioned medium accelerates distraction osteogenesis through multiple regenerative mechanisms. Bone. 2014;61:82–90.

99. Zhou B-R, Xu Y, Guo S-L, Xu Y, Wang Y, Zhu F, et al. The effect of conditioned media of adipose-derived stem cells on wound healing after ablative fractional carbon dioxide laser resurfacing. Biomed Res Int. 2013;2013:519126.

100. Katagiri W, Osugi M, Kawai T, Hibi H. First-in-human study and clinical case reports of the alveolar bone regeneration with the secretome from human mesenchymal stem cells. Head Face Med. 2016;12(1):1–10.

101. Katagiri W, Watanabe J, Toyama N, Osugi M, Sakaguchi K, Hibi H. Clinical study of bone regeneration by conditioned medium from mesenchymal stem cells after maxillary sinus floor elevation. Implant Dent. 2017;26(4):607–12.

102. Ragni E, Perucca Orfei C, De Luca P, Mondadori C, Viganò M, Colombini A, et al. Inflammatory priming enhances mesenchymal stromal cell secretome potential as a clinical product for regenerative medicine approaches through secreted factors and EV-miRNAs: the example of joint disease. Stem Cell Res Ther. 2020;11(1):1863.

103. Grigorian-Shamagian L, Liu W, Fereydooni S, Middleton RC, Valle J, Cho JH, et al. Cardiac and systemic rejuvenation after cardiosphere-derived cell therapy in senescent rats. Eur Heart J. 2017;38(39):2957–67.

104. Mianehsaz E, Mirzaei HR, Mahjoubin-Tehran M, Rezaee A, Sahebnasagh R, Pourhanifeh MH, et al. Mesenchymal stem cell-derived exosomes: a new therapeutic approach to osteoarthritis? Stem Cell Res Ther. 2019;10(1):1233.

105. Chamberlain C, Clements A, Kink J, Choi U, Baer G, Halanski M, et al. Extracellular vesicle-educated macrophages promote early Achilles tendon healing. Stem Cells. 2019;37(5):652–62.

106. Shen H, Yoneda S, Abu-Amer Y, Guilak F, Gelberman RH. Stem cell-derived extracellular vesicles attenuate the early inflammatory response after tendon injury and repair. J Orthop Res. 2019;38(1):117–27.

107. Nakamura Y, Miyaki S, Ishitobi H, Matsuyama S, Nakasa T, Kamei N, et al. Mesenchymal-stem-cell-derived exosomes accelerate skeletal muscle regeneration. FEBS Lett. 2015;589(11):1257–65.

108. Wu J, Kuang L, Chen C, Yang J, Zeng W-N, Li T, et al. miR-100-5p-abundant exosomes derived from infrapatellar fat pad MSCs protect articular cartilage and ameliorate gait abnormalities via inhibition of mTOR in osteoarthritis. Biomaterials. 2019;206:87–100.

109. Liu X, Yang Y, Li Y, Niu X, Zhao B, Wang Y, et al. Integration of stem cell-derived exosomes with in situ hydrogel glue as a promising tissue patch for articular cartilage regeneration. Nanoscale. 2017;9(13):4430–8.

110. Tofiño-Vian M, Guillén MI, Pérez del Caz MD, Castejón MA, Alcaraz MJ. Extracellular vesicles from adipose-derived mesenchymal stem cells downregulate senescence features in osteoarthritic osteoblasts. Oxidative Med Cell Longev. 2017;2017(6):1–12.

111. Ragni E, Orfei CP, De Luca P, Lugano G, Viganò M, Colombini A, et al. Interaction with hyaluronan matrix and miRNA cargo as contributors for in vitro potential of mesenchymal stem cell-derived extracellular vesicles in a model of human osteoarthritic synoviocytes. Stem Cell Res Ther. 2019;10(1):1–17.

112. Kim M, Steinberg DR, Burdick JA, Mauck RL. Extracellular vesicles mediate improved functional outcomes in engineered cartilage produced from MSC/chondrocyte cocultures. Proc Natl Acad Sci U S A. 2019;116(5):1569–78.

113. Ren L, Song Z-J, Cai Q-W, Chen R-X, Zou Y, Fu Q, et al. Adipose mesenchymal stem cell-derived exosomes ameliorate hypoxia/serum deprivation-induced osteocyte apoptosis and osteocyte-mediated osteoclastogenesis in vitro. Biochem Biophys Res Commun. 2019;508(1):138–44.

114. Hu Y, Xu R, Chen C-Y, Rao S-S, Xia K, Huang J, et al. Extracellular vesicles from human umbilical cord blood ameliorate bone loss in senile osteoporotic mice. Metabolism. 2019;95:93–101.

115. Li W, Liu Y, Zhang P, Tang Y, Zhou M, Jiang W, et al. Tissue-engineered bone immobilized with human adipose stem cells-derived exosomes promotes bone regeneration. ACS Appl Mater Interfaces. 2018;10(6):5240–54.

116. Furuta T, Miyaki S, Ishitobi H, Ogura T, Kato Y, Kamei N, et al. Mesenchymal stem cell-derived exosomes promote fracture healing in a mouse model. Stem Cells Transl Med. 2016;5(12):1620–30.

117. Lu Z, Chen Y, Dunstan C, Roohani-Esfahani S, Zreiqat H. Priming adipose stem cells with tumor necrosis factor-alpha preconditioning potentiates their exosome efficacy for bone regeneration. Tissue Eng Part A. 2017;23(21–22):1212–20.

118. Casado JG, Blázquez R, Vela FJ, Álvarez V, Tarazona R, Sánchez-Margallo FM. Mesenchymal stem cell-derived exosomes: immunomodulatory evaluation in an antigen-induced synovitis porcine model. Front Vet Sci. 2017;4(3):249.

119. Liu Y, Lin L, Zou R, Wen C, Wang Z, Lin F. MSC-derived exosomes promote proliferation and inhibit apoptosis of chondrocytes via lncRNA-KLF3-AS1/miR-206/GIT1 axis in osteoarthritis. Cell Cycle. 2018;17(21–22) 2411–22.

120. Zavatti M, Beretti F, Casciaro F, Bertucci E, Maraldi T. Comparison of the therapeutic effect of amniotic fluid stem cells and their exosomes on monoiodoacetate-induced animal model of osteoarthritis. Biofactors. 2019;46(1):106–17.

121. Qi X, Zhang J, Yuan H, Xu Z, Li Q, Niu X, et al. Exosomes secreted by human-induced pluripotent stem cell-derived mesenchymal stem cells repair critical-sized bone defects through enhanced angiogenesis and osteogenesis in osteoporotic rats. Int J Biol Sci. 2016;12(7):836–49.

122. Qin Y, Wang L, Gao Z, Chen G, Zhang C. Bone marrow stromal/stem cell-derived extracellular vesicles regulate osteoblast activity and differentiation in vitro and promote bone regeneration in vivo. Sci Rep. 2016;6(1):1–11.

123. Xiang C, Yang K, Liang Z, Wan Y, Cheng Y, Ma D, et al. Sphingosine-1-phosphate mediates the therapeutic effects of bone marrow mesenchymal stem

cell-derived microvesicles on articular cartilage defect. Transl Res. 2018;193:42–53.

124. Gallina C, Turinetto V, Giachino C. A new paradigm in cardiac regeneration: the mesenchymal stem cell secretome. Stem Cells Int. 2015;2015(19):1–10.

125. Borges FT, Convento MB, Schor N. Bone marrow-derived mesenchymal stromal cell: what next? Stem Cells Cloning. 2018;11:77–83.

126. Chen J, Li Y, Hao H, Li C, Du Y, Hu Y, et al. Mesenchymal stem cell conditioned medium promotes proliferation and migration of alveolar epithelial cells under septic conditions in vitro via the JNK-P38 signaling pathway. Cell Physiol Biochem. 2015;37(5):1830–46.

127. Clabaut A, Grare C, Léger T, Hardouin P, Broux O. Variations of secretome profiles according to conditioned medium preparation: the example of human mesenchymal stem cell-derived adipocytes. Electrophoresis. 2015;36(20):2587–93.

128. Siegel G, Kluba T, Hermanutz-Klein U, Bieback KR, Northoff H, Schäfer R. Phenotype, donor age and gender affect function of human bone marrow-derived mesenchymal stromal cells. BMC Med. 2013;11(1):146.

129. Katsara O, Mahaira LG, Iliopoulou EG, Moustaki A, Antsaklis A, Loutradis D, et al. Effects of donor age, gender, and in vitro cellular aging on the phenotypic, functional, and molecular characteristics of mouse bone marrow-derived mesenchymal stem cells. Stem Cells Dev. 2011;20(9):1549–61.

130. Tsiapalis D, O'Driscoll L. Mesenchymal stem cell derived extracellular vesicles for tissue engineering and regenerative medicine applications. Cell. 2020;9(4):991.

131. Headland SE, Jones HR, Norling LV, Kim A, Souza PR, Corsiero E, et al. Neutrophil-derived microvesicles enter cartilage and protect the joint in inflammatory arthritis. Sci Transl Med. 2015;7(315):315ra190.

132. Cosenza S, Ruiz M, Toupet K, Jorgensen C, Noël D. Mesenchymal stem cells derived exosomes and microparticles protect cartilage and bone from degradation in osteoarthritis. Sci Rep. 2017;7(1):1–12.

133. Zhang S, Chuah SJ, Lai RC, Hui JHP, Lim S-K, Toh WS. MSC exosomes mediate cartilage repair by enhancing proliferation, attenuating apoptosis and modulating immune reactivity. Biomaterials. 2018;156:16–27.

134. Liu Y, Zou R, Wang Z, Wen C, Zhang F, Lin F. Exosomal KLF3-AS1 from hMSCs promoted cartilage repair and chondrocyte proliferation in osteoarthritis. Biochem J. 2018;475(22):3629–38.

135. Mao G, Zhang Z, Hu S, Zhang Z, Chang Z, Huang Z, et al. Exosomes derived from miR-92a-3p-overexpressing human mesenchymal stem cells enhance chondrogenesis and suppress cartilage degradation via targeting WNT5A. Stem Cell Res Ther. 2018;9(1):1697.

136. Mao G, Huang Z, Chen W, Huang G, Meng F, Zhang Z, et al. MicroRNA-92a-3p regulates the expression of cartilage-specific genes by directly targeting histone deacetylase 2 in chondrogenesis and degradation. Osteoarthr Cartil. 2017;25(4):521–32.

Part II

Injectable Orthobiologics: Methods and Results Based on Anatomy and Pathology

Rotator Cuff Tendinopathy: Cell Therapy

Philippe Hernigou and Jacques Hernigou

13.1 Introduction

Shoulder pain is the third most common musculoskeletal complaint. Some commonly diagnosed shoulder problems include impingement of the rotator cuff (RC) tendons or subdeltoid bursa, bicipital tendinopathy, frozen shoulder, and gleno-humeral and acromio-clavicular arthritis. However, the most frequent cause of pain is tendinopathy or partial RC tears. Tendinopathy (tendon disorder) is a condition that occurs in the general population and more frequently in athletes [1], with pain and functional limitations of the affected joint involving the tendon insertion [2]. Many therapies [3] have been proposed: eccentric exercises, corticosteroids, anti-inflammatory non-steroidal drugs, shockwaves, platelet-rich plasma (PRP), and surgery. Despite positive results for some of these therapies, recalcitrant cases remain, and these may have an evolution to a RC tear.

If a diagnosis of tendinopathy has been made, it is important to take the diagnosis a step further and decipher whether the tendinopathy is from extrinsic causes, intrinsic causes, or a combination of the two. They may also have pain at rest, night pain, or a painful arc. Upon evaluation, the clinician may find weak external rotators, a weak supraspinatus, and signs of impingement. Signs of impingement may include painful overhead reaching, an inflamed subdeltoid bursa, or positive special tests meant to provoke symptoms. When a patient partially tears his RC, it is common that he/she presents with reduced shoulder function (i.e., dyskinesis, weakness, pain, and stiffness); in patients over the age of 60 with two out of three of the aforementioned symptoms (i.e., weak external rotators, weak supraspinatus, and impingement signs previously listed), there is a 98% chance of a RC tear, although this is not always true and asymptomatic tears exist.

13.2 Rotator Cuff: From Mechanical to Biological Improvement

13.2.1 Early History

The first description of RC injury appeared in *Edwin Smith Papyrus* (1600 BC) describing a RC injury in Egypt. But from a historical point of view [4], the first real description [5] was in 1788, when Alexander Monro depicted a tear in supraspinatus and infraspinatus in his book *A Description of All the Bursal Mucosae of the Human Body*. Five decades later, John Smith [6] published the first series of RC tears. In 1855, the

P. Hernigou (✉)
Department of Orthopaedic and Traumatology Surgery, Hospital Henri Mondor; University of Paris, Paris, France

J. Hernigou
Department of Orthopaedic and Traumatology Surgery, EpiCURA Hospital, Baudour/Hornu, Belgium

© ISAKOS 2022
G. Filardo et al. (eds.), *Orthobiologics*, https://doi.org/10.1007/978-3-030-84744-9_13

French surgeon Jean-François Malgaigne [7] discussed the role of dislocations in the cause of RC avulsions identified in cadaver specimens. Between 1860 and 1980, other etiologies for shoulder pain were investigated, including impingement of the thickened bursa under the acromion, entrapment or dislocation of the long head of the biceps, and subacromial stiffness and adhesions. After the introduction of X-rays in 1895, calcifications were identified as a cause in 1907 [8].

According to Perthes [9], in 1870 Karl Hüter firstly re-attached torn cuff tendons to the humeral diaphysis after a humeral head resection in chronic dislocation, and in 1898 Wilhelm Müller repaired RC tears to the humeral head during a surgical intervention for shoulder stabilization. The new century represented the beginning of a new era for RC surgery. Perthes reported in 1906 [9] a small series of RC repair with suture anchors, and in 1911, Codman [10] in the USA described the technique to repair the supraspinatus tendon tears.

Important steps of open RC repair surgery are represented by McLaughlin's [11] technique in 1944, Jean Debeyre and Patte's [12] supraspinatus advancement technique, and Neer's [13] tendon transfers, previously used to treat neurological palsies, which were first performed for RC tears by Cofield [14] and Gerber et al. [15]. In the 1990s, arthroscopic surgery was proposed to repair RC tears by Raymond Thal who published in 1993 a technical note on arthroscopic sutures [16].

13.2.2 Biological Enhancement of Rotator Cuff Repair

Tendinopathy is based on tendon degeneration [17] associated with presence of inflammatory cells [18]. Histological findings [19] demonstrated necrotic and apoptotic tenocytes, neovascularization, and collagen disarray. The repair process [20] is based first on the possibility of cells' differentiating into new tenocytes to generate new tendon tissue and, second, on the paracrine effects of these cells [21] to modulate the inflammatory reaction

and to stimulate repair by the production cytokines. Until the end of the twentieth century, most of the therapeutic improvements consisted in improving the mechanics of the surgical technique either by means of anchors or by means of two rows of fixation; it was not until the beginning of the twenty-first century that the biological improvement of treatment was considered.

In the beginning of the twenty-first century, Goutallier [22] reported the influence of biologic factors on the results with a classification of cuff muscle fatty degeneration. McElvany et al. [23] highlighted that the introduction of new surgical techniques generally did not result in an improvement of results and that novel biological strategies to enhance RC healing should be investigated.

To enhance tendon tissue regeneration, new biological solutions are being investigated, including growth factors, PRP, as well as both freshly isolated tissue derived cells and culture-expanded cell populations (such as mesenchymal stromal cells (aka MSCs)). The use of PRP as a biological solution to improve RC tendon healing has gained popularity over the last several years. The first report of a clinical use of PRP in RC surgery was published by Randelli et al. in 2008 [24], but evidence on the real benefit of this approach is still inconclusive [25].

Cell-based approaches were suggested to enhance tendon healing by the author, professor Hernigou, who has an experience of 30 years with freshly isolated preparations of bone marrow-derived cells [26], particularly in hip osteonecrosis. The first RC tears operated with cellular adjunction were performed by the author in open surgery in 1996, and arthroscopic technique results with cellular adjunction [27] were reported in 2014.

These preparations of freshly isolated cells include a small but important population of stem and progenitor cells [28–30] that are capable of proliferation in vitro, commonly referred to as connective tissue progenitors (CTPs). The concentration and prevalence of CTPs in a sample of tissue-derived cells can be estimated based on the number of colonies that are formed when those cells are placed into tissue culture. Tissue-derived cell populations can also be expanded through tissue culture in vitro [31] to generate homogeneous populations of plastic adherent cells that have been defined as mesenchymal stromal cells (MSCs).

13.3 The Theoretical Benefits of Cell Transplantation on Enthesis Healing

13.3.1 Reparative Process of the Enthesis

The enthesis has been divided into four zones (Fig. 13.1): the tendon, the non-mineralized fibrocartilage, the mineralized fibrocartilage, and the cancellous bone [32]. In the tendon area (zone 1), there is a predominance of type I collagen fibers together with a small amount of decorin which is a small cellular or pericellular matrix proteoglycan; in the non-mineralized fibrocartilage area (zone 2), type II and III collagen fibers are predominant and small amounts of type I, IX, and X collagen fibers have also been detected. Aggrecans and decorin are also present. Zone 3 is constituted by the mineralized fibrocartilage, with a highly specialized mineralized content and type I collagen fibers. Lastly, zone 4 is characterized by a bone-like composition, as it corresponds to the bony insertional area. As previously

mentioned, it has been demonstrated that this specialized tissue does not regenerate after injury and repair. The fibrovascular tissue that replaces the native enthesis is characterized by a predominance of type III collagen due to the excessive formation of scar tissue and the absence of fibrocartilage.

The reparative process can be divided into three phases (inflammatory, reparative, and remodeling), and numerous cells and cytokines have been implicated. Some authors have pointed out that the inability to regenerate the native enthesis could be caused by the incomplete expression of the genes implicated in its formation [32]. The main problem with failure in RC repair is probably biologic, as it is well-known that the delicate and highly specialized fibrocartilaginous transition zone between the RC and the bone does not regenerate after repair [33]. Standard tendon to bone repair techniques attain only a fibro-vascular scar tissue (Fig. 13.2) that has relatively poor mechanical properties [34]. Thus, the focus in research has changed from mechanical improvement of the repair techniques to finding ways to improve the biological environment around that repair. The most important histopathological findings of ruptured tendons consist mainly in disarray (i.e., loss of structural

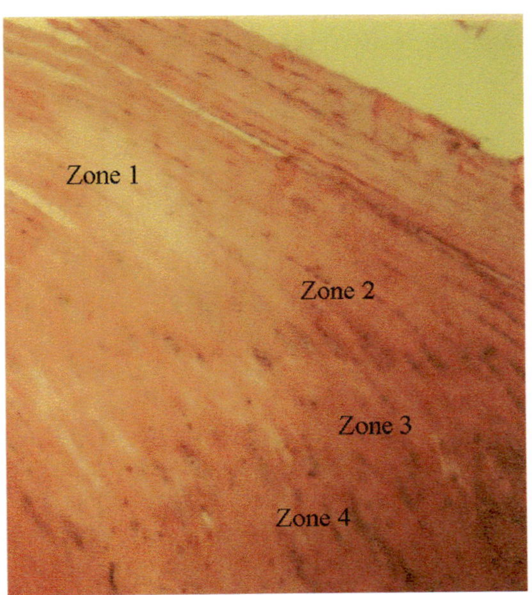

Fig. 13.1 The four zones of the normal enthesis

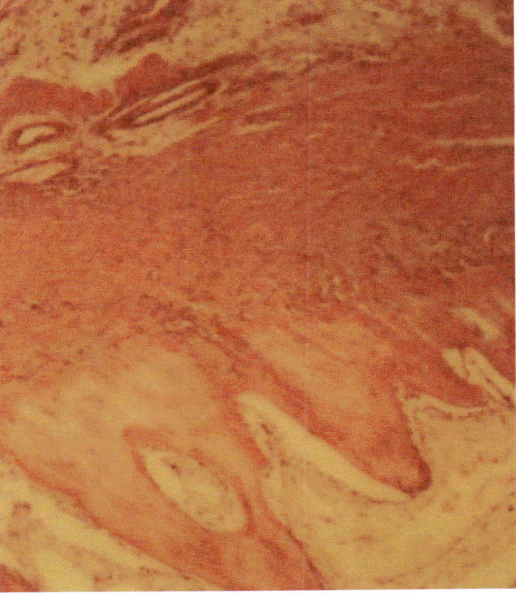

Fig. 13.2 Fibrocartilage healing of a tear

organization), poor or absent neoangiogenesis, chondral metaplasia, and fibrosis. All these features seem to give to tendon tissue a low healing capability, so these aspects may therefore explain why the lesions of the RC are at high risk of re-tear. In fact, as reported in the literature [35, 36], RC tears recurrence varies from 39 to 89% of cases. Considering these observations, it seems possible to refer to a condition of primary healing failure rather than re-tear, so we would like to introduce the new concept of "non-healing." In contrast, the histopathological features observed in the subacromial bursa consist of absence of disarray, presence of neoangiogenesis, absence of chondral metaplasia, hyperplasia/hypertrophy, and absence of necrosis. These aspects could be interpreted as an attempt to repair starting from the bursal tissue. Unfortunately, as time passes bursal tissue inflammation decreased so that the results of this "reparative burst" stops. In conclusion, in the small lesion, the repair process could start from the tendon tissue, but in large lesions the attempt to repair starts with the activation of the bursal tissue. So, the main goal of cellular therapy could be to boost the repair process.

13.3.2 Rationale for the Use of Cells to Treat Tendon Disorders

The association between biological abnormalities and RC tears has been investigated by Hernigou. Hernigou et al. [37] demonstrated that there is a reduced level of colony founding CTPs at the tendon-bone interface tuberosity in patients with tendinopathy and RC tears.

Cells were isolated from the bone marrow at the tendon-bone interface tuberosity of the humeral head of patients with and without RC tears and placed into a colony formation assay. In patients with tendinopathy, and after a RC tear, there is a significant decrease in the prevalence of colony founding CTPs in the bone marrow of the proximal humerus at the tendon-bone interface tuberosity. Overall, the lag time between the onset of RC symptoms and RC repair played a

major role in the decrease of CTPs in the affected tuberosity. In the case of traumatic tears, they observed that the decrease was moderate in patients with small tears that were diagnosed and treated early after the onset of symptoms. A severe decrease was present in patients for whom 1 year or more had elapsed between the initial incident/onset of symptoms and surgery, particularly in elderly patients. Overall, the findings in this study provided a context for surgeons to consider use of cellular augmentation in treating tendinopathies.

Marrow-derived cells have demonstrated therapeutic benefit in a wide range of orthopedic pathologies. As a first result of this observation (decrease of CTPs in the tuberosity of tendinopathies), it is potentially counterproductive to remove bone marrow from a patient's injured shoulder. Therefore, other sites of aspiration, like patient's iliac crest, were used as a source of bone marrow-derived CTPs.

Note that culture-expanded populations of plastic adherent cells can be generated by placing a mixture of cells including CTPs into culture and expanding the progeny of these cells in competition with one another. The result can be a culture-expanded population of adherent cells that are capable of limited self-renewal and of differentiating into other cell types [38]. These populations of culture-expanded cells are defined as mesenchymal stromal cells (MSCs) [31].

13.3.3 Preclinical Studies

Many preclinical studies suggest that the use of cell-based therapies in subjects with RC tears can lead to improved enthesis healing. Recent animal studies have been reported on the role of bone marrow-derived cells for the reconstruction of RC ruptures. The progeny of some marrow-derived CTPs has the potential to become both tenocytes and osteoblast, as well as to provide multiple growth factors to establish an environment conducive to soft and hard tissue regeneration. The principal source of cells for cell-enhanced healing of the RC has been autologous bone marrow. Gulotta et al. [39] performed an experimental unilateral detachment of supraspinatus tendon and a trans-osseous repair in rats. They showed that transplanted cells were present at the repair site and that they were metabolically active. Although they did not find significant differences in between the treated and untreated groups, at 4 weeks, there was a higher amount of fibrocartilage formation and better orientation of fibrocartilage fibers.

In order to reproduce RC surgery, Kida et al. [40] designed a study in which they performed additional drilling to the greater tuberosity to release bone marrow and allow bone marrow cells to migrate into the suture zone. They tested chimeric rats that expressed green fluorescent protein in the bone marrow cells and looked for the expression of this protein after a period of 2, 4, and 8 weeks. It seems that drilling and the subsequent migration of stem/stromal cells might improve maximum load to failure at 4 and 8 weeks.

More recently, Gulotta et al. [41] have used genetically modified culture-expanded MSCs in order to express scleraxis and produce MIT1 and BMP-13 with promising results. MSCs genetically modified to overexpress MT1-MMP might be useful for augmenting suture as it has demonstrated improved biomechanical strength at 4 weeks based on a higher presence of fibrocartilage.

Culture-expanded adipose tissue-derived mesenchymal stromal cells (AMSC) have also shown multipotentiality in vitro [42]. Oh et al. [43] have published the first study in a RC model using AMSCs. Four groups were compared for a suture of the subscapularis tendon in rabbit using saline, saline and AMSCs, only AMSCs, and only suture. They found better healing properties in animals treated with AMSCs and a capacity for regeneration after fatty infiltration of the muscle. Recently, Viganò et al. [44] used autologous micro-fragmented adipose tissue (µFAT) to improve tenocyte proliferation. Micro-fragmented adipose tissue exerted an anti-inflammatory action on supraspinatus tendon cells resulting in the reduction of catabolic and inflammatory marker expression. These observations potentially support the use of µFAT as adjuvant therapy in the treatment of RC disease.

13.4 Benefit of Cell Transplantation for Rotator Cuff Repair: What Is the Current Evidence?

MEDLINE/PubMed, EMBASE, and trial registers were searched to identify human trials on RC tears treatment. Studies investigating any type of cell-based therapy for patients with tendon disorders were eligible if they included patient-reported outcome measures or assessed tendon healing. We identified eight studies that examined the efficacy of stem/stromal cell injections in human RC tendon tears.

13.4.1 The First Trial: The Experience of the Senior Author

The first trial (level 2 evidence; prospective, controlled matched trial was completed with MRI and ultrasonography control) to evaluate the efficacy of an injection of bone marrow-derived cells [27] is an adjunct to single-row RC arthroscopy in comparison with single-row RC arthroscopy alone in patients diagnosed with RC tears (Fig. 13.3). The study group was composed of 45

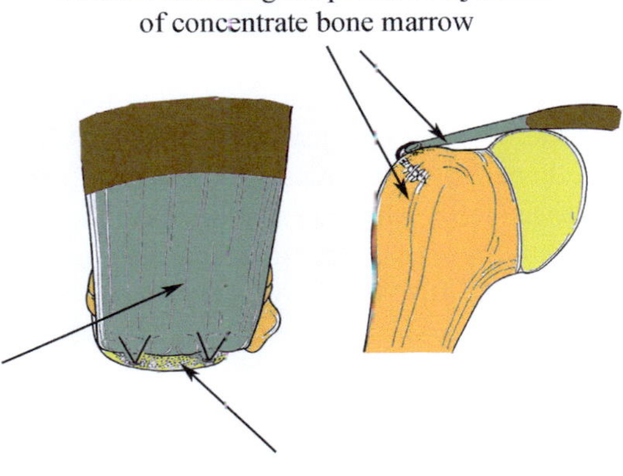

Fig. 13.3 Points of injection of cells after rotator cuff repair

patients with symptomatic rupture (and tear size from 1.5 to 3.5 cm) of the RC that had undergone surgical repair with an arthroscopic protocol and adjunct therapy using autologous marrow-derived cells during a period of 5 years between 2000 and 2004. Injection of processed marrow aspirates was performed at the end of rotator cuff tendon fixation. MSCs were injected in the tendon at the junction between the bone and tendon (4 mL) and in the bone at the site of the footprint (8 mL). Each patient receiving cell augmentation received a total of 12 mL of bone marrow concentrate processed using a centrifuge. Each bone marrow concentrate (BMC) sample was assayed in vitro to quantify the number of colony founding CTPs. The mean concentration of CTPs returned to the 45 patients was 4300 + 1800 per mL and the total number of CTPs returned to these patients was 51,600 ± 25,000 cells. In addition to CTPs, these injections also delivered a total of 1824 ± 648 million mononucleated cells (non-CTPs) at a concentration of 152 ± 53 million cells per ml. Therefore, the overall prevalence of CTPs among all cells transplanted was one in 36 thousand mononucleated cells.

Outcomes of patients receiving BMC during their repair were compared to those of a matched control group of 45 patients who did not receive BMC. The total follow-up was 10 years, and the results were published in 2014. In the BMC treatment group, significantly fewer re-tears were reported after 10 years of follow-up as measured by MRI and ultrasonography (13% vs. 56%, respectively; $P < 0.05$). The efficiency of BMC augmentation of the standard of care rotator cuff repair was analyzed for the healing time, the quality of the healing surface on the footprint, and the absence of re-tear. Injection of BMC as an adjunctive therapy during rotator cuff repair enhanced the healing rate and improved the quality of the repaired surface as determined by ultrasound and MRI. Forty-five (100%) of the 45 repairs with BMC augmentation had healed by 6 months versus 30 (67%) of the 45 repairs without MSC treatment by 6 months. BMC injection also prevented further ruptures during the next 10 years. At the most recent 10-year follow-up, intact rotator cuffs were found in 39 (87%) of the

45 patients in the BMC-treated group, but just 20 (44%) of the 45 patients in the control group. The number of transplanted CTPs was determined to be the most relevant to the outcome in the study group, since patients with a loss of tendon integrity at any time up to the 10-year follow-up milestone received fewer CTPs as compared with those who had maintained a successful repair during the same interval. There was level 2 evidence for improved outcome of the arthroscopic RC repair with BMC augmentation in comparison with arthroscopic RC repair for re-tears as measured by MRI and ultrasonography.

Although not included in the article, the authors had checked as reported in another study with hip surgery [45] that these bone marrow-derived cells when "culture expanded" as MSCs were meeting the standard criteria [31] from the International Society for Cellular Therapy (i.e., cells that adhere to tissue culture plastic; retain the capability for tri-lineage differentiation (bone, cartilage, and adipose); express CD105, CD73, and CD90 (with 95% prevalence); and lack expression of CD45, CD34, CD14 or CD11b, CD79 alpha or CD19, and HLA-DR surface molecules). No severe adverse events related to the injection of culture-expanded MSCs after treatment were observed: no shoulder ossification or calcification, no local tumor or remote tumors after a follow-up that is now of 10–20 years.

13.4.2 The Other Seven Trials

The second case series (level 4 evidence; absence of control group) examined the efficacy of another preparation of marrow-derived cells that the authors referred to as bone marrow mononuclear cells (BMMC) in patients undergoing surgery for complete RC tears. Fourteen patients with RC tears entered the study from 2009 to 2010. After a minimum of 12-month follow-up period, the series [46] was reported in 2012. In that series, 14 consecutive patients were treated with trans-osseous stitches through mini-open incision and subsequent BMMC injection obtained from the iliac crest. Samples were assayed based on the number of CD34 positive cells, rather than colony formation. CD34, a marker of immature hematopoietic stem and progenitor cells, was used as a surrogate measure of the relative quality of the marrow aspirates. Ten mL were injected. The mean number of mononucleated cells implanted was 380 million per mL, and among these cells the mean number of CD34+ cells transplanted was 5.65 million per mL, making the mean prevalence of CD34+ cells one in 67 mononucleated cells. The University of California (UCLA) score was used as main outcome; this score increased, but no statistical analysis was performed in this study. After a minimum 12-month follow-up period, the UCLA score increased from 12 ± 3.0 to 31 ± 3.2. Clinical findings remained unaltered in the following year in all but one patient (13/14). MRI analysis after a 12-month follow-up period demonstrated tendon integrity in all cases (14/14), presence of low-signal intensity areas along the supraspinatus tendon and distal muscle belly in eight cases (8/14), and high-intensity blooming small round artifact at the bursal and tendon topography in 11 cases (11/14). Six patients (6/14) showed formation of a high-signal intensity zone at the critical zone. There was a level 4 evidence for a superior effect of BMMC on RC tears compared with non-surgery control group (with exercise therapy) for clinical score and VAS pain score.

In the third trial (level 4 evidence; nonsurgical controlled trial), a total of 81 shoulders were treated with autologous BMC injections for RC tears [47]. Nucleated cell count data were available for 81 joints. CTP assays were not performed. The mean cell count was 470 million cells (standard deviation = 3.1, range = 0.6–22.7). Ultrasound or fluoroscopy was used to guide intra-articular or rotator cuff tear needle placement. Clinical outcomes were assessed serially over time using the disabilities of the arm, shoulder, and hand score (DASH), the numeric pain scale (NPS), and a subjective improvement rating scale. Baseline scores were compared to the most recent outcome scores at the time of the analysis and adjusted for demographic differences. At the most current follow-up assessment after treatment, the average subjective score improved from 36.1 to 17.1 ($P = 0.001$), and the average numeric pain scale value decreased (improved) from 4.3 to 2.4 ($P = 0.001$). These changes were associated with an average subjective improvement of 48.8%. There was a level 4 evidence for a positive effect of bone marrow-derived cells on RC tears when comparisons were done between pre- and posttreatment scores.

The fourth trial (level 4 evidence; nonsurgical controlled trial) used culture-expanded human autologous bone marrow stromal cells [48] in repair of a RC tear and reported the preliminary results of a safety study. Ten patients were included in the study. Marrow was harvested at 3–4 weeks before surgery. Subsequently, an arthroscopic repair of the rotator cuff tear was performed, and a suspension of culture-expanded MSCs was applied to the suture site at the end of the procedure. The clinical assessment of these patients included the visual analogue scale (VAS) and subjective questionnaires for Constant and UCLA scores. All patients underwent MRI examination at 6 postoperative months. The average values at 6 months postoperatively were 0 points for the VAS score, 32 for the UCLA score, and 84 for the Constant score. The MRI findings at 6 months after surgery showed fully healed and well-integrated tissue of the rotator cuff tendon attachment in eight patients.

The fifth trial (level 2 evidence; prospective, surgical control group, with MRI control) determined the efficacy of one injection of allogenic adipose-derived culture-expanded MSCs loaded in fibrin glue (injection group: allo-ASC injection group) during arthroscopic RC repair (35 patients), comparing the double-row suture bridge technique with injection versus the double-row suture bridge technique alone in 35 matched patients with RC tears [49]. A dose of 4.46 million MScs loaded in 2 mL of fibrin glue was used. The mean VAS score at rest and during motion improved significantly in both groups after surgery. At minimum of 12 months after intervention, a re-tear rate of 28.5% was found with MRI in the control group versus 14.3% in the MSC injection group ($P < 0.001$). Complete healing of the tendon, as measured with MRI, was observed in 85.7% of patients in the intervention group versus 71.4% of patients in the control group. There was level 3 evidence for a superior effect of allo-ASC injection augmentation for RC tears compared with arthroscopic double-row suture bridge technique alone on tendon healing and re-tears as measured by MRI.

The sixth trial (level 3 evidence; nonsurgical controlled trial) studied the efficacy of one injection with bone marrow aspirate concentration (BMAC) and PRP injection versus exercise therapy in patients diagnosed with a RC tear with a total follow-up of 3 months [50]. Cell count and CTPs assay were not performed. The mean American Shoulder and Elbow Surgeons (ASES) score changed from 39.4 ± 13.0 to 54.5 ± 11.5 at 3 weeks and 74.1 ± 8.5 at 3 months in the BMAC-PRP group and changed from 45.9 ± 12.4 to 56.3 ± 12.3 at 3 weeks and 62.2 ± 12.2 at 3 months in the control group. The change in the ASES score differed significantly between groups at 3 months ($P = 0.011$) but not at 3 weeks ($P = 0.712$). There was level 3 evidence for a superior effect of BMAC-PRP on RC tears compared with exercise therapy.

The seventh trial (level 2 evidence; prospective, controlled matched trial with MRI and arthroscopic control) assessed the safety and efficacy of one intratendinous injection of culture-expanded autologous adipose tissue-derived MSCs (AD MSCs) in patients with RC disease with three dose escalation cohorts [51]. Nineteen patients were treated, in two phases. With an initial dose escalation from 10 million cells to 100 million cells and the last ten patients treated with 100 million cells. The primary outcomes were the safety and the Shoulder Pain and Disability Index (SPADI). Secondary outcomes included clinical, radiological, and arthroscopic evaluations. Intratendinous injections of AD MSCs were not associated with adverse events. It significantly decreased the SPADI scores by 80% and 77% in the mid- and high-dose groups, respectively. Shoulder pain was significantly alleviated by 71% in the high-dose group. MRI examination showed that volume of the bursal side defect significantly decreased by 90% in the high-dose group.

Arthroscopic examination demonstrated that volume of the articular and bursal side defects decreased by 83 and 90% in the mid- and high-dose groups, respectively. There was level 2 evidence that one intratendinous injection of autologous AD MSCs in patient with a partial-thickness RC tear did not cause adverse events, but improved shoulder function, and relieved pain through regeneration of RC tendon.

The eighth trial (level of evidence: level 1; prospective, randomized, controlled trial) tested the hypothesis that treatment of symptomatic, partial-thickness RC tears (PT-RCT) with fresh, point of care, uncultured, unmodified, autologous adipose-derived from lipoaspirate was safe and more effective than corticosteroid injection [52]. A mean dose of 11.4 million cells was delivered in 5 mL. This pilot study suggested that the use of adipose-derived cells can lead to improved shoulder function without adverse effects. This level 1 study evidenced the following key findings: no severe adverse events in the 12 months after treatment and no greater risks than those connected with treatment of PT-RCT with corticosteroid injection. Despite the small number of subjects in this pilot study, those 12 in the autologous adipose-derived group showed statistically significantly higher mean ASES total scores at Week 24 and Week 52 posttreatment than the 8 subjects in the corticosteroid group ($p < 0.05$).

13.5 Conclusion

Many gaps in our knowledge remain with regard to indications, dosage, cell source, cell preparation, cell composition, cell concentration, and method of cell delivery. Tendon tears represent more severe injuries than tendinopathy, and outcomes of tendon tears are often poorer, particularly for rotator cuff tears of the shoulder. According to our review and to our experience, we are currently able to recommend cellular therapies for some patients with rotator cuff tendon disorders. Our current practice is not to recommend cell therapy for every surgery but rather to propose this treatment for patients who have a risk of a re-tear after arthroscopic repair of the rotator cuff. Those patients with risk of re-tear show advanced degeneration of the tendons, which are thinner and atrophic.

However, most of the evidence of efficacy is based on injections done during surgery and associated with surgical repair of rotator cuff tears; more studies are needed on the conservative cell-based injective treatment of RC tendinopathy without rupture. Several other therapies, such as eccentric exercises, nonsteroidal anti-inflammatory drugs, shockwave therapy, corticosteroids, platelet-rich plasma, and surgery before rupture, have been proposed as treatments for tendinopathy, with moderate to low variable levels of evidence and success. Based on current evidence, it is not possible to judge whether cellular therapy using either freshly isolated tissue-derived cell preparations or culture-expanded cell preparations should be preferred over other conventional treatments in the management of tendon disorders.

> **Take-Home Messages**
> - Cell therapy is an emerging treatment for tendon disorders.
> - In the current literature, there are only a few trials that studied the efficacy of bone marrow-derived cell therapy for rotator cuff tears and of adipose-derived cells.
> - All controlled trials reported superior outcomes and/or better tendon healing when marrow-derived or adipose-derived cells were transplanted.
> - This treatment approach could be proposed for patients who have a risk of a re-tear after arthroscopic repair of the rotator cuff.
> - Several gaps in our knowledge remain with regard to dosage, cell source, cell preparation, cell composition, cell concentration, and method of cell delivery; these factors could influence the effects and the final results.

References

1. Ackermann PW, Renström P. Tendinopathy in sport. Sports Health. 2012;4:193–201.
2. Maffulli N, Wong J, Almekinders LC. Types and epidemiology of tendinopathy. Clin Sports Med. 2003;22:675–92.
3. Andres BM, Murrell GA. Treatment of tendinopathy: what works, what does not, and what is on the horizon. Clin Orthop Relat Res. 2008;466:1539–54.
4. Randelli P, Cucchi D, Ragone V, de Girolamo L, Cabitza P, Randelli M. History of rotator cuff surgery. Knee Surg Sports Traumatol Arthrosc. 2015;23(2):344–62.
5. Monro A. A description of all the bursae mucosae of the human body. Edinburgh: Elliot C, Kay T & Co.; 1788.
6. Smith JG. Pathological appearances of seven cases of injury of the shoulder-joint: with remarks. 1834. Lond Med Gaz. 1834;14:280–5.
7. Malgaigne JF. Traité des fractures et des luxations. Paris: Baillière; 1855.
8. Painter CF. Subdeltoid Bursitis. Boston Med Surg J. 1907;156(12):345–9.
9. Perthes GC. Über Operationen bei habitueller Schulterluxation. Deutsche Zeitschrift f Chirurgie. 1906;85(1):199–227.
10. Codman EA. Complete rupture of the supraspinatus tendon; operative treatment with report of two successful cases. Boston Med Surg J. 1911;164:708–10.
11. McLaughlin HL. Lesions of the musculotendinous cuff of the shoulder. The exposure and treatment of tears with retraction. Clin Orthop Relat Res. 1994;304:3–9.
12. Debeyre J, Patte D. A technic of repair of the musculotendinous cuff of the shoulder. Transacromial route approach and advancement of the body of the supraspinosus muscle. Presse Med. 1961;69:2019–20.
13. Neer CS II. Anterior acromioplasty for the chronic impingement syndrome in the shoulder: a preliminary report. J Bone Joint Surg Am. 1972;54(1):41–50.
14. Cofield RH. Subscapular muscle transposition for repair of chronic rotator cuff tears. Surg Gynecol Obstet. 1982;154(5):667–72.
15. Gerber C. Latissimus dorsi transfer for the treatment of irreparable tears of the rotator cuff. Clin Orthop Relat Res. 1992;275:152–60.
16. Thal R. A technique for arthroscopic mattress suture placement. Arthroscopy. 1993;9(5):605–7.
17. Xu Y, Murrell GA. The basic science of tendinopathy. Clin Orthop Relat Res. 2008;466:1528–38.
18. Rees JD, Stride M, Scott A. Tendons—time to revisit inflammation. Br J Sports Med. 2014;48:1553–7.
19. Lui PP. Histopathological changes in tendinopathy—potential roles of BMPs? Rheumatology (Oxford). 2013;52:2116–26.

20. Nixon AJ, Watts AE, Schnabel LV. Cell- and gene-based approaches to tendon regeneration. J Shoulder Elb Surg. 2012;21:278–94.

21. Caplan AI, Dennis JE. Mesenchymal stem cells as trophic mediators. J Cell Biochem. 2006;98: 1076–84.

22. Goutallier D, Postel JM, Gleyze P, Leguilloux P, Van Driessche S. Influence of cuff muscle fatty degeneration on anatomic and functional outcomes after simple suture of full-thickness tears. J Shoulder Elb Surg. 2003;12(6):550–4.

23. McElvany MD, McGoldrick E, Gee AO, Neradilek MB, Matsen FA 3rd. Rotator cuff repair: published evidence on factors associated with repair integrity and clinical outcome. Am J Sports Med. 2015;43(2):491–500.

24. Randelli PS, Arrigoni P, Cabitza P, Volpi P, Maffulli N. Autologous platelet rich plasma for arthroscopic rotator cuff repair. A pilot study. Disabil Rehabil. 2008;30(20–22):1584–9.

25. Filardo G, Di Matteo B, Kon E, Merli G, Marcacci M. Platelet-rich plasma in tendon-related disorders: results and indications. Knee Surg Sports Traumatol Arthrosc. 2018;26(7):1984–99.

26. Hernigou P, Dubory A, Homma Y, Guissou I, Flouzat Lachaniette CH, Chevallier N, Rouard H. Cell therapy versus simultaneous contralateral decompression in symptomatic corticosteroid osteonecrosis: a thirty year follow-up prospective randomized study of one hundred and twenty five adult patients. Int Orthop. 2018;42(7):1639–49.

27. Hernigou P, Flouzat Lachaniette CH, Delambre J, Zilber S, Duffiet P, Chevallier N, Rouard H. Biologic augmentation of rotator cuff repair with mesenchymal stem cells during arthroscopy improves healing and prevents further tears: a case-controlled study. Int Orthop. 2014;38(9):1811–8.

28. Muschler GF, Midura R. Connective tissue progenitors: practical concepts for clinical applications. Clin Orthop Rel Res. 2002;395:66–80.

29. Muschler GF, Midura R, Nakamoto C. Practical modeling concepts for connective tissue stem cell and progenitor compartment kinetics. J Biomed Biotechnol. 2003;3:170–93.

30. Patterson TE, Boehm C, Nakamoto C, Rozic R, Walker E, Piuzzi NS, Muschler GF. The efficiency of bone marrow aspiration for the harvest of connective tissue progenitors from the human iliac crest. J Bone Joint Surg Am. 2017;99(19):1673–82.

31. Dominici M, Le Blanc K, Mueller I, et al. Minimal criteria for defining multipotent mesenchymal stromal cells. The International Society for Cellular Therapy position statement. Cytotherapy. 2006;8:315–7.

32. Thomopoulos S, Genin GM, Galatz LM. The development and morphogenesis of the tendon-to-bone insertion—what development can teach us about healing. J Musculoskelet Neuronal Interact. 2010;10:35–45.

33. Galatz LM, Sandell LJ, Rothermich SY, Das R, Mastny A, Havlioglu N, Silva MJ, Thomopoulos S. Characteristics of the rat supraspinatus tendon during tendon-to-bone healing after acute injury. J Orthop Res. 2006;24:541–50.

34. Kovacevic D, Rodeo SA. Biological augmentation of rotator cuff tendon repair. Clin Orthop Relat Res. 2008;466:622–33.

35. Bishop J, Klepps S, Lo IK, Bird J, Gladstone JN, Flatow EL. Cuff integrity after arthroscopic versus open rotator cuff repair: a prospective study. J Shoulder Elb Surg. 2006;15:290–9.

36. Tashjian RZ, Hollins AM, Kim HM, Teefey SA, Middleton WD, Steger-May K, Galatz LM, Yamaguchi K. Factors affecting healing rates after arthroscopic double-row rotator cuff repair. Am J Sports Med. 2010;38:2435–42.

37. Hernigou P, Merouse G, Duffiet P, Chevalier N, Rouard H. Reduced levels of mesenchymal stem cells at the tendon-bone interface tuberosity in patients with symptomatic rotator cuff tear. Int Orthop. 2015;39(6):1219–25.

38. Phinney DG, Prockop DJ. Concise review: mesenchymal stem/multipotent stromal cells: the state of transdifferentiation and modes of tissue repair—current views. Stem Cells. 2007;25:2896–902.

39. Gulotta LV, Kovacevic D, Ehteshami JR, Dagher E, Packer JD, Rodeo SA. Application of bone marrow-derived mesenchymal stem cells in a rotator cuff repair model. Am J Sports Med. 2009;37: 2126–33.

40. Kida Y, Morihara T, Matsuda K, Kajikawa Y, Tachiiri H, Iwata Y, Sawamura K, Yoshida A, Oshima Y, Ikeda T, Fujiwara H, Kawata M, Kubo T. Bone marrow-derived cells from the footprint infiltrate into the repaired rotator cuff. J Shoulder Elb Surg. 2013;22:197–205.

41. Gulotta LV, Kovacevic D, Packer JD, Deng XH, Rodeo SA. Bone marrow-derived mesenchymal stem cells transduced with scleraxis improve rotator cuff healing in a rat model. Am J Sports Med. 2011;39:1282–9.

42. Zuk PA, Zhu M, Ashjian P, De Ugarte DA, Huang JI, Mizuno H, Alfonso ZC, Fraser JK, Benhaim P, Hedrick MH. Human adipose tissue is a source of multipotent stem cells. Mol Biol Cell. 2002;13:4279–95.

43. Oh JH, Chung SW, Kim SH, Chung JY, Kim JY. 2013 Neer Award: effect of the adipose-derived stem cell for the improvement of fatty degeneration and rotator cuff healing in rabbit model. J Shoulder Elb Surg. 2014;23:445–55.

44. Viganò M, Lugano G, Perucca Orfei C, Menon A, Ragni E, Colombini A, De Luca P, Randelli P, de Girolamo L. Autologous microfragmented adipose tissue reduces inflammatory and catabolic markers in supraspinatus tendon cells derived from patients affected by rotator cuff tears. Int Orthop. 2020;45(2):419–26.

45. Lebouvier A, Poignard A, Coquelin-Salsac L, Léotot J, Homma Y, Jullien N, Bierling P, Galactéros F, Hernigou P, Chevallier N, Rouard H. Autologous bone marrow stromal cells are promising candidates

for cell therapy approaches to treat bone degeneration in sickle cell disease. Stem Cell Res. 2015;15(3): 584–94.

46. Ellera Gomes JL, da Silva RC, Silla LMR, Abreu MR, Pellanda R. Conventional rotator cuff repair complemented by the aid of mononuclear autologous stem cells. Knee Surg Sports Traumatol Arthrosc. 2012;20(2):373–7.

47. Centeno CJ, Al-Sayegh H, Bashir J, Goodyear S, Freeman MD. A prospective multi-site registry study of a specific protocol of autologous bone marrow concentrate for the treatment of shoulder rotator cuff tears and osteoarthritis. J Pain Res. 2015;8:269–76. eCollection 2015.

48. Havlas V, Kotaška J, Koníček P, Trč T, Konrádová Š, Kočí Z, Syková E. Pouziti kultivovanych lidskych autolognich kmenovych bunek kostni drene pri rekonstrukci ruptury rotatorove manzety—studie bezpecnosti metody, predbezne vysledky [Use of cultured human autologous bone marrow stem cells in repair of a rotator cuff tear: preliminary results of a safety study] [Article in Czech]. Acta Chir Orthop Traumatol Cechoslov. 2015;82(3): 229–34.

49. Kim YS, Sung CH, Chung SH, Kwak SJ, Koh YG. Does an injection of adipose derived mesenchymal stem cells loaded in fibrin glue influence rotator cuff repair outcomes? A clinical and magnetic resonance imaging study. Am J Sports Med. 2017;45(9):2010–8.

50. Kim SJ, Kim EK, Kim SJ, Song DH. Effects of bone marrow aspirate concentrate and platelet-rich plasma on patients with partial tear of the rotator cuff tendon. J Orthop Surg Res. 2018;13(1):1.

51. Jo CH, Chai JW, Jeong EC, Oh S, Kim PS, Yoon JY, Yoon KS. Intratendinous injection of autologous adipose tissue-derived mesenchymal stem cells for the treatment of rotator cuff disease: a first-in-human trial. Stem Cells. 2018;36(9):1441–50.

52. Hurd JL, Facile TR, Weiss J, Hayes M, Hayes M, Furia JP, Maffulli N, Winnier GE, Alt C, Schmitz C, Alt EU, Lundeen M. Safety and efficacy of treating symptomatic, partial-thickness rotator cuff tears with fresh, uncultured, unmodified, autologous adipose-derived regenerative cells (UA-ADRCs) isolated at the point of care: a prospective, randomized, controlled first-in-human pilot study. J Orthop Surg Res. 2020;15(1):122.

Rotator Cuff Tendinopathy: Biologics

14

Pietro Simone Randelli, Chiara Fossati, Marianna Vitale, Francesca Pedrini, and Alessandra Menon

14.1 Introduction

Shoulder pain is a very common musculoskeletal problem in the general population with a prevalence between 6.9% and 34%, and rotator cuff tendinopathy represents the leading cause of this symptom, increasing as a function of age [1]. The patho-etiology of rotator cuff tendinopathy is still poorly defined, but it is surely multifactorial including extrinsic and intrinsic pathogenetic mechanisms. Extrinsic factors cause a compression on the bursal side of the rotator cuff tendons. Acromion, subacromial, and acromioclavicular joint spurs, alterations in scapular or humeral kinematics, postural abnormalities, rotator cuff and scapular muscle performance deficits, and decreased extensibility of the pectoralis minor are mainly responsible for tendon compression [2]. Also, overuse is considered one of the most important extrinsic factor especially in individuals performing repetitive overhead activities, including throwing sports, like baseball or volleyball, and heavy works [2]. Intrinsic factors, including alterations in biology, mechanical properties, morphology, and vascularity, contribute to rotator cuff tendon degeneration with tensile or shear overload. Rotator cuff tendinosis may be the result of a disorganization in collagen fibril morphology and tendon ultrastructure. The loss of cellularity, followed by thinning and disorganization of tendon fibers, lead to the formation of a granulation tissue and fibrocartilaginous changes [3, 4]. All these changes increase the risk of rotator cuff tears.

The intrinsic factors could explain the rationale for the use of biologics in the treatment of rotator cuff tendinopathy, having demonstrated an influence on tendon morphology and function. Depending on the patients' comorbidities, age, activity level, symptoms, and findings on physical examination and imaging, different biologic therapies have been described. Injection therapy of biologics can be applied as conservative

P. S. Randelli
Laboratory of Applied Biomechanics, Department of Biomedical Sciences for Health, Università degli Studi di Milano, Milan, Italy

U.O.C. 1° Clinica Ortopedica, ASST Centro Specialistico Ortopedico Traumatologico Gaetano Pini-CTO, Milan, Italy

Research Center for Adult and Pediatric Rheumatic Diseases (RECAP-RD), Department of Biomedical Sciences for Health, Università degli Studi di Milano, Milan, Italy
e-mail: pietro.randelli@unimi.it

C. Fossati · A. Menon
Laboratory of Applied Biomechanics, Department of Biomedical Sciences for Health, Università degli Studi di Milano, Milan, Italy

U.O.C. 1° Clinica Ortopedica, ASST Centro Specialistico Ortopedico Traumatologico Gaetano Pini-CTO, Milan, Italy

M. Vitale (✉) · F. Pedrini
U.O.C. 1° Clinica Ortopedica, ASST Centro Specialistico Ortopedico Traumatologico Gaetano Pini-CTO, Milan, Italy

© ISAKOS 2022
G. Filardo et al. (eds.), *Orthobiologics*, https://doi.org/10.1007/978-3-030-84744-9_14

treatment or adjuvant in surgical procedures. Historically, the injection therapy of choice was represented by the use of corticosteroids that can reduce pain and, consequently, improve functional outcomes but without a tendon healing stimulation. Recently, an increasing interest in the use of platelet-rich plasma (PRP) injections for nonoperative management of rotator cuff tendinopathy or in association with arthroscopic rotator cuff repair has emerged. At the same time, also hyaluronic acid (HA) injections, among the biologicals, are being widely studied. Meanwhile, the use of selected cytokines and growth factors is still restricted to animal studies.

14.2　Platelet-Rich Plasma (PRP)

PRP is one of the most studied adjuvants in rotator cuff tendinopathy and tears, able to stimulate healing at the tendon-bone interface and decrease pain and inflammation [5, 6]. PRP is rich in growth factors including transforming growth factor-β (TGF-β), fibroblast growth factor (FGF), platelet-derived growth factor (PDGF), and others that are involved in the tendon healing. Several in vitro and in vivo preclinical studies have been conducted in order to define the mechanism of action of PRP and the histological and mechanical effects on the healed tendons. TGF-b increases the expression of procollagen types I and III and mechanical properties. PDGF-BB, IGF-1, VEGF, and B-FGF promote tendon cell proliferation and tendon healing. These factors stimulate the gene expression of the matrix molecules and tendon cell proliferation and promote the synthesis of angiogenic and other growth factors and also activate circulation-derived cells [7, 8].

The impossibility of reproducing chronic tendinopathy in animal models means that most of the studies are based on the effectiveness of PRP in surgically created tendon tears. Several studies analyzing different tendons (Achilles tendon, patellar tendon, flexor digitorum tendon) in different animals (rats, rabbits, horses) found that PRP accelerated healing time and led to superior quality tendons than control groups, with better organization of fibroblasts and collagen bundles. PRP treatment groups also had earlier regression of granulation tissue than the control groups, indicating an increased rate of repair. Tendons treated with PRP showed an increased load to failure [7, 8]. Animal studies confirmed these results on rotator cuff repair with intraoperative injection of PRP [9]. Furthermore augmentation with a sponge carrier or a dermal graft did not influence the effects of PRP on rotator cuff healing [10, 11], while the use of scaffolds for PRP may have adverse effects compared to the injectable preparation. In an animal study, platelet-rich fibrin matrix (PRFM) did not recapitulate the native enthesis, but rather induced an exuberant and disordered healing response that was characterized by fibrovascular scar tissue [12]. Similarly, in humans there is no evidence to support the use of PRFM applied to the bone-tendon interface [13–15].

Currently, there are several PRP preparation protocols, which result in different formulations with varying platelet concentration and other components. There is a high heterogeneity even among the randomized controlled trials (RCTs) because disease severity, treatment formulation, and methodological quality vary widely, making it difficult to reach a definitive conclusion on the efficacy of PRP [5, 6].

In conservative treatment, the procedure is performed in an outpatient setting. Patients are in supine position with their arms placed on the superior part of the iliac wing with the palm up and the elbow flexed, or they are seated with the arm relaxed along the trunk and the elbow flexed. PRP is aseptically injected into the subacromial space. The approach could be anterior, lateral, or posterior (Fig. 14.1a–c) [16].

Compared to a landmark-guided approach, the subacromial ultrasound-guided injection is more accurate, allowing to deliver PRP exactly into the tendon lesion. After the visualization of the hypoechoic lesion, approximately 2 mL of PRP are injected via a 22 gauge needle. Some authors suggest the use of a peppering technique which consists of multiple intratendinous injections of little amounts of PRP solution withdrawing without emerging from the skin, but slightly

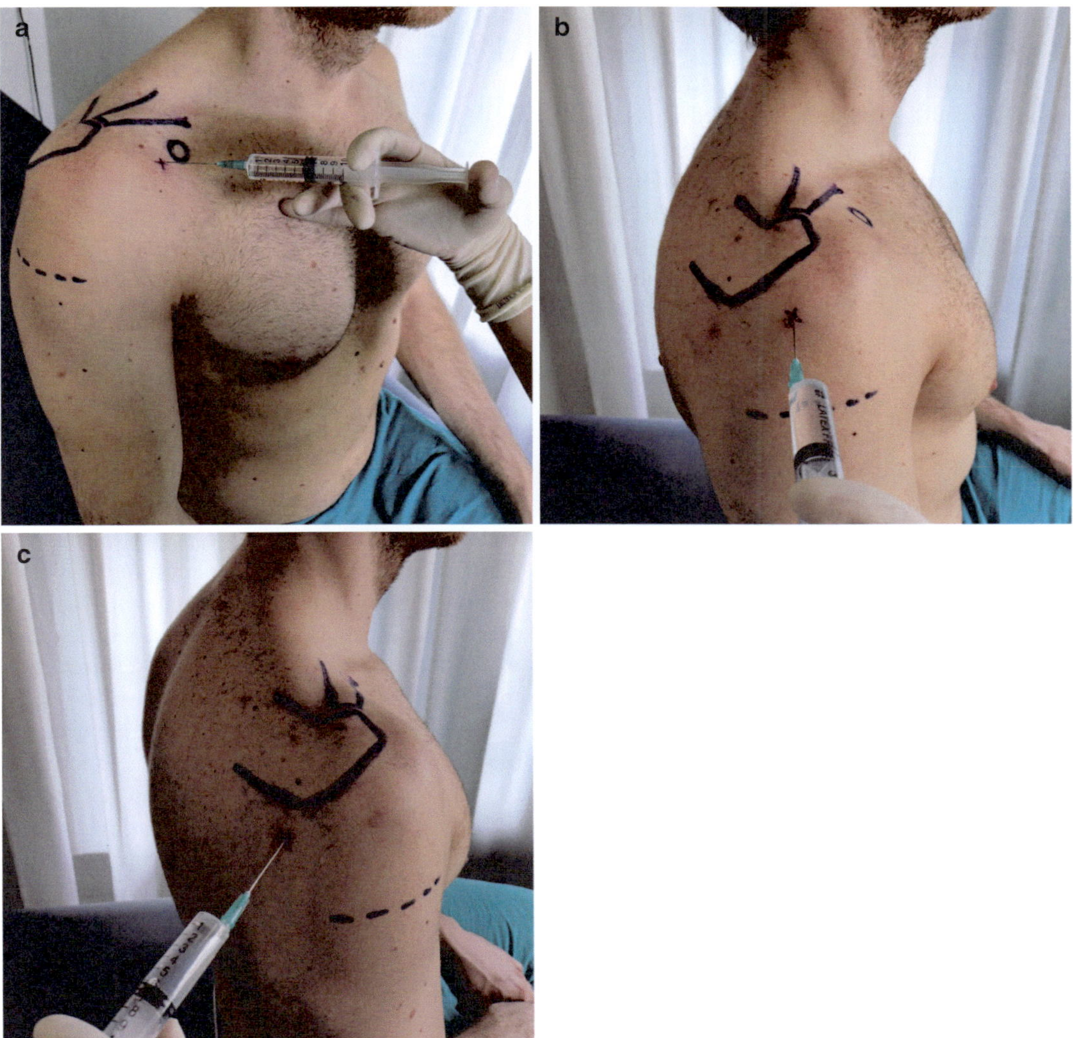

Fig. 14.1 Landmark-guided approaches in shoulder subacromial injections. (**a**) Anterior, (**b**) lateral, and (**c**) posterior approaches

redirecting and reinserting the syringe needle. In fact, intratendinous PRP injections are retained within the tendon and primarily distribute longitudinally with minimal cross-sectional spread. Meanwhile, intratendinous injections with this peppering technique may alter tendon morphology and mechanics [17]. To our knowledge there are no comparative studies on the difference in the effects of subacromial and intratendinous injections.

Several studies have reported potential positive effects of PRP in the treatment of acute and chronic tendinopathy, but in rotator cuff ones the results are still inconsistent [18–20]. In 2019 a systematic review of RCTs by Harley et al. found that at short-term follow-up, PRP (mainly as single injection protocol) may not be a beneficial therapy in nonoperative management for partial-thickness rotator cuff tears [18]. However, in a comparison with other injectable agents, in absence of physical therapy, PRP seemed to have a potential effect in rotator cuff tendinopathy [18]. In the RCT performed by Shams et al., 20 patients treated with a single PRP injection were compared with 20 patients treated with a corticosteroid injection [19]. Patients were assessed at 6 weeks and 3 and 6 months of follow-up. The patients in the PRP group showed significantly greater improvements in American Shoulder and Elbow Surgeons (ASES) score, Constant-Murley score (CMS), Simple Shoulder Test (SST), and Visual Analogue Scale (VAS) at 3 months of follow-up compared to the corticosteroid group. However, no significant difference was found in terms of clinical outcomes and tendinopathy grade on magnetic resonance imaging (MRI) at final follow-up [19].

PRP in addition or compared to exercise therapy did not show clear results [20]. Kesikburun et al. found that, in patients who are undergoing a 6-week standard exercise program, a single injection of PRP did not result in greater clinical improvements compared with placebo injection at 1 year of follow-up [20]. However, in other studies, PRP seemed to show a potential, although not significant, efficacy in terms of improvement of tendinopathy grade on MRI compared to exercise therapy. Therefore, PRP may have a role alongside exercise therapy to promote tendon healing, which may be beneficial in the long-term [19–21]. Probably the beneficial effect of physical exercise was due to its cumulative effect maintained over time. Therefore, multiple injections of PRP could increase these effects [22].

In surgical rotator cuff repairs, despite the resolution of pain and the improvement in subjective outcomes, there is a high rate of incomplete healing or re-tearing of the tendons [23]. Several studies have demonstrated that the failure of tendon repair occurs relatively early after surgery, between 6 and 12 months postoperatively [23]. Moreover, the failure did not seem to be caused by commercially available sutures, but by the weakness of the tendon-suture interface. The reason for not healing was the poor tendon tissue quality due to degenerative changes, increasing with age [24, 25]. For all these reasons, augmentation with PRP may be considered in patients with high risk of repair failure, such as older age (mostly > 65 years), multiple tendon involvement, small/moderate tear size (> 2 cm), retraction (> 2 cm), and high-grade fatty infiltration of the muscles (Goutallier grade > 2) [26, 27].

Comparative clinical studies on the use of PRP after rotator cuff repair have demonstrated conflicting results, making it now difficult to draw definitive conclusions [6]. Literature data suggested a beneficial effect on the healing of arthroscopically repaired small and medium rotator cuff lesions (re-tear rate 7.9% among patients treated with PRP, compared to 26.8% of those treated without PRP) [28].

However, we report here our preliminary results, not already published, of a RCT which compared clinical (University of California at Los Angeles (UCLA) Shoulder Score, VAS, SST, CMS, and shoulder external rotation strength) and radiological outcomes of 53 patients who underwent arthroscopic rotator cuff repair with or without the addition of PRP at the 10-year follow-up. The addition of PRP consisted of an injection between the bone and the repaired rotator cuff and then performing a dry arthroscopic check of the clot formation. Compared with the previous radiological control at the 2-year follow-up, new re-lesions occurred in 6% of the patients that received PRP treatment, whereas in the control group, the percentage raises to 14% ($p = 0.61$). The clinical and radiological outcomes at the 10-year follow-up showed a substantial uniformity of results between the two groups. The minor differences observed in previous follow-ups tended to converge over time.

Even a large number of systematic reviews and meta-analyses did not point to a consensus. The recent meta-analysis of Chen et al. reviewed exclusively level 1 RCTs to assess the efficacy of PRP for rotator cuff-related abnormalities and evaluate how specific tendon involvement, the use of gel/non-gel formulations, and inclusion of leukocytes affect pain and functional outcomes and compared quantitative results with the minimal clinically important difference (MCID) [5, 29]. This study showed that long-term re-tear rates were significantly decreased in patients who received PRP in different manner (i.e., intraoperative injections, augmentation with autologous platelet-rich fibrin matrix or gel formulation); significant improvements in PRP-treated patients were noted for multiple functional outcomes, but none reached their respective MCIDs [5].

A recent meta-analysis by Harley et al. of 18 RCTs analyzed separately the effect of PRP and platelet-rich fibrin on rotator cuff repair [30]. They found that PRP improved the structural integrity of tendon compared with the controls (82.8% versus 69.5%; $p < 0.05$), in small to medium tears (93.3% versus 73.5%; $p < 0.05$). In addition, the use of PRP leads to significantly improved visual analogue score at 30 days and at final follow-up, as well as improved CMS and UCLA score. They also investigated the differences between leukocyte-rich end leukocyte-poor PRP formulation finding that leukocyte-poor formulation had better healing rates [30].

Lastly, in a systematic review and meta-analysis including 16 RCTs or prospective cohort studies, Cavendish et al. found that intraoperative PRP injection resulted in a 25% reduction in the risk of repair failure (as defined by postoperative imaging) regardless of tear size, with low heterogeneity among the included studies [6].

14.3 Hyaluronic Acid (HA)

HA is a high molecular weight glycosaminoglycan constituted by a repetition of N-acetyl-glucosamine and a β-glucuronic acid [31]; its physiochemical properties of retaining water with a very high hydration ratio and viscoelasticity are very well studied. HA seems to be effective for the treatment of shoulder osteoarthritis [32]; however, no consensus is present in literature about the clinical indication for rotator cuff tendinopathy [33, 34].

In vitro studies showed how HA enhances viability and proliferation, counteracting apoptosis, in rotator cuff tendon-derived cells, in a dose-dependent manner [35]. This biological property may be helpful as in tendinopathy a higher incidence of tenocyte apoptosis and a decrease in collagen synthesis have been reported [36]. Mitsui et al. proved HA significantly and dose-dependently inhibited the expression of proinflammatory cytokine mRNA (IL-1b, IL-6, and TNF-alpha) in subacromial-synovium fibroblasts

(SSF) frequently dysregulated in rotator cuff disease [37]. In both in vitro and in vivo trials, high concentration of HA increased collagen type I, stimulated endogenous growth and cell-cell interaction, resulting in a faster recovery, accelerating healing processes after tendon repair and decreasing scar formation within the tendons [38]. Animal studies confirmed results of in vitro studies: a rabbit rotator cuff tear model showed that an intraoperative injection of 1 mL HA accelerated tendon-to-bone healing, enhancing the biomechanical strength and increasing chondroid formation and tendon maturity at the tendon-bone interface [39, 40]. In addition, Nakamura et al. proved the safety of HA in term of cell viability of tendon fibroblasts when compared to corticosteroids in an in vivo study [41]. Nevertheless, in vivo studies had the limit of the acute injury model, where weight-bearing and biochemical reactions differ from humans with chronic tendinopathy [41].

In 2018, Lin et al. [34] performed the first meta-analysis on injection therapies for rotator cuff diseases, extracting all data from RCTs published up to September 2017. The study combined both direct and indirect evidences into the same statistical framework, giving the more consistent results of a level I study. In this meta-analysis, the authors reported that HA did not show a greater pain reduction in the short-, medium-, and long-term follow-up, compared to placebo. The main limitation of this study was that RCTs included in the meta-analysis considered several causes of rotator cuff tendinopathy, both degenerative or overuse disorders and impingement syndrome. On the contrary, three RCTs published just after this meta-analysis reported more encouraging results [42–44]. In the trial of Flores et al., 84 patients, suffering for persistent supraspinatus tendinopathy without rotator cuff tear, were randomized to receive either physical therapy in association with subacromial HA injections (two injections with an interval of 1 week) or physical therapy only. The results supported the use of subacromial HA injections as adjuvant treatment to physical therapy in the management of supraspinatus tendinopathy thanks to an earlier return to pre-injury activity and lower number of rehabilitation sessions [42]. Moreover, Jeong et al. demonstrated that a hyaluronate/carboxymethyl cellulose intraoperative subacromial injection in 80 patients with full-thickness rotator cuff tear improved gliding motion in the sub-deltoid space at 8 weeks of follow-up, with a not statistically significant tendency of a faster recovery [43]. Finally, in their RCT on 184 patients with a partial-thickness rotator cuff tear, Cai et al. compared subacromial injections of normal saline (NS), saline hyaluronate (SH), PRP, and a combination of SH and PRP at several follow-ups. Each patient received subacromial injections consecutively once a week for 4 weeks. SH, and even more SH + PRP, showed to improve the function of the shoulder reducing pain at every follow-up more than NS [44].

Clinical trials showed less promising results compared to in vitro and animal studies probably because of the different concentration in nutrients and oxygen that it is very different in the pathological environment. Moreover, the effect of HA on the main proteins of extracellular matrix tendons remains unsolved [35].

14.4 Cytokines and Growth Factors

Cytokines play an important role in cell chemotaxis, proliferation, matrix synthesis, and cell differentiation having the potential to improve rotator cuff tendon healing via autocrine and paracrine signaling [4]. As highlighted by Sundman et al., these factors can vary across various preparations of PRP, which is why researchers have looked to specific cytokines to increase healing rates with rotator cuff repairs [45]. The enthesis of a tendon includes tendon, fibrocartilage, and bone, so early research investigated osteoinductive factors as possible adjuvants.

Kim et al., studying the effect of a single intra-tendinous PRP injection on the degenerative rotator cuff tendinopathy according to compositions, found that PRP subgroup above IL-1B or TGF-B1 cutoff values (>5.19 pg/mL and >61.79 μg/mL, respectively) showed significant differences in all examined clinical outcomes compared with the exercise group, while the PRP subgroup below IL-1B or TGF-B1 cutoff values did not [22]. IL-1β is a major cytokine that induces a catabolic action on tendon fibroblasts through the upregulation of inflammatory mediators and plays a role in the tendon's degenerative changes in tendinopathy or regenerative capacity [22]. TGF-β1 inhibits matrix metalloprotease (MMP)-9 and MMP-13 expression preventing the degradation of collagen and enhancing the formation of tough fibrous tissues and is known to improve tendon strength and tendon healing [22]. Therefore MMP enzymes, due to their role in degrading collagen and other extra-cellular matrix proteins, have also been studied in soft tissue repairs to prevent re-tears [22]. A significantly higher levels of MMP-3 were found in patients with rotator cuff re-tears in a study by Gotoh et al. [46].

Doxycycline, a common antibiotic, has been widely studied because it inhibits MMP. In a rat study, doxycycline (130 mg/kg/day) showed promising results in preventing tendons re-tears. One hundred and eighty-three rats underwent repair of the supraspinatus tendon, and the animals were divided into four groups. In experimental groups, an identical surgery was performed, and doxycycline was started orally at preoperative day 1 (group 1 = 66 animals), postoperative day 5 (group 2 = 28 animals), or postoperative day 14 (group 3 = 23 animals) and administered every day until the time of sacrifice. In the control group (group 4 = 66 animals), the supraspinatus was repaired to its anatomical footprint, and no dose of doxycycline was administered. They found that in the first postoperative day, doxycycline-treated animals demonstrated greater metachromasia and improved collagen organization at the healing enthesis with the MMP-13 activity significantly reduced [47].

Specific growth factors, such as FGF and PDGF, have also been studied. Recombinant PDGF showed to improve the biomechanical strength of repair and increase the bone-tendon interdigitation histologically when used with a collagen matrix [48], but only gain the histological characteristics of repairs when embedded in sutures alone [49]. Similarly, in rat models, FGF-2 showed the same properties when applied to rotator cuff repairs [50–52] probably due to its role in tenogenic progenitor cells stimulation that can improve tendon to bone healing [53].

Despite numerous animal studies, these cytokines and growth factors have not been studied in a human model yet, and new rigorous trials are necessary to evaluate the safety and efficacy of these adjuvants in rotator cuff tendinopathy and repair.

14.5 Conclusions

Orthobiologics injection therapies represent one of the new frontiers in the treatment of each stage of rotator cuff disease. The use of PRP and HA injections in the conservative treatment of rotator cuff tendinopathy or tears has been widely studied. Although results of in vitro and in vivo studies are very encouraging, clinical trials have shown inconsistent evidences, making it difficult to reach definitive conclusions. In order to support the routine use in clinical practice of PRP and HA injections for the treatment of rotator cuff tendinopathy, new studies should be performed to clarify the best timing, doses, injection intervals, and type of PRP and HA formulations.

Finally, the promising results of growth factors and cytokines in in vitro and animal trials will allow to better understand the structural and compositional deficiencies of the injured rotator cuff tissue to identify the biological needs and create a targeted injection therapy in the future.

> **Take-Home Messages**
> - Historically, corticosteroids represented the injection therapy for rotator cuff tendinopathy that can reduce pain and, consequently, improve functional outcomes but without a tendon healing stimulation.
> - The use of PRP and HA injections in the conservative treatment of rotator cuff tendinopathy or tears has been widely studied in humans, while the use of cytokines and growth factors is limited to in vitro and animal studies.
> - In basic science studies, PRP accelerated healing time and led to superior quality tendons.
> - In humans, PRP is injected for nonoperative management of rotator cuff tendinopathy or in association with arthroscopic rotator cuff repair with a potential efficacy in terms of reduction of pain and improvement of clinical outcomes and tendinopathy grade on MRI.
> - Currently, it is difficult to reach a definitive conclusion on the efficacy of PRP for rotator cuff tendinopathy due to different formulations and high heterogeneity among randomized controlled trials.
> - Likewise, the promising results of HA in vitro and animal studies are not confirmed in clinical trials.

References

1. Chard MD, Hazleman R, Hazleman BL, King RH, Reiss BB. Shoulder disorders in the elderly: a community survey. Arthritis Rheum. 1991;34(6):766–9.
2. Factor D, Dale B. Current concepts of rotator cuff tendinopathy. Int J Sports Phys Ther. 2014;9(2):274.
3. Abat F, Alfredson H, Cucchiarini M, Madry H, Marmotti A, Mouton C, et al. Current trends in tendinopathy: consensus of the ESSKA basic science committee. Part I: biology, biomechanics, anatomy and an exercise-based approach. J Exp Orthopaed. 2017;4(1):18.
4. Bedi A, Maak T, Walsh C, et al. Cytokines in rotator cuff degeneration and repair. J Shoulder Elb Surg. 2012;21(2):218–27.
5. Chen X, Jones IA, Togashi R, Park C, Vangsness CT Jr. Use of platelet-rich plasma for the improvement of pain and function in rotator cuff tears: a systematic review and meta-analysis with bias assessment. Am J Sports Med. 2019:363546519881423.
6. Cavendish PA, Everhart JS, DiBartola AC, Eikenberry AD, Cvetanovich GL, Flanigan DC. The effect of perioperative platelet-rich plasma injections on postoperative failure rates following rotator cuff repair: a systematic review with meta-analysis. J Shoulder Elb Surg. 2020;29(5):1059–70.

7. Kon E, Filardo G, Di Martino A, Marcacci M. Platelet-rich plasma (PRP) to treat sports injuries: evidence to support its use. Knee Surg Sports Traumatol Arthrosc. 2011;19(4):516–27.

8. Baksh N, Hannon CP, Murawski CD, Smyth NA, Kennedy JG. Platelet-rich plasma in tendon models: a systematic review of basic science literature. Arthroscopy. 2013;29(3):596–607.

9. Beck J, Evans D, Tonino PM, Yong S, Callaci JJ. The biomechanical and histologic effects of platelet-rich plasma on rat rotator cuff repairs. Am J Sports Med. 2012;40(9):2037–44.

10. Ersen A, Demirhan M, Atalar AC, Kapicioğlu M, Baysal G. Platelet-rich plasma for enhancing surgical rotator cuff repair: evaluation and comparison of two application methods in a rat model. Arch Orthop Trauma Surg. 2014;134(3):405–11.

11. Chung SW, Song BW, Kim YH, Park KU, Oh JH. Effect of platelet-rich plasma and porcine dermal collagen graft augmentation for rotator cuff healing in a rabbit model. Am J Sports Med. 2013;41(12):2909–18.

12. Hasan S, Weinberg M, Khatib O, Jazrawi L, Strauss EJ. The effect of platelet-rich fibrin matrix on rotator cuff healing in a rat model. Int J Sports Med. 2016;37(1):36–42.

13. Rodeo SA, Delos D, Williams RJ, Adler RS, Pearle A, Warren RF. The effect of platelet-rich fibrin matrix on rotator cuff tendon healing: a prospective, randomized clinical study. Am J Sports Med. 2012;40(6):1234–41.

14. Castricini R, Longo UG, De Benedetto M, Panfoli N, Pirani P, Zini R, Maffulli N, Denaro V. Platelet-rich plasma augmentation for arthroscopic rotator cuff repair: a randomized controlled trial. Am J Sports Med. 2011;39(2):258–65.

15. Barber FA, Hrnack SA, Snyder SJ, Hapa O. Rotator cuff repair healing influenced by platelet-rich plasma construct augmentation. Arthroscopy. 2011;27(8):1029–35.

16. Ogbeivor C. Needle placement approach to subacromial injection in patients with subacromial impingement syndrome: a systematic review. Musculoskeletal Care. 2019;17(1):13–22.

17. Wilson JJ, Lee KS, Chamberlain C, DeWall R, Baer GS, Marcus Greatens and Nicole Kamps. Intratendinous injections of platelet-rich plasma: feasibility and effect on tendon morphology and mechanics. J Exp Orthopaed. 2015;2:5.

18. Hurley ET, Hannon CP, Pauzenberger L, Fat DL, Moran CJ, Mullett H. Nonoperative treatment of rotator cuff disease with platelet-rich plasma: a systematic review of randomized controlled trials. Arthroscopy. 2019;35(5):1584–91.

19. Shams A, El-Sayed M, Gamal O, Ewes W. Subacromial injection of autologous platelet-rich plasma versus corticosteroid for the treatment of symptomatic partial rotator cuff tears. Eur J Orthop Surg Traumatol. 2016;26:837–42.

20. Kesikburun S, Tan AK, Yilmaz B, Yasar E, Yazicioglu K. Platelet-rich plasma injections in the treatment of chronic rotator cuff tendinopathy: a randomized controlled trial with 1-year follow-up. Am J Sports Med. 2013;41:2609–16.

21. Rha DW, Park GY, Kim YK, Kim MT, Lee SC. Comparison of the therapeutic effects of ultrasound-guided platelet-rich plasma injection and dry needling in rotator cuff disease: a randomized controlled trial. Clin Rehabil. 2013;27:113–22.

22. Kim SJ, Yeo SM, Noh SJ, Ha CW, Lee BC, Lee HS, Kim SJ. Effect of platelet-rich plasma on the degenerative rotator cuff tendinopathy according to the compositions. J Orthop Surg Res. 2019;14(1):408.

23. Mirzayan R, Weber AE, Petrigliano FA, Chahla J. Rationale for biologic augmentation of rotator cuff repairs. J Am Acad Orthop Surg. 2019;27(13):468–78.

24. Ponce BA, Hosemann CD, Raghava P, Tate JP, Sheppard ED, Eberhardt AW. A biomechanical analysis of controllable intraoperative variables affecting the strength of rotator cuff repairs at the suture-tendon interface. Am J Sports Med. 2013;41(10):2256–61.

25. Miller BS, Downie BK, Kohen RB, et al. When do rotator cuff repairs fail? Serial ultrasound examination after arthroscopic repair of large and massive rotator cuff tears. Am J Sports Med. 2011;39(10):2064–70.

26. Wylie JD, Baran S, Granger EK, Tashjian RZ. A comprehensive evaluation of factors affecting healing, range of motion, strength, and patient-reported outcomes after arthroscopic rotator cuff repair. Orthop J Sports Med. 2018;6(1).

27. Mall NA, Tanaka MJ, Choi LS, Paletta GA Jr. Factors affecting rotator cuff healing. J Bone Joint Surg Am. 2014;96(9):778–88.

28. Chahal J, Van Thiel GS, Mall N, et al. The role of platelet-rich plasma in arthroscopic rotator cuff repair: a systematic review with quantitative synthesis. Arthroscopy. 2012;28(11):1718–27.

29. Sedaghat AR. Understanding the minimal clinically important difference (MCID) of patient-reported outcome measures. Otolaryngol Head Neck Surg. 2019;161(4):551–60.

30. Hurley ET, Lim Fat D, Moran CJ, Mullett H. The efficacy of platelet-rich plasma and platelet-rich fibrin in arthroscopic rotator cuff repair: a meta-analysis of randomized controlled trials. Am J Sports Med. 2019;47(3):753–61.

31. Meyer K. Chemical structure of hyaluronic acid. Fed Proc. 1958;17(4):1075–7.

32. Kwon YW, Eisenberg G, Zuckerman JD. Sodium hyaluronate for the treatment of chronic shoulder pain associated with glenohumeral osteoarthritis: a multicenter, randomized, double-blind, placebo-controlled trial. J Shoulder Elb Surg. 2013;22(5):584–94.

33. Osti L, Buda M, Buono AD, Osti R, Massari L. Clinical evidence in the treatment of rotator cuff tears with hyaluronic acid. Muscles Ligaments Tendons J. 2016;5(4):270–5.

34. Lin MT, Chiang CF, Wu CH, Huang YT, Tu YK, Wang TG. Comparative effectiveness of injection therapies in rotator cuff tendinopathy: a systematic review, pairwise and network meta-analysis of

randomized controlled trials. Arch Phys Med Rehabil. 2019;100(2):336–349.e15.

35. Gallorini M, Berardi AC, Berardocco M, Gissi C, Maffulli N, Cataldi A, Oliva F. Hyaluronic acid increases tendon derived cell viability and proliferation in vitro: comparative study of two different hyaluronic acid preparations by molecular weight. Muscles Ligaments Tendons J. 2017;7(2):208–14.

36. Via AG, De Cupis M, Spoliti M, Oliva F. Clinical and biological aspects of rotator cuff tears. Muscles Ligaments Tendons J. 2013;3(2):70–9. Erratum in: Muscles Ligaments Tendons J. 2014 Oct;3(4):359.

37. Mitsui Y, Gotoh M, Nakama K, Yamada T, Higuchi F, Nagata K. Hyaluronic acidinhibits mRNA expression of proinflammatory cytokines and cyclooxygenase-2/prostaglandin E(2) production via CD44 in interleukin-1-stimulated subacromial synovial fibroblasts from patients with rotator cuff disease. J Orthop Res. 2008;26(7):1032–7.

38. Abate M, Schiavone C, Salini V. The use of hyaluronic acid after tendon surgery and in tendinopathies. Biomed Res Int. 2014;2014:783632.

39. Honda H, Gotoh M, Kanazawa T, Ohzono H, Nakamura H, Ohta K, Nakamura KI, Fukuda K, Teramura T, Hashimoto T, Shichijo S, Shiba N. Hyaluronic acid accelerates tendon-to-bone healing after rotator cuff repair. Am J Sports Med. 2017;45(14):3322–30.

40. Li H, Chen Y, Chen S. Enhancement of rotator cuff tendon-bone healing using bone marrow-stimulating technique along with hyaluronic acid. J Orthop Translat. 2019;17:96–102.

41. Nakamura H, Gotoh M, Kanazawa T, Ohta K, Nakamura K, Honda H, Ohzono H, Shimokobe H, Mitsui Y, Shirachi I, Okawa T, Higuchi F, Shirahama M, Shiba N, Matsueda S. Effects of corticosteroids and hyaluronic acid on torn rotator cuff tendons in vitro and in rats. J Orthop Res. 2015;33(10):1523–30.

42. Flores C, Balius R, Álvarez G, Buil MA, Varela L, Cano C, Casariego J. Efficacy and tolerability of peritendinous hyaluronic acid in patients with supraspinatus tendinopathy: a multicenter, randomized, controlled trial. Sports Med Open. 2017;3(1):22.

43. Jeong JY, Chung PK, Yoo JC. Effect of sodium hyaluronate/carboxymethyl cellulose (Guardixsol) on retear rate and postoperative stiffness in arthroscopic rotator cuff repair patients: a prospective cohort study. J Orthop Surg (Hong Kong). 2017;25(2):2309499017718908.

44. Cai YU, Sun Z, Liao B, Song Z, Xiao T, Zhu P. Sodium hyaluronate and platelet-rich plasma for partial-thickness rotator cuff tears. Med Sci Sports Exerc. 2019;51(2):227–33.

45. Sundman EA, et al. Growth factor and catabolic cytokine concentrations are influenced by the cellular composition of platelet-rich plasma. Am J Sports Med. 2011;39(10):2135–40.

46. Gotoh M, Mitsui Y, Shibata H, et al. Increased matrix metalloprotease-3 gene expression in ruptured rotator cuff tendons is associated with postoperative tendon retear. Knee Surg Sports Traumatol Arthrosc. 2013;21:1807–181

47. Bedi A, Fox AJS, Kovacevic D, Deng X, Warren RF, Rodeo SA. Doxycycline-mediated inhibition of matrix metalloproteinases improves healing after rotator cuff repair. Am J Sports Med. 2010;38(2):308–17.

48. Hee CK, Dines JS, Dines DM, Roden CM, Wisner-Lynch LA, Turner AS, Santoni BG. Augmentation of a rotator cuff suture repair using rhPDGF-BB and a type i bovine collagen matrix in an ovine model. Am J Sports Med. 2011;39(8):1630–40.

49. Uggen C, Dines J, McGarry M, Grande D, Lee T, Limpisvasti O. The effect of recombinant human platelet-derived growth factor BB-coated sutures on rotator cuff healing in a sheep model. Arthroscopy. 2010;26(11):1456–62.

50. Ide J, Kikukawa K, Hirose J, Iyama K, Sakamoto H, Fujimoto T, et al. The effect of a local application of fibroblast growth factor-2 on tendon-to-bone remodelling in rats with acute injury and repair ofthe supraspinatus tendon. J Shoulder Elb Surg. 2009;18(3):391–8.

51. Ide J, Kikukawa K, Hirose J, Iyama K, Sakamoto H, Mizuta H. The effects of fibroblast growth factor-2 on rotator cuff reconstruction with acellular dermal matrix grafts. Arthroscopy. 2009;25(6):608–16.

52. Tokunaga T, Karasugi T, Arimura H, Yonemitsu R, Sakamoto H, Ide J, et al. Enhancement of rotator cuff tendon-bone healing with fibroblast growth factor 2 impregnated in gelatin hydrogel sheets in a rabbit model. J Shoulder Elb Surg. 2017;26(10):1708–17.

53. Tokunaga T, Shukunami C, Okamoto N, Taniwaki T, Oka K, Sakamoto H, et al. FGF-2 stimulates the growth of Tenogenic progenitor cells to facilitate the generation of Tenomodulin-positive tenocytes in a rat rotator cuff healing model. Am J Sports Med. 2015;43(10):2411–22.

Orthobiologics for the Treatment of Tennis Elbow

William D. Murrell, Sharmila Tulpule,
Nagib Atallah Yurdi, Agnes Ezekwesili,
Nicola Maffulli, and Gerard A. Malanga

15.1 Introduction

Lateral epicondylitis (LE), also known as tennis elbow, is a prevalent, common, usually self-limiting disorder of the dorsolateral aspect of the elbow. It affects men and women equally, predominantly between the ages of 45 and 54 years [1]. If the local healing response fails, LE can evolve into a chronic condition involving the origins of the wrist extensor muscles (especially the extensor carpi radialis brevis—ECRB) where they attach to the lateral epicondyle of the humerus [2]. LE commonly affects individuals involved in repetitive tasks, including trade/office personnel. Despite the moniker, LE is less common in athletes.

The etiology of LE is not known, but excessive use, microtrauma, age, smoking, and obesity have been defined as risk factors [1]. LE is more prevalent than its medial counterpart, medial epicondylitis, affecting 1.3% of the general population [3, 4], and its incidence seems to be decreasing [5]. Historically, LE was classified as an inflammatory disorder. Conservative plans have been characterized by the use of oral anti-inflammatory drugs, physical therapy (stretching, range of motion, strengthening), activity modification, rest, and steroid injections with variable long-term success [6]. However, the modern descriptions classify the condition as a tendinopathy.

While the specific causes of LE are not always clear, the process of mechanically induced tendinopathy is followed by a failed healing response, which leads to some feature of chronic inflammation, calcification, fibrosis, vascular proliferation, and hyaline degeneration of the affected tendons. Although acute inflammatory cell infiltration is not typically seen in biopsies of the tendon lesion, it is now well-accepted that "molecular inflammation" and the production of numerous inflam-

W. D. Murrell (✉)
Healthpoint Hospital,
Abu Dhabi, United Arab Emirates

411th HC, Jacksonville, FL, USA

S. Tulpule
Saudi German Hospital, Dubai, United Arab Emirates

N. A. Yurdi
411th HC, Jacksonville, FL, USA

A. Ezekwesili
Perelman School of Medicine, University of
Pennsylvania, Philadelphia, PA, USA

N. Maffulli
Department of Musculoskeletal Disorders, University
of Salerno School of Medicine and Dentistry,
Salerno, Italy

Queen Mary University of London, Barts and the
London School of Medicine and Dentistry Centre for
Sports and Exercise Medicine, Mile End Hospital,
London, UK

G. A. Malanga
New Jersey Regenerative Institute LLC,
Cedar Knolls, NJ, USA

Department of Physical Medicine and Rehabilitation,
Rutgers University, New Jersey Medical School,
Newark, NJ, USA

© ISAKOS 2022
G. Filardo et al. (eds.), *Orthobiologics*, https://doi.org/10.1007/978-3-030-84744-9_15

matory mediators play a critical role in the progression and chronicity of tendinopathy and probably the initiation phase as well [7–9].

Given this new understanding of the physiopathology of LE, treatment approaches have shifted toward novel biological therapies to restore tendon biology through reactivation of a healing response that may stimulate local tendon repair and/or regeneration, rather than just treating inflammation and pain. The developing treatment strategies include local injection of platelet-rich plasma (PRP), collagen-producing tenocyte-like cells, mixed cell preparations containing connective tissue progenitor cells from different sources (mostly bone marrow and adipose tissue), and autologous conditioned serum (ACS).

15.2 Pathophysiology

Describing the lateral epicondyle, the inferior surface of this bony structure serves as the origin for the lateral ulnar collateral ligament (LUCL) and the radial collateral ligament (RCL) [10]. Superficial to the RCL is the common extensor tendon. This tendon has been extensively investigated in the diagnosis of LE, with detailed study defining the extent and severity of the abnormalities seen within the tendon [11]. The ECRB, a part of the common extensor tendon origin, is the most commonly affected tendon and the structure where the pathology of LE is located [12].

Many different activities that involve the repetitive use of these muscles can lead to the condition, including, but not limited to, manual labor, playing a musical instrument, typing, and playing golf, tennis, or other racquet sports. The strength of tendons typically exceeds physiological forces. However, like other materials, they also have a limit of fatigue resistance at forces that do not result in immediate rupture. The tendons' fatigue limit of loading exceeds the capacity for local repair and remodeling to heal microtears; such microtears can then accumulate, weakening the regional structural integrity to the point that ongoing injury continues to exceed the

rate of repair even at reduced levels of activity. This can lead to failure of healing and subsequently degeneration of the tendon which defines a tendinopathy. Although often called a "tendonitis," at histopathology LE presents with few acute inflammatory cells and is instead more associated with hypertrophy of fibroblasts, abundant disorganized collagen, hyperplasia of vascular elements, and eventually apoptosis and increase turnover of disorganized extracellular matrix [13, 14]. Various inflammatory mediators have been identified in biopsies of tendinopathic tendons. The pathophysiology of tendinopathy is complex, with variability in the cellular and molecular mediators based on the stage of the condition, inducing several abnormalities in tenocytes, including abnormalities in cell proliferation, matrix synthesis, production of matrix-degrading metalloproteinases, and apoptosis [8]. The end result is the failure of tendon homeostasis and/or imbalance of tissue injury and repair/remodeling.

At histology, in patients with LE, the ECRB tendon presents apoptosis of tendon cells and signs of fibrotic regeneration [15], together with angiofibroblastic degeneration secondary to a failure of natural tendon repair mechanisms, with low numbers of inflammatory cells seen in patients with chronic LE [4, 16]. The underuse of the tendon (stress deprivation) can also predispose to LE [17]. Underuse can cause the tendon structure to weaken, increasing the risk of full or partial rupture/failure [18]. Underuse can also decrease the vascularization of the tendon causing sustained use to lead to ischemia of the tendon and hyperthermic injuries that will greatly weaken the muscle-tendon-bone structure [17].

The etiology of LE is not always known and likely to be multifactorial. Repetitive overuse/microtrauma, age, smoking, and obesity have been defined as risk factors. There are no consistent associations between clinical presentation, imaging findings, and common histological findings of vascular proliferation, hyaline degeneration, and fibroblastic proliferation with calcific debris [19].

15.3 Evaluation of Tennis Elbow: Lateral Epicondylitis

There is great variance in the presentation of LE with varying peripheral nerve irritation and pain response. Clinically, LE can present with intermittent, low-grade pain to escalate to chronic and continuous severe pain which disturbs sleep [17]. The pain is generally worse with activities that involve grasping and wrist extension or those that can result in eccentric loading of the tendon such as in racquet sports.

At initial investigation, tenderness to palpation of the lateral epicondyle and common extensor tendon are indicative of LE [12]. As the condition progresses, calcifications can be detected over the lateral epicondyle. Special maneuvers such as Maudsley's test, Thomson's maneuver, diminished grip strength, and the "chair" test can help make the diagnosis [20, 21]. Additionally, ultrasonography is often used to try to analyze the alterations of the tendons involved in this condition. Using this technique in various ultrasound images, intratendinous and peritendinous evaluations can be made to differentiate between normal and pathologic tendons [11].

The diagnosis of LE, in general, is performed following clinical examination, with lateral elbow pain that increases on palpation and/or through resisted wrist dorsiflexion and resisted extension of the long finger [22]. Various other indicators used and tools help to evaluate the presence and extent of the condition [23].

The tenderness of the lateral epicondyle and the triggering of pain by resisted wrist extension and/or finger extensors with the elbow extended or flexed are key clinical findings to make the diagnosis [24]. The evaluation and diagnosis of tennis elbow is largely clinical, but imaging studies such as radiographs, ultrasonography, or MRI can be performed, when the diagnosis is not clear. Current ultrasound technology can provide high-resolution images [25]. The ultrasonographic findings classically included the following findings: (1) enthesopathy, (2) tendinitis, (3) peritendinitis, (4) bursitis, (5) intramuscular hematoma, and (6) mixed findings [26]. With modern devices and high frequency transducers, real-time, high-definition images have been able to more clearly define the condition. However, it should be kept in mind that anatomical abnormalities can be identified in both symptomatic and asymptomatic subjects, as many of the tendon alterations seen in imaging studies represent typical age-related changes [27]. The use of color flow Doppler can be very useful in determining a painful, pathologic tendon which would demonstrate neovascularization.

15.4 Orthobiologic Treatment Options

15.4.1 Basic Mechanism of Action

The biologic treatment options in the management of LE are designed to intervene and change the balance of injury and repair, generally be inducing a new injury or inflammatory process that may activate and accelerate a new cycle of repair and remodeling. These include injection of hyperosmolar dextrose for prolotherapy, platelet-rich plasma (PRP), culture-expanded autologous tenocytes, various types of progenitor cells (including BMAC, adipose derivatives, culture expanded mesenchymal stromal cells (MSCs)), and incubated autologous conditioned serum such as gold-induced cytokines at the site of the tendon lesion [28–36].

In the case of PRP, the potential active components include a multitude of growth factors contained in the alpha granules of the platelets. Following local injection, in aggregate, these promote a local inflammatory response followed by a local proliferation of intrinsic or recruited connective tissue progenitors. This is followed by a remodeling phase [18] (Fig. 15.1). However, PRP is a complex device. In addition to cytokines produced that have positive (desirable) immuno-regulatory and anti-inflammatory functions, there may also be components that have negative undesirable effects. The precise composition of individual preparations of PRP varies markedly from one preparation to another, and there are currently insufficient data to define an optimal composition.

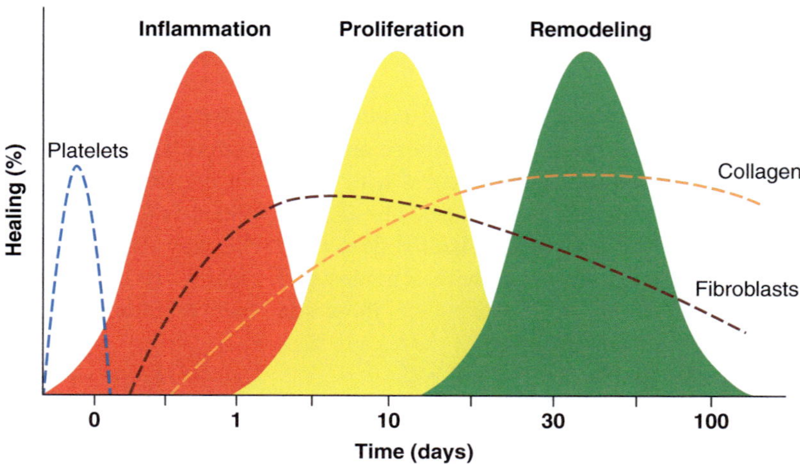

Fig. 15.1 Platelet administration is believed to initiate a healing response in tendons, similar to what is seen in cutaneous wound healing

The specific growth factors believed to play a positive role in PRP include platelet-derived growth factor (PDGF), vascular endothelial growth factor (VEGF), transforming growth factor (TGF)-beta-1, insulin-like growth factor (IGF), epidermal growth factor (EGF), and fibroblast growth factor (FGF) [37]. PDGF and EGF are involved in activating chemotaxis, activation and proliferation of local connective tissue progenitors (CTPs), and stimulation of other growth factor cascades. VEGF stimulates angiogenesis in the injured tendon. TGF-beta-1 and IGF factor are associated with an increased collagen synthesis. FGF has multiple roles: angiogenesis, cell migration, CTP proliferation, and collagen synthesis [38]. PDGF and EGF activate proliferation of CTPs in local tissues, likely including local perivascular cells (e.g., pericytes) that amplify the healing response.

Culture-expanded mesenchymal stromal cells (MSCs) have been delivered by local injection. These cells are not found at the site of activation for much time. However, MSCs are possibly phagocytized by local monocytes that prolong the healing potential by activation and induction of a specific population of T-cells, the T-regulatory cells, which continue to favor a healing environment long after MSCs are removed [39].

Exogenous injected culture-expanded autologous tenocytes appear to also be immunomodulatory with very little engraftment. If effective, they are thought to primarily work through a paracrine mechanism, stimulating local and perhaps distant host cells. Injected cells produce PDGF-alpha, FGF-beta, and TGF-beta in a similar fashion to PRP to restore the normal biologic and mechanical properties to the tendon by promoting collagen synthesis, extracellular matrix, and tendon repair [31, 32, 40].

Processed autologous serum (aka autologous conditioned serum (ACS)) is a variant on the processing of blood for PRP. Incubation at 37 °C with specially designed glass beads for 24 h upregulates the Interleukin-1 receptor antagonist (IL-1ra) up to 100-fold and increases other anti-inflammatory cytokines including IL-4, IL-10, and IL-13 [41].

The mechanism of action of gold-induced autologous conditioned serum injected into the local tissue is thought to be mediated by local change in cytokines to tip the balance of local injury and wound healing/regeneration [42]. The process specifically increases the concentration of cytokines and growth factors in autologous blood, with significant increases in p-Gelsolin, granulocyte colony-stimulating factor (G-CSF). The following proteins are also upregulated: IL-8, macrophage chemotactic protein (MCP-3), stromal-derived protein-alpha (SDF-α), tumor necrosis factor-alpha (TNF-α), leukemia inhibitory factor (LIF), IL-10, macrophage inflammatory protein (MIP-1α), and MIP-1ß. Macrophage colony-stimulating factor (M-CSF), IL-15, IL-17, granulocyte-macrophage colony-stimulating factor (GM-CFS), hepatocyte growth factor (HGF), IL-2Ra, IL-12p40, chemokine (C-C motif) ligand 11 (Eotaxin/CCL11), fibroblast growth factor-basic (bFGF), and interferon-gamma (IFN-γ). Gelsolin has proven to be a particularly important cytokine at the epigenetic, transcriptional, and translational levels [36, 43] and is found in many cells in the human body and in the circulation as plasma-gelsolin (p-GSN). Its molecular structure is well conserved between mammalian species as an actin-binding protein which is critically important for the maintenance of cellular structure and homeostasis [44, 45]. Gelsolin is currently being investigated as a biomarker for several different disease processes as decreased levels are associated with the onset of the various disease states in some conditions and has been found to have impacts on cancer, cellular apoptosis, infection and inflammation, cardiac injury, pulmonary diseases, and aging [43, 44, 46].

15.4.2 Sites of Harvests and Source Materials

PRP and processed serum preparations are most commonly obtained from peripheral blood. As described in other chapters, bone marrow aspirate (BMA) can be collected from the pelvis, spine, and the sternum. BMA contains a heterogeneous mixture of immature and adult hematopoietic cells that include lymphoid and myeloid precursors, rare megakaryocytes, and a small population of CTPs, erythrocytes, and platelets [47–49]. Adipose tissue can be harvested by percutaneous liposuction, direct excision, or Coleman's technique from the abdomen, inner thigh, outer thigh, and flank (the highest quantity and quality of cells yielded from the abdomen [50]). Adipose tissue harvested in this way will contain a mixture of adipocytes, vascular and perivascular cells, fibrous tissue fragments, blood, serum, clot material, and debris resulting from cell lysis. As a result, both BMA and adipose-derived starting materials contain a mixture of cell populations, growth factors, and cytokines, with both pro-inflammatory and anti-inflammatory properties. Some form of processing is needed in both cases to remove undesirable components and concentrate, and possibly even select for, desirable components.

15.4.3 Biologic Selection

Once the diagnosis of LE is confirmed, and conservative strategies have failed, treatment is determined by a combination of provider recommendation and patients' choice. Many options are available spanning from supervised neglect and conservative management to surgery. Since this book is dedicated to the review of injectable biological therapies, we will quickly review other nonoperative therapies. Generally, at least 6 weeks of conservative therapy are recommended before considering more invasive modalities. A review of available conventional therapies with timeframe of treatment and efficacy is provided in Table 15.1.

15.5 Clinical Results

15.5.1 PRP

Many comparators to PRP have been used as controls in studies examining the role of PRP in the management of LE. This results in heterogeneity in the available data, which complicates systematic reviews and meta-analyses. For the purpose of this chapter, for comparison studies we will mainly focus on reviews and articles that compare PRP to corticosteroid (CS) injection, although this in itself is a controversial point, as CS offer at best short-term benefits and may possibly be detrimental.

Huang et al., in a systematic review and meta-analysis of nine randomized controlled trials (RCTs) and 565 subjects, quantitatively compared PRP to CS in the management of LE and did not demonstrate that PRP treatment was superior to conventional management in terms of improvement in visual analog scale (VAS) (8/9 trials) and Patient-Rated Tennis Elbow Evaluation (PRTEE) (1/9 trials). The 9 RCTs reported 11

Table 15.1 Schematic summary of available conventional therapies

Treatment	How instituted	Special equipment/personnel	Efficacy
Rest	+/− Initiated healthcare professional +/− oral medication (analgesic or anti-inflammatory) +/− topical medication (analgesic or anti-inflammatory)	None	Likely short-term benefit
Stretching/exercise therapy	Self-directed or supervised	Therabands. Ice packs	Unknown
Manual therapy	Supervised therapy	Various soft tissue manipulation devices	Unknown
Modalities	Supervised therapy	Iontophoresis, electrical stimulation, ultrasound, TENS, and low intensity light therapy	Likely short-term benefit
Shockwave	Supervised treatment	Shockwave device with both radial and focal applicators	Unknown
Corticosteroid injections	Clinician administered	Ultrasound, corticosteroid	Short-term benefit
Autologous blood products	Clinician administered	Ultrasound, commercial kits/centrifuge	Intermediate-term benefit

short- and 5 long-term comparisons of pain and 6 short-term comparisons of Disability of the Arm, Shoulder, and Hand (DASH) scores that were used for data pooling. The short-term data analysis showed a statistically significant medium effect size of CS over PRP for pain relief (SMD, 0.56; 95% CI, 0.14–0.99; $I^2 = 86\%$; $P = 0.009$) with moderate quality of evidence. There were no differences in the short-term DASH scores (SMD, −0.18; 95% CI, −0.88 to 0.51; $I^2 = 88\%$; $P = 0.6$) with low quality of evidence. In contrast to the short-term analyses, the improvement in pain scores reversed long-term: PRP provided significantly better pain relief than CS, with a very large effect size (SMD, −1.3; 95% CI, −1.9 to −0.7; $I^2 = 85\%$; $P < 0.0001$), although the quality of evidence was low [28].

Mi et al. conducted a systematic review and meta-analysis of 8 RCTs (511 patients) published between 1980 and 2016, comparing PRP vs CS injections in LE patients. CS improved pain and function in the short-term (2–8 weeks), but PRP was more effective in the intermediate (12 weeks) and longer term (6 and 12 months). Li et al. conducted a similar systematic review with meta-analysis of 7 RCTs to compare LE treatment of PRP vs CS: local CS fared better at short-term (4–8 weeks) follow-up, but beyond 24 weeks, PRP produced a significantly better outcome according to DASH score, Mayo Elbow Performance Score (MEPS), and VAS score [51]. The short-term efficacy of CS supports the notion of the presence of "molecular inflammation" [52]. Increasing evidence has demonstrated that inflammatory mechanisms likely are activated within the tendon during the symptomatic phase of the condition and in many cases results in a dysregulated homeostasis of the tissue [53].

The type of PRP injection administered is another area of active investigation to determine the optimal preparation. Currently, studies have used different types of PRP to treat LE (leukocyte rich or leukocyte poor or other), as well as different platelet concentrations, and activation status. Currently, there are no clear recommendations based on evidence for the choice of leukocyte-rich PRP (LR-PRP) over leukocyte-poor PRP (LP-PRP), as both LR- and LP-PRP preparations have demonstrated comparable benefit [54]. A further limiting factor is that many studies have relatively short follow-up, which contributes to uncertainty concerning the duration of treatment effectiveness.

Omar et al. also reviewed available RCTs comparing PRP to CS and did not find any significant difference between PRP and CS after a follow-up of 6 weeks [55]. Analyzing individual studies, Mishra et al. investigated the clinical value of PRP by comparing it with a control group who received a bupivacaine injection in

230 patients. At 6 months, the PRP group reported a significant improved VAS score and an improved PRTEE score when compared to the control [56]. Krogh et al. compared the reduction in the intensity of pain in 60 patients divided into the 3 different groups, LR-PRP, CS, and saline, with no significant difference between the 3 groups at the primary end-point of 3 months. At 3 months, PRP demonstrated continued improvement, whereas CS demonstrated a recurrence of pain after an initial improvement [57]. Yadav et al. compared LR-PRP with CS in 65 patients, with PRP displaying a significantly improved quick Disability of the Arm, Shoulder, and Hand (qDASH) score, higher VAS score, and greater grip strength at all time points, 15 days, 1 month, and 3 months [58]. Gosens et al. investigated the efficacy of an ultrasound-guided injection of LR-PRP against CS in a RCT including 100 patients. Significant improvement was observed in the VAS score at 6 months and 1 and 2 years in the PRP group. The CS group showed early improvement, but VAS and DASH scores returned to baseline levels at 2-year follow-up [59]. Gautam et al. looked at CS and PRP, injected in a non-ultrasound-guided manner, in 30 patients [4]. There was no significant difference observed in the VAS score, MEPS, Oxford Elbow Score (OES), DASH score, or grip strength, but CS demonstrated a relative decrease in outcomes at 6 months as compared to 3 months. Palacio et al. performed a RCT in 60 patients with 90- and 180-day follow-ups, comparing LR-PRP to 0.5% neocaine to dexamethasone: no difference was seen at both the 90- and 180-day follow-ups [60]. Seetharamaiah et al. randomly compared and evaluated 90 elbows with equal numbers in each of the three treatment groups (LR-PRP, saline, and CS) [61]. Both the PRP and the triamcinolone groups produced better pain relief at 3 and 6 months as compared to normal saline group ($P < 0.05$). At 6 months, pain relief in the PRP group was significantly better than the triamcinolone group. Varshney et al. carried out a randomized study of 83 patients (50 treated with CS and 33 with LR-PRP): there was no difference between the groups at 2 months after treatment [62]. At 6-month follow-up, the CS injection

group returned to baseline VAS, but the PRP group continued to improve. In a randomized study, Lebiedzinski et al. administered autologous conditioned plasma (ACP) to 53 patients, and 46 patients received CS with lignocaine [63]. The DASH score was significantly better in the CS group at 6 weeks and 6 months, but significantly improved in the ACP group at 1 year. Gupta et al. conducted a randomized trial which included a total of 80 patients divided into 2 groups, group A-PRP versus group B-CS and local anesthetic mixture in LE treatment [64]. Results were compared using DASH, VAS, and MEPS. Though the results with CS were better at 6 weeks, PRP patients fared significantly better at 3 and 12 months.

Overall, these studies and reviews suggest that PRP therapies for LE are both safe and effective. They are at least as effective as traditional CS injections and appear to provide better long-term outcomes. Further investigation in prospective studies and registry data collection will be needed to better refine the optimal timing, dose, and composition of PRP therapy in the clinical setting of LE.

15.5.2 Bone Marrow Aspirate (BMA) or Bone Marrow Aspirate Concentrate (BMAC)

A single report details the use of BMA or BMAC in LE. Singh et al. reported on a cohort of 30 patients who had received no previous treatment for LE [33]. They were evaluated with the PRTEE score prior to and following the treatment of a single administration of BMAC. BMA (10 mL) was aspirated from the iliac crest, anticoagulated with heparin (1 mL) and centrifuged for 20–30 min at 2000 RPM. A total volume of 4–5 mL was injected. Unfortunately, cell count and viability of the injectate was not reported. Patients were evaluated at 2, 6, and 12 weeks after administration. Overall, patients demonstrated a baseline pre-injection mean PRTEE score of 72.8 ± 6.97 which decreased to a mean PRTEE score of 40.93 ± 5.94, $P < 0.0001$ at 2 weeks. The mean PRTEE score was significantly improved at the 6-week and 12-week follow-ups, reported at 24.46 ± 4.58 and 14.86 ± 3.48, respectively; both $P < 0.0001$. Since this work did not provide a comparison to control group, further studies will be needed before BMA or BMAC preparations can be considered as therapy options for LE.

15.5.3 Adipose Tissue-Derived Cells

Evidence on using freshly isolated populations of fat-derived cells for LE therapy is lacking.

Lee et al. evaluated the safety and efficacy of allogenic adipose-derived culture-expanded MSCs (AD-MSCs) in treating LE [34]. In this investigation, allogeneic AD-MSCs combined with fibrin glue were injected into the hypoechoic common extensor tendon lesions of 12 patients with chronic LE; 6 patients each received a dose of 10^6 or 10^7 cells in 1 mL. Efficacy was analyzed by VAS score for elbow pain, modified Mayo Clinic Performance Index for the Elbow, and evaluating ultrasound images of tendon lesions after 6, 12, 26, and 52 weeks. From baseline through 52 weeks of periodic follow-up, VAS scores progressively decreased, elbow performance scores improved, and the size of the tendon lesions also decreased. This provided evidence of safety and potential efficacy for allogenic AD-MSCs. However, some adverse effects were reported during the study, including mild swelling and joint effusion. Further studies in larger cohorts with appropriate controls will be needed before consideration of adipose-derived cells for LE therapy.

15.5.4 Autologous Culture-Expanded Fibroblasts

Another potential treatment is culture-expanded autologous fibroblast injection, particularly for more severe, chronic, and resistant forms of LE. In a pilot study of 12 patients with refractory LE, Connell et al. evaluated culture-expanded fibroblasts derived from autologous skin biopsies [35]. A dose of 10×10^6 cells in 2 mL was injected in one arm, concentrated plasma (2 mL) in the other arm. Patients were followed for 6 months. Clinical and structural improvements were noted. Specifically, the median PRTEE score decreased from 78 before the procedure to 47 at 6 weeks, 35 at 3 months, and 12 at 6 months after the procedure ($P = 0.05$). The healing response on ultrasonography showed median decrease in (1) number of tears, from 5 to 2; (2) number of new vessels, from 3 to 1; and (3) tendon thickness, from 4.35 to 4.2 ($P = 0.05$). Of the 12 patients, 11 had a satisfactory outcome, and only 1 patient proceeded to surgery after failure of treatment at the end of 3 months.

More recently, Wang et al. published a study with 4.5-year follow-up regarding the use of culture-expanded fibroblastic cells derived from autologous patellar tendon [40]. A dose of 2 mL ($2–5 \times 10^6$ cells/mL) was injected under ultrasound guidance near the lateral epicondyle for treatment of severe refractory LE in 17 patients. The implantation was ultrasound guided. No adverse effects were reported. VAS scores, qDASH scores, and grip strength all improved significantly at 12 months.

Further studies using culture-expanded fibroblastic cells derived from skin, tendon tissue, bone marrow, or fat tissue are justified, particularly for patients with LE who fail less invasive methods.

15.5.5 Gold-Induced Cytokines Injection (Autologous Conditioned Serum)

Clinically resistant LE in 22 patients was treated from 2010 to 2015 to investigate the safety and efficacy of injection of gold-induced cytokines for the treatment of LE. The composition of the injection is described elsewhere [36]. Primary outcome measure was improvement in VAS score at 1 month, 3 months, 6 months, and 1 year. At baseline, median VAS score was 6. Fourteen of 22 patients were available at final follow-up at 1-year, all 14 patients reported no pain (Fig. 15.2). Prospective controlled studies are planned with longer follow-up.

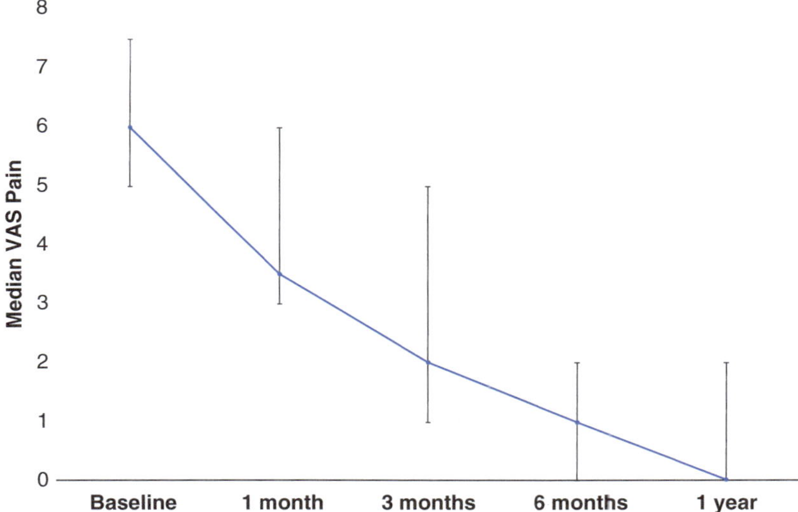

Fig. 15.2 Gold-induced cytokines LE visual analog scale (VAS) ($N = 14$) up to 1-year follow-up

15.6 Delivery of Orthobiologics

Injections in LE patients can be performed using anatomical landmarks only (Fig. 15.3, LR-PRP injections), however, as much as 30% of these injections may not be placed intralesionally [65]. This may be true because of the significant variations in anatomy [27]. For this reason, many practitioners who utilize orthobiologic preparations often insist that ultrasound guidance of injection is critical: (1) to ensure optimal localization of the injectate and (2) to optimize the patient experience and confidence. Because the site of injection can be particularly painful, a radial nerve block prior to the procedure can be considered (Figs. 15.4 and 15.5). This can minimize the patient pain and anxiety and eliminate the need for local anesthetics that may result in undesirable distortion of local anatomy and compromise the concentration, delivery, and therefore the efficacy of the biologic injectate.

Treatment is best administered with the patient sitting or supine, with a pillow to support the forearm, so that the elbow is freely accessible for injection proximal to distal or distal to proximal (Fig. 15.6). Many prefer distal to proximal along the muscle fibers of the ECRB, although a trans-

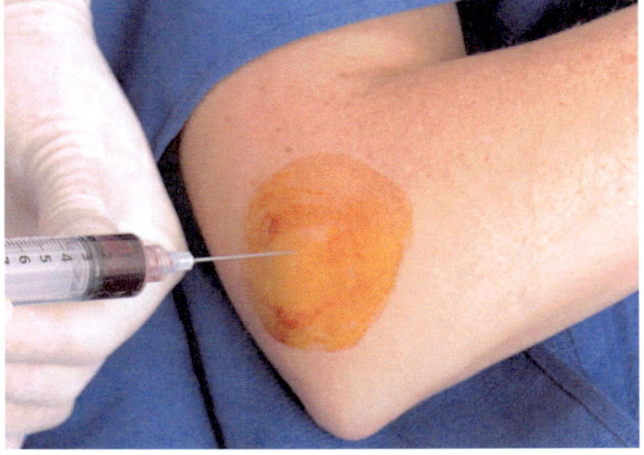

Fig. 15.3 LR-PRP injection, landmark technique. Courtesy Allan Mishra, MD

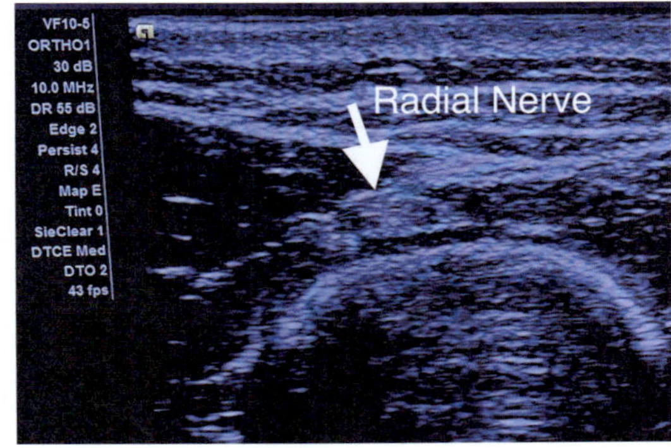

Fig. 15.4 Ultrasound image of radial nerve

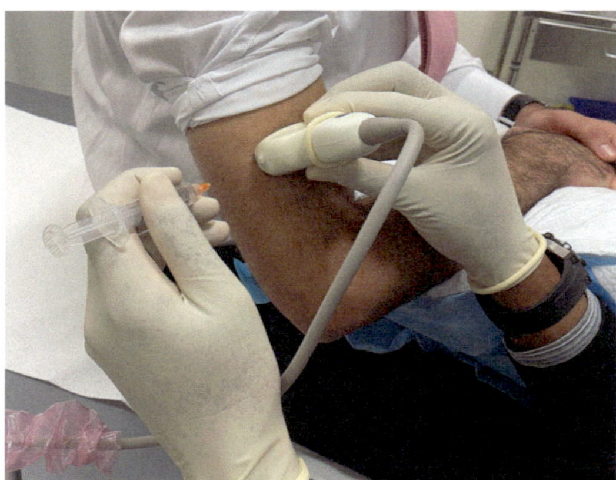

Fig. 15.5 Ultrasound-guided radial nerve block

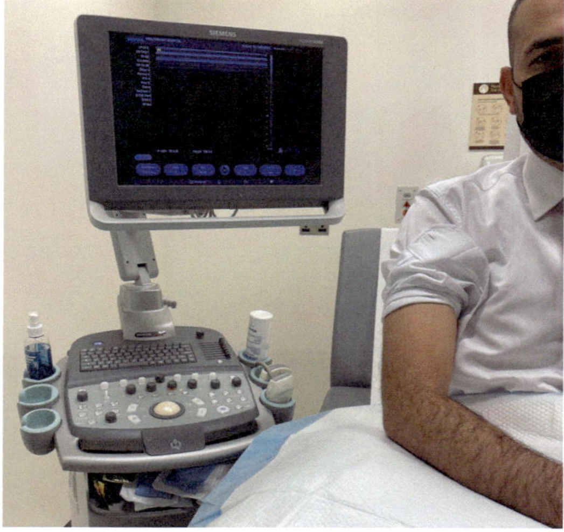

Fig. 15.6 Preferred patient positioning for ultrasound-guided injection

verse approach to the tendon can also be performed. If ultrasound is available, obtaining a pre-treatment image of the pathology is helpful, as it is always advantageous to have pre- and posttreatment images to document the clinical findings. Most practitioners will use a high frequency linear probe with depth set to 2.5 or 3 cm. As the structure is quite superficial, the focus does not need to be below 1 cm.

Although most of these procedures can be performed with a small amount of dilute lidocaine restricted to the subcutaneous tissues only, a radial nerve block may also be used for patient comfort. A field or radial nerve block can be administered; 3–5 mL of a mixture of equal parts 1% lidocaine and 0.5% ropivicaine can be used to anesthetize the patient. The field block is performed in the soft tissue 1–2 cm proximal to the most proximal portion of the injection site. The radial nerve block is performed under ultrasound guidance and is located in the lateral arm approximately 5 cm proximal to the elbow joint (Figs. 15.4 and 15.5). The block can be performed distal to the branching point of the posterior interosseous nerve to spare motor function to a portion of the forearm. The injection around the nerve is performed on the long axis of the ultrasound probe in a lateral to medial direction, and care is taken not to inject the nerve, but to inject around it. The depth is not great, so using a 1.5 inch 25 or 27 gauge needle can easily complete the task.

Once the patient is adequately anesthetized, the PRP or other orthobiologic can be administered, with or without ultrasound guidance. The results of treatment can be seen at ultrasound before (Fig. 15.7a) as early as 3 months following the procedure (Fig. 15.7b). Often, several passes are performed into the hypoechoic areas as in performing a needle tenotomy procedure.

Post-injection, especially if a block has been administered, it is helpful to use a sling to support the hand and elbow for a few hours until the block has worn off. Many rehabilitation protocols have been described, but many are not based on empirical data, but experience derived from conservative treatment of LE. There is debate on whether to start with stretching versus immediate loading, with no agreement or data to support either of these approaches. One area of agreement is that eccentric strengthening exercises should not be performed until the later phases of rehabilitation. In general, aggressive strengthening is initiated at 4–6 weeks after injection. It is suggested that the patient refrains from all NSAIDs for a minimum of 2 weeks but many clinicians will restrict these medications for up to 4 weeks.

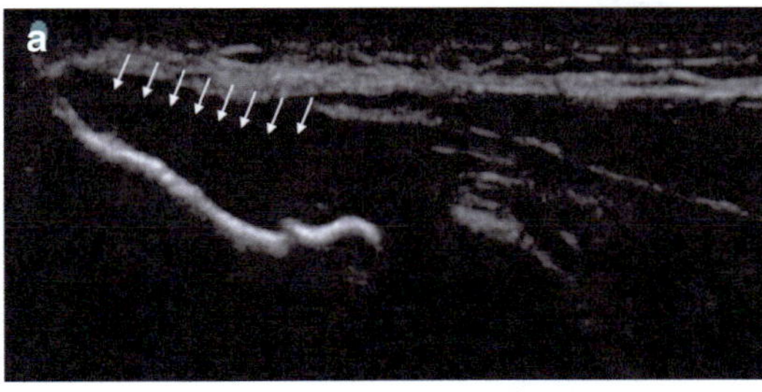
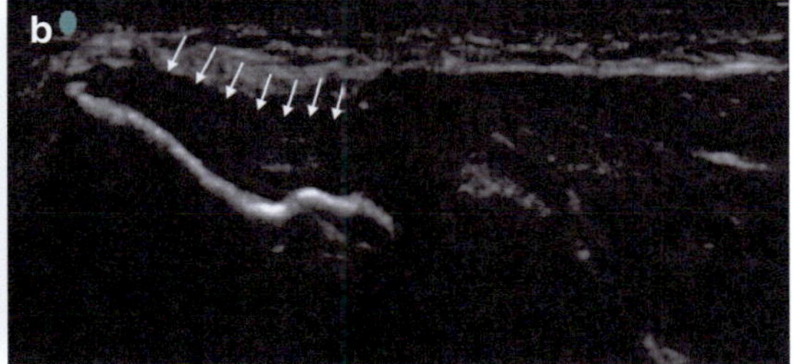

Fig. 15.7 Ultrasound image: (**a**) Pre-treatment. (**b**) Three months after LR-PRP injection. LE, note the increased tissue density and increased echogenicity. Courtesy Allan Mishra, MD

15.7 Conclusion

LE is a common and disabling condition that leads to pain, time away from work and sport, and difficulty with activities of daily living. The goals of treatments (both orthobiologic and non-orthobiologic) are to completely relieve pain and rapidly restore long-term function.

After reviewing the wide range of available treatments and their results, there is emerging evidence for the effectiveness of orthobiologic therapies for LE. PRP is the most commonly used and best documented of these therapies. PRP appears to be superior to traditional steroid injection therapies. However, the optimal timing, dose, and composition of PRP injections remain to be established.

Other therapies are promising, but remain much less well developed, and their efficacy is not yet firmly established. These include injection of culture-expanded autologous fibroblasts (from skin, tendon, bone marrow, or adipose origin) processed autologous serum or injection of gold-induced cytokines. Preliminary evaluation of these alternative therapies shows that they are safe and exhibit minimal side effects. These orthobiologic alternatives must be considered investigational, as the body of available studies is insufficient to provide definitive evidence of clinical efficacy and incremental value beyond the current efficacy of PRP.

A fundamental limitation in the current literature is the lack of reporting of the composition of the orthobiologic administered, as well as frequent absence of details related to the processing of the material, technique of adminis-tration, and post-injection rehabilitation protocol. We encourage authors to report the "Minimum Information for Studies Evaluating Biologics in Orthopaedics" (MIBO) criteria for all clinical studies reporting on orthobiologics [66]. Another distinct limitation in the current literature is the paucity of follow-up imaging studies to evaluate changes in the structure and composition of the treated tendon. Future goals should focus on the completion of additional long-term controlled studies that further highlight safety and adverse events, cost-effectiveness, long-term efficacy, indications, patient selection, preparation and procedure standardization, characterization, and quality management, as this will likely help to eliminate the great variability seen in the current results.

Take-Home Messages
- The mainstay of treatment of LE is conservative and can include NSAIDs, paracetamol (acetaminophen), physiotherapy, and extracorporeal shockwave therapy as initial treatment and should be carried out for at least 6 weeks before considering alternative approaches.
- Corticosteroids can be considered after failing noninvasive nonoperative therapy; however, this option likely only provides short-term benefit and is possibly harmful.
- PRP seems to provide better results than other injectable options such as saline,

corticosteroids, or local anesthesia, with longer-lasting duration.

- Orthobiologic approaches can be an alternative option to reduce symptoms and possibly delay or avoid surgery in patients with LE if all other conservative measures fail.
- Ultrasound diagnosis and delivery of orthobiologic injectable therapies can be helpful for the treatment of LE.
- Several variables could influence the effects of PRP, but current literature cannot clearly identify the optimal dose, timing, or composition of PRP for LE.
- Alternative preparations such as processing cells from bone marrow or fat; culture expanding fibroblastic cells from the skin, marrow, fat, or tendon; and preparations of autologous conditioned serum appear to be safe. However, long-term comparative studies have not been carried out to confirm the safety and efficacy of these treatment options.

Acknowledgments The authors would like to acknowledge Pam Jackson, PhD for her contribution to this work.

References

1. Waseem M, Nuhmani S, Ram C, Sachin Y. Lateral epicondylitis: a review of the literature. J Back Musculoskelet Rehabil. 2012;25:131–42.
2. Nirschl R, Ashman E. Tennis elbow tendinosis (epicondylitis). Instr Course Lect. 2004;53:587–98.
3. Ciccotti MG, Charlton WP. Epicondylitis in the athlete. Clin Sports Med. 2001;20(1):77–93.
4. Gautam VK, Verma S, Batra S, Bhatnagar N, Arora S. Platelet-rich plasma versus corticosteroid injection for recalcitrant lateral epicondylitis: clinical and ultrasonographic evaluation. J Orthop Surg. 2015;23:1–5.
5. Sanders TL Jr, Maradit Kremers H, Bryan AJ, Ransom JE, Smith J, Morrey BF. The epidemiology and health care burden of tennis elbow: a population-based study. Am J Sports Med. 2015;43(5):1066–71.
6. Nirschl RP, Pettrone F. Tennis elbow. The surgical treatment of lateral epicondylitis. J Bone Joint Surg Am. 1979;61(6A):832–9.
7. Cook J, Purdam CR. Is tendon pathology a continuum? A pathology model to explain the clinical presentation of load-induced tendinopathy. Br J Sports Med. 2009;43:409–16.
8. Aicale R, Tarantino D, Maffulli N. Overuse injuries in sport: a comprehensive overview. J Orthop Surg Res. 2018;13:309.
9. Chisari E, Rehak L, Khan WS, Maffulli N. Tendon healing in presence of chronic low-level inflammation: a systematic review. Br Med Bull. 2019;132:97–116.
10. Karbach LE, Elfar J. Elbow instability: anatomy, biomechanics, diagnostic maneuvers, and testing. J Hand Surg Am. 2017;42:118–26.
11. Tagliafico AS, Bignotti B, Martinoli C, Elbow US. Anatomy, variants, and scanning technique. Radiology. 2015;275:636–50.
12. Taylor SA, Hannafin JA. Evaluation and management of elbow tendinopathy. Sports Health. 2012;4:384–93.
13. Kraushaar BS, Nirschl RP. Tendinosis of the elbow (Tennis elbow): clinical features and findings of histological, immunohistochemical, and electron microscopy studies. J Bone Joint Surg Am. 1999;81-A:259–78.
14. Chen J, Wang A, Xu J, Zheng M. In chronic lateral epicondylitis, apoptosis and autophagic cell death occur in the extensor carpi radialis brevis tendon. J Shoulder Elb Surg. 2010;19:355–62.
15. Alfredson H, Ljung BO, Thorsen K, Lorentzon R. In vivo investigation of ECRB tendons with microdialysis technique—no signs of inflammation but high amounts of glutamate in tennis elbow. Acta Orthop Scand. 2000 Oct;71:475–9.
16. Doran A, Gresham GA, Rushton N, Watson C. Tennis elbow. A clinicopathologic study of 22 cases followed for 2 years. Acta Orthop Scand. 1990;61:535–8.
17. Vaquero-Picado A, Barco R, Antuña SA. Lateral epicondylitis of the elbow. EFORT Open Rev. 2017;1:391–7.
18. Ahmad Z, Siddiqui N, Malik SS, et al. Lateral epicondylitis: a review of pathology and management. Bone Joint J [Br]. 2013;95-B:1158–64.
19. Regan W, Wold LE, Coonrad R, Morrey BF. Microscopic histopathology of chronic refractory lateral epicondylitis. Am J Sports Med. 1992;20:746–9.
20. McCallum SDA, Paoloni JA, Murrell GAC. Five-year prospective comparison study of topical glyceryl trinitrate treatment of chronic lateral epicondylosis at the elbow. Br J Sports Med. 2011;45:416–20.
21. Hsu SH, Moen TC, Levine WN, Ahmad CS. Physical examination of the athlete's elbow. Am J Sports Med. 2012;40:699–708.
22. Bisset L, Paungmali A, Vicenzino B, Beller E. A systematic review and meta-analysis of clinical trials on physical interventions for lateral epicondylalgia. Br J Sports Med. 2005;39:411–22.
23. Arrigoni P, Cucchi D, Menon A, Randelli P. It's time to change perspective! New diagnostic tools for lateral elbow pain. Musculoskelet Surg. 2017;101(Suppl 2):175–9.
24. Cho YT, Hsu WY, Lin LF, Lin YN. Kinesio taping reduces elbow pain during resisted wrist extension in patients with chronic lateral epicondylitis: a ran-

domized, double-blinded, cross-over study. BMC Musculoskelet Disord. 2018;19:193.

25. Dones VC 3rd, Grimmer K, Thoirs K, Suarez CG, Luker J. The diagnostic validity of musculoskeletal ultrasound in lateral epicondylalgia: a systematic review. BMC Med Imaging. 2014;14:10.

26. Maffulli N, Regine R, Carrillo F, Capasso G, Minelli S. Tennis elbow: an ultrasonographic study in tennis players. Br J Sports Med. 1990;24:151–5.

27. Keijsers R, Koenraadt KLM, Turkenburg JL, Beumer A, Bertram T, Eygendaal D. Ultrasound measurements of the ECRB tendon shows remarkable variations in patients with lateral epicondylitis. Arch Bone Jt Surg. 2020;8:168–72.

28. Huang K, Giddins G, Wu LD. Platelet-rich plasma versus corticosteroid injections in the management of elbow epicondylitis and plantar fasciitis: an updated systematic review and meta-analysis. Am J Sports Med. 2020;48:2572–85.

29. Murrell WD, Anz AW, Badsha H, Bennett WF, Boykin RE, Caplan AI. Regenerative treatments to enhance orthopedic surgical outcome. PM R. 2015 Apr;7(4 Suppl):S41–52.

30. Scarpone M, Rabago DP, Zgierska A, Arbogast G, Snell E. The efficacy of prolotherapy for lateral epicondylosis: a pilot study. Clin J Sport Med. 2008;18:248–54.

31. Wang A, Breidahl W, Mackie KE, Lin Z, Qin A, Chen J, et al. Autologous tenocyte injection for the treatment of severe, chronic resistant lateral epicondylitis: a pilot study. Am J Sports Med. 2013;41:2925–32.

32. Bucher TA, Ebert JR, Smith A, Breidahl W, Fallon M, Wang T, Zheng MH, Janes GC. Autologous tenocyte injection for the treatment of chronic recalcitrant gluteal tendinopathy: a prospective pilot study. Orthop J Sports Med. 2017;5:2325967116688866.

33. Singh A, Gangwar DS, Singh S. Bone marrow injection: a novel treatment for tennis elbow. J Nat Sci Biol Med. 2014;5:389–91.

34. Lee SY, Kim W, Lim C, Chung SG. Treatment of lateral epicondylosis by using allogenic adipose-derived mesenchymal stem cells: a pilot study. Stem Cells. 2015;33:2995–3005.

35. Connell D, Datir A, Alyas F, Curtis M. Treatment of lateral epicondylitis using skin-derived tenocyte-like cells. Br J Sport Med. 2009;43:293–8.

36. Schneider U, Wallich R, Felmet G, Murrell WD. Gold-induced autologous cytokine treatment in Achilles tendinopathy. In: Canata G, d'Hooghe P, Hunt K, editors. Muscle and tendon injuries. Berlin, Heidelberg: Springer; 2017.

37. de Mos M, van der Windt AE, Jahr H, van Schie HT, Weinans H, Verhaar JA, van Osch GJ. Can platelet-rich plasma enhance tendon repair? A cell culture study. Am J Sports Med. 2008;36:1171–8.

38. Yan Z, Yin H, Nerlich M, Pfeifer CG, Docheva D. Boosting tendon repair: interplay of cells, growth factors and scaffold-free and gel-based carriers. J Exp Orthop. 2018;5:1.

39. de Witte SFH, Luk F, Sierra Parraga JM, Gargesha M, Merino A, Korevaar SS, Shankar AS, O'Flynn L, Elliman SJ, Roy D, Betjes MGH, Newsome PN, Baan CC, Hoogduijn MJ. Immunomodulation by therapeutic mesenchymal stromal cells (MSC) is triggered through phagocytosis of MSC by monocytic cells. Stem Cells. 2018;36:602–15.

40. Wang A, Mackie K, Breidahl W, Wang T, Zheng MH. Evidence for the durability of autologous tenocyte injection for treatment of chronic resistant lateral epicondylitis: mean 4.5-year clinical follow-up. Am J Sports Med. 2015;43:1775–83.

41. Alvarez-Camino JC, Vázquez-Delgado E, Gay-Escoda C. Use of autologous conditioned serum (Orthokine) for the treatment of the degenerative osteoarthritis of the temporomandibular joint. Review of the literature. Med Oral Patol Oral Cir Bucal. 2013;18: e433–8.

42. Kopecki Z, Cowin AJ. The role of actin remodeling proteins in wound healing and tissue regeneration. In: Alexandrescu VA, editor. Wound healing—new insights into ancient challenges. IntechOpen; 2020.

43. Li GH, Arora PD, Chen Y, McCulloch CA, Liu P. Multifunctional roles of gelsolin in health and diseases. Med Res Rev. 2012;32:999–1025.

44. Piktel E, Levental I, Durnaś B, Janmey PA, Bucki R. Plasma gelsolin: indicator of inflammation and its potential as a diagnostic tool and therapeutic target. Int J Mol Sci. 2018;19:2516.

45. Schneider U, Kumar A, Murrell W, Ezekwesili A, Yurdi NA, Maffulli N. Intra-articular gold induced cytokine (GOLDIC®) injection therapy in patients with osteoarthritis of knee joint: a clinical study. Int Orthop. 2021 Feb;45:497–507.

46. DiNubile MJ. Plasma gelsolin as a biomarker of inflammation. Arthritis Res Ther. 2008;10:124.

47. Muschler GF, Boehm C, Easley K. Aspiration to obtain osteoblast progenitor cells from human bone marrow: the influence of aspiration volume. J Bone Joint Surg Am. 1997;79:1699–709.

48. Piuzzi NS, Mantripragada VP, Sumski A, Selvam S, Boehm C, Muschler GF. Bone marrow-derived cellular therapies in orthopaedics: part i: recommendations for bone marrow aspiration technique and safety. JBJS Rev. 2018;6:e4.

49. Patterson TE, Boehm C, Nakamoto C, Rozic R, Walker E, Piuzzi NS, Muschler GF. The efficiency of bone marrow aspiration for the harvest of connective tissue progenitors from the human iliac crest. J Bone Joint Surg Am. 2017;99:1673–82.

50. Iyyanki T, Hubenak J, Liu J, Chang EI, Beahm EK, Zhang Q. Harvesting technique affects adipose-derived stem cell yield. Aesthet Surg J 2015;35:467–76.

51. Mi B, Liu G, Zhou W, Lv H, Liu Y, Wu Q, Liu J. Platelet rich plasma versus steroid on lateral epicondylitis: meta-analysis of randomized clinical trials. Phys Sportsmed. 2017;45:97–104.

52. Dean BJF, Dakin SG, Millar NL, Carr AJ. Review: emerging concepts in the pathogenesis of tendinopathy. Surgeon. 2017;15:349–54.

53. Millar NL, Silbernagel KG, Thorborg K, Kirwan PD, Galatz LM, Abrams GD, Murrell GAC, McInnes IB, Rodeo SA. Tendinopathy. Nat Rev Dis Primers. 2021;7:1. Erratum in: Nat Rev Dis Primers. 2021;7:10.
54. Le ADK, Enweze L, DeBaun MR, Dragoo JL. Current clinical recommendations for use of platelet-rich plasma. Curr Rev Musculoskelet Med. 2018;11:624–34.
55. Omar AS, Ibrahim ME, Ahmed AS, Said M. Local injection of autologous platelet rich plasma and corticosteroid in treatment of lateral epicondylitis and plantar fasciitis: randomized clinical trial. Egypt Rheumatol. 2012;34:43–9.
56. Mishra AK, Skrepnik NV, Edwards SG, Jones GL, Sampson S, Vermillion DA, et al. Efficacy of platelet-rich plasma for chronic tennis elbow: a double-blind, prospective, multicenter, randomized controlled trial of 230 patients. Am J Sports Med. 2014;42: 463–71.
57. Krogh TP, Fredberg U, Stengaard-Pedersen K, Christensen R, Jensen P, Ellingsen T. Treatment of lateral epicondylitis with platelet-rich plasma, glucocorticoid, or saline: a randomized, double-blind, placebo-controlled trial. Am J Sports Med. 2013;41:625–35.
58. Yadav R, Kothari SY, Borah D. Comparison of local injection of platelet rich plasma and corticosteroids in the treatment of lateral epicondylitis of humerus. J Clin DIAGNOSTIC Res. 2015;9(7):RC05–7.
59. Gosens T, Peerbooms JC, van Laar W, den Oudsten BL. Ongoing positive effect of platelet-rich plasma versus corticosteroid injection in lateral epicondylitis: a double-blind randomized controlled trial with 2-year follow-up. Am J Sports Med. 2011;39:1200–8.
60. Palacio EP, Schiavetti RR, Kanematsu M, Ikeda TM, Mizobuchi RR, Galbiatti JA. Effects of platelet-rich plasma on lateral epicondylitis of the elbow: prospective randomized controlled trial. Rev Bras Ortop. 2016;51:90–5.
61. Seetharamaiah V, Gantaguru A, Basavarajanna S. A comparative study to evaluate the efficacy of platelet-rich plasma and triamcinolone to treat tennis elbow. Indian J Orthop. 2017;51(3):304.
62. Varshney A, Maheshwari R, Juyal A, Agrawal A, Hayer P. Autologous platelet-rich plasma versus corticosteroid in the management of elbow epicondylitis: a randomized study. Int J Appl Basic Med Res. 2017;7:125–8.
63. Lebiedziński R, Synder M, Buchcic P, Polguj M, Grzegorzewski A, Sibiński M. A randomized study of autologous conditioned plasma and steroid injections in the treatment of lateral epicondylitis. Int Orthop. 2015;39:2199–203.
64. Gupta PK, Acharya A, Khanna V, Roy S, Khillan K, Sambandam SN. PRP versus steroids in a deadlock for efficacy: long-term stability versus short-term intensity-results from a randomised trial. Musculoskelet Surg. 2020;104:285–94.
65. Keijsers R, van den Bekerom MPJ, Koenraadt KLM, Bleys RLAW, van Dijk CN, Eygendaal D, Elbow Study Collaborative. Injection of tennis elbow: hit and miss? A cadaveric study of injection accuracy. Knee Surg Sports Traumatol Arthrosc. 2017;25:2289–92.
66. Murray IR, Geeslin AG, Goudie EB, Petrigliano FA, LaPrade RF. Minimum information for studies evaluating biologics in orthopaedics (MIBO): platelet-rich plasma and mesenchymal stem cells. J Bone Joint Surg Am. 2017;99:809–19.

Patellar Tendinopathy: Cell Therapy

16

Chris H. Jo and Sanghoon Oh

16.1 Introduction

Patellar tendinopathy causes pain and tenderness, which lead to a decrease in function at sports activity [1]. It occurs frequently in the people of young age, especially in athletes who are subject to repeated stress on the extensor mechanism [2]. Its prevalence is about 14% of all athletes; 45% of volleyball players and 35% of basketball players have experienced this disease [2, 3].

Whereas patellar tendinopathy was considered as an inflammatory tendinitis before the 1990s, it has been understood as a chronic, non-inflammatory, histologically degenerative condition in the 2000s [4]. When 4~8% strain is repeatedly loaded to the tendon, microtrauma occurs via failure of cross-linked structure [5]. Accumulated microtrauma which is not repaired adequately results in the formation of a degenerative zone within the tendon [4]. However, with advance of immunohistochemistry and molecular biology technologies, many studies showed that an inflammatory mechanism has become apparent in the initiation and progress of tendinopathy [6, 7]. These studies suggest that key inflammatory interactions occur in the early stages (~12 weeks) of repetitive tendon microtrauma when patients may be asymptomatic [8]. At these early stages, changes in tissue microenvironment and activation of the innate immune system interact at a crossroads between reparative and degenerative inflammatory healing. These recent studies provide strong evidence that inflammation is a key component of tendinopathy that should not be overlooked in the development of therapeutic strategies in tendinopathy. Tenocytes play a major role to regulate tendon matrix and maintain the homeostasis of the tendon-cell environment [9]. If balance between the pro-inflammatory system and the pro-resolving system is lost, tenocytes develop to inflammatory phenotypes and undergo apoptotic changes [9, 10]. Through these changes, dysregulation occurs in the remodeling of the extracellular matrix that tenocytes are responsible for. Then, the tendon loses the reparative capacity and progresses to the degenerative inflammatory phase [10]. If microtrauma occurs continuously in this state, damage accumulates beyond the reparative capacity in the tendon which results in entering into a vicious cycle [1, 4]. Thus, in order to treat tendinopathy that is entering the chronic degenerative phase, the reparative capacity of the tendon must be restored with the treatment of inflammation and degeneration, and reinforcing pro-resolving system would be important. Through advance of molecular biology technologies, many studies have found out pro-resolving pathways [10]. Utilizing the potential of pro-resolving response represents a new approach to

C. H. Jo (✉) · S. Oh
Department of Orthopedic Surgery, SMG-SNU Boramae Medical Center, Seoul National University College of Medicine, Seoul, South Korea
e-mail: chrisjo@snu.ac.kr

© ISAKOS 2022
G. Filardo et al. (eds.), *Orthobiologics*, https://doi.org/10.1007/978-3-030-84744-9_16

treat inflammatory and degenerative tendon disease [11].

Recently, many studies suggested that stem/stromal cells have an immunomodulatory effect and can orchestrate the inflammatory environment by reconstituting/repopulating the injured tendon [12]. The ability of stem/stromal cells to promote switching from pro-inflammatory to pro-resolving cellular response can be also used to modulate the degenerative environment of tendons, so stem/stromal cell therapy is expected to be a novel promising therapeutic approach [12].

16.2   Current Treatments

Nonsteroidal anti-inflammatory drugs (NSAID) may relieve early acute pain. However, there is no conclusive evidence that they could prevent or treat the progression of patellar tendinopathy, neither that they are effective in the treatment of chronic patellar tendinopathy entering a degenerative phase [4]. Furthermore, NSAIDs may rather mask symptoms and disturb proper treatment [4]. So, until conclusive evidence on NSAIDs for treating tendinopathy, they should rather be used for simple pain control.

Eccentric exercise is known to have the effect of increasing the remodeling process of the collagen fibers of patellar tendon and is commonly used as the main treatment of patellar tendinopathy [13]. Several studies suggested that eccentric exercise has the most high-level evidence among the treatments used to date [13–17]. The most frequently used method for the treatment of patellar tendinopathy is a decline board squat training that repeats the knee bending and stretching exercise on the decline board while standing by the foot of the affected leg [13]. Although this treatment has high-level evidence, the therapeutic outcomes are not optimal yet, and there is no unified protocol [17, 18].

Corticosteroid injection has been used widely for patellar tendinopathy. Nevertheless, the therapeutic effect of corticosteroid injection is controversial [4, 15, 16, 19]. Paavola et al. reported that corticosteroid injection was helpful for tendinopathy by altering the release of noxious and painful chemicals [20]. However, Kongsgaard et al. reported that corticosteroid injection was effective in VISA-p score or VAS at short-term follow-up, but not at long-term follow-up [21]. Moreover, its effect was inferior to eccentric exercise at 6 months, and it may increase the likelihood of tendon rupture. Fredberg et al. also reported that corticosteroid injection only has short-term effect on pain in patellar tendinopathy and does not have long-term effects [22]. Furthermore, Dean et al. (2014) argued that the local administration of corticosteroids disturbs collagen synthesis of tendon and reduces their mechanical properties [23]. Thus, corticosteroid injection on patellar tendinopathy should be carefully performed for specific conditions.

Extracorporeal shock wave therapy (ESWT) has been suggested as a treatment option for patellar tendinopathy [4]. The effect of ESWT on patellar tendinopathy is based on three theories [24]. The first is the relief of pain caused by hyperstimulation analgesia. Overstimulation of pain areas reduces signal transmission in the brain stem. The second is that the mechanical load generated by mechanical stimulation of EWST increases tendon regeneration. The third is the removal of tendon calcification through ESWT. ESWT has been attempted in patients who had no results with other conservative treatments, but its effectiveness and role are controversial [13].

Platelet-rich plasma (PRP) is a platelet concentrate which can help the repair and regeneration by delivering cytokines and growth factors to tendons [25]. Several author reported conflicting results [26, 27]. The optimal number of PRP injections for patellar tendinopathy is controversial among several studies [27–29]. PRP injection for patellar tendinopathy is expected to be promising, but still further mechanistic studies should be necessary. Further details on PRP use for patellar tendinopathy are reported in this chapter.

Dry needling is based on the following assumptions; that is, by repeatedly stabbing the tendinosis site, disruption and internal bleeding of the collagen fibers are caused, and inflammatory processes are activated. Autologous blood rich in growth factors can therefore reach the site to accelerate collagen regeneration, cell proliferation, and tendon healing [25]. However, there is no research showing that dry needling alone has an effect on patellar tendinopathy neither if performed with autologous blood injection [2]. James et al. reported that dry needling significantly improved the VISA score and the thickness and tendinosis zone decreased on ultrasound evaluation [30]. However, there is no sufficient evidence about the effect of dry needling for general clinical uses [25].

Injections of a sclerosing agent such as polidocanol were suggested to treat patellar tendinopathy by preventing excessive neovascularization and destroying new vessels and vasa nervorum. In a study by Alfredson and Ohberg, polidocanol injection resulted in pain improvement and reduction of neovascularization under ultrasound evaluation and promoted return to activity as prior to symptom development [31]. However, according to a study by Hoksrud and Bahr, sclerosing agent injection was effective at short-term follow-up, but not at long-term follow-up. More than 1/3 of the patients showed symptoms worsening and eventually received surgical treatment [32]. So far, the therapeutic effect of sclerosing agent injections is unclear, and further research is needed.

16.3 Cell Therapy

Currently conservative therapies are still the mainstay treatment for patellar tendinopathy. However, as scar tissue formation and consequent inferior biomechanical properties cannot be managed with these conservative treatments, alternative approaches that could improve histological and biomechanical properties, in addition to symptoms, through regeneration have been proposed [33].

Cell therapies could be grouped into one of three categories; culture-expanded undifferentiated cells, culture-expanded differentiated cells, and minimally manipulated heterogeneous native cells including native connective tissue progenitors (CTPs) (Table 16.1). In order to investigate

Table 16.1 Cell therapy options

Culture-expanded undifferentiated cells	Culture-expanded differentiated cells	Minimally manipulated heterogeneous native cells
Mesenchymal stromal cells (MSC) Tendon-derived stromal cells (TDSC) Embryonic stem cells (ESC) Induced pluripotent stem cells (iPSC)	Tenocytes Dermal fibroblasts	Bone marrow-derived mononuclear cells (BM MNC) Stromal vascular fraction (SVF) Human placenta-derived cells

the effect of cells on the regeneration of impaired tendons, studies on cell characterization, mode of action (mechanism), and adverse events are needed. To date, several studies have been carried out to establish in vitro, in vivo, and clinical evidence for cell injection therapy. To this end, these studies will be introduced and evidence for the usefulness of cell therapy will be presented.

16.4 In Vitro Isolation and Preparation of Culture-Expanded Cell Populations with Potential Value in Treating Tendinopathy

Mesenchymal stromal cells (MSCs) are defined to have adherence to plastic, specific cell surface antigen and multipotent differentiation potential [34]. MSCs can be culture expanded readily from a variety of tissue sources and are sub-classified by the tissue of cell origin and the harvest and processing methods used [35]. Bone marrow-derived MSCs (BM MSCs) and adipose tissue-derived MSCs (AD MSCs) are the most widely utilized cells in tendon tissue engineering. For clinical use, MSC differentiation toward a target tissue-specific lineage is considered to be important [36]. Tenogenic differentiation of MSCs can be induced by biochemical and mechanical stimulation [37, 38]. Several in vitro studies reported that tenogenesis of BM MSCs can be facilitated by various growth factors such as transforming growth factor-β3 (TGF-β3), bone morphogenic protein-12, 14 (BMP-12, BMP-14), and ascorbic acid [36, 39]. TGF-β3 was reported to induce the expression of tendon-specific markers such as scleraixs in BM MSCs, and the effect of TGF-β3 was upregulated by the presence of BMP-12. In addition, with media containing ascorbic acid, the expression of tendon matrix markers such as collagen was upregulated [36]. BMP-14 was also revealed to increase the expression of scleraxis and tenomodulin [39]. Increased expression of scleraxis and tenomodulin led to the activation of Sirt1-JNK/Smad1-PPARγ signaling pathway and then induce the tenogenic differentiation of BM MSCs [39]. Tenogenic differentiation of AD MSCs can also be promoted by biochemical materials. Like in BM MSCs, TGF-β3, BMP-12, and BMP-14 induce tenogenic differentiation of AD MSCs [36]. These growth factors increased proliferation and expression of tendon-specific markers in AD MSCs [40]. Mechanical stimulation is another factor to promote tenogenic differentiation of MSCs. There were some studies that argued mechanical stretch could promote proliferation and tenogenic differentiation of MSCs [37, 38].

The potential of tendon-derived stromal cells (TDSCs) has been studied by various authors. Guo et al. reported the spontaneous tenogenic differentiation of TDSCs [41]. Spontaneous tenogenic differentiation occurs in vitro with formation of a 3D layer with abundant ECM resembling normal tendon. CD90 and nucleostemin decreased, whereas tenogenic markers of scleraxis, early growth response factor 1 (EGR1), and eyes absent homolog transcriptional coactivator and phosphatase 1 (EYA1) as well as matrix markers of collagen type I, tenomodulin, decorin, and fibromodulin increased. In addition, they showed that transforming growth factor-β (TGF-β) promoted tenogenic differentiation of TDSCs. Xu et al. reported that transfection of BMP-12 and connective tissue growth factor (CTGF) into TDSCs led to tenogenic differentiation of TDSCs, whereas it prevented differentiation of other lineages with downregulation of osteogenic, adipogenic, and chondrogenic markers [42]. Bi et al. reported that biglycan and fibromodulin are fac-

tors required for the differentiation of TDSCs [43]. The use of CD 146⁺ endogenous tendon-derived cells recruited with CTGF is another strategy for tendon regeneration without exogenous cell transplantation [44].

Embryonic stem cells (ESCs) are pluripotent cells found in early developmental stages [45]. Some authors reported tenogenic differentiation of ESCs using several growth factors [46]. But there is ethical concern about the clinical use of ESCs due to the destruction of a human embryo in cell preparing [47]. Induced pluripotent stem cells (iPSCs) are an alternative cell population, with very similar properties to ESCs but bypassing this ethical concern. However, there are still limitations of ESCs and iPSCs in light of the potential of carcinogenesis due to unlimited pluripotency [47, 48].

Dermal fibroblasts are another potential option for cell therapy. Similar to tenocytes, dermal fibroblasts can form tendon-like tissue with abundant collagen fibers, and the maximum tensile strength of tissue was similar to normal tendon [49]. Based on these results, it was argued that dermal fibroblasts could have a useful effect on tendon healing. However, Zhang et al. and Evans and Trail showed that fibroblasts produced both collagen types I and III, whereas tenocytes only produced collagen type I [50, 51]. Moreover, it was found that the arrangement of collagen fibers was more regular with tenocytes than with fibroblasts suggesting that dermal fibroblasts may have a lower tendon healing potential than tenocytes. In addition, no studies have characterized the details of tenogenic markers in dermal fibroblasts, so these cells have insufficient evidence as a treatment option for tendinopathy [52]. There are some studies using dermal fibroblasts, but many problems exist on this cell group, and there are some studies that were retracted. In addition, scar tissue formation may be a problem in tendon healing. Fibroblasts can induce such scar tissue formation and thus have limitations in cell therapy for tendinopathy compared to other cell groups.

Minimally manipulated cells also have been widely used in tendon tissue engineering. These non-cultured cells do not require cell expansion;

physicians can prepare and use them ad hoc. Mononuclear cells from BM and SVF from adipose tissue are the most commonly used minimally manipulated cells, [53, 54] and some authors suggested that they would promote tendon healing via expression and secretion of growth factors [55].

16.5 Preclinical In Vivo Evidence for Cellular Therapy for Tendinopathy

BM MSCs implanted with collagen gel or sponges or fibrin carrier in a patellar tendon defect in a rabbit model have been extensively studied, and most of them reported beneficial effects of BM MSCs for tendon regeneration [56, 57]. Awad et al. reported that autologous BM MSCs in collagen type I gel implanted into a patellar tendon defect in a rabbit model showed significant increases in biomechanical properties but produced no visible improvement in its histological microstructure [56]. In a subsequent study, they demonstrated that MSC-collagen composites implanted in a patellar tendon defect significantly improved the biomechanical properties of tendon repair tissues up to 26 weeks. However, they also found that greater MSC concentrations produced no additional histological or biomechanical improvement [58]. Moreover, unexpected ectopic bone formation was discovered in 28% of the MSC treated group. This complication suggested that preimplantation tenogenic differentiation may be desirable to reduce the risk of differentiation of transplanted cells to osteogenic or chondrogenic lineages. Yin et al. [59] reported that a stepwise tenogenic differentiation approach, by first using TGF-b1 stimulation followed by combination with CTGF, would prompt tenogenic differentiation of BM MSCs while suppressing differentiation into other lineages.

Culture-expanded fibroblastic MSC-like cells from tendon tissue (TDSCs) are an alternative to BM MSCs that may reduce risk of osteogenic and chondrogenic differentiation. TDSCs can also be cultured readily, because the prevalence of colony founding connective tissue progenitors (CTPs) is higher in tendon tissue than in native bone marrow [60–62]. Ni et al. analyzed the effects of culture-expanded TDSC on tendon healing by injecting TDSCs into patellar tendon defects in rats and performed TDSCs tracking [63]. Good results were obtained from gross observation, histologic, and biomechanical studies. TDSCs implantation was reported to aid in earlier and better recovery of injured tendons. Lui et al. investigated the medium- and long-term tendon regeneration effects and inflammatory responses on tendons by implanting allogeneic TDSCs targeting patellar tendon window defects in rats [64]. TDSCs were helpful for tendon healing including histological and biomechanical results without increasing the risk of ectopic bone formation. In addition, TDSCs implantation showed weaker immunoreaction compared to the control group, so it was also inferred that the TDSCs implantation had an anti-inflammatory effect. Lui et al. reported that CTGF and ascorbic acid pre-treated TDSCs with fibrin glue showed better results in histologic, biomechanical, and ultrasound evaluation as well as lower ectopic bone formation in a rat model of patellar tendon defect [62]. Similarly, Xu et al. also demonstrated that the implantation of TDSCs transfected with BMP-12 and CTGF in a rat patellar tendon defect promoted patellar tendon regeneration [42]. Scleraxis is a basic helix-loop-helix transcription factor observed in tendons from the condensation stage to adulthood. Tan et al. compared the effects of scleraxis-transduced TDSCs and naïve TDSCs in a patellar tendon defect of a rat model [65]. Scleraxis-transduced cells resulted in improved tendon repair and did not increase side effects such as ectopic cartilage and bone formation. As such, TDSCs have potentials for tendon regeneration like BM MSCs, but reported to have fewer side effects such as ectopic bone formation.

The mechanism behind successful cell implantation for tendon regeneration is not clearly understood. Becerra et al. carried out an intralesional injection of ten million technetium 99 m (Tc 99 m)-labeled autologous MSCs into a natural tendinopathy lesion of a horse [66]. Cell tracking showed that the number of MSCs in the lesion was decreased gradually over time. In other cell tracking studies, Ni et al. [63], Lui et al. [64], and Tan et al. [65] observed that the number of transplanted TDSCs decreased during the healing process gradually in the patellar tendon defects in rats. These studies indicated that MSCs disappeared gradually within the tendon without migrating to other organs. Thus, it can be inferred that the intralesional injection of cells helped tendon healing through one or more mechanisms that did not involve direct tenogenic differentiation. Many researchers have focused on the healing process of tendon through the paracrine effects. In the future, much more research will need to focus on the paracrine or anti-inflammatory effects.

iPSC-derived cells have also been examined. Xu et al. observed the recovery of patellar tendon after the implantation of human iPSC-derived neural crest stem cells (iPSC-NCSC) in patellar tendon defects of a rat model [48]. The group treated with stem cells showed significantly better recovery in terms of macroscopic observation, histologic examination, and biomechanical analysis. Transplanted iPSC-NCSC also increased the host ECM deposit and thereby upregulated the endogenous repair system.

In a collagenase-induce tendinopathy of patellar tendon of a rat model, Ma et al. showed that the intratendinous injection of culture-expanded human placenta-derived MSC-like cells (PLX-PAD) resulted in better load to failure and stiffness in comparison with those in the control group. Gene expression analysis demonstrated higher levels of interleukin-1β (IL-1β) and IL-6 early in the healing process in the PLX-PAD-treated tendons. The authors concluded that a transient beneficial effect on tendon failure load would be expected because of the induction of an early inflammatory response with PLX-PAD. While cells derived from human extraembryonic tissues, such as the placenta, could emerge as source of cells for musculoskeletal repair, they need further preclinical and clinical research.

Lee at al. [44] confirmed that proliferation and tenogenic differentiation of culture-expanded CD146+ tendon-derived CTPs is regulated in vitro by CTGF and experimented with delivering CTGF to transection defects of rat patellar tendon. As a result, CTGF increased the prevalence of CD146+ cells in the tendon, and CD146+ cells differentiated into tenocyte-like cells over time. The positive results were confirmed also by the histological and biomechanical analysis. This experiment supported the potential of targeting endogenous CD146+ cells to induce differentiation into tenocytes and promote tendon regeneration through CTGF delivery. Targeting native progenitor populations of cells, rather than injecting cells, is a yet untapped option for cell therapy by promoting the endogenous repair.

16.6 Clinical Evidence for Cellular Therapy for Patellar Tendinopathy

Clarke et al. carried out a randomized controlled trial in 46 patients with chronic patellar tendinopathy to investigate the effect of culture-expanded autologous skin-derived tenocyte-like cells (SDTCs) injections [19]. SDTCs were obtained and cultured by harvesting fibroblasts from skin biopsies. SDTCs showed collagen type I and III expression in an ex vivo linear stretching model and showed tenocyte-like behavior. SDTCs injected with autologous plasma showed symptom improvement, more quickly than the autologous plasma injection only in the VISA score at 6 months after the injection. Ultrasonography demonstrated reduced hypoechogenicity and tear size in all patients compared to those before the injection, whereas the tendon thickness only decreased in patients treated with SDTCs injected with the autologous plasma. However, fibroblasts from the skin and tenocytes from tendon are different and have distinct biological characteristics [67, 68]. Skin injury results in scar formation in most cases. Skin fibroblasts have limited ability to repair or regenerate damaged skin and are associated with excessive formation of extracellular matrix by fibrocytes and myofibroblasts suggesting a less probable role in tendon regeneration [69].

In a study of Pascual-Garrido et al., eight patients with refractory patellar tendinopathy were treated with an ultrasound-guided injection of freshly isolated and heterogeneous autologous cells from a bone marrow aspirate (BM MNCs) that were concentrated using a centrifuge. Patients were evaluated for clinical symptoms at 2 and 5 years after injection [70]. Some functional scores and ultrasound evaluation of tendons improved significantly after the injection. All of eight patients were satisfied with the clinical symptoms, and seven out of eight patients said that they were willing to receive the same treatment if symptoms relapse in the future. While this provides evidence of potential safety, this cohort was not compared to a control group, so superiority to conventional therapy could not be established.

Finally, Rodas et al. published a study protocol comparing the effects of culture-expanded autologous BM MSCs and PRP injection in 20 football players with chronic patellar tendinopathy [71]. Evidence comparing BM MSCs and PRP injections would provide useful information on cell therapy for patellar tendinopathy, but currently data are still missing.

16.7 Conclusions

Despite the potential promise of cell therapy strategies for treatment of patellar tendinopathy, and a plethora of potential alternatives for cell sourcing and processing, both preclinical and clinical evidence are no sufficient to make a recommendation for the use of cellular therapy in this setting. Further basic research is needed to identify the potential mechanism of successful cell therapy, tracking injected cells to determine their fate, and quality attributes for injected cells that both improve performance and avoid the risk of ectopic cartilage and bone formation. It is highly desirable to explore scarless regeneration of tendon using methods that could recapitulate the formation of native tendon with cells. Clinical research must invest in high-quality clinical trial design, with adequate sample size, controls, pre- and posttreatment functional assessment, nonin-

vasive tissue assessment, and sufficient follow-up. By solid evidence, we hope that cell therapy for tendinopathy will open a promising new era of musculoskeletal regeneration.

Take-Home Messages
- Most of the current treatment options for patellar tendinopathy are only for symptomatic relief, and even outcomes are still controversial suggesting the need for alternative strategies including cell therapy.
- Current cell therapies are grouped in one of three categories: stem/stromal cells, differentiated cells, and minimally manipulated cells. Among them, only minimally manipulated cells are currently clinically available, while the others are still experimental.
- Stem/stromal cells include a variety of different immature cells such as embryonic stem cells, tissue-specific stem/stromal cells, and induced pluripotent cells. Lots of experimental studies showed their potential as a promising treatment option for patellar tendinopathy. Nonetheless, clinical translation is slow and much to be elucidated remains before widespread clinical use.
- Differentiated cells include dermal fibroblast and tenocytes. Relatively few experimental and clinical studies have investigated their potential for tendinopathy. Considering that fibroblasts form scar tissue and that cultured tenocytes are prone to lose their original phenotype and could generate scar tissue rather than tendon matrix, they have limited usefulness as cell therapy for tendinopathy.
- Some recent studies reported good outcomes with minimally manipulated cells from bone marrow in patients with refractory patellar tendinopathy. Nonetheless, more clinical high-quality evidence is necessary for clinical use.

References

1. Schwartz A, Watson JN, Hutchinson MR. Patellar tendinopathy. Sports Health. 2015;7(5):415–20.
2. Vander Doelen T, Jelley W. Non-surgical treatment of patellar tendinopathy: a systematic review of randomized controlled trials. J Sci Med Sport. 2020;23(2):118–24.
3. Lian OB, Engebretsen L, Bahr R. Prevalence of jumper's knee among elite athletes from different sports: a cross-sectional study. Am J Sports Med. 2005;33(4):561–7.
4. Peers KH, Lysens RJ. Patellar tendinopathy in athletes: current diagnostic and therapeutic recommendations. Sports Med (Auckland, NZ). 2005;35(1):71–87.
5. Kannus P, Natri A. Etiology and pathophysiology of tendon ruptures in sports. Scand J Med Sci Sports. 1997;7(2):107–12.
6. Rees JD, Stride M, Scott A. Tendons—time to revisit inflammation. Br J Sports Med. 2014;48(21):1553–7.
7. Tang C, Chen Y, Huang J, Zhao K, Chen X, Yin Z, et al. The roles of inflammatory mediators and immunocytes in tendinopathy. J Orthopaed Translat. 2018;14:23–33.
8. Battery L, Maffulli N. Inflammation in overuse tendon injuries. Sports Med Arthrosc Rev. 2011;19(3):213–7.
9. D'Addona A, Maffulli N, Formisano S, Rosa D. Inflammation in tendinopathy. Surgeon. 2017;15(5):297–302.
10. Millar NL, Murrell GA, McInnes IB. Inflammatory mechanisms in tendinopathy—towards translation. Nat Rev Rheumatol. 2017;13(2):110–22.
11. Dakin SG, Dudhia J, Smith RK. Resolving an inflammatory concept: the importance of inflammation and resolution in tendinopathy. Vet Immunol Immunopathol. 2014;158(3–4):121–7.
12. Costa-Almeida R, Calejo I, Gomes ME. Mesenchymal stem cells empowering tendon regenerative therapies. Int J Mol Sci. 2019;20(12).
13. Figueroa D, Figueroa F, Calvo R. Patellar tendinopathy: diagnosis and treatment. J Am Acad Orthop Surg. 2016;24(12):e184–e92.
14. Larsson ME, Kall I, Nilsson-Helander K. Treatment of patellar tendinopathy—a systematic review of randomized controlled trials. Knee Surg Sports Traumatol Arthrosc. 2012;20(8):1632–46.
15. van Ark M, Zwerver J, van den Akker-Scheek I. Injection treatments for patellar tendinopathy. Br J Sports Med. 2011;45(13):1068–76.
16. Everhart JS, Cole D, Sojka JH, Higgins JD, Magnussen RA, Schmitt LC, et al. Treatment options for patellar tendinopathy: a systematic review. Arthroscopy. 2017;33(4):861–72.
17. Gaida JE, Cook J. Treatment options for patellar tendinopathy: critical review. Curr Sports Med Rep. 2011;10(5):255–70.
18. Bahr R, Fossan B, Loken S, Engebretsen L. Surgical treatment compared with eccentric training for patellar tendinopathy (Jumper's Knee). A randomized, controlled trial. J Bone Joint Surg Am. 2006;88(8):1689–98.
19. Clarke AW, Alyas F, Morris T, Robertson CJ, Bell J, Connell DA. Skin-derived tenocyte-like cells for the treatment of patellar tendinopathy. Am J Sports Med. 2011;39(3):614–23.
20. Paavola M, Kannus P, Jarvinen TA, Jarvinen TL, Jozsa L, Jarvinen M. Treatment of tendon disorders. Is there a role for corticosteroid injection? Foot Ankle Clin. 2002;7(3):501–13.
21. Kongsgaard M, Kovanen V, Aagaard P, Doessing S, Hansen P, Laursen AH, et al. Corticosteroid injections, eccentric decline squat training and heavy slow resistance training in patellar tendinopathy. Scand J Med Sci Sports. 2009;19(6):790–802.
22. Fredberg U, Bolvig L, Pfeiffer-Jensen M, Clemmensen D, Jakobsen BW, Stengaard-Pedersen K. Ultrasonography as a tool for diagnosis, guidance of local steroid injection and, together with pressure algometry, monitoring of the treatment of athletes with chronic jumper's knee and Achilles tendinitis: a randomized, double-blind, placebo-controlled study. Scand J Rheumatol. 2004;33(2):94–101.
23. Dean BJ, Lostis E, Oakley T, Rombach I, Morrey ME, Carr AJ. The risks and benefits of glucocorticoid treatment for tendinopathy: a systematic review of the effects of local glucocorticoid on tendon. Semin Arthritis Rheum. 2014;43(4):570–6.
24. van der Worp H, van den Akker-Scheek I, van Schie H, Zwerver J. ESWT for tendinopathy: technology and clinical implications. Knee Surg Sports Traumatol Arthrosc. 2013;21(6):1451–8.
25. Nuhmani S. Injection therapies for patellar tendinopathy. Phys Sportsmed. 2019:1–6.
26. Ferrero G, Fabbro E, Orlandi D, Martini C, Lacelli F, Serafini G, et al. Ultrasound-guided injection of platelet-rich plasma in chronic Achilles and patellar tendinopathy. J Ultrasound. 2012;15(4):260–6.
27. Charousset C, Zaoui A, Bellaiche L, Bouyer B. Are multiple platelet-rich plasma injections useful for treatment of chronic patellar tendinopathy in athletes? A prospective study. Am J Sports Med. 2014;42(4):906–11.
28. Zayni R, Thaunat M, Fayard JM, Hager JP, Carrillon Y, Clechet J, et al. Platelet-rich plasma as a treatment for chronic patellar tendinopathy: comparison of a single versus two consecutive injections. Muscles Ligaments Tendons J. 2015;5(2):92–8.
29. Kaux JF, Croisier JL, Forthomme B, Le Goff C, Buhler F, Savanier B, et al. Using platelet-rich plasma to treat jumper's knees: exploring the effect of a second closely-timed infiltration. J Sci Med Sport. 2016;19(3):200–4.
30. James SL, Ali K, Pocock C, Robertson C, Walter J, Bell J, et al. Ultrasound guided dry needling and autologous blood injection for patellar tendinosis. Br J Sports Med. 2007;41(8):518–21; discussion 22.
31. Alfredson H, Ohberg L. Neovascularisation in chronic painful patellar tendinosis—promising results after sclerosing neovessels outside the tendon challenge

the need for surgery. Knee Surg Sports Traumatol Arthrosc. 2005;13(2):74–80.

32. Hoksrud A, Bahr R. Ultrasound-guided sclerosing treatment in patients with patellar tendinopathy (jumper's knee). 44-month follow-up. Am J Sports Med. 2011;39(11):2377–80.

33. Rees JD, Wilson AM, Wolman RL. Current concepts in the management of tendon disorders. Rheumatology (Oxford). 2006;45(5):508–21.

34. Osborne H, Anderson L, Burt P, Young M, Gerrard D. Australasian College of Sports Physicians-position statement: the place of mesenchymal stem/stromal cell therapies in sport and exercise medicine. Br J Sports Med. 2016;50(20):1237–44.

35. Luo Q, Song G, Song Y, Xu B, Qin J, Shi Y. Indirect co-culture with tenocytes promotes proliferation and mRNA expression of tendon/ligament related genes in rat bone marrow mesenchymal stem cells. Cytotechnology. 2009;61(1):1.

36. Perucca Orfei C, Viganò M, Pearson JR, Colombini A, De Luca P, Ragni E, et al. In vitro induction of tendon-specific markers in tendon cells, adipose- and bone marrow-derived stem cells is dependent on TGFβ3, BMP-12 and ascorbic acid stimulation. Int J Mol Sci. 2019;20(1).

37. Lui PPY, Rui YF, Ni M, Chan KM. Tenogenic differentiation of stem cells for tendon repair—what is the current evidence? J Tissue Eng Regen Med. 2011;5(8):e144–e63.

38. Zhang B, Luo Q, Halim A, Ju Y, Morita Y, Song G. Directed differentiation and paracrine mechanisms of mesenchymal stem cells: potential implications for tendon repair and regeneration. Curr Stem Cell Res Ther. 2017;12(6):447–54.

39. Wang D, Jiang X, Lu A, Tu M, Huang W, Huang P. BMP14 induces tenogenic differentiation of bone marrow mesenchymal stem cells in vitro. Exp Ther Med. 2018;16(2):1165–74.

40. Park A, Hogan MV, Kesturu GS, James R, Balian G, Chhabra AB. Adipose-derived mesenchymal stem cells treated with growth differentiation factor-5 express tendon-specific markers. Tissue Eng Part A. 2010;16(9):2941–51.

41. Guo J, Chan KM, Zhang JF, Li G. Tendon-derived stem cells undergo spontaneous tenogenic differentiation. Exp Cell Res. 2016;341(1):1–7.

42. Xu K, Sun Y, Kh Al-Ani M, Wang C, Sha Y, Sung KP, et al. Synergistic promoting effects of bone morphogenetic protein 12/connective tissue growth factor on functional differentiation of tendon derived stem cells and patellar tendon window defect regeneration. J Biomech. 2018;66:95–102.

43. Bi Y, Ehirchiou D, Kilts TM, Inkson CA, Embree MC, Sonoyama W, et al. Identification of tendon stem/progenitor cells and the role of the extracellular matrix in their niche. Nat Med. 2007;13(10):1219–27.

44. Lee CH, Lee FY, Tarafder S, Kao K, Jun Y, Yang G, et al. Harnessing endogenous stem/progenitor cells for tendon regeneration. J Clin Invest. 2015;125(7):2690–701.

45. Guevara-Alvarez A, Schmitt A, Russell RP, Imhoff AB, Buchmann S. Growth factor delivery vehicles for tendon injuries: mesenchymal stem cells and platelet rich plasma. Muscles Ligaments Tendons J. 2014;4(3):378–85.

46. Dale TP, Mazher S, Webb WR, Zhou J, Maffulli N, Chen GQ, et al. Tenogenic differentiation of human embryonic stem cells. Tissue Eng Part A. 2018;24(5–6):361–8.

47. Volarevic V, Markovic BS, Gazdic M, Volarevic A, Jovicic N, Arsenijevic N, et al. Ethical and safety issues of stem cell-based therapy. Int J Med Sci. 2018;15(1):36–45.

48. Xu W, Wang Y, Liu E, Sun Y, Luo Z, Xu Z, et al. Human iPSC-derived neural crest stem cells promote tendon repair in a rat patellar tendon window defect model. Tissue Eng Part A. 2013;19(21–22):2439–51.

49. Deng D, Liu W, Xu F, Wu XL, Wei X, Zhong B, et al. In vitro tendon engineering using human dermal fibroblasts. Zhonghua Yi Xue Za Zhi. 2008;88(13):914–8.

50. Zhang Q, Yang Z, Peng W. [Experimental study on biological characteristics of tenocyte and fibroblast in rabbit]. Zhongguo Xiu Fu Chong Jian Wai Ke Za Zhi. 1997;11(1):46–8.

51. Evans CE, Trail IA. Fibroblast-like cells from tendons differ from skin fibroblasts in their ability to form three-dimensional structures in vitro. J Hand Surg (Edinburgh, Scotland). 1998;23(5):633–41.

52. Gaspar D, Spanoudes K, Holladay C, Pandit A, Zeugolis D. Progress in cell-based therapies for tendon repair. Adv Drug Deliv Rev. 2015;84:240–56.

53. Sampson S, Botto-van Bemden A, Aufiero D. Autologous bone marrow concentrate: review and application of a novel intra-articular orthobiologic for cartilage disease. Phys Sportsmed. 2013;41(3):7–18.

54. Steinert AF, Rackwitz L, Gilbert F, Noth U, Tuan RS. Concise review: the clinical application of mesenchymal stem cells for musculoskeletal regeneration: current status and perspectives. Stem Cells Transl Med. 2012;1(3):237–47.

55. Polly SS, Nichols AEC, Donnini E, Inman DJ, Scott TJ, Apple SM, et al. Adipose-derived stromal vascular fraction and cultured stromal cells as trophic mediators for tendon healing. J Orthop Res. 2019;37(6):1429–39.

56. Awad HA, Butler DL, Boivin CP, Smith FN, Malaviya P, Huibregtse B, et al. Autologous mesenchymal stem cell-mediated repair of tendon. Tissue Eng. 1999;5(3):267–77.

57. Hankemeier S, Hurschler C, Zeichen J, van Griensven M, Miller B, Meller R, et al. Bone marrow stromal cells in a liquid fibrin matrix improve the healing process of patellar tendon window defects. Tissue Eng Part A. 2009;15(5):1019–30.

58. Awad HA, Boivin GP, Dressler MR, Smith FN, Young RG, Butler DL. Repair of patellar tendon injuries using a cell-collagen composite. J Orthop Res. 2003;21(3):420–31.

59. Yin Z, Guo J, Wu TY, Chen X, Xu LL, Lin SE, et al. Stepwise differentiation of mesenchymal stem cells augments tendon-like tissue formation and defect repair in vivo. Stem Cells Transl Med. 2016;5(8):1106–16.

60. Tan Q, Lui PP, Rui YF, Wong YM. Comparison of potentials of stem cells isolated from tendon and bone marrow for musculoskeletal tissue engineering. Tissue Eng Part A. 2012;18(7–8):840–51.

61. Stanco D, Vigano M, Perucca Orfei C, Di Giancamillo A, Peretti GM, Lanfranchi L, et al. Multidifferentiation potential of human mesenchymal stem cells from adipose tissue and hamstring tendons for musculoskeletal cell-based therapy. Regen Med. 2015;10(6):729–43.

62. Lui PP, Wong OT, Lee YW. Transplantation of tendon-derived stem cells pre-treated with connective tissue growth factor and ascorbic acid in vitro promoted better tendon repair in a patellar tendon window injury rat model. Cytotherapy. 2016;18(1):99–112.

63. Ni M, Lui PP, Rui YF, Lee YW, Lee YW, Tan Q, et al. Tendon-derived stem cells (TDSCs) promote tendon repair in a rat patellar tendon window defect model. J Orthop Res. 2012;30(4):613–9.

64. Lui PP, Kong SK, Lau PM, Wong YM, Lee YW, Tan C, et al. Allogeneic tendon-derived stem cells promote tendon healing and suppress immunoreactions in hosts: in vivo model. Tissue Eng Part A. 2014;20(21–22):2998–3009.

65. Tan C, Lui PP, Lee YW, Wong YM. Scx-transduced tendon-derived stem cells (TDSCS) promoted better tendon repair compared to mock-transduced cells in a rat patellar tendon window injury model. PLoS One. 2014;9(5):e97453.

66. Becerra P, Valdes Vazquez MA, Dudhia J, Fiske-Jackson AR, Neves F, Hartman NG, et al. Distribution of injected technetium(99m)-labeled mesenchymal stem cells in horses with naturally occurring tendinopathy. J Orthop Res. 2013;31(7):1096–102.

67. Mackley Jennifer R, Ando J, Herzyk P, Winder SJ. Phenotypic responses to mechanical stress in fibroblasts from tendon, cornea and skin. Biochem J. 2006;396(2):307–16.

68. Evans CE, Trail IA. Fibroblast-like cells from tendons differ from skin fibroblasts in their ability to form three-dimensional structures in vitro. J Hand Surg Br. 1998;23(5):633–41.

69. Jackson WM, Nesti LJ, Tuan RS. Mesenchymal stem cell therapy for attenuation of scar formation during wound healing. Stem Cell Res Ther. 2012;3(3):20.

70. Pascual-Garrido C, Rolon A, Makino A. Treatment of chronic patellar tendinopathy with autologous bone marrow stem cells: a 5-year-followup. Stem Cells Int. 2012;2012:953510.

71. Rodas G, Soler R, Balius R, Alomar X, Peirau X, Alberca M, et al. Autologous bone marrow expanded mesenchymal stem cells in patellar tendinopathy: protocol for a phase I/II, single-centre, randomized with active control PRP, double-blinded clinical trial. J Orthop Surg Res. 2019;14(1):441.

Patellar Tendinopathy: Biologics

17

Rahman Kandil and Jason Dragoo

17.1 Introduction

Patellar tendinopathy is a common cause of anterior knee pain. It is often an overuse injury caused by excessive activities that require jumping, running, or rapid changes in direction. The reported prevalence ranges from 14 to 32% in basketball players and 45% in volleyball players [1]. This disease process is more common in males and has been reported in all activity levels from recreational to professional athletes [2].

Patellar tendinopathy is a disease process characterized by the degeneration of the proximal patellar tendon. It is not an inflammatory process, and therefore patellar tendinitis is no longer commonly used to describe this disease. Histologically, patellar tendinopathy is characterized by progressive degeneration of the tendinous tissue, an inability of the tissue to repair itself, and the absence of inflammatory cells. To the naked eye, the affected region of tendinopathy appears yellow and commonly is described as "mucoid degeneration" [3].

17.2 Evaluation and Diagnostic Workup

The common classification system for patellar tendinopathy is based on its clinical features [4]. This classification system includes four progressive stages graded according to disease severity (see Table 17.1).

Blazina stage 1 patellar tendinopathy is categorized by pain only after participation in sports. Patients in stage 1 do not have pain at baseline. Blazina stage 2 is categorized by pain at the beginning of the sport activity, slowly disappearing after warm-up, but then reappearing after fatigue. Blazina stage 3 is categorized by constant pain at rest and with activity. Finally, Blazina stage 4 is a complete rupture of the patellar tendon.

Patients with patellar tendinopathy often complain of anterior knee pain and tenderness to palpation of the proximal patellar tendon and inferior pole of the patella [4]. The decline squat test, in which a single leg squat is performed in 30° of knee flexion, reproduces the pain associated with patellar tendinopathy [5].

Table 17.1 Blazina classification of patellar tendinopathy

Stage	Description
1	Pain only after sports
2	Pain at the beginning of sports, disappearing after warm-up, but reappearing after fatigue
3	Constant pain at rest and with activity
4	Complete rupture of the patellar tendon

R. Kandil
The Orthopedic Group, Leesburg, VA, USA

J. Dragoo (✉)
Department of Orthopaedic Surgery, University of Colorado, Denver, CO, USA
e-mail: J.DRAGOO@cuanschutz.edu

© ISAKOS 2022
G. Filardo et al. (eds.), *Orthobiologics*, https://doi.org/10.1007/978-3-030-84744-9_17

Typical findings on magnetic resonance imaging (MRI) include focal T2 hyperintensity signal changes in the proximal patellar tendon, larger tendon cross-sectional area, proximal thickening, partial tearing, and abnormalities of the posterior border of the patellar tendon and infrapatellar fat pad [5]. While there have been multiple proposed MRI classification systems for patellar tendinopathy, no single grading system is commonly and consistently used in the literature.

17.3 Treatment Options

17.3.1 Nonoperative Management

The mainstay of treatment for patellar tendinopathy is conservative management. Many patients report improvement in symptoms and return to play after a period of nonoperative management. A variety of treatment options have been reported in the literature with physical therapy being the foundation of conservative management. Other conservative treatment options include dry needling and extracorporeal shock wave therapy (ESWT) [6].

An eccentric exercise physical therapy program is considered the standard treatment for patellar tendinopathy. Eccentric exercises have been proposed to help in patellar tendinopathy by increasing the remodeling process of the collagen fibers in the diseased portion of the patellar tendon.

One landmark study compared eccentric to concentric exercises in patients with patellar tendinopathy and found a significant improvement in visual analog scores (VAS) and Victorian Institute of Sports Assessment (VISA) scores within the eccentric group but no improvement within the concentric group [7]. A recent systematic review found that eccentric exercises are the only high-level evidence treatment option for patients with patellar tendinopathy [8].

Despite being an important component of most treatment programs for patellar tendinopathy, many patients fail to improve with eccentric exercises only and require additional treatment [9].

17.4 Surgical Treatment

While most cases of patellar tendinosis can be managed non-surgically, some recalcitrant cases may benefit from surgical management. Surgery is indicated in patients with persistent symptoms and functional impairment that have failed at least 6 months of conservative management.

If patients continue to have pain interfering with physical activity after exhausting a course of physical therapy, injections, and other nonoperative interventions, then surgery can be considered. About 10% of patients treated for patellar tendinosis undergo surgery [10].

Goals of surgery include removal of fibrous tissue, stimulation of biologic healing through debridement, and re-establishing vascularity to the diseased tendon. There is strong evidence to support surgical treatment for cases of chronic patellar tendinosis in patients with severe baseline symptoms and those who failed conservative management [11]. The overall success rate of surgical management is greater than 80% [12].

Multiple surgical options exist for the treatment of patellar tendinosis including the following:

- Open longitudinal tenotomy with excision of pathology +/− resection of the inferior pole of the patella
- Open, multiple longitudinal tenotomies
- Arthroscopic debridement and/or patellar tenotomy
- Arthroscopic vs open drilling or resection of the inferior pole of the patella

17.5 Role of Injectable Biologics

17.5.1 Overview

Different types of injections have been studied for patellar tendinopathy, including corticosteroids or biologic injections such as platelet-rich plasma (PRP), hyaluronic acid (HA), and autologous bone marrow stromal cells.

Corticosteroid injections are widely used in orthopedic surgery for a variety of conditions including arthritis, bursitis, and tendinopathy. Since patellar tendinopathy is not an inflammatory condition, the effect of anti-inflammatories on the pathogenesis of patellar tendinopathy is dubious. The evidence is limited on the efficacy of corticosteroid injections in patient with patellar tendinopathy, and one significant drawback is the risk of patellar tendon rupture. A study by Chen et al. found that a series of seven weightlifters with patellar tendon rupture were all found to have had a corticosteroid injection in the patellar tendon prior to the rupture [13]. In addition, some studies raise concerns regarding the effect of corticosteroids on tendon strength [14]. Finally, a newer study found that dexamethasone induced stem cells to differentiate into non-tenocytes including chondrocytes and adipocytes, suggesting that injections of dexamethasone into a tendon may lead to the formation of non-tendon tissue within the tendon, ultimately weakening the tendon [15]. HA has been studied in the treatment of multiple orthopedic conditions, but the primary indication is osteoarthritis. One study by Fogli et al. looked at the effect of HA injections in patients with patellar tendinopathy and other tendinopathies [16]. They found that patients receiving ultrasound-guided HA peritendinous injections had significant pain relief and reduction in tendon thickness and neovascularization. Autologous bone marrow stromal cells are another biologic treatment option for patellar tendinopathy. Pascual-Garrido et al. performed a case series looking at the clinical outcomes of autologous bone marrow stromal cell injections in patients with chronic patellar tendinopathy refractory to conservative treatment [17]. They found that statistically significant improvement in most clinical scores at 5-year follow-up.

Beside these attempts to investigate the role of injectable therapies, the majority of the current studies specifically look at biologic injections with a focus on the effect of PRP on patellar tendinopathy. The level of evidence ranges from case series to randomized control trials. PRP is prepared by centrifuging anticoagulated autologous whole blood obtained by phlebotomy (see Fig. 17.1). PRP contains a hyper-physiological concentration of autologous platelets, 3–8 times

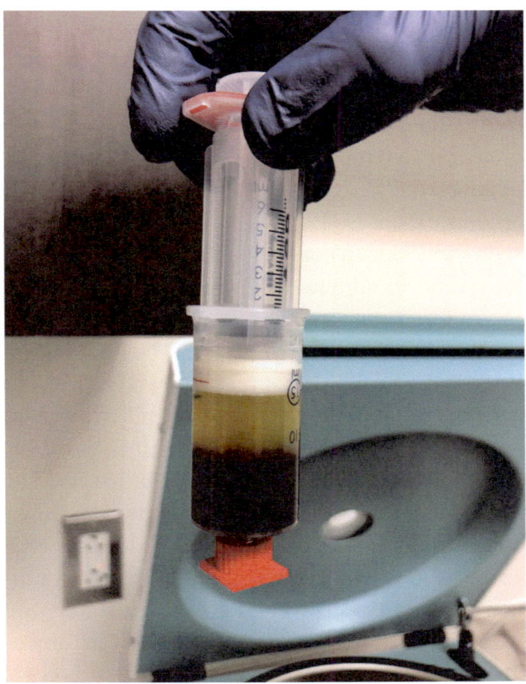

Fig. 17.1 PRP preparation

the concentration of platelets in whole blood [18]. Platelets are one of the first responders arriving at the site of tissue injury and help release growth factors that play a critical role in mediating healing [19]. PRP is therefore thought to be capable of enhancing tissue repair because of its high concentration of growth factors.

There is no standard protocol for obtaining PRP and variability in the number of injections needed for optimal results. In addition, the exact definition of PRP has not been determined in terms of the concentration of platelets, and most published reports differ on PRP concentrations [20]. One systematic review by Jeong et al. reported that PRP seems to have a positive effect in treating patellar tendinopathy, but the available evidence at that time was deemed to be low quality, and, thus, their finding was not considered definitive [21].

Another systematic review showed that studies comparing PRP with other treatments had inconsistent results and that none of the studies showed marked differences between PRP and other treatments [22]. A comprehensive and recent systematic review meta-analysis included a total of 70 studies involving 2530 patients. Eccentric exercise therapies obtained the best results at short-term. At long-term follow-up greater than 6 months, multiple injections of PRP obtained the best results, followed by ESWT and eccentric exercise [23].

A randomized controlled trial by Vetrano et al. looked at 46 athletes with patellar tendinopathy and randomized into two homogeneous treatment groups with the treatment arms being PRP vs ESWT. Both treatment groups showed significant improvement of symptoms at all follow-up assessments. There were no significant differences between groups at 2-month follow-up. The PRP group showed significantly better improvement than the ESWT group in VISA-P, VAS scores at 6- and 12-month follow-up, and modified Blazina scale score at 12-month follow-up. In conclusion, this study showed that PRP injections lead to better mid-term clinical results compared with focused ESWT in the treatment of patellar tendinopathy in athletes [24].

17.5.2 Number of Injections

There is variability in the number of injections used for patellar tendinopathy. Some providers administer a single PRP injection for patellar tendinopathy, while others may provide multiple injections. One of the first prospective case series looking at PRP and patellar tendinopathy evaluated the efficacy of multiple PRP injections on the healing of chronic refractory patellar tendinopathy. The study documented good and stable results up to 4 years' follow-up with high patient satisfaction rates and return to sports. They concluded that three ultrasound-guided intra-tendinous injections of 5 milliliters of PRP, 2 weeks apart from each other, provided a good clinical outcome for the treatment of chronic recalcitrant patellar tendinopathy with stable results up to medium-term follow-up [25].

A high-quality prospective series found that the application of three consecutive US-guided PRP injections, 1 week apart from each other,

significantly improved symptoms and function in athletes with chronic patellar tendinopathy and allowed fast recovery to their pre-symptom sporting level. In addition, this study looked at the effect of PRP on imaging findings and found that PRP treatment permitted a return to a normal architecture of the tendon as assessed by MRI [26].

A randomized controlled trial compared clinical outcomes in patellar tendinopathy after a single ultrasound-guided injection of leukocyte-rich PRP vs dry needling. The authors concluded that a regimen of leukocyte-rich PRP injection and a standardized eccentric exercise program accelerates the recovery from patellar tendinopathy relative to exercise and ultrasound-guided dry needling alone [9].

Regarding location of injection, there is variability in available studies. Some studies describe peritendinous injections, while most authors describe injections into the tendon and area of tendinopathy using ultrasound guidance for confirmation. Our preferred patellar tendinopathy injection technique is to localize the area of tendinopathy by ultrasound and patient feedback and then injecting local anesthetic subcutaneously using sterile technique. Care is taken not to anesthetize the tendon or tendon sheath. The patellar tendon is then injected with the needle oriented in cranio-caudal direction at an angle of 45°.

17.6 PRP Formulations

With regard to various PRP formulations, one study found that leukocyte-rich PRP (LR-PRP) induces a greater short-term inflammatory and fibrotic response than leukocyte-poor PRP (LP-PRP) [27]. Dragoo et al. suggested that the inclusion of the white blood cell fraction in PRP preparations may increase growth factor yield but may also lead to increased inflammation and possibly a delayed healing response [22].

A recent high-quality randomized controlled trial looked at the effect of a single ultrasound-guided injection of LR-PRP, LP-PRP, or saline in patients with advanced patellar tendinopathy. There was no significant difference in mean change in VISA-P score, pain, or global rating of change among the three treatment groups at 12 weeks or any other time point. The authors found that when combined with an exercise-based rehabilitation program, a single injection of LR-PRP or LP-PRP was no more effective than saline for the improvement of patellar tendinopathy symptoms [28].

17.7 Conclusion

Injectable PRP has promising results for the treatment of patellar tendinopathy. Numerous high-quality studies show improvement in symptoms, and many show a significant improvement compared to other conservative options including physical therapy, ESWT, dry needling, and more. There is room for improvement with regard to uniformity of administration protocols, and more high-quality controlled studies are needed to determine its true efficacy.

Take-Home Messages
- Patellar tendinopathy is a common cause of anterior knee pain characterized by the degeneration of the proximal patellar tendon.
- Injectable biologics are an emerging treatment option in patients with patellar tendinopathy.
- Numerous high-quality studies show PRP leading to an improvement in symptoms and many show a significant improvement compared to other conservative options including physical therapy, ESWT, and dry needling.
- There is room for improvement with regard to uniformity of PRP administration protocols, and more high-quality controlled studies are needed to determine its true efficacy.

References

1. Lian OB, Engebretsen L, Bahr R. Prevalence of jumper's knee among elite athletes from different sports: a cross-sectional study. Am J Sports Med. 2005;33(4):561–7.
2. Zwerver J, Bredeweg SW, van den Akker-Scheek I. Prevalence of jumper's knee among nonelite athletes from different sports: a cross-sectional survey. Am J Sports Med. 2011;39(9):1984–8.
3. Khan KM, Maffulli N, Coleman BD, Cook JL, Taunton JE. Patellar tendinopathy: some aspects of basic science and clinical management. Br J Sports Med. 1998;32(4):346–55.
4. Blazina ME, Kerlan RK, Jobe FW, Carter VS, Carlson GJ. Jumper's knee. Orthop Clin North Am. 1973;4(3):665–78.
5. Warden SJ, Kiss ZS, Malara FA, Ooi AB, Cook JL, Crossley KM. Comparative accuracy of magnetic resonance imaging and ultrasonography in confirming clinically diagnosed patellar tendinopathy. Am J Sports Med. 2007;35(3):427–36.
6. van Leeuwen MT, Zwerver J, van den Akker-Scheek I. Extracorporeal shockwave therapy for patellar tendinopathy: a review of the literature. Br J Sports Med. 2009;43(3):163–8.
7. Jonsson P, Alfredson H. Superior results with eccentric compared to concentric quadriceps training in patients with jumper's knee: a prospective randomised study. Br J Sports Med. 2005;39(11):847–50.
8. Larsson ME, Kall I, Nilsson-Helander K. Treatment of patellar tendinopathy: a systematic review of randomized controlled trials. Knee Surg Sports Traumatol Arthrosc. 2012;20(8):1632–46.
9. Dragoo JL, Wasterlain AS, Braun HJ, et al. Platelet-rich plasma as a treatment for patellar tendinopathy: a double-blind, randomized controlled trial. Am J Sports Med. 2014;42(3):610–8.
10. Ogon P, Maier D, Jaeger A, Suedkamp NP. Arthroscopic patellar release for the treatment of chronic patellar tendinopathy. Arthroscopy. 2006;22(4):462.e1–5.
11. Everhart JS, Cole D, Sojka JH, et al. Treatment options for patellar tendinopathy: a systematic review. Arthroscopy. 2017;33(4):861–72.
12. Coleman BD, Khan KM, Maffulli N, Cook JL, Wark JD. Studies of surgical outcome after patellar tendinopathy: clinical significance of methodological deficiencies and guidelines for future studies. Victorian Institute of Sport Tendon Study Group. Scand J Med Sci Sports. 2000;10(1):2–11.
13. Chen SK, Lu CC, Chou PH, Guo LY, Wu WL. Patellar tendon ruptures in weight lifters after local steroid injections. Arch Orthop Trauma Surg. 2009;129(3):369–72.
14. Paavola M, Kannus P, Järvinen TA, Järvinen TL, Józsa L, Järvinen M. Treatment of tendon disorders. Is there a role for corticosteroid injection? Foot Ankle Clin. 2002;7(3):501–13.

15. Zhang J, Keenan C, Wang JH. The effects of dexamethasone on human patellar tendon stem cells: implications for dexamethasone treatment of tendon injury. J Orthop Res. 2013;31(1):105–10.
16. Fogli M, Giordan N, Mazzoni G. Efficacy and safety of hyaluronic acid (500-730kDa) ultrasound-guided injections on painful tendinopathies: a prospective, open label, clinical study. Muscles Ligaments Tendons J. 2017;7(2):388–95.
17. Pascual-Garrido C, Rolón A, Makino A. Treatment of chronic patellar tendinopathy with autologous bone marrow stem cells: a 5-year-followup. Stem Cells Int. 2012;2012:953510.
18. Middleton KK, Barro V, Muller B, Terada S, Fu FH. Evaluation of the effects of platelet-rich plasma (PRP) therapy involved in the healing of sports-related soft tissue injuries. Iowa Orthop J. 2012;32:150–63.
19. Creaney L, Hamilton B. Growth factor delivery methods in the management of sports injuries: the state of play. Br J Sports Med. 2008;42(5):314–20.
20. Nguyen RT, Borg-Stein J, McInnis K. Applications of platelet-rich plasma in musculoskeletal and sports medicine: an evidence-based approach. PM & R. 2011;3(3):226–50.
21. Jeong DU, Lee CR, Lee JH, et al. Clinical applications of platelet-rich plasma in patellar tendinopathy. Biomed Res Int. 2014;2014:249498.
22. Liddle AD, Rodriguez-Merchin EC. Platelet-rich plasma in the treatment of patellar tendinopathy: a systematic review. Am J Sports Med. 2015;43(10):2583–2590.25524323.
23. Andriolo L, Altamura SA, Reale D, Candrian C, Zaffagnini S, Filardo G. Nonsurgical treatments of patellar tendinopathy: multiple injections of platelet-rich plasma are a suitable option: a systematic review and meta-analysis. Am J Sports Med. 2019;47(4):1001–18.
24. Vetrano M, Castorina A, Vulpiani MC, Baldini R, Pavan A, Ferretti A. Platelet-rich plasma versus focused shock waves in the treatment of jumper's knee in athletes. Am J Sports Med. 2013;41(4):795–803.
25. Filardo G, Kon E, Di Matteo B. Platelet-rich plasma for the treatment of patellar tendinopathy: clinical and imaging findings at medium-term follow-up. International Orthopaedics (SICOT). 2013;37:1583–9.
26. Charousset C, Zaoui A, Bellaiche L, Bouyer B. Are multiple platelet-rich plasma injections useful for treatment of chronic patellar tendinopathy in athletes? A prospective study. The Am J Sports Med. 2014;42(4):906–11.
27. Dragoo JL, Braun HJ, Durham JL, et al. Comparison of the acute inflammatory response of two commercial platelet-rich plasma systems in healthy rabbit tendons. Am J Sports Med. 2012;40(6):1274–81.
28. Scott A, LaPrade RF, Harmon KG, Filardo G, Kon E, Della Villa S, Bahr R, Moksnes H, Torgalsen T, Lee J, Dragoo JL, Engebretsen L. Platelet-rich plasma for patellar tendinopathy: a randomized controlled trial of leukocyte-rich PRP or leukocyte-poor PRP versus saline. Am J Sports Med. 2019;47(7):1654–61.

Orthobiologics for the Treatment of Achilles Tendinopathy

Joseph D. Lamplot, Cort D. Lawton, and Scott A. Rodeo

18.1 Introduction

Achilles tendinopathy (AT) is a chronic degenerative process associated with diminished vascularity, repetitive microtrauma, and aging [1, 2]. Although surgical specimens rarely contain typical inflammatory cell populations, the current understanding is that "molecular inflammation" plays an important role in the underlying pathologic process. Inflammatory mediators produced by both intrinsic tendon stromal cells and infiltrating immune cells play a fundamental role in the initiation and regulation of tendinopathy [3, 4]. Achilles tendinopathy can occur at its insertion on the calcaneus or midsubstance, typically 2–6 cm from its bony insertion [5]. Insertional Achilles tendinopathy (IAT) accounts for approximately 20–25% of Achilles tendon disorders, whereas midsubstance Achilles tendinopathy (MAT) accounts for 66% [6, 7]. AT typically presents with pain occurring at the beginning of activity, with a decrease in discomfort as activity continues [8, 9]. In advanced cases, pain may occur in all phases of activity and interfere with activities of daily living [8, 9]. Although AT typically results from overuse, it may present insidiously in middle-aged overweight patients with no history of increased physical activity [10]. Risk factors associated with AT include hypertension, diabetes, and obesity [2, 11]. However, in many cases, the etiology is multifactorial [12, 13]. Changes in activity level, previous tendon injuries, poorly fitting footwear, and environmental factors such as training on hard or uneven surfaces may predispose patients to AT [7]. Multiple studies have reported that AT accounts for up to 18% of all injuries in runners [14–16]. Presenting signs and symptoms include pain and swelling over a region of nodular tendon thickening [5]. While the diagnosis of AT is largely clinical, radiographs may show calcifications within the tendon or insertional osteophytes, and ultrasound (US) or magnetic resonance imaging (MRI) may demonstrate tendon thickening (Fig. 18.1) [5, 7].

Initial management of Achilles tendinopathy is non-operative and may include eccentric exercises, cryotherapy, extracorporeal shockwave therapy, orthotics, splints, injections, and nonsteroidal anti-inflammatory drugs (NSAIDs) [17–20]. There is a consensus that progressive tendon loading in the form of eccentric strengthening is an important component of a non-operative management strategy [21–23]. Non-operative management has been reported to be effective in approximately 75% of patients, with failure of non-operative management correlated with

J. D. Lamplot (✉)
Department of Orthopaedics, Emory University School of Medicine, Flowery Branch, GA, USA
e-mail: joseph.daniel.lamplot@emory.edu; lamplotj@hss.edu

C. D. Lawton
Sports Medicine Institute, Hospital for Special Surgery, New York, NY, USA

S. A. Rodeo
OrthoIllinois, Algonquin, IL, USA

© ISAKOS 2022
G. Filardo et al. (eds.), *Orthobiologics*, https://doi.org/10.1007/978-3-030-84744-9_18

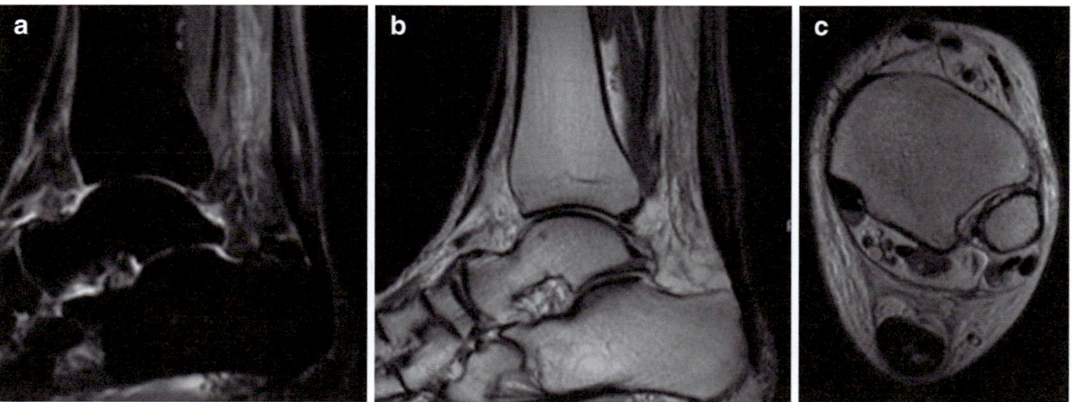

Fig. 18.1 Magnetic resonance imaging (MRI) of Achilles tendinopathy. (**a**) Sagittal inversion recovery (IR), (**b**) sagittal proton density (PD), and (**c**) axial PD images demonstrate severe Achilles tendinosis with profound intrasubstance degeneration and interstitial tearing without partial thickness defect. Tendinosis extends distally to the Achilles insertion

severity of tendinosis, advanced patient age, and duration of symptoms [5, 21]. Several different types of injection therapies have been described including corticosteroids, high-volume saline, prolotherapy, autologous blood products (ABP), platelet-rich plasma (PRP), aprotinin, botulinum toxin, sodium hyaluronate (HA), polysulfated glycosaminoglycan, sclerosing agents, and cell-based therapies [20, 22, 24–31]. This chapter specifically focuses on injectable biologic therapies for Achilles tendon pathology. While the main focus will be on AT, results of injectable biologics in the setting of Achilles tendon repair will briefly be discussed.

18.2 Pathophysiology

The Achilles tendon originates from the aponeuroses of the gastrocnemius, soleus, and plantaris muscles and is primarily composed of type I collagen [32]. It inserts 2 cm distal to the posterosuperior calcaneal prominence and has an anterior-posterior diameter of 5–6 mm [33, 34]. The insertion may be predisposed to tendinopathy due to high shear and compressive forces in this location [24]. Meanwhile, MAT

has been attributed to decreased vascularity as the tendon spirals through a 90° lateral turn toward its insertion [35]. When the tendon experiences repetitive submaximal forces, microscopic damage occurs to the extracellular matrix (ECM), and collagen cross-links begin to fail across the length of the tendon. The tendon then undergoes attempted remodeling and repair [24]. Tendinopathy results when there is an imbalance between repetitive injury and repair. Histologically, this process is characterized by a disorganized proliferation of tenocytes, degeneration of mature tendon cells, and disrupted collagen fibers with an increase in ECM, with an abnormal increase in matrix glycosaminoglycans [8–10, 36–38]. As the parallel orientation of collagen fibers is disrupted, there is a decrease in collagen fiber diameter and density of collagen with an increase in type III collagen [8–10, 36–39]. In addition to a loss of parallel collagen structure and loss of fiber integrity, tendon degeneration is characterized by capillary proliferation and, in later phases, fatty infiltration [38, 40]. These changes in tendon microstructure and composition lead to symptoms and may predispose to Achilles tendon rupture [35, 41].

18.3 Role of Biologic Therapies

Biologic therapies for AT aim to restore the properties of the native tendon, thereby reducing pain, improving function, and reducing the risk of rupture [42]. The reparative process that occurs following tendon injury leads to the formation of a fibrovascular scar comprised mainly of type III collagen rather than the native type I collagen. This results in a tendon that is mechanically weaker than the uninjured tendon [43]. Biologic therapies may help facilitate a more robust healing response compared to other commonly used nonsurgical treatment modalities, thereby resulting in a tendon with biomechanical properties that are more similar to the native uninjured tendon. These therapies may also have a role as an adjuvant treatment in the surgical setting [7].

18.4 Corticosteroids

Corticosteroid injections have largely fallen out of favor for treatment of AT, mainly out of concern for an increased risk of iatrogenic tendon rupture [20, 44]. In a randomized controlled trial (RCT) in which AT patients received either three corticosteroid or placebo peritendinous injection, Fredberg et al. [28] reported a significant reduction in Achilles tendon diameter in the corticosteroid group compared with placebo. In another RCT of patients with AT, DaCruz et al. [26] randomized patients to either a single peritendinous injection of methylprednisolone with 0.25% Marcaine or Marcaine alone, reporting no difference in return to normal activity between the two groups.

With regard to complications related to corticosteroid injections, reversible atrophy of the Achilles tendon has been described in approximately 50% of patients, representing a higher rate when compared to placebo [45]. While Fredberg et al. [28] reported that only 8% of patients treated with corticosteroids required future surgery, DaCruz [26] et al. reported that 39% eventually required operative intervention. One study reported a single episode of Achilles tendon rupture [28]. Altogether, current evidence does not support the use of corticosteroid injection for AT.

18.5 Sodium Hyaluronate

Although the effects of sodium hyaluronate (hyaluronic acid [HA]) on Achilles tendon healing have been explored in the setting of both AT and Achilles tendon repair, most studies have been preclinical investigations in animal models. In one recent clinical trial, Lyren et al. [46] performed an RCT comparing extracorporeal shock wave therapy (ESWT) and peritendinous HA injections in treatment of MAT, reporting a clinically relevant improvement in terms of VAS pain scores and VISA-A scores at 4 weeks and 6 months. The authors concluded that two peritendinous HA injections led to greater treatment success than three ESWT applications at weekly intervals in the treatment of MAT. Halici et al. [47] evaluated the effects of HA on vascular endothelial growth factor (VEGF) and type IV collagen expression during the healing process in a rabbit Achilles rupture model, finding a decrease in tendon adhesions and accelerated tissue repair in the HA group at 6 and 12 weeks, with strongly positive VEGF immunostaining at the 6th week persisting into the 12th week. In a rat Achilles tendon rupture model, Tosun et al. [48] reported that rats treated with either HA or combination therapy with HA and chondroitin sulfate demonstrated a significantly lower number of adhesions in the combination group and a higher maximum stress compared to HA alone or control. In a rat Achilles tendinopathy model in which degeneration was induced with local corticosteroid injections, Tatari [49] reported improved histopathologic scores in the HA group compared to the control group at 75 days post-treatment. However, Wu et al. [50] performed intratendinous HA or phosphate buffered saline (PBS) injection in a rat model, demonstrating more severe tendon injury and inflammatory changes as indicated by histopathology, IL-1β expression, and macrophage density in the HA group. Altogether, the role of HA in the treatment of Achilles tendon injury remains poorly defined. While Lynen [47] et al. reported promising results, additional randomized studies with larger numbers will be necessary to better elucidate its role in Achilles tendon injury. Additionally, further evidence will be necessary to elucidate the role of peritendinous and intratendinous injections of HA in this setting.

18.6 Platelet-Rich Plasma

PRP is defined as a sample of plasma with an increase in platelet concentration above baseline levels or platelet levels greater than $1.1 \times 10^6/\mu L$ [51]. When the platelets degranulate, growth factors are released including platelet-derived growth factor (PDGF), transforming growth factor-β (TGF-β), VEGF, platelet-derived angiogenesis factor (PDAF), epidermal growth factor (EGF), platelet-derived endothelial growth factor (PDEGF), epithelial cell growth factor (ECGF), and insulin-like growth factor (IGF) [52, 53]. As these growth factors have been shown to enhance tissue healing, their use may be of particular interest in areas with low intrinsic healing potential such as the Achilles tendon [54]. Like other biologic therapies, the goal of PRP is to restore the properties of the injured tendon. Due to its autologous nature, PRP is inherently safe [55]. However, current evidence summarized below does not appear to support its use in AT [54, 56–58].

The few published Level I RCTs on the use of PRP in patients with AT have not reported that PRP decreases pain or improves function compared with placebo [14, 27]. Zhang et al. [23] performed a meta-analysis on the use of PRP for chronic AT including only Level I studies in which patients underwent a peritendinous PRP injection or saline injection. Both groups also underwent eccentric training. The authors noted that they had sufficient statistical power analysis to determine a clinically meaningful difference in the primary outcome (VISA-A score). Meta-analysis demonstrated no difference between the PRP and saline groups when examining VISA-A score, ultrasonographic evaluation of tendon

thickness, or color Doppler activity. Only one RCT reported that four injections of PRP at 2-week intervals improved VISA-A scores for AT compared with saline injection [59]. The authors of that study postulated that repetition of injections may prolong the exposure of the tendon matrix to cytokines and anti-inflammatory mediators, thereby enhancing improvement in symptoms and possibly tendon healing.

In a meta-analysis by Filardo et al. [60] examining PRP in tendon-related disorders including chronic AT, a discrepancy was observed between the results described in RCTs and case series. All of the non-controlled studies reported significant improvements in clinical outcomes, with good return to sport rates and positive effects lasting up to mid-term evaluation. Conversely, consistent with the findings of Zhang et al. [23], the RCTs which featured a control group did not favor PRP. Similar results were reported in an RCT by Krogh et al. [61] which showed no difference in patient-reported outcomes (VISA-A and VAS pain scores) between those receiving peritendinous injection of PRP or saline solution 3 months posttreatment (Fig. 18.2). PRP was found to significantly increase tendon thickness as assessed by US at 3 months posttreatment. Kearney et al. [62] performed an RCT of only 20 patients comparing a PRP injection with a rehabilitation program, reporting a trend toward improvement in the PRP group which did not meet statistical significance, possibly in part due to an underpowered sample size. Most recently, Liu et al. [63] performed a meta-analysis of PRP in the treatment of chronic AT, also finding no significant difference in VISA-A improvement or VAS pain scores following PRP injection compared to control. To date, the majority of studies using PRP for AT have been in patients with either MAT or mixed cohorts, and results have been inconsistent. Two retrospective case series specifically examining the use of PRP for IAT have recently been presented but not yet published [64, 65], reporting patient satisfaction rates of just 53% (10/19) [62], and 57% (8/14) [63] at 6-month follow-up.

PRP has also been evaluated in the setting of Achilles tear and as an adjuvant to surgical repair

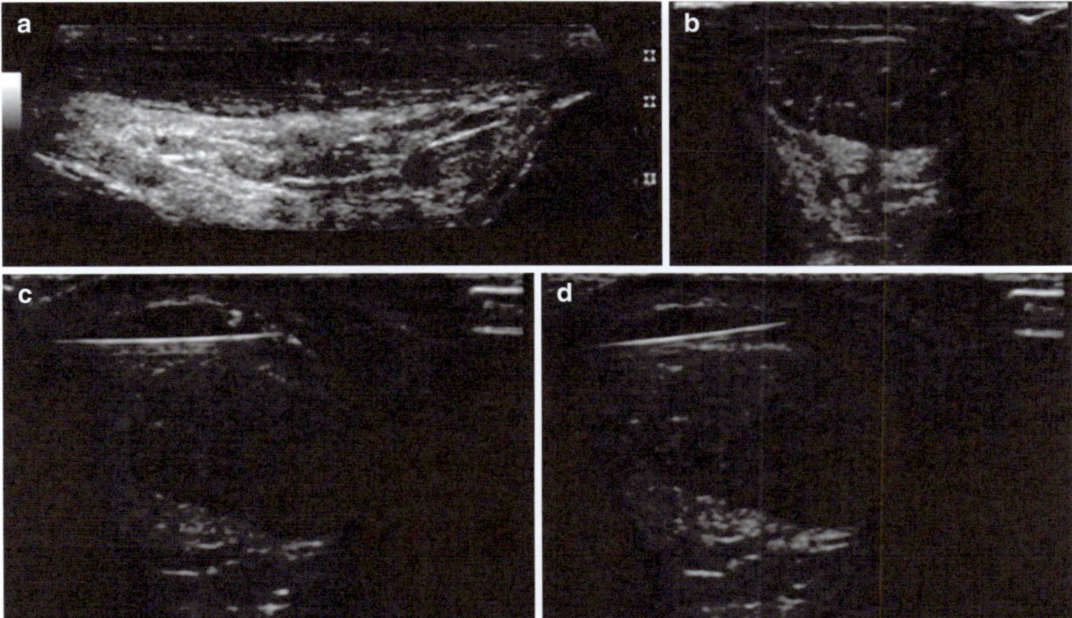

Fig. 18.2 Ultrasound (US)-guided peri-Achilles injection. (**a**) Long-axis and (**b**) short-axis images demonstrating moderate to severe Achilles tendinosis with tendon thickening of the paratenon without tear at the critical (watershed) zone. (**c**) Short-axis image demonstrating short 25-guage needle advancing into peritendinous tissues of Achilles tendon at the critical zone. (**d**) After confirming proper needle positioning, the injection was placed around the superficial and deep margins of the Achilles at the critical zone

[53]. In a rat Achilles tendon injury model, the early phases of Achilles tendon healing following rupture were enhanced following PRP injections, with early increases in fibrillary collagen deposition [66]. Rat Achilles tendons injected with PRP following surgical repair have also demonstrated improved maturation of tendon callus, stronger biomechanical properties, enhanced neovascularization, and improved histological quality [66–68]. While few human studies have investigated PRP in Achilles tendon tear, case series have reported encouraging outcomes [58, 69]. A case series of six athletes with torn Achilles tendons treated with PRP following repair demonstrated earlier recovery of range of motion and return to sport compared to athletes who underwent surgery without PRP treatment. These findings were corroborated by a review of ten studies demonstrating a strong positive effect of PRP when administered following Achilles tendon repair [58]. However, another recent review identified four studies investigating the use of PRP in the treatment of Achilles tendon rupture [60], finding no beneficial effects of PRP administration during or immediately after tendon suturing. In fact, one study reported inferior outcomes when PRP was added to Achilles tendon repair compared to repair alone, concluding that the addition of PRP may be detrimental to tissue healing in this setting [70].

Other variations of PRP have been explored in an effort to optimize the therapeutic benefits of autologous platelets. Plasma rich in growth factors (PRGF) is a type of leukocyte-poor PRP which has demonstrated positive effects on tenocyte growth and migration in vitro [71]. Platelet-rich clot releasate (PRCR) is an acellular serum product with activated platelets derived from PRP. In an in vitro model of rat tendon stem/stromal cells, PRCR was shown to induce tenocyte differentiation while inhibiting the differentiation of other cell types thought to impede tendon healing [72]. At this time, there is no data on these variations of PRP in human subjects. Altogether, current evidence does not appear to support the use of PRP for Achilles tendon injuries.

18.7 Autologous Blood Products

Autologous blood alone has also been investigated as a treatment for AT. Similar to PRP, autologous blood has an excellent safety profile but does not require any processing following a peripheral blood draw. A recent meta-analysis of seven studies sought to compare the efficacy of autologous blood-derived product (ABP) injection with placebo (sham injection, no injection, or physical therapy alone) in patients with AT [73]. The authors reported no between-group difference in VISA-A score at any time point. Furthermore, meta-regression demonstrated no association between change in VISA-A score and duration of symptoms. In another study comparing outcome of peritendinous autologous blood injection with placebo for chronic AT, Pearson et al. [29] reported no between-group difference in VISA-A score but a 21% rate of post-injection discomfort in the autologous blood group which subsequently resolved in all cases. Current data does not support the use of autologous blood in the treatment of AT.

18.8 Peripheral Blood Mononuclear Cells

Several recent studies have suggested that peripheral blood mononuclear cells (PBMNCs), which include monocytes, macrophage, and lymphocyte, may have a role in the treatment of Achilles tendon injuries [74, 75]. Monocytes and macrophages have known roles in promoting tissue repair and regeneration [74, 76]. Monocytes and macrophages are multipotent, with the ability to differentiate into different macrophage phenotypes depending on environmental signals [7]. These cells promote angiogenesis by inducing the release of VEGF [7, 77], facilitate tissue regeneration by stimulating the release of growth factors and cytokines [7, 78], and activate the connective tissue progenitor cell (CTP) popula-

tion that are resident in native tendon via paracrine signaling [7, 79].

Inflammation within tendons in early phases of tendinopathy is characterized by an infiltration of immune cells including neutrophils and macrophages [80, 81]. Initially, macrophages release cytokines at the repair site which promote ECM degradation, inflammation, and apoptosis. In later stages of tendon healing, macrophages release anti-inflammatory cytokines which promote tendon remodeling [82–84]. The balance between pro-inflammatory (M1) and anti-inflammatory (M2) macrophages during the healing process has a large impact on the resolution of inflammation and ultimate tendon healing. To illustrate this, Sugg et al. [85] monitored changes in the macrophage phenotype (M1/M2) following rat Achilles tendon repair, confirming the sequential transition of macrophages between the M1 and M2 phenotypes during the healing process. These findings support the dual function of macrophages in the early degradation and later repair of damaged tendon tissue. In a rat Achilles tendon tear model, Daher et al. [86] found that, when seeded on a biodegradable scaffold, allogeneic cells from peripheral blood improved biomechanical properties and histological organization in healing tendons when used as an augment to suture repair.

Although no studies have examined the role of macrophages or other PBMNCs in the treatment of human AT or Achilles repair, the injection of monocytes harvested from peripheral blood may represent a novel therapeutic option with a strong scientific rationale [80]. Further preclinical studies will be necessary to determine the efficacy of PBMNCs in this setting.

18.9 Bone Marrow Aspirate and Bone Marrow Aspirate Concentrate

Bone marrow aspirates (BMA) from the iliac crest contain a mixed population of marrow-derived cells. This includes hematopoietic cells, but also includes connective tissue progenitors (CTPs), with the capacity to proliferate and generate progeny of cells with the capacity to form one or more connective tissue phenotypes in culture. The prevalence of CTPs, even in healthy marrow, is low (range 1 in 10,000 to 1 in 100,000 nucleated cells, so CTPs with colony founding potential comprise less than 0.01% of all nucleated cells in a BMA sample [87–90]}.

An anticoagulated bone marrow aspirate can be processed using a centrifuge to remove red cells based on density and concentrate nucleated cells four to sixfold and provide what is commonly called a bone marrow aspirate concentrate (BMAC) [7]. Bone marrow aspirates are most commonly obtained from the iliac crest. Marrow can also be aspirated from other sites including the proximal tibia, distal femur, and proximal humerus [7, 42], but the yield of CTPs is generally lower and more variable. This method of centrifugation concentrates the mononucleated cells, hematopoietic stem cells, and platelets in one layer and both acellular serum and the red blood cells in other layers, which are typically discarded [7]. While the main function of BMAC is to deliver nucleated cells to target tissues, BMAC processing can be modified to also concentrate platelets and platelet concentrations similar to PRP, i.e., platelet-rich BMAC [87]. The presumed effect of BMAC in the treatment of tendinopathy appears to be to reduce inflammation and fibrosis and to facilitate activation of native CTPs and their potential to differentiate into tenocytes to via a paracrine mechanism [91].

An in vitro study demonstrated an increase in in situ cell proliferation within Achilles tendon scaffolds seeded with BMAC compared to culture-expanded MSC-like human bone marrow stromal cells [92]. In a rat Achilles injury model, Okamato et al. [93] peritendinously injected transected Achilles tendons with either processed bone marrow cells ("BMCs"), bone marrow-derived culture-expanded MSCs, or control vehicle and observed the highest load to failure in the BMC group. The load to failure for tendons treated with BMCs was equal to that of an uninjured tendon at 28 days postoperatively [93].

Only one study to date has evaluated the effect of BMAC injections in the setting of human Achilles tendon injury [45]. In a case series of 27 patients (28 tendons), Stein et al. [45] performed open Achilles tendon repair augmented with BMAC injection, reporting a 92% rate of return-to-sport at a mean of 5.9 months postoperative. There were no re-ruptures and one superficial wound dehiscence. Although these results are encouraging, there was no comparative group. Further investigation into the use of BMAC in Achilles tendon injuries is warranted.

18.10 Processing of Adipose Tissue as a Cellular Preparation

Adipose tissue can be harvested and processed using mechanical dissociation or enzymatic digestion to prepare cells in a form that allows injection in a "same-day point of care" procedure.

One recent RCT compared intratendinous cells released by enzymatic digestion (the stromal vascular fraction [SVF)] and PRP injections for the treatment of midsubstance Achilles tendinopathy (MAT), utilizing the within adipose tissue [94]. The authors found no clinical differences in VAS pain, VISA-A, American Orthopaedic Foot and Ankle Society Ankle-Hindfoot Score, Short Form-36 [SF-36]), or image-based metrics (MRI and ultrasonography [US]) between the two groups at 60, 120, and 180-day follow-up [94]. Both treatments provided improved clinical and image-based outcomes at all time points. However, the SVF group had improved VAS pain, AOFAS, and VISA-A scores at 15 and 30 days post-injection. While early clinical results are encouraging, there is currently limited evidence to support the use of SVF in treatment of MAT. Additional clinical studies, and particularly high-quality RCTs, will be necessary to support their use in this setting.

18.11 Culture-Expanded Mesenchymal Stromal Cells

Culture-expanded mesenchymal stromal cells (MSCs), primarily derived from bone marrow, have been investigated in the treatment of various

Achilles tendon injuries [95, 96]. Culture-expanded MSC populations include plastic adherent culture-expanded cells expressing appropriate in vitro markers that are derived from bone marrow and from adipose tissue (sometimes called ADSCs), which are addressed separately below.

MSC populations are thought to have the potential to enhance tissue healing by differentiation into tenocytes and by signaling anti-inflammatory and anti-apoptotic pathways [97]. There is evidence suggesting that the most important effects of injection of culture-expanded cells occur in the early phases of the healing process [42]. In a rabbit Achilles tendon tear model, tendons treated with culture-expanded bone marrow-derived MSCs demonstrated improved collagen fiber organization at 3 weeks posttreatment, but not later in the healing process [98].

Large variation exists between batches of MSCs from various sources. There is also variation that depends on processing technique and the environment use during expansion [42]. For example, MSCs cultured under physiological oxygen tension conditions (e.g., 3–7% FiO_2) have been shown to exhibit an increased proliferation rate and differentiation potential along with improved cell migration [42]. MSCs cultured under physiological oxygen conditions also secrete more cytokines and growth factors compared to MSCs grown under hyperoxic atmospheric (20% FiO_2) conditions, which place culture cells under increased oxidative stress [99]. One study in a rat Achilles tendon tear model demonstrated that the ultimate load to failure was significantly higher when the tendons were treated with MSCs cultured with a lower oxygen tension compared to MSC cultures at 20% O_2 at 2 and 4 weeks postoperatively [99].

It is possible that genetically modified MSCs may be utilized to enhance tendon repair. In a rat Achilles tendon tear model, treatment of injured tendons with MSCs modified to produce the zinc finger transcription factor EGR1 led to an increased formation of tendon-like tissue [100]. In another study, transected rat Achilles tendons treated with Smad8/BMP-2-engineered MSCs demonstrated an earlier recovery of biomechanical properties compared to control [101]. Future investigation of genetically engineered MSCs may seek to determine if a single growth factor or combination of growth factors is needed to optimize tendon healing and repair. Altogether, further studies will be necessary to better elucidate the role of culture-expanded MSCs in the treatment of Achilles tendon injury.

18.12 Culture-Expanded Adipose-Derived Stromal Cells (ADSCs)

Adipose-derived stromal cells (ADSCs) have been investigated for their role in the treatment of tendinopathies and have several potential advantages as a cell source, including the low morbidity in harvest (like BMA) and a relatively high prevalence of tissue resident CTPs. ADSCs have a higher prevalence of colony founding CTPS (mean ~0.2% vs. 0.005%) compared to bone marrow [42, 102]. Like marrow-derived MSCs, adipose-derived MSCs (ADSCs) have the capacity to differentiate into multiple cellular lines, including adipocytes, chondrocytes, osteoblasts, and muscle cells and may have comparable or even better efficacy to marrow-derived MSCs in some clinical settings [94, 103] [42].

The use of culture-expanded ADSCs for the treatment of tendinopathy has been investigated in multiple animal models with encouraging results [7, 94, 103]. ADSCs have been shown to induce tenocyte differentiation by overexpressing bone morphogenetic protein 12 (BMP12) [103–105]. ADSCs have also been shown to increase the expression of cartilage oligomeric matrix protein (COMP), an ECM protein critical for the formation of organized collagen fibrils [106, 107]. Additionally, local delivery of ADSCs have been shown to increase collagen type I, VEGF, and fibroblast growth factor (FGF) as well as the tensile strength of healing rabbit Achilles tendons [7].

18.13 Growth Factors

Growth factors including TGF-β, VEGF, PDGF, and IGF-I are involved in proliferation, differentiation, chemotaxis, and synthesis of ECM, all of which occur during tissue healing. In the setting of tendon healing, growth factors are produced by tenocytes and various leukocyte populations and are released from platelets during degranulation. While there is minimal data in human subjects to support the use of injectable growth factors in Achilles tendon injury, an understanding of some of the key growth factors in tendon healing and regeneration may be important in the development of future injectable therapies [108].

18.13.1 Transforming Growth Factor-β

TGF-β is a family of proteins present in three different isoforms (TGF-β1, TGF-β2, and TGF-β3) involved in several cellular processes including proliferation and cell migration [109]. TGF-β has an important role in tissue healing [109]. Collagen fibers are largely responsible for the mechanical strength of the healing tendon, and as such, TGF-β therapy may be an important target for tendon healing applications [110]. TGF-β therapy may increase the mechanical strength of the healing Achilles tendon via upregulation of collagen synthesis, increased cross-link formation, and enhanced ECM remodeling [111, 112]. TGF-β1 increases the synthesis of collagen type I and III in tenocytes and tenocyte progenitors and is overexpressed within the tendon in the early post-injury period. Promising results have been reported following TGF-β injections in a rat Achilles tear model [113]. As early as 2 weeks postoperative, Achilles tendon injected with TGF-β following repair had a histological appearance similar to that of native tendon [114]. Additionally, a dose-dependent increase in the expression of type I and type III procollagen was observed.

18.13.2 Vascular Endothelial Growth Factor

The VEGF family consists of several isoforms (VEGF-A, VEGF-B, VEGF-C, VEGF-D, and VEGF-E, and placenta growth factor) [7]. While VEGF-A is a critical regulator of neovascularization which is required for tissue healing, excessive VEGF production can lead to scar formation [115]. VEGF appears to be important during early phases of tendon healing, as a canine model of flexor tendon healing demonstrated peak mRNA levels at days 7–10 post-injury followed by a return to baseline by day 14 [116]. In an Achilles tendon repair model, exogenous VEGF injection at the repair site was found to increase TGF-β gene expression and lead to early increases in tensile strength [117]. In a rat Achilles tendon defect model, Tempfer et al. [118] utilized a local injection of the monoclonal antibody bevacizumab at the defect site to block VEGF-A signaling. In the bevacizumab-treated tendon defects, angiogenesis was found to be significantly reduced. This was accompanied by a significantly reduced cross-sectional tendon area, improved matrix organization, increased stiffness, increased maximum load to failure, and an improved gait pattern [118]. While the findings of these studies suggest that VEGF plays a key role in tissue healing in the setting of Achilles tendon injury, additional studies will be necessary to determine how both overexpression or suppression of VEGF affect Achilles tendon healing, structure, and function.

18.13.3 Bone Morphogenetic Proteins (BMPs)

Bone morphogenetic protein-12 (BMP-12), also known as growth and differentiation factor-7 (GDF-7), promotes the differentiation of stem/stromal cells into tenocytes and induces the formation of tendon and ligament-like tissue [119]. Treatment of transected rat Achilles tendons with a genetically modified skeletal muscle flap transduced with adenovirus carrying BMP-12 demonstrated an early increase in the maximum load to failure, persistently increased stiffness, and improved collagen organization and collagen fiber diameter at all time points compared to controls [119]. In a rat Achilles tendon rupture model, concomitant administration of BMP-12 and BMP-13 resulted in an increased rate of cellular infiltration, increased tissue volume, and altered levels of mRNA consistent with tendon healing [120]. Conversely, genetically engineered BMP-12 knockout mice demonstrated no significant effects on the composition or structure of Achilles tendons [121]. As such, the role of BMP-12 in tendon healing and regeneration remains poorly defined.

Bone morphogenetic protein-14 (BMP-14), also known as growth and differentiation factor-5 (GDF-5), is also a member of the transforming growth factor superfamily. This growth factor plays a role in tendon collagen organization [122]. The results of several animal studies have suggested a role in promoting Achilles tendon healing [42]. When injected ectopically, BMP-14 has been shown to induce neotendon formation [42]. Bolt et al. [123] reported that transected rat Achilles tendons that were treated with BMP-14-transfected adenovirus following repair had a 70% greater tensile strength than control at 2 weeks following repair, less visible gapping, and no ectopic bone or cartilage within the tendons treated with BMP-14. When rat Achilles tendon fibroblasts were treated with BMP-14, Keller et al. [122] demonstrated that ECM synthesis and cellular proliferation were increased. The use of BMP-14-coated sutures also has been shown to lead to improved early tendon healing at 3 weeks postoperative but not at later time points [124]. Further research will be necessary to determine the role of BMPs as adjuvants in the treatment of Achilles tendon pathologies.

### 18.13.4	Interleukin-6 (IL-6)

The ability of interleukin-6 (IL-6) to stimulate collagen synthesis has led investigators to explore its role in the treatment of AT. Concentrations of IL-6 in the peritendinous region surrounding the Achilles increase markedly following exercise or mechanical loading [42]. As such, IL-6 may allow for the transformation of mechanical loading into collagen synthesis. In a patellar tendon repair model, Andersen et al. [125] found that when recombinant human IL-6 was injected into the peritendinous tissues surrounding the Achilles tendon in healthy volunteers, it resulted in increased local collagen synthesis compared to the contralateral side with or without exercise. Additional preclinical studies will be needed to determine the efficacy of IL-6 in the setting of Achilles tendon injury.

### 18.13.5	Combination Treatment with Growth Factors

In addition to those mentioned above, several individual growth factors including PDGF, cartilage-derived morphogenetic protein-2 (CDMP-2), IGF-1, and hepatocyte growth factor (HGF) have shown positive effects in terms of tendon healing when applied to Achilles tendons using a variety of delivery methods [42, 67, 126–131]. Konerding et al. [132] hypothesized that the application of multiple mitogenic and angiogenic growth factors would lead to improvement of tendon healing in a rabbit Achilles tendon tear model. However, the concomitant peritendinous injection of VEGF, bFGF, and PDGF did not result in significant improvements in any outcomes including blood vessel density, collagen I/III ratio, or mechanical strength. Further research will be necessary to determine whether the exogenous administration of growth factors in isolation or in combination enhances healing in the setting of Achilles tendon injury.

## 18.14	Proteinase Inhibitors

Aprotinin is a serine proteinase inhibitor obtained from the bovine lung which may function as a collagenase inhibitor in the setting of tendinopathy [25]. Brown et al. [25] reported results of using a series of three once-weekly injections of aprotinin for the treatment of Achilles tendinopathy, finding no significant differences between the treatment group and placebo with respect to the primary outcome measure (VISA-A score) or any secondary outcome measure (number of hops to pain, return to sport, single-leg heel raises to pain, and patient pain rating). Aprotinin is no longer commonly used for the treatment of AT, and data does not support its use.

18.15 Conclusions

At this time, no definite treatment recommendations can be made on the currently available evidence, and biologic therapies should be used judiciously, tailoring the treatment to the needs of the patient while setting appropriate expectations for treatment outcomes. While various injectable biologic treatments for Achilles tendon disorders hold promise, further scientific investigation is needed before evidence-based protocols can be developed, tested, or widely adopted. As most cases of AT are successfully treated non-operatively [5, 21], a goal of any biologic therapy in this setting should be to improve early healing, thereby expediting recovery. As several risk factors are associated with AT [2, 11], personalized approaches to biologic therapy based on demographic and patient-specific risk factors of AT should be explored. Corticosteroids have largely fallen out of favor due to adverse tissue effects and concern for tendon rupture [28]. The role of HA remains poorly defined, and the scientific rationale underlying its potential efficacy in this setting is not as strong as some other biologic therapies. While the results of multiple case series have been encouraging, PRP has been shown in multiple high-level RCTs to be ineffective in the treatment of AT, and as such, its use cannot be recommended at this time [23].

Preclinical and early human studies of cell-based biologic therapies have demonstrated more success than corticosteroids and autologous blood products. Adipose-derived cells have shown great promise in preclinical and early clinical studies of AT, and future randomized controlled studies will need to corroborate these promising findings [7, 94]. Similarly, preclinical and early clinical studies of BMAC and BMCs in the treatment of AT have been encouraging, and further investigation is warranted before widespread adoption can be recommended [45, 93]. Although culture-expanded MSCs including those derived from bone marrow and peripheral blood have not yet been investigated in human AT, preclinical studies have demonstrated good outcomes [86, 99].

Various growth factors are known to have important roles in tendon healing and regeneration, and while exogenous injection of various growth factors including VEGF, TGF-β [115], BMP-12 [114], and BMP-14 [123] have improved tendon healing in preclinical models, no human studies have yet been performed.

Take-Home Messages
- The reparative process following Achilles tendon injury leads to the formation of a fibrovascular scar that is mechanically weaker than the native uninjured tendon.
- Biologic therapies may facilitate a more robust healing response with biomechanical properties more similar to the native uninjured tendon.
- Current evidence does not support the use of hyaluronic acid (HA), platelet-rich plasma (PRP), or autologous blood products in the treatment of Achilles tendinopathy.
- While bone marrow aspirate concentrate (BMAC) has demonstrated promising outcomes in preclinical models of Achilles injury and in the setting of surgical repair of the Achilles rupture, larger comparative studies are necessary to determine the efficacy of BMAC in this setting.
- While injectable biologics including culture-expanded mesenchymal stromal cells (MSCs), adipose-derived stromal cells (ADSCs), and exogenous growth factors hold promise, the currently

available evidence suggests that no definite treatment recommendations can be made at this time.
- Biologic therapies should be used judiciously, tailoring the treatment to the needs of the patient while setting appropriate expectations for treatment outcomes.

References

1. Hattrup SJ, Johnson KA. A review of ruptures of the Achilles tendon. Foot Ankle. 1985;6(1):34–8.
2. Holmes GB, Lin J. Etiologic factors associated with symptomatic Achilles tendinopathy. Foot Ankle Int. 2006;27(11):952–9.
3. Dakin SG, Newton J, Martinez FO, Hedley R, Gwilym S, Jones N, et al. Chronic inflammation is a feature of Achilles tendinopathy and rupture. Br J Sports Med. 2018;52(6):359–67.
4. Mosca MJ, Rashid MS, Snelling SJ, Kirtley S, Carr AJ, Dakin SG. Trends in the theory that inflammation plays a causal role in tendinopathy: a systematic review and quantitative analysis of published reviews. BMJ Open Sport Exerc Med. 2018;4(1):e000332.
5. Gross CE, Hsu AR, Chahal J, Holmes GB Jr. Injectable treatments for noninsertional Achilles tendinosis: a systematic review. Foot Ankle Int. 2013;34(5):619–28.
6. Kvist M. Achilles tendon injuries in athletes. Sports Med. 1994;18(3):173–201.
7. Indino C, D'Ambrosi R, Usuelli FG. Biologics in the treatment of Achilles tendon pathologies. Foot Ankle Clin. 2019;24(3):471–93.
8. Longo UG, Ronga M, Maffulli N. Achilles tendinopathy. Sports Med Arthrosc Rev. 2009;17(2):112–26.
9. Longo UG, Ronga M, Maffulli N. Achilles tendinopathy. Sports Med Arthrosc Rev. 2018;26(1):16–30.
10. Maffulli N, Khan KM, Puddu G. Overuse tendon conditions: time to change a confusing terminology. Arthroscopy. 1998;14(8):840–3.
11. Puddu G, Ippolito E, Postacchini F. A classification of Achilles tendon disease. Am J Sports Med. 1976;4(4):145–50.
12. Kraemer R, Wuerfel W, Lorenzen J, Busche M, Vogt PM, Knobloch K. Analysis of hereditary and medical risk factors in Achilles tendinopathy and Achilles tendon ruptures: a matched pair analysis. Arch Orthop Trauma Surg. 2012;132(6):847–53.
13. van Sterkenburg MN, van Dijk CN. Mid-portion Achilles tendinopathy: why painful? An evidence-based philosophy. Knee Surg Sports Traumatol Arthrosc. 2011;19(8):1367–75.
14. de Jonge S, de Vos RJ, Weir A, van Schie HT, Bierma-Zeinstra SM, Verhaar JA, et al. One-year follow-up of platelet-rich plasma treatment in chronic Achilles tendinopathy: a double-blind randomized placebo-controlled trial. Am J Sports Med. 2011;39(8):1623–9.
15. de Jonge S, van den Berg C, de Vos RJ, van der Heide HJ, Weir A, Verhaar JA, et al. Incidence of midportion Achilles tendinopathy in the general population. Br J Sports Med. 2011;45(13):1026–8.
16. Van Middelkoop M, Kolkman J, Van Ochten J, Bierma-Zeinstra SM, Koes B. Prevalence and incidence of lower extremity injuries in male marathon runners. Scand J Med Sci Sports. 2008;18(2):140–4.
17. Andres BM, Murrell GA. Treatment of tendinopathy: what works, what does not, and what is on the horizon. Clin Orthop Relat Res. 2008;466(7):1539–54.
18. Kearney R, Costa ML. Insertional Achilles tendinopathy management: a systematic review. Foot Ankle Int. 2010;31(8):689–94.
19. Sussmilch-Leitch SP, Collins NJ, Bialocerkowski AE, Warden SJ, Crossley KM. Physical therapies for Achilles tendinopathy: systematic review and meta-analysis. J Foot Ankle Res. 2012;5(1):15.
20. Coombes BK, Bisset L, Vicenzino B. Efficacy and safety of corticosteroid injections and other injections for management of tendinopathy: a systematic review of randomised controlled trials. Lancet. 2010;376(9754):1751–67.
21. Alfredson H, Cook J. A treatment algorithm for managing Achilles tendinopathy: new treatment options. Br J Sports Med. 2007;41(4):211–6.
22. Atkins D, Best D, Briss PA, Eccles M, Falck-Ytter Y, Flottorp S, et al. Grading quality of evidence and strength of recommendations. BMJ. 2004;328(7454):1490.
23. Zhang YJ, Xu SZ, Gu PC, Du JY, Cai YZ, Zhang C, et al. Is platelet-rich plasma injection effective for chronic Achilles tendinopathy? A meta-analysis. Clin Orthop Relat Res. 2018;476(8):1633–41.
24. Kearney RS, Parsons N, Metcalfe D, Costa ML. Injection therapies for Achilles tendinopathy. Cochrane Database Syst Rev. 2015;26(5):CD010960.
25. Brown R, Orchard J, Kinchington M, Hooper A, Nalder G. Aprotinin in the management of Achilles tendinopathy: a randomised controlled trial. Br J Sports Med. 2006;40(3):275–9.
26. DaCruz DJ, Geeson M, Allen MJ, Phair I. Achilles paratendonitis: an evaluation of steroid injection. Br J Sports Med. 1988;22(2):64–5.
27. de Vos RJ, Weir A, van Schie HT, Bierma-Zeinstra SM, Verhaar JA, Weinans H, et al. Platelet-rich plasma injection for chronic Achilles tendinopathy: a randomized controlled trial. JAMA. 2010;303(2):144–9.
28. Fredberg U, Bolvig L, Pfeiffer-Jensen M, Clemmensen D, Jakobsen BW, Stengaard-Pedersen K. Ultrasonography as a tool for diagnosis, guidance of local steroid injection and, together with pressure algometry, monitoring of the treatment of athletes with chronic jumper's knee and Achilles tendinitis: a

randomized, double-blind, placebo-controlled study. Scand J Rheumatol. 2004;33(2):94–101.

29. Pearson J, Rowlands D, Highet R. Autologous blood injection to treat Achilles tendinopathy? A randomized controlled trial. J Sport Rehabil. 2012;21(3):218–24.

30. Willberg L, Sunding K, Ohberg L, Forssblad M, Fahlstrom M, Alfredson H. Sclerosing injections to treat midportion Achilles tendinosis: a randomised controlled study evaluating two different concentrations of Polidocanol. Knee Surg Sports Traumatol Arthrosc. 2008;16(9):859–64.

31. Yelland MJ, Sweeting KR, Lyftogt JA, Ng SK, Scuffham PA, Evans KA. Prolotherapy injections and eccentric loading exercises for painful Achilles tendinosis: a randomised trial. Br J Sports Med. 2011;45(5):421–8.

32. Phisitkul P. Endoscopic surgery of the Achilles tendon. Curr Rev Musculoskelet Med. 2012;5(2):156–63.

33. Lohrer H, Arentz S, Nauck T, Dorn-Lange NV, Konerding MA. The Achilles tendon insertion is crescent-shaped: an in vitro anatomic investigation. Clin Orthop Relat Res. 2008;466(9):2230–7.

34. Syed TA, Perera A. A proposed staging classification for minimally invasive management of Haglund's syndrome with percutaneous and endoscopic surgery. Foot Ankle Clin. 2016;21(3):641–64.

35. Riley G. Tendinopathy—from basic science to treatment. Nat Clin Pract Rheumatol. 2008;4(2):82–9.

36. Longo UG, Franceschi F, Ruzzini L, Rabitti C, Morini S, Maffulli N, et al. Histopathology of the supraspinatus tendon in rotator cuff tears. Am J Sports Med. 2008;36(3):533–8.

37. Longo UG, Franceschi F, Ruzzini L, Rabitti C, Morini S, Maffulli N, et al. Light microscopic histology of supraspinatus tendon ruptures. Knee Surg Sports Traumatol Arthrosc. 2007;15(11):1390–4.

38. Movin T, Gad A, Reinholt FP, Rolf C. Tendon pathology in long-standing achillodynia. Biopsy findings in 40 patients. Acta Orthop Scand. 1997;68(2):170–5.

39. Jarvinen M, Jozsa L, Kannus P, Jarvinen TL, Kvist M, Leadbetter W. Histopathological findings in chronic tendon disorders. Scand J Med Sci Sports. 1997;7(2):86–95.

40. Klauser AS, Miyamoto H, Tamegger M, Faschingbauer R, Moriggl B, Klima G, et al. Achilles tendon assessed with sonoelastography: histologic agreement. Radiology. 2013;267(3):837–42.

41. Narici MV, Maffulli N, Maganaris CN. Ageing of human muscles and tendons. Disabil Rehabil. 2008;30(20–22):1548–54.

42. Shapiro E, Grande D, Drakos M. Biologics in Achilles tendon healing and repair: a review. Curr Rev Musculoskelet Med. 2015;8(1):9–17.

43. Gott M, Ast M, Lane LB, Schwartz JA, Catanzano A, Razzano P, et al. Tendon phenotype should dictate tissue engineering modality in tendon repair: a review. Discov Med. 2011;12(62):75–84.

44. Metcalfe D, Achten J, Costa ML. Glucocorticoid injections in lesions of the Achilles tendon. Foot Ankle Int. 2009;30(7):661–5.

45. Stein BE, Stroh DA, Schon LC. Outcomes of acute Achilles tendon rupture repair with bone marrow aspirate concentrate augmentation. Int Orthop. 2015;39(5):901–5.

46. Lynen N, De Vroey T, Spiegel I, Van Ongeval F, Hendrickx NJ, Stassijns G. Comparison of peritendinous hyaluronan injections versus extracorporeal shock wave therapy in the treatment of painful Achilles' tendinopathy: a randomized clinical efficacy and safety study. Arch Phys Med Rehabil. 2017;98(1):64–71.

47. Halici M, Karaoglu S, Canoz O, Kabak S, Baktir A. Sodium hyaluronate regulating angiogenesis during Achilles tendon healing. Knee Surg Sports Traumatol Arthrosc. 2004;12(6):562–7.

48. Tosun HB, Gumustas SA, Kom M, Uludag A, Serbest S, Eroksuz Y. The effect of sodium hyaluronate plus sodium chondroitin sulfate solution on peritendinous adhesion and tendon healing: an experimental study. Balkan Med J. 2016;33(3):258–66.

49. Tatari H, Skiak E, Destan H, Ulukus C, Ozer E, Satoglu S. Effect of hylan G-F 20 in Achilles' tendonitis: an experimental study in rats. Arch Phys Med Rehabil. 2004;85(9):1470–4.

50. Wu PT, Jou IM, Kuo LC, Su FC. Intratendinous injection of hyaluronate induces acute inflammation: a possible detrimental effect. PLoS One. 2016;11(5):e0155424.

51. Fortier LA, Barker JU, Strauss EJ, McCarrel TM, Cole BJ. The role of growth factors in cartilage repair. Clin Orthop Relat Res. 2011;469(10):2706–15.

52. Dhillon MS, Behera P, Patel S, Shetty V. Orthobiologics and platelet rich plasma. Indian J Orthop. 2014;48(1):1–9.

53. Soomekh DJ. Current concepts for the use of platelet-rich plasma in the foot and ankle. Clin Podiatr Med Surg. 2011;28(1):155–70.

54. Vannini F, Di Matteo B, Filardo G, Kon E, Marcacci M, Giannini S. Platelet-rich plasma for foot and ankle pathologies: a systematic review. Foot Ankle Surg. 2014;20(1):2–9.

55. Randelli P, Randelli F, Ragone V, Menon A, D'Ambrosi R, Cucchi D, et al. Regenerative medicine in rotator cuff injuries. Biomed Res Int. 2014;2014:129515.

56. Moraes VY, Lenza M, Tamaoki MJ, Faloppa F, Belloti JC. Platelet-rich therapies for musculoskeletal soft tissue injuries. Cochrane Database Syst Rev. 2014;29(4):CD010071.

57. Owens RF Jr, Ginnetti J, Conti SF, Latona C. Clinical and magnetic resonance imaging outcomes following platelet rich plasma injection for chronic mid-substance Achilles tendinopathy. Foot Ankle Int. 2011;32(11):1032–9.

58. Sadoghi P, Rosso C, Valderrabano V, Leithner A, Vavken P. The role of platelets in the treat-

ment of Achilles tendon injuries. J Orthop Res. 2013;31(1):111–8.

59. Boesen AP, Hansen R, Boesen MI, Malliaras P, Langberg H. Effect of high-volume injection, platelet-rich plasma, and sham treatment in chronic midportion Achilles tendinopathy: a randomized double-blinded prospective study. Am J Sports Med. 2017;45(9):2034–43.

60. Filardo G, Di Matteo B, Kon E, Merli G, Marcacci M. Platelet-rich plasma in tendon-related disorders: results and indications. Knee Surg Sports Traumatol Arthrosc. 2018;26(7):1984–99.

61. Krogh TP, Ellingsen T, Christensen R, Jensen P, Fredberg U. Ultrasound-guided injection therapy of Achilles tendinopathy with platelet-rich plasma or saline: a randomized, blinded, placebo-controlled trial. Am J Sports Med. 2016;44(8):1990–7.

62. Kearney RS, Parsons N, Costa ML. Achilles tendinopathy management: a pilot randomised controlled trial comparing platelet-richplasma injection with an eccentric loading programme. Bone Joint Res. 2013;2(10):227–32.

63. Liu CJ, Yu KL, Bai JB, Tian DH, Liu GL. Platelet-rich plasma injection for the treatment of chronic Achilles tendinopathy: a meta-analysis. Medicine (Baltimore). 2019;98(16):e15278.

64. Murawski CN NH, Kennedy JG. Platelet-rich plasma injection for the treatment of chronic insertional Achilles tendinopathy. American Orthopaedic Foot & Ankle Society Annual Meeting; 2011.

65. O'Malley M. PRP shows potential for treating Achilles tendinosis. American Academy of Orthopaedic Surgeons; 2010.

66. Kaux JF, Drion PV, Colige A, Pascon F, Libertiaux V, Hoffmann A, et al. Effects of platelet-rich plasma (PRP) on the healing of Achilles tendons of rats. Wound Repair Regen. 2012;20(5):748–56.

67. Aspenberg P, Virchenko O. Platelet concentrate injection improves Achilles tendon repair in rats. Acta Orthop Scand. 2004;75(1):93–9.

68. Lyras DN, Kazakos K, Verettas D, Polychronidis A, Tryfonidis M, Botaitis S, et al. The influence of platelet-rich plasma on angiogenesis during the early phase of tendon healing. Foot Ankle Int. 2009;30(11):1101–6.

69. Filardo G, Presti ML, Kon E, Marcacci M. Nonoperative biological treatment approach for partial Achilles tendon lesion. Orthopedics. 2010;33(2):120–3.

70. Schepull T, Kvist J, Norrman H, Trinks M, Berlin G, Aspenberg P. Autologous platelets have no effect on the healing of human Achilles tendon ruptures: a randomized single-blind study. Am J Sports Med. 2011;39(1):38–47.

71. Tohidnezhad M, Varoga D, Wruck CJ, Brandenburg LO, Seekamp A, Shakibaei M, et al. Platelet-released growth factors can accelerate tenocyte proliferation and activate the anti-oxidant response element. Histochem Cell Biol. 2011;135(5):453–60.

72. Chen L, Dong SW, Tao X, Liu JP, Tang KL, Xu JZ. Autologous platelet-rich clot releasate stimulates proliferation and inhibits differentiation of adult rat tendon stem cells towards nontenocyte lineages. J Int Med Res. 2012;40(4):1399–409.

73. Lin MT, Chiang CF, Wu CH, Hsu HH, Tu YK. Meta-analysis comparing autologous blood-derived products (including platelet-rich plasma) injection versus placebo in patients with Achilles tendinopathy. Arthroscopy. 2018;34(6):1966–75. e5.

74. Ogle ME, Segar CE, Sridhar S, Botchwey EA. Monocytes and macrophages in tissue repair: implications for immunoregenerative biomaterial design. Exp Biol Med (Maywood). 2016;241(10):1084–97.

75. Zhang M, Huang B. The multi-differentiation potential of peripheral blood mononuclear cells. Stem Cell Res Ther. 2012;3(6):48.

76. Forbes SJ, Rosenthal N. Preparing the ground for tissue regeneration: from mechanism to therapy. Nat Med. 2014;20(8):857–69.

77. Barnett FH, Rosenfeld M, Wood M, Kiosses WB, Usui Y, Marchetti V, et al. Macrophages form functional vascular mimicry channels in vivo. Sci Rep. 2016;6:36659.

78. Caplan AI. New era of cell-based orthopedic therapies. Tissue Eng Part B Rev. 2009;15(2):195–200.

79. Pajarinen J, Lin T, Gibon E, Kohno Y, Maruyama M, Nathan K, et al. Mesenchymal stem cell-macrophage crosstalk and bone healing. Biomaterials. 2019;196:80–9.

80. Dean BJ, Gettings P, Dakin SG, Carr AJ. Are inflammatory cells increased in painful human tendinopathy? A systematic review. Br J Sports Med. 2016;50(4):216–20.

81. Millar NL, Hueber AJ, Reilly JH, Xu Y, Fazzi UG, Murrell GA, et al. Inflammation is present in early human tendinopathy. Am J Sports Med. 2010;38(10):2085–91.

82. John T, Lodka D, Kohl B, Ertel W, Jammrath J, Conrad C, et al. Effect of pro-inflammatory and immunoregulatory cytokines on human tenocytes. J Orthop Res. 2010;28(8):1071–7.

83. Manning CN, Havlioglu N, Knutsen E, Sakiyama-Elbert SE, Silva MJ, Thomopoulos S, et al. The early inflammatory response after flexor tendon healing: a gene expression and histological analysis. J Orthop Res. 2014;32(5):645–52.

84. Marsolais D, Cote CH, Frenette J. Neutrophils and macrophages accumulate sequentially following Achilles tendon injury. J Orthop Res. 2001;19(6):1203–9.

85. Sugg KB, Lubardic J, Gumucio JP, Mendias CL. Changes in macrophage phenotype and induction of epithelial-to-mesenchymal transition genes following acute Achilles tenotomy and repair. J Orthop Res. 2014;32(7):944–51.

86. Daher RJ, Chahine NO, Razzano P, Patwa SA, Sgaglione NJ, Grande DA. Tendon repair augmented

with a novel circulating stem cell population. Int J Clin Exp Med. 2011;4(3):214–9.

87. Smyth NA, Murawski CD, Haleem AM, Hannon CP, Savage-Elliott I, Kennedy JG. Establishing proof of concept: platelet-rich plasma and bone marrow aspirate concentrate may improve cartilage repair following surgical treatment for osteochondral lesions of the talus. World J Orthop. 2012;3(7):101–8.

88. Muschler GF, Boehm C, Easley K. Aspiration to obtain osteoblast progenitor cells from human bone marrow: the influence of aspiration volume. J Bone Joint Surg Am. 1997;79(11):1699–709.

89. Muschler GF, Midura RJ. Connective tissue progenitors: practical concepts for clinical applications. Clin Orthop Relat Res. 2002;395:66–80.

90. Muschler GF, Midura RJ, Nakamoto C. Practical modeling concepts for connective tissue stem cell and progenitor compartment kinetics. J Biomed Biotechnol. 2003;2003(3):170–93.

91. Imam MA, Holton J, Horriat S, Negida AS, Grubhofer F, Gupta R, et al. A systematic review of the concept and clinical applications of bone marrow aspirate concentrate in tendon pathology. SICOT J. 2017;3:58.

92. Broese M, Toma I, Haasper C, Simon A, Petri M, Budde S, et al. Seeding a human tendon matrix with bone marrow aspirates compared to previously isolated hBMSCs—an in vitro study. Technol Health Care. 2011;19(6):469–79.

93. Okamoto N, Kushida T, Oe K, Umeda M, Ikehara S, Iida H. Treating Achilles tendon rupture in rats with bone-marrow-cell transplantation therapy. J Bone Joint Surg Am. 2010;92(17):2776–84.

94. Usuelli FG, Grassi M, Maccario C, Vigano M, Lanfranchi L, Alfieri Montrasio U, et al. Intratendinous adipose-derived stromal vascular fraction (SVF) injection provides a safe, efficacious treatment for Achilles tendinopathy: results of a randomized controlled clinical trial at a 6-month follow-up. Knee Surg Sports Traumatol Arthrosc. 2018;26(7):2000–10.

95. Ju YJ, Muneta T, Yoshimura H, Koga H, Sekiya I. Synovial mesenchymal stem cells accelerate early remodeling of tendon-bone healing. Cell Tissue Res. 2008;332(3):469–78.

96. Nourissat G, Diop A, Maurel N, Salvat C, Dumont S, Pigenet A, et al. Mesenchymal stem cell therapy regenerates the native bone-tendon junction after surgical repair in a degenerative rat model. PLoS One. 2010;5(8):e12248.

97. Selek O, Buluc L, Muezzinoglu B, Ergun RE, Ayhan S, Karaoz E. Mesenchymal stem cell application improves tendon healing via anti-apoptotic effect (Animal study). Acta Orthop Traumatol Turc. 2014;48(2):187–95.

98. Chong AK, Ang AD, Goh JC, Hui JH, Lim AY, Lee EH, et al. Bone marrow-derived mesenchymal stem cells influence early tendon-healing in a rabbit Achilles tendon model. J Bone Joint Surg Am. 2007;89(1):74–81.

99. Huang TF, Yew TL, Chiang ER, Ma HL, Hsu CY, Hsu SH, et al. Mesenchymal stem cells from a hypoxic culture improve and engraft Achilles tendon repair. Am J Sports Med. 2013;41(5):1117–25.

100. Guerquin MJ, Charvet B, Nourissat G, Havis E, Ronsin O, Bonnin MA, et al. Transcription factor EGR1 directs tendon differentiation and promotes tendon repair. J Clin Invest. 2013;123(8):3564–76.

101. Pelled G, Snedeker JG, Ben-Arav A, Rigozzi S, Zilberman Y, Kimelman-Bleich N, et al. Smad8/BMP2-engineered mesenchymal stem cells induce accelerated recovery of the biomechanical properties of the Achilles tendon. J Orthop Res. 2012;30(12):1932–9.

102. Mantripragada VP, Piuzzi NS, Bova WA, Boehm C, Obuchowski NA, Lefebvre V, et al. Donor-matched comparison of chondrogenic progenitors resident in human infrapatellar fat pad, synovium, and periosteum—implications for cartilage repair. Connect Tissue Res. 2019;60(6):597–610.

103. Usuelli FG, D'Ambrosi R, Maccario C, Indino C, Manzi L, Maffulli N. Adipose-derived stem cells in orthopaedic pathologies. Br Med Bull. 2017;124(1):31–54.

104. Akpancar S, Tatar O, Turgut H, Akyildiz F, Ekinci S. The current perspectives of stem cell therapy in orthopedic surgery. Arch Trauma Res. 2016;5(4):e37976.

105. Yan Z, Yin H, Nerlich M, Pfeifer CG, Docheva D. Boosting tendon repair: interplay of cells, growth factors and scaffold-free and gel-based carriers. J Exp Orthop. 2018;5(1):1.

106. Fuoco C, Petrilli LL, Cannata S, Gargioli C. Matrix scaffolding for stem cell guidance toward skeletal muscle tissue engineering. J Orthop Surg Res. 2016;11(1):86.

107. Longo UG, Lamberti A, Petrillo S, Maffulli N, Denaro V. Scaffolds in tendon tissue engineering. Stem Cells Int. 2012;2012:517165.

108. Sharma P, Maffulli N. Tendon injury and tendinopathy: healing and repair. J Bone Joint Surg Am. 2005;87(1):187–202.

109. Leksa V, Godar S, Schiller HB, Fuertbauer E, Muhammad A, Slezakova K, et al. TGF-beta-induced apoptosis in endothelial cells mediated by M6P/IGFII-R and mini-plasminogen. J Cell Sci. 2005;118(Pt 19):4577–86.

110. Hou Y, Mao Z, Wei X, Lin L, Chen L, Wang H, et al. The roles of TGF-beta1 gene transfer on collagen formation during Achilles tendon healing. Biochem Biophys Res Commun. 2009;383(2):235–9.

111. Benjamin M, Ralphs JR. Tendons and ligaments—an overview. Histol Histopathol. 1997;12(4):1135–44.

112. Hou Y, Mao Z, Wei X, Lin L, Chen L, Wang H, et al. Effects of transforming growth factor-beta1 and vascular endothelial growth factor 165 gene transfer on Achilles tendon healing. Matrix Biol. 2009;28(6):324–35.

113. Kashiwagi K, Mochizuki Y, Yasunaga Y, Ishida O, Deie M, Ochi M. Effects of transforming growth factor-beta 1 on the early stages of healing of the Achilles tendon in a rat model. Scand J Plast Reconstr Surg Hand Surg. 2004;38(4):193–7.

114. Majewski M, Porter RM, Betz OB, Betz VM, Clahsen H, Fluckiger R, et al. Improvement of tendon repair using muscle grafts transduced with TGF-beta1 cDNA. Eur Cell Mater. 2012;23:94–101; discussion-2.

115. Wilgus TA, Ferreira AM, Oberyszyn TM, Bergdall VK, Dipietro LA. Regulation of scar formation by vascular endothelial growth factor. Lab Investig. 2008;88(6):579–90.

116. Boyer MI, Watson JT, Lou J, Manske PR, Gelberman RH, Cai SR. Quantitative variation in vascular endothelial growth factor mRNA expression during early flexor tendon healing: an investigation in a canine model. J Orthop Res. 2001;19(5):869–72.

117. Zhang F, Liu H, Stile F, Lei MP, Pang Y, Oswald TM, et al. Effect of vascular endothelial growth factor on rat Achilles tendon healing. Plast Reconstr Surg. 2003;112(6):1613–9.

118. Tempfer H, Kaser-Eichberger A, Lehner C, Gehwolf R, Korntner S, Kunkel N, et al. Bevacizumab improves Achilles tendon repair in a rat model. Cell Physiol Biochem. 2018;46(3):1148–58.

119. Majewski M, Betz O, Ochsner PE, Liu F, Porter RM, Evans CH. Ex vivo adenoviral transfer of bone morphogenetic protein 12 (BMP-12) cDNA improves Achilles tendon healing in a rat model. Gene Ther. 2008;15(16):1139–46.

120. Jelinsky SA, Li L, Ellis D, Archambault J, Li J, St Andre M, et al. Treatment with rhBMP12 or rhBMP13 increase the rate and the quality of rat Achilles tendon repair. J Orthop Res. 2011;29(10):1604–12.

121. Heisterbach PE, Todorov A, Fluckiger R, Evans CH, Majewski M. Effect of BMP-12, TGF-beta1 and autologous conditioned serum on growth factor expression in Achilles tendon healing. Knee Surg Sports Traumatol Arthrosc. 2012;20(10): 1907–14.

122. Keller TC, Hogan MV, Kesturu G, James R, Balian G, Chhabra AB. Growth/differentiation factor-5 modulates the synthesis and expression of extracellular matrix and cell-adhesion-related molecules of rat Achilles tendon fibroblasts. Connect Tissue Res. 2011;52(4):353–64.

123. Bolt P, Clerk AN, Luu HH, Kang Q, Kummer JL, Deng ZL, et al. BMP-14 gene therapy increases tendon tensile strength in a rat model of Achilles tendon injury. J Bone Joint Surg Am. 2007;89(6):1315–20.

124. Dines JS, Weber L, Razzano P, Prajapati R, Timmer M, Bowman S, et al. The effect of growth differentiation factor-5-coated sutures on tendon repair in a rat model. J Shoulder Elb Surg. 2007;16(5 Suppl):S215–21.

125. Andersen MB, Pingel J, Kjaer M, Langberg H. Interleukin-6: a growth factor stimulating collagen synthesis in human tendon. J Appl Physiol (1985). 2011;110(6):1549–54.

126. Virchenko O, Fahlgren A, Skoglund B, Aspenberg P. CDMP-2 injection improves early tendon healing in a rabbit model for surgical repair. Scand J Med Sci Sports. 2005;15(4):260–4.

127. Thomopoulos S, Das R, Silva MJ, Sakiyama-Elbert S, Harwood FL, Zampiakis E, et al. Enhanced flexor tendon healing through controlled delivery of PDGF-BB. J Orthop Res. 2009;27(9):1209–15.

128. Thomopoulos S, Zaegel M, Das R, Harwood FL, Silva MJ, Amiel D, et al. PDGF-BB released in tendon repair using a novel delivery system promotes cell proliferation and collagen remodeling. J Orthop Res. 2007;25(10):1358–68.

129. Wurgler-Hauri CC, Dourte LM, Baradet TC, Williams GR, Soslowsky LJ. Temporal expression of 8 growth factors in tendon-to-bone healing in a rat supraspinatus model. J Shoulder Elb Surg. 2007;16(5 Suppl):S198–203.

130. Cui Q, Wang Z, Jiang D, Qu L, Guo J, Li Z. HGF inhibits TGF-beta1-induced myofibroblast differentiation and ECM deposition via MMP-2 in Achilles tendon in rat. Eur J Appl Physiol. 2011;111(7):1457–63.

131. Zhang J, Middleton KK, Fu FH, Im HJ, Wang JH. HGF mediates the anti-inflammatory effects of PRP on injured tendons. PLoS One. 2013;8(6):e67303.

132. Konerding MA, Arlt F, Wellmann A, Li V, Li W. Impact of combinatory growth factor application on rabbit Achilles tendon injury with operative versus conservative treatment: a pilot study. Int J Mol Med. 2010;25(2):217–24.

Orthobiologics for the Treatment of Plantar Fasciitis

Filippo Rosati Tarulli, Cristian Aletto, and Nicola Maffulli

19.1 Introduction

Plantar fasciitis (PF) is characterized by plantar medial heel pain, usually presenting in the morning at the first few steps. Obese individuals, who stand for prolonged periods and who walk on hard surfaces [1, 2], typically suffer from PF, the commonest cause of plantar heel pain in adults [3].

The diagnosis can be achieved through patient clinical history and clinical findings. Stretching exercises, activity modification, and use of several analgesics resolve symptoms in over 80% of patients, while biomechanical factors can be corrected by insoles or various kinds of orthotics or night splints [4, 5]. In the small group of patients who develop intractable PF, other available strategies are extracorporeal shock wave therapy and corticosteroid injections [6]. Surgical management of PF consists of plantar fascia release, but

F. R. Tarulli · C. Aletto
Department of Musculoskeletal Disorders, School of Medicine and Surgery, University of Salerno, Salerno, Italy

N. Maffulli (✉)
Department of Musculoskeletal Disorders, School of Medicine and Surgery, University of Salerno, Salerno, Italy

Centre for Sports and Exercise Medicine, Queen Mary University of London, Barts and the London School of Medicine and Dentistry, Mile End Hospital, London, England
e-mail: n.maffulli@qmul.ac.uk

efficacy is still debated [7, 8]. In recent years, biological treatments have been getting popularity in many orthopaedic conditions [9].

19.2 Platelet-Rich Plasma (PRP)

Platelet-rich plasma (PRP) is a device used for several chronic degenerative soft tissue conditions, including PF. To prepare PRP, the patient's own blood is centrifuged to obtain an increased platelet concentration. Platelet alpha-granules contain growth factors and mediators [vascular endothelial growth factor (VEGF), transforming growth factor-beta 1 (TGF-β1), EGF, platelet-derived growth factor (PDGF), bFGF, IGF-1], which are concentrated through a single- or double-centrifugation process. Supraphysiological amounts of these cytokines and growth factors are injected into the injury site and promote the physiological healing process [10]. PRP is postulated to promote native tissue regeneration [11].

The efficacy of PRP injections in the management of chronic PF has been evaluated in several randomized controlled trials. Platelet-rich plasma is not associated with the complications of corticosteroid injections, such as plantar fascia rupture or fat pad atrophy [12].

Platelet-rich plasma injections were compared to corticosteroid injections in two recent meta-analyses: PRP injections are a valid alternative to corticosteroid injections with some studies demonstrating superiority of PRP [13–15].

© ISAKOS 2022

G. Filardo et al. (eds.), *Orthobiologics*, https://doi.org/10.1007/978-3-030-84744-9_19

Ragab and Othman assessed 25 patients managed with a single injection of PRP. The average visual analog scale (VAS) pain decreased from 9.1 to 2.1 by 1 year after the injection [16].

After 1 year, a marked improvement in terms of VAS after PRP injection (from 7.1 ± 1.1 to 1.9 ± 1.5) was reported in a prospective uncontrolled study by Martinelli et al. [17] (Table 19.1).

Sami et al. compared the use of PRP injections under ultrasonography guidance to physiotherapy. They prospectively recruited patients suffering from chronic PF and divided them into two treatment groups (PRP group vs physiotherapy group). All patients were evaluated using the American Orthopaedic Foot and Ankle Society (AOFAS) before and after treatment. The AOFAS score improved significantly in the PRP group. Ultrasonography was performed before and 4 weeks after treatment, fascial echogenicity was significantly changed in most of the patients after PRP injection, and fascial thickness was statistically decreased in the PRP group compared to the physiotherapy group [20].

A comprehensive systematic review analysed the use of PRP in the treatment of PF [21]. Most the analysed studies mentioned a significantly larger improvement in symptoms between the first visit and the last follow-up assessment.

Platelet-rich plasma injections are an effective strategy to decrease pain and enhance function in chronic plantar fasciitis and may be superior to corticosteroids, especially considering its safety profile [16] (Fig. 19.1).

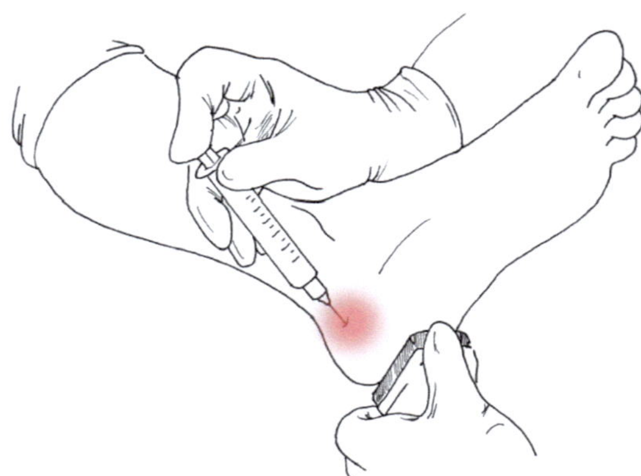

Fig. 19.1 Ultrasound-guided injection of platelet-rich plasma

19.3 AWB vs PRP

Autologous whole blood (AWB), unlike PRP, is not obtained separating the blood components. AWB contains red blood cells, white blood cells, and platelets rich in growth factor. After extracting the patient's blood from a vein, the doctor injects this blood into the heel. This whole blood with its growth factors can help trigger a healing response. Autologous whole blood cells seem to be as effective as PRP, but more randomized comparative clinical trials are needed to establish efficacy of AWB compared to PRP injections [22].

19.4 Corticosteroid vs PRP

Different studies analysed the use of corticosteroids injections vs PRP in patients with PF using functional evaluation and pain scales. Monto recruited 40 individuals with chronic unilateral PF who had failed traditional conservative treatment. They were randomized into two groups [14]. Group 1 was managed with one ultrasound-guided injection of 40 mg DepoMedrol (methylprednisolone), and group 2 was treated with one ultrasound-guided injection of autologous PRP.

The average AOFAS score before treatment was 52 in the corticosteroid group, improving to 81 at 3 months posttreatment and decreasing to 74 at 6 months; at 12 months it dropped to

Table 19.1 Success rate of PRP injection therapy

References	Success definition	Success rate (%)
Kumar et al. (2013) [18]	Patients satisfaction at the question "would have the procedure again"	64
Martinelli et al. (2013) [17]	Roles and Maudsley score: excellent and good	78.6
Ragab and Othman (2012) [16]	Satisfaction at patient's questionnaire	88
Aksahin et al. (2012) [19]	Roles and Maudsley score: excellent and good	33.3

58, with a final score of 56 at 24 months. Conversely, average AOFAS score was 37 in the PRP group before treatment and raised to 95 at 3 months, remaining elevated at 94 at 6 and 12 months, and reaching a final score of 92 at 24 months. In patients with severe chronic PF who have not obtained the desired result following traditional conservative management, PRP is able to provide successful benefits in the long-term, being more efficacious than corticosteroid injections, and appearing safer than surgical alternatives [23].

In the controlled, randomized, blinded clinical study by Acosta-Olivo et al., patients were randomized into two groups. Administration of dexamethasone 8 mg plus 2 mL of lidocaine was adopted in the steroid treatment group, while 3 mL of PRP activated with 0.45 mL of 10% calcium gluconate was used in the PRP treatment group. The VAS, Foot and Ankle Disability Index (FADI), and American Orthopaedic Foot and Ankle Society (AOFAS) scale were proposed to all patients at the beginning of the study and at 2, 4, 8, 12, and 16 weeks posttreatment. In conclusion, none of the cores was statistically different between the two groups, but PRP was more efficacious than corticosteroid injections in patients with PF failed to respond to non-operative treatment [24]. In the end, the efficacy of PRP is comparable to that of steroids injections, without complications associated with steroid use and, in addition, promoting the regeneration of damaged tissue, especially soft tissues, such as muscles and tendons, thanks to its regenerative properties. On the contrary, there are some disadvantages: PRP is more costly than steroids; the process to obtain PRP is time expensive for patients and physicians.

19.5 Prolotherapy

Prolotherapy is an injection-based treatment used in chronic musculoskeletal conditions, such as PF. In this procedure, a natural irritant (such as hyperosmolar dextrose) is injected into the soft tissues of the plantar fascia to cause osmotic desiccation and potential death of local cells and trigger a healing response. Prolotherapy injections can be effective in patients with chronic PF [25–27]. The efficacy of a prolotherapy injection has been reported to be superior to that of corticosteroids, as it allows tissue healing similar to PRP [28]. Prolotherapy injections are simpler to prepare than PRP, noninvasive, and more cost-effective [29]. This procedure is considered safe and effective, with only minor reported adverse effects mainly in the site of injections [30]. Uğurlar et al. compared the use of extracorporeal shock wave therapy (ESWT) and corticosteroids, PRP, and prolotherapy injections for the treatment of plantar fasciitis through a randomized controlled prospective clinical study. The clinical outcomes were assessed using the visual analog scale (VAS) and Revised Foot Function Index. The corticosteroid injection reduced foot pain in the first 3 months, while ESWT had similar results in the first 6 months. The result of prolotherapy and PRP was seen during the follow-up period, while the corticosteroid injection lost its effectiveness. Nevertheless, at the 36-month follow-up point, no significant difference was noted in terms of VAS and Revised Foot Function Index score among the four treatments [31].

19.6 Bone Marrow Aspirate Concentrate (BMAC)

One promising new non-surgical treatment is bone marrow concentrate (BMC) or bone marrow aspirate concentrate (BMAC) therapy [32]. BMAC is based on autologous bone marrow aspiration, followed by centrifugation to concentrate nucleated cells from bone marrow. These cells (20–100 nucleated cells/mL) include a small population of connective tissue stem cells and progenitors (CTPs) (generally <0.05%), hematopoietic stem cells (HSCs) and progenitors (<1%), and growth factors and platelets [33]. The percentage of CTPs in BMAC varies from 0.001% to 0.01% of mononuclear cells after centrifugation. However, BMAC is also a source of growth factors, including transforming growth factor-beta (TGF-b), vascular endothelial growth factor (VEGF), platelet-derived growth factor (PDGF), bone morphogenetic protein (BMP)-2, and BMP-7, which are potentially valuable as anabolic and anti-inflammatory agents [34, 35].

Recently, there has been a spike in interest in the use of bone marrow aspirates (BMAs) and BMAC in musculoskeletal ailments. Much of the early hope of cell therapy has been placed on the potential for CPTs in native tissues [36] [37]. However, these hopes of new tissue formation from transplanted cells have been elusive. The progeny of transplanted cells has not been shown to survive in new tissues. Paracrine effects appear more likely, whereby secretory products of one or more of the many cell types that are present in a BMAC preparation contribute to increase the activation of local CTPs resident in tendon tissue where they may proliferate and generate new tissue and organized extracellular matrix [38]. To date, there are no published clinical studies on efficacy of BMAC in plantar fasciitis.

19.7 Placenta Tissue Extracts

Foetal tissues consist of the chorionic membrane, amniotic membrane, and umbilical cord. These tissues are known for their healing characteristics, from numerous growth factors, cytokines, and matrix components [39]. Fresh human amniotic membrane is not used in the clinical setting because of his risk of disease transmission, but, thanks to the PURION process, it can be cleaned, sterilized, and dried to obtain human amnion/chorion membrane (dHACM) allograft [40]. Furthermore, the dHACM can be refined using a micronization process to produce a powder; dispersion of this powder into suspension with sterile 0.9% saline solution can be injected in injured soft tissues to promote regeneration and reduce inflammation. In vivo and in vitro studies have shown that the amniotic membrane does not only help to modulate inflammation and enhance soft tissue healing but also has antibacterial and pain reduction properties [41]. Cazzel et al. evaluated, in a prospective, single-blind, randomized controlled trial, one injection of micronized dHACM or 0.9% sodium chloride placebo in 145 subjects with chronic PF. Visual analog scale (VAS), Foot

Function Index-Revised (FFI-R) score, and presence/absence of adverse events were assessed at 4 weeks, 8 weeks, 3 months, 6 months, and 12 months post injection. Pain reduction (VAS) and functional improvement outcomes (FFI-R) were statistically significant and clinically relevant in patients treated with micronized dHAMC compared to patients managed with placebo [42]. A prospective, randomized, single-centre clinical trial was performed by Zelen et al. examining the efficacy of micronized dehydrated human amniotic/chorionic membrane (mdHACM) injections. Follow-up visits occurred over 8 weeks to measure pain, function, health, and well-being. Significant improvement in terms of American Orthopaedic Foot and Ankle Society (AOFAS) Hindfoot scores was observed in patients receiving 0.5 mL or 1.25 mL mdHACM versus controls [43]. Hanselman et al. compared the treatment with cryopreserved human amniotic membrane to a traditional corticosteroid injection in a controlled randomized double-blind trial. Total follow-up was 12 weeks in 23 patients. Cryopreserved human amniotic membrane injection is considered safe and comparable to corticosteroid injection in terms of VAS and Foot Health Status Questionnaire (FHSQ), but sample size was small, and additional comparative effectiveness studies are required [44]. Treatment with dHACM injection can be used to manage chronic PF, but larger multicentre randomized controlled trials are required to assess its efficacy compared to corticosteroids and other biologics treatments.

19.8 Conclusions

Conservative treatment is successful in the majority of patients with PF. Physicians need to bear in mind that early recognition and treatment usually lead to a shorter course of treatment as well as increased probability of success with conservative measures.

Stretching and strengthening programmes play an important role in the management of plantar fasciitis and must be advised in addition to biological treatments. In general, plantar fasciitis is a self-limiting condition. Unfortunately, the time until resolution is often 6 to 18 months, which can lead to frustration for patients and physicians.

Ignoring plantar fasciitis may sometimes result in chronic heel pain that hinders patients' regular activities, and therefore using the right conservative measure can make the difference. In some cases, after 6 or more months of conservative treatment, it is necessary to try more invasive options. Surgical release of the plantar fascia is effective in the small proportion of patients who do not respond to biological measures. New techniques, such as endoscopic or percutaneous plantar fascia release, may have a role.

The current widespread use of corticosteroids must be discouraged because of the adverse reactions and the limitation of their prolonged use.

Biological treatments are becoming a viable management option given their low apparent risk and potential for new mechanisms for tissue restoration.

Additional well-designed randomized double-blinded controlled trials are needed to distinguish between improvement related to intervention and improvement that would take place in the natural course of resolution.

Take-Home Messages
- Platelet-rich plasma injections are an effective strategy to decrease pain and enhance function in chronic plantar fasciitis.
- PRP, compared to steroids injections, has no complications associated with steroid use and, in addition, promotes the regeneration of the damaged tissue.
- Autologous whole blood cells seem to be as effective as PRP, but more randomized comparative clinical trials are needed to establish efficacy.
- Prolotherapy injection is considered safe and effective, with only minor reported adverse effects in the site of injections, and it allows tissue healing similar to PRP.

- Bone marrow aspirate concentrate is potentially effective, but more clinical studies are needed.
- Treatment with placental tissue extracts injection can be used to manage chronic PF, but more and larger multicentre trials are required.

References

1. Lapidus PW, Guidotti FP. Painful heel: report of 323 patients with 364 painful heels. Clin Orthop. 1965;39:178–86.
2. Riddle DL, Pulisic M, Pidcoe P, Johnson RE. Risk factors for plantar fasciitis: a matched case-control study. J Bone Joint Surg Am. 2003;85(5):872–7.
3. Riddle DL, Schappert SM. Volume of ambulatory care visits and patterns of care for patients diagnosed with plantar fasciitis: a national study of medical doctors. Foot Ankle Int. 2004;25(5):303–10.
4. Levy JC, Mizel MS, Clifford PD, Temple HT. Value of radiographs in the initial evaluation of nontraumatic adult heel pain. Foot Ankle Int. 2006;27(6):427–30.
5. American Orthopaedic Foot and Ankle Society. Position statement: endoscopic and open heel surgery; 2010.
6. Porter MD, Shadbolt B. Intralesional corticosteroid injection versus extracorporeal shock wave therapy for plantar fasciopathy. Clin J Sport Med Off J Can Acad Sport Med. 2005;15(3):119–24.
7. Sammarco GJ, Helfrey RB. Surgical treatment of recalcitrant plantar fasciitis. Foot Ankle Int. 1996;17(9):520–6.
8. Davies MS, Weiss GA, Saxby TS. Plantar fasciitis: how successful is surgical intervention? Foot Ankle Int. 1999;20(12):803–7.
9. Sampson S, Gerhardt M, Mandelbaum B. Platelet rich plasma injection grafts for musculoskeletal injuries: a review. Curr Rev Musculoskelet Med. 2008;1(3–4):165–74.
10. Boswell SG, Cole BJ, Sundman EA, Karas V, Fortier LA. Platelet-rich plasma: a milieu of bioactive factors. Arthrosc J Arthrosc Relat Surg Off Publ Arthrosc Assoc N Am Int Arthrosc Assoc. 2012;28(3):429–39.
11. Lee KS, Wilson JJ, Rabago DP, Baer GS, Jacobson JA, Borrero CG. Musculoskeletal applications of platelet-rich plasma: fad or future? AJR Am J Roentgenol. 2011;196(3):628–36.
12. Neufeld SK, Cerrato R. Plantar fasciitis: evaluation and treatment. J Am Acad Orthop Surg. 2008;16(6):338–46.
13. Mahindra P, Yamin M, Selhi HS, Singla S, Soni A. Chronic plantar fasciitis: effect of platelet-rich

plasma, corticosteroid, and placebo. Orthopedics. 2016;39(2):e285–9.

14. Monto RR. Platelet-rich plasma efficacy versus corticosteroid injection treatment for chronic severe plantar fasciitis. Foot Ankle Int. 2014;35(4):313–8.

15. Say F, Gürler D, İnkaya E, Bülbül M. Comparison of platelet-rich plasma and steroid injection in the treatment of plantar fasciitis. Acta Orthop Traumatol. 2014;48(6):667–72.

16. Ragab EMS, Othman AMA. Platelets rich plasma for treatment of chronic plantar fasciitis. Arch Orthop Trauma Surg. 2012;132(8):1065–70.

17. Martinelli N, Marinozzi A, Carnì S, Trovato U, Bianchi A, Denaro V. Platelet-rich plasma injections for chronic plantar fasciitis. Int Orthop. 2013;37(5):839–42.

18. Kumar V, Millar T, Murphy PN, Clough T. The treatment of intractable plantar fasciitis with platelet-rich plasma injection. Foot (Edinb). 2013;23(2–3):74–7.

19. Akşahin E, Doğruyol D, Yüksel HY, Hapa O, Doğan Ö, Çelebi L, et al. The comparison of the effect of corticosteroids and platelet-rich plasma (PRP) for the treatment of plantar fasciitis. Arch Orthop Trauma Surg. 2012;132(6):781–5.

20. Sami M, Nassr MH, Hamdy M, Khalil A. Preliminary study reveals the efficiency of platelet rich plasma injection over physiotherapy for chronic plantar fasciitis treatment. Int J Clin Rheumatol. 2019;14(3):120.

21. Franceschi F, Papalia R, Franceschetti E, Paciotti M, Maffulli N, Denaro V. Platelet-rich plasma injections for chronic plantar fasciopathy: a systematic review. Br Med Bull. 2014;112(1):83–95.

22. Kirmani T, Gul I, Waris Q, Kangoo K. Autologous whole blood injection in chronic plantar fasciitis: a prospective clinical study. Int J Res Orthop. 2018;4:634.

23. Tabrizi A, Dindarian S, Mohammadi S. The effect of corticosteroid local injection versus platelet-rich plasma for the treatment of plantar fasciitis in obese patients: a single-blind, randomized clinical trial. J Foot Ankle Surg Off Publ Am Coll Foot Ankle Surg. 2020;59(1):64–8.

24. Acosta-Olivo C, Elizondo-Rodriguez J, Lopez-Cavazos R, Vilchez-Cavazos F, Simental-Mendia M, Mendoza-Lemus O. Plantar fasciitis-a comparison of treatment with intralesional steroids versus platelet-rich plasma a randomized, blinded study. J Am Podiatr Med Assoc. 2017;107(6):490–6.

25. Ryan MB, Wong AD, Gillies JH, Wong J, Taunton JE. Sonographically guided intratendinous injections of hyperosmolar dextrose/lidocaine: a pilot study for the treatment of chronic plantar fasciitis. Br J Sports Med. 2009;43(4):303–6.

26. Chen C-M, Chen J-S, Tsai W-C, Hsu H-C, Chen K-H, Lin C-H. Effectiveness of device-assisted ultrasound-guided steroid injection for treating plantar fasciitis. Am J Phys Med Rehabil. 2013;92(7):597–605.

27. Ersen Ö, Koca K, Akpancar S, Seven MM, Akyıldız F, Yıldız Y, et al. A randomized-controlled trial of prolotherapy injections in the treatment of plantar fasciitis. Turk J Phys Med Rehabil. 2018;64(1):59–65.

28. Seven M, Koca K, Akpancar S, Turkkan S, Uysal B, Yildiz Y, et al. Prolotherapy injections in the treatment of overuse injuries. Balk Mil Med Rev. 2016;19:1.

29. Rabago D, Slattengren A, Zgierska A. Prolotherapy in primary care practice. Prim Care. 2010;37(1):65–80.

30. Dorman TA. Prolotherapy: a survey. J Orthop Med. 1993;15(2):49–50.

31. Uğurlar M, Sönmez MM, Uğurlar ÖY, Adıyeke L, Yıldırım H, Eren OT. Effectiveness of four different treatment modalities in the treatment of chronic plantar fasciitis during a 36-month follow-up period: a randomized controlled trial. J Foot Ankle Surg Off Publ Am Coll Foot Ankle Surg. 2018;57(5):913–8.

32. Harford JS, Dekker TJ, Adams SB. Bone marrow aspirate concentrate for bone healing in foot and ankle surgery. Foot Ankle Clin. 2016;21(4):839–45.

33. Cavinatto L, Hinckel BB, Tomlinson RE, Gupta S, Farr J, Bartolozzi AR. The role of bone marrow aspirate concentrate for the treatment of focal chondral lesions of the knee: a systematic review and critical analysis of animal and clinical studies. Arthrosc J Arthrosc Relat Surg Off Publ Arthrosc Assoc N Am Int Arthrosc Assoc. 2019;35(6):1860–77.

34. Chahla J, Dean CS, Moatshe G, Pascual-Garrido C, Serra Cruz R, LaPrade RF. Concentrated bone marrow aspirate for the treatment of chondral injuries and osteoarthritis of the knee: a systematic review of outcomes. Orthop J Sports Med. 2016;4(1):2325967115625481.

35. Courneya J-P, Luzina IG, Zeller CB, Rasmussen JF, Bocharov A, Schon LC, et al. Interleukins 4 and 13 modulate gene expression and promote proliferation of primary human tenocytes. Fibrogenesis Tissue Repair. 2010;3:9.

36. Dominici M, Le Blanc K, Mueller I, Slaper-Cortenbach I, Marini F, Krause D, et al. Minimal criteria for defining multipotent mesenchymal stromal cells. The International Society for Cellular Therapy position statement. Cytotherapy. 2006;8(4):315–7.

37. Pittenger MF, Mackay AM, Beck SC, Jaiswal RK, Douglas R, Mosca JD, et al. Multilineage potential of adult human mesenchymal stem cells. Science. 1999;284(5411):143–7.

38. Cottom JM, Plemmons BS. Bone marrow aspirate concentrate and its uses in the foot and ankle. Clin Podiatr Med Surg. 2018;35(1):19–26.

39. Werber B. Amniotic tissues for the treatment of chronic plantar fasciosis and achilles tendinosis. J Sports Med. 2015;2015:e219896.

40. Fetterolf D, Snyder R. Scientific and clinical support for the use of dehydrated amniotic membrane in wound management. Wounds Compend Clin Res Pract. 2012;24:299–307.

41. Niknejad H, Peirovi H, Jorjani M, Ahmadiani A, Ghanavi J, Seifalian AM. Properties of the amniotic membrane for potential use in tissue engineering. Eur Cell Mater. 2008;15:88–99.

42. Cazzell S, Stewart J, Agnew PS, Senatore J, Walters J, Murdoch D, et al. Randomized controlled trial of micronized dehydrated human amnion/chorion membrane (dHACM) injection compared to placebo for the treatment of plantar fasciitis. Foot Ankle Int. 2018;39(10):1151–61.
43. Zelen CM, Poka A, Andrews J. Prospective, randomized, blinded, comparative study of injectable micronized dehydrated amniotic/chorionic membrane allograft for plantar fasciitis—a feasibility study. Foot Ankle Int. 2013;34(10):1332–9.
44. Hanselman AE, Tidwell JE, Santrock RD. Cryopreserved human amniotic membrane injection for plantar fasciitis: a randomized, controlled, double-blind pilot study. Foot Ankle Int. 2015;36(2):151–8.

Robert S. Dean, Nicholas N. DePhillipo, and Robert F. LaPrade

20.1 Introduction

Cell therapies provide a broad range of promising strategies as potential treatments and/or augmentation procedures for current orthopedic pathologies. In concept, they may include use of undifferentiated cells that can either be induced into forming target tissues/structures or both differentiated and undifferentiated cells that can enhance the native healing process through direct or indirect mechanisms. While there is a relative paucity in the literature regarding clinical outcomes using cell therapies in ligament injuries, the preclinical studies have been encouraging. Success will depend on integration of specific knowledge of the healing process that occurs in damaged ligaments, the local anatomy, with the intrinsic biological attributes and potential of cellular therapy options.

20.1.1 Native Ligament Healing

There are three stages of ligament healing (Fig. 20.1). The first stage is the *inflammatory phase*, which is initiated immediately following the inciting injury and lasts for 2–3 days. The

R. S. Dean · R. F. LaPrade (✉)
Twin City Orthopedics, Edina, MN, USA

N. N. DePhillipo
Twin City Orthopedics, Edina, MN, USA

Oslo Sports Trauma Research Center, Oslo, Norway

hallmark event that occurs during this phase is the adhesion and activation of platelets which initiate a clot formation. The activated platelets release products from their alpha granules, including platelet-derived growth factor (PDGF), transforming growth factor-β (TGF-ß), vascular endothelial growth factor (VEGF), and fibroblast growth factor (FGF), among others. These growth factors may help stimulate local growth including angiogenesis and collagen synthesis, initiate cellular differentiation, and are crucial to the progression of the healing process [1].

The *hypercellularity/proliferative phase* is the second phase of ligament healing and typically begins several days after the inciting ligament injury (Fig. 20.1). It is defined by cell proliferation, neovascularization, and matrix synthesis, among other metabolic processes which aid in remodeling and organization of a healing ligament. During this phase, growth factors, chemoattractants, and mitogenic agents help induce native connective tissue progenitors (CTPs) in local injured and adjacent tissues to proliferate and differentiate into myofibroblasts [2]. There are also increased levels of insulin-like growth factor-I (IGF-I) (a proliferative growth factor) and insulin-like growth factor-II (IGF-II) (a collagen synthesis stimulator) locally and an increased expression of transforming growth factor-ß1 (TGF-ß1-a collagen and non-collagen protein stimulator). During the hypercellularity/proliferative phase, there is a decreased level of basic fibroblast growth factor (bFGF) which has an important role

G. Filardo et al. (eds.), *Orthobiologics*, https://doi.org/10.1007/978-3-030-84744-9_20

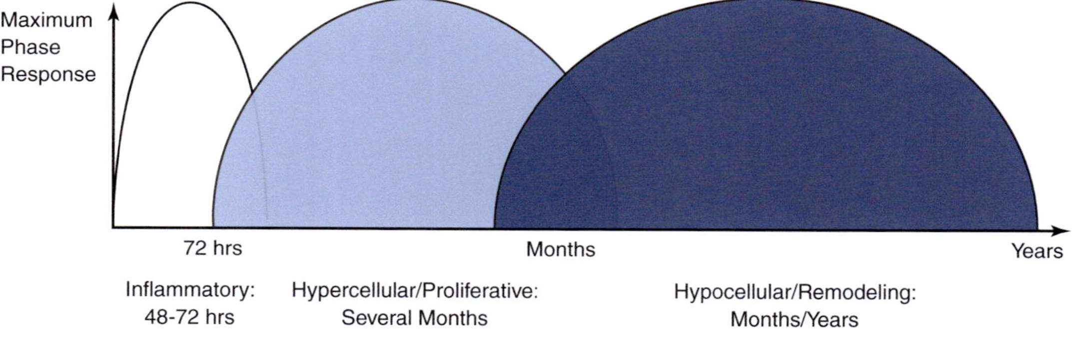

Fig. 20.1 The three phases of ligament healing

in inhibiting angiogenesis [3]. By the conclusion of this phase of the inflammatory process, several months after the inciting event, rat medial collateral ligaments (MCLs) have been shown to regain approximately 80% of their strength [4].

The third and final stage of healing is the *hypocellular/remodeling phase*. This stage starts several weeks after the inciting injury and can continue for months to years. During this phase, infiltration of the acute inflammatory products including fibroblasts, inflammatory cells, and endothelial cells diminish to pre-injury levels. Furthermore, collagen fibers and ligament matrix components undergo nearly continuous remodeling to promote strong ligamentous growth. This phase is primarily marked by collagen maturation and strengthening of the injured ligament [1]. In the end, ligaments heal with fibrovascular scar, which possess inferior biomechanical and mechanical properties compared to the native structure [5].

20.1.2 Relevant Osseous and Soft Tissue Anatomy

The ACL is an intra-articular, extra-synovial ligament consisting of two separate bundles that primarily serve to resist anterior translation of the tibia relative to the femur. The anteromedial and posterolateral bundles attach proximally at the posterior aspect of the medial wall of the lateral femoral condyle, just lateral to the intercondylar ridge, and distally to the tibia in between the medial and lateral tibial eminences (Figs. 20.2 and 20.3) [6]. The ACL is made primarily of type

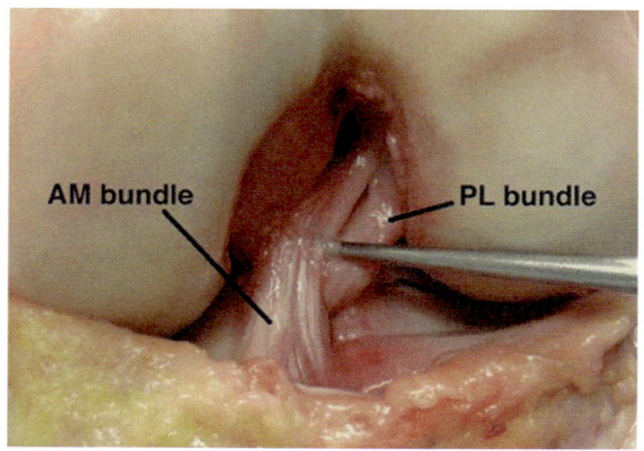

Fig. 20.2 Anatomical dissection from an anterior to posterior perspective. This dissection shows the intimate, yet distinct relationship of the two bundles of the anterior cruciate ligament. *AM* Anteromedial bundle, *PL* Posterolateral bundle

I collagen and has an elastic characteristic which helps maintain its stability with sudden movements and prolonged strain [7]. It has a characteristically poor blood supply. It is coated in synovial tissue and surrounded by synovial fluid which likely contribute to the intrinsic healing capacity of this ligament [8].

The bony anatomy of the medial aspect of the knee is formed by the medial femoral condyle and the medial tibial plateau which articulate in a convex-on-concave fashion, respectively [9], whereas the opposing bony surfaces of the lateral aspect of the knee articulate in a convex-on-convex manner, creating an inherently unstable region of the knee (Fig. 20.4) [10, 11]. Animal model studies have demonstrated that injuries to the lateral aspect of the knee heal poorly due to this inherent bony instability, leaving the knee

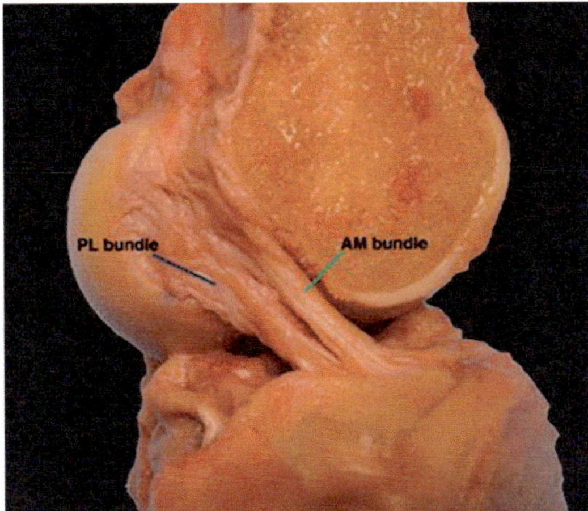

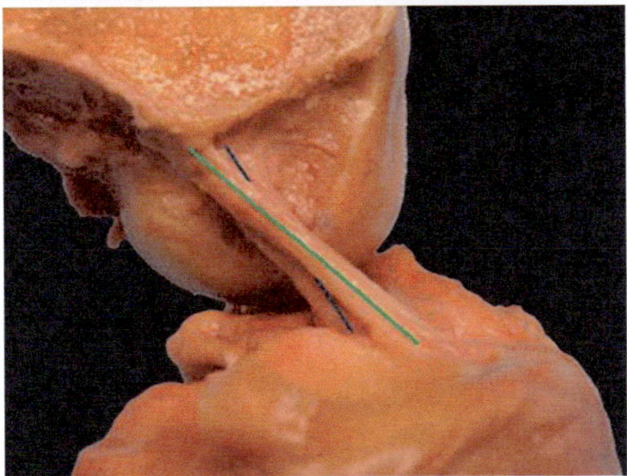

Fig. 20.3 Anatomical dissection of the ACL from a medial to lateral perspective of the knee in full extension (top) and approximately 90 degrees of flexion (bottom). The green line corresponds to the anteromedial (AM) bundle, and the blue line corresponds to the posterolateral (PL) bundle

susceptible to medial compartment osteoarthritis, medial meniscus injury, instability, and lateral compartment gapping [10–14].

20.1.3 Differences in Healing Between the Collateral Ligaments and Cruciate Ligaments

Comparisons between ACL and MCL have revealed unique properties of the respective ligaments which may contribute to the differences in their intrinsic healing capacity. Zhang et al. [15] demonstrated that the MCL contained a cell population with elevated natural levels of putative stem cell markers STRO-1 and OCT-4. Further, the MCL demonstrated a higher prevalence of colony founding connective tissue progenitor cells, which demonstrated higher proliferation rate when cultured in vitro and a greater potential for multi-lineage differentiation than the ACL [15]. This difference in the native intrinsic population of colony founding CTPs may be related to the difference in healing capacity between the ACL and MCL in both partial and complete tears [8, 16]. Another potential explanation for the differences in repair potential is related to the differences in mechanical stabilization and the microenvironment surrounding each ligament. While the MCL is found extra-articularly, the ACL is surrounded by synovial fluid [2, 8, 16]. The synovial fluid prevents clot formation at the injured ACL, which limits the body's ability to form a provisional scaffold to initiate self-repair, and restricts the release of crucial growth factors [16–19].

20.2 Types of Cellular Therapies

Mesenchymal stem cells (MSCs) are defined by the International Society for Cellular Therapy as culture-expanded cells that exhibit plastic adherence and possess specific cell surface markers, i.e., cluster of differentiation (CD), CD73, CD90, and CD105, and lack the expression of CD14, CD34, CD45, and human leukocyte antigen-DR. Additionally, these cells must have the ability to differentiate into adipocytes, chondrocytes, and osteoblasts [20]. The original MSCs were derived from bone marrow, but they have since been derived from many other types of tissues including adipose, skin, and synovial fluid, among many others [21–23]. The potential benefits of culture-expanded MSCs can be grouped into the following general categories: immunomodulation, anti-apoptosis, angiogenesis, support of stem/progenitor cells, anti-scarring, and chemoattraction [24]. Specifically, for use in orthopedics, clinicians take advantage of the immunomodulatory paracrine effects of MSCs to limit the inflammatory microenvironment in acute ligamentous injury [25]. The most commonly used MSCs come from bone marrow or

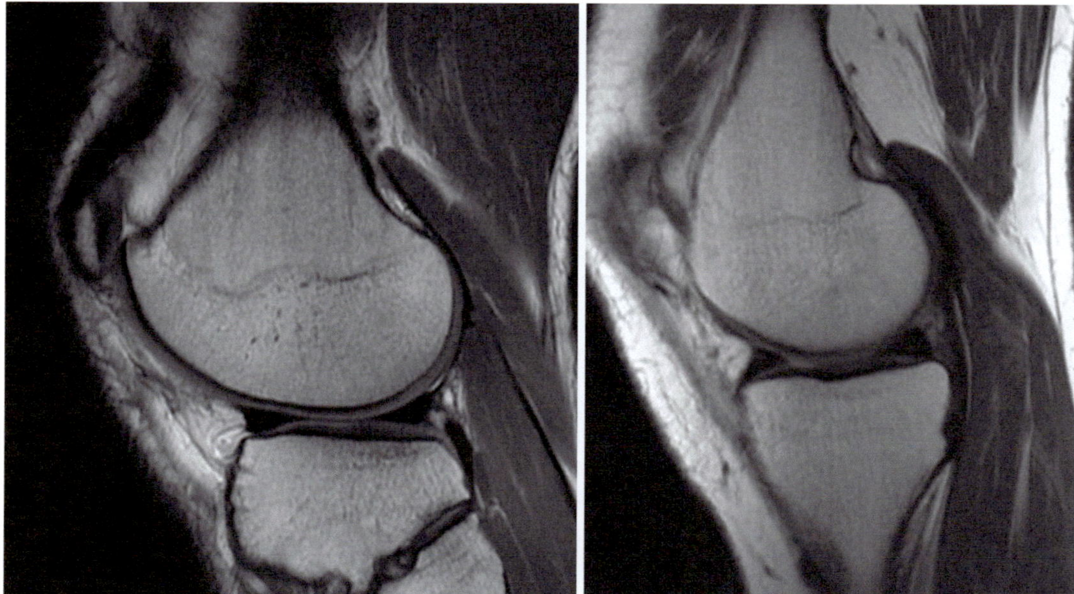

Fig. 20.4 Magnetic resonance image of the lateral knee (left) and medial knee (right). This image demonstrates the inherently unstable convex-on-convex articulation of the femur and tibia at the lateral aspect of the knee (left) and the more stable convex-on-concave articulation of the joint at the medial side of the knee (right)

from adipose tissue, and the iliac crest, metaphysis of the distal femur, proximal humerus, and proximal tibia are the most common sites to harvest bone marrow-derived MSCs [26].

It should be noted that cell populations that are present in native tissues or cells that can be extracted from native tissues are heterogenous. Native tissues contain a predominance of mature differentiated cells, along with a mixed population of tissue-specific CTPs, endothelial progenitors (EPCs), and, in the case of bone marrow, hematopoietic stem cells and progenitors (HSCs). True stem or progenitor cells are expected to be the least common cells among those extracted from native tissues. Moreover, mixtures of freshly isolated progenitors exhibit large variation in biological attributes and potential. As a result, culture-expanded MSC populations and laboratory purified cell lines, which can be fabricated in vastly greater cell concentration, are felt to offer a greater degree of potential potency [27].

Induced pluripotent stem cells (iPSCs) are reprogrammed, or dedifferentiated, cells that have pluripotent differentiation capacity. Essentially, these iPSC populations, once isolated, allow scientists to reprogram mature cells into any cell type other than trophoblast (i.e., placenta, amnion, and chorion). While the research regarding iPSCs and their use in osteogenesis, chondrogenesis, and myogenesis is more robust, research regarding tenogenesis and ligament regeneration are in the very early, preclinical stages [28]. iPS-derived cell populations are not currently available for clinical use in any musculoskeletal application.

Mishra et al. provided preclinical evidence that suggested that there is a synergistic relationship between MSCs and platelet-rich plasma (PRP) therapy [29]. This study confirmed that PRP-enhanced chondrogenic differentiation of MSCs in vitro. This relationship shows promise with respect to clinical outcomes in the treatment of ligament lesions.

20.3 Outcomes and Reasons for Cell Therapy

Historically, direct ACL repair using native tissues alone has been associated with sub-optimal clinical outcomes, with more than half of patients suffering re-tears of the repair by 5 years postoperation [30]. It has been theorized that this is because of the poor intrinsic healing capacity of the ACL which has been linked to the decreased amount of local growth factors, cellularity, and expression of molecules compared to other ligaments and tendons [19, 31–35]. Additional studies have suggested that the poor success following ACL repair is also correlated with the unfavorable environmental factors including the intrinsic progenitor cell populations, mechanical environment, inflammatory condition, blood supply, and nutrient and growth factor supply [8, 18, 36]. ACL reconstruction techniques have been more successful than primary repairs with success rates approaching 80–90% [37]. However, there is still room for improvement, including faster recovery and even higher success rates.

PCL tears are associated with meniscus tears, osteoarthritis, and subsequent total knee replacement [38]. Historically, PCL injuries have been managed nonoperatively. Shelbourne et al. reported on 68 patients with acute PCL injuries that were managed conservatively for an average of 17.6 years [39]. They reported an average IKDC of 73.4 and reported moderate to severe osteoarthritis in 11% of patients [39]. Surgical reconstruction techniques of the PCL are separated into single and double bundle reconstructions. Randomized controlled trials have shown that single bundle reconstructions have a greater amount of posterior tibial translation at 2 years postoperatively [40, 41]. Additionally, LaPrade et al. [42] concluded that double bundle PCL reconstructions had a failure rate of 1% and mean side-to-side difference of only 1.6 mm at average of 3 years postoperation [42].

Studies have reported that after healing of a torn MCL, the remodeled MCL is ultimately weaker and less stiff than the native MCL [43–46]. Furthermore, injured MCLs in the remodeling phase of healing are less elastic and are slower to regain their native length after being stretched compared to an uninjured MCL [46]. Increased ligamentous laxity leaves the ligament at an increased risk of re-tear and the knee at an increased risk of mechanical instability [46, 47]. However, outcomes after conservative management of grade III MCL tears have been relatively encouraging, with an average return to competitive contact sports at 9.2 weeks [48]. Surgical treatment of these injuries using non-anatomic reconstruction of the superficial MCL and posterior oblique ligament have been shown to subjectively reduce medial compartment laxity to <5 mm and have yielded IKDC subjective scores of A or B in 74% of patients [49]. Additionally, LaPrade and Wijdicks [50] reported on 28 patients who had single-stage anatomic reconstruction of the sMCL and posterior oblique ligament, with concurrent cruciate ligament reconstruction. Patients in this cohort reported improved subjective IKDC scores, and all patients demonstrated resolution of side-to-side medial instability at 2-year follow-up. On valgus stress radiographs there was improvement from 6.3 to 1.3 mm in side-to-side medial compartment opening [50].

With respect to lateral-sided knee injuries, there is a relative paucity in the literature regarding conservative management. Of the few studies that exist, conservative management has demonstrated favorable results with minimal radiographic changes and minimal symptomatology at 8-year follow-up in grade I and grade II posterolateral corner (PLC) injuries [51, 52]. However, in patients with acute grade III FCL injuries without clinical evidence of cruciate ligament damage, there was a high incidence (50%) of posttraumatic osteoarthritis and relatively poor outcome scores (Lysholm: 55, Marshall/HSS: 28) at 8-year follow-up [51]. Two studies have demonstrated positive outcomes for isolated FCL tears diagnosed on magnetic resonance imaging. A case series reported 4/5 patients were able to return to professional football in under 2 weeks after injury, and a case report described return to competitive sport at 15 weeks in one patient [53]. Of note, neither of these studies utilized stress radiographs to aid in diagnosis, and neither conducted follow-up imaging of any kind to confirm FCL healing. Furthermore, operative management using anatomic-based reconstruction of isolated grade III FCL injuries has provided encouraging outcomes (Lysholm: 84, WOMAC: 8) [54, 55].

20.3.1 Cell Therapies for Anterior Cruciate Ligament

The use of cell therapy agents in the setting of knee ligament injury has yielded promising results in preclinical in vitro models, but there is a lack of clinical studies. Moreover, the understanding of the specific mechanisms of action, dosing recommendations, and reproducible preparation of biologic agents are currently lacking in the orthopedic literature.

20.3.1.1 Preclinical Evidence

There have been several preclinical studies that have examined the effects of culture-expanded cells (embryonic, iPSC, and MSC) on ACL healing in animal models. Collectively, these studies have shown enhanced ligament recovery at 8–12 weeks following injury [56–67]. These improvements were defined by histological scores and biomechanical performance of the injured ligaments [56–67].

Tissue-engineered ACLs seeded with culture-expanded MSCs have demonstrated increased production of collagen [68, 69]. Furthermore, a study demonstrated that culture-expanded porcine adipose-derived MSCs cells were able to stimulate procollagen and ACL-fibroblast proliferation [70]. To date, experiments involving the induction of human adipose-derived stem cells into ligaments have not been successful, and in one experiment, injection of adipose-derived stem cells inhibited chondrogenesis in ligaments [71–73].

Additionally, one study compared MSCs generated from old and young donors during ACL reconstruction or arthroplasty [74]. They concluded that the regenerative potential of ACL MSCs was not significantly different based upon age and that cells from both sources had adequate and equal regenerative potential [74].

Lange-Consiglio et al. [75] compared the use of amniotic membrane MSCs to bone marrow-derived MSCs in a horse model with 95 horses with injury to either the suspensory ligament or the superficial digital flexor tendon. The horses that received the amniotic derived stem cells had a lower reinjury rate at 2 years (4% vs. 23.1%) and were able to return to activity at earlier times [75].

20.3.1.2 Clinical Evidence

One clinical case series considered ten patients that underwent percutaneous injection of autologous bone marrow-derived nucleated cells in both partial and complete ACL tears [76]. Approximately 10–15 mL of bone marrow aspirate was obtained from the patients' iliac crest. Ultimately the injectate consisted of 2–3 mL of bone marrow aspirate concentrate, PRP, and platelet lysate. The average nucleated cell count of the included patients was 694 million cells (range, 376–123 million cells). This study reported improved imaging metrics at 3.7 months after injury in grade 1 and 2 grade tears. All patients in this study reported improvements in functionality scores, while there was not a significant improvement in pain [76].

To date, one randomized controlled clinical trial has evaluated the use of adult marrow-derived cells in ACL injuries [77]. In this trial on ACL reconstructions, Silva et al. [77] collected 30 mL of bone marrow harvested from the anterior iliac crest into the femoral end of the graft and around the femoral tunnel during the initial reconstruction. The harvested bone marrow was centrifuged for 15 min at 3200 RPM, and 3 mL of concentrate was obtained; 1.5 mL was injected inside the femoral end of the graft, and the other 1.5 mL was placed in the tunnel around the graft. According the manufacturer of the equipment utilized in this study, this concentrating technique produces a concentrate with six times the number of stem cells compared to baseline. This study found no significant difference in graft to bone healing assessed on MRI at 3 months after the reconstruction.

Similarly, one prospective randomized trial has examined the effects of a single injection of culture-expanded allogenic MSCs during ACL reconstruction with 17 patients. The study group received 75 million allogenic MSC suspended in hyaluronan, while the control received only hyaluronan. This study concluded that injection of MSCs was safe with minimal side effects and that they may improve symptoms and structural outcomes [78]. They reported that the study group has significant improvements in KOOS pain, activities of daily living, and SF-36 bodily pain scores. The study also found that the group that received culture-expanded allogenic MSCs had reduced medial and lateral tibiofemoral joint space narrowing, less bone expansion, and a trend toward reduced tibial cartilage volume loss compared to the hyaluronan control group [78].

Finally, a recent surgical technique has described infusion of an autogenous bone marrow aspirate concentrate (BMAC) within allograft ACL reconstructions [79]. While outcomes for this technique are yet to be described, surgeons have theorized that infusion of BMAC, either from the intercondylar notch or the iliac crest, may improve the microenvironment for healing following surgical reconstruction.

20.3.2 Cell Therapies for Collateral Ligaments

20.3.2.1 Preclinical Evidence

In a preclinical trial, culture-expanded MSCs were introduced locally in the setting of grade III MCL tears in an animal model. The study found that this led to a significant increase in type II macrophages, increased early endothelization (day 5), and procollagen 1α matrix deposition, which collectively resulted in a limited inflammatory microenvironment post injury [25]. Both low concentrations (1×10^6) and high (4×10^6) concentrations of MSCs have been shown to provide beneficial anti-inflammatory effects on ligament healing at 14 days following introduction [80]. However, the samples from the rats treated with the lower concentration had decreased M1 macrophages and decreased numbers of pro-inflammatory cytokines compared to the samples treated with the higher dose. Moreover, the high dose was shown to cause decreased early inflammation after 5 days but was less advantageous than the lesser dose of MSCs at 14 days [80]. At 14 days, lower doses demonstrated the most significant improvement in failure load and stiffness [80].

Jiang et al. [81] reported that the addition of MCL-derived stromal cells and umbilical-cord-blood-derived CD34+ cells to rat MCL tears leads to improved healing, including diminished swelling, improved tensile strength at 2 and 4 weeks, increased type I collagen deposition, and increased local angiogenesis [81]. An additional preclinical study determined that CD34+ promoted local vasculogenesis and was ultimately associated with improved histological appearance of the MCL at 2 and 4 weeks post injury [82].

Nishimori et al. [83] found that murine-derived stromal cells transduced to overexpress the *VEGF* gene in the setting of MCL rupture led to increased capillary density, without improvement in biomechanical properties. Interestingly, this study also demonstrated that inhibition of angiogenesis yielded significantly decreased biomechanical properties in MCL injuries [83].

20.3.2.2 Clinical Evidence

Currently, there are no known clinical studies which report the efficacy of cell therapies in collateral ligament injuries.

20.3.3 Cell Therapies for Ulnar Collateral Ligament

A 2015 case report described a 25-year-old professional pitcher with ulnar collateral ligament (UCL) instability and ulnar neuritis who underwent UCL reconstruction with ulnar nerve decompression. The procedure utilized a dermal allograft, PRP, in addition to one milliliter of MSCs (Ovation, Osiris Therapeutics,

Columbia, MD). By 4 months postoperation he returned to a throwing program and did not report complications by 21 months [84].

20.4 Current Indications and Contraindications

There are currently no scientifically validated indications for the use of cell therapy in ligament pathologies. However, recent advancements in preclinical and clinical research have been encouraging, and it seems likely that the use of cellular therapy may be scientifically supported in the future. Furthermore, well-designed clinical trials are needed to determine the efficacy and potential side effects associated with cell therapy in the clinical setting [85]. There are currently no known absolute contraindications to cell therapy in orthopedics.

20.5 Conclusions

The clinical use of cellular therapies to treat orthopedic pathologies has outpaced the relatively limited clinical evidence. Specifically, there is no compelling evidence as yet in the orthopedic literature supporting the use of cell therapies for ligament injuries.

A majority of the existing studies on this topic describes preclinical trials considering the use of cell therapies in MCL tears, with only a few clinical case series or case reports (level V evidence) that describe the use of BMAC or culture-expanded MSCs in human ACL or UCL injuries. Further, at this time there are no studies, clinical or preclinical, which have examined the effects of cell therapies on PCL or FCL injuries.

Considering this lack of information, the current authors recommend caution with the use of cell therapies in the setting of ligament injuries until further clinical evidence is provided. We urge further preclinical and clinical research to investigate the utility of cellular therapies in acute ligament injury and reconstructive procedures too. Promising science should not be blindly implemented in the clinic setting without clinical trials demonstrating successful, reproducible outcomes.

Take-Home Messages
- Given the relative paucity of literature on the utility of cell therapies with respect to ligament injuries, further research is needed to examine the clinical outcomes using these agents prior to widespread implementation.
- Preclinical and anecdotal studies have reported encouraging results with the use of culture-expanded and bone marrow-derived nucleated cells, respectively, in the setting of ACL injuries.
- Cell therapies have yielded improved early tensile strength, increased type I collagen deposition, and increased local angiogenesis in MCL injuries using animal models. There have been no clinical studies on cell therapies in collateral ligament injuries to date.

Conflicts of Interest Robert S. Dean declares that he has no conflict of interest.

Nicholas N. DePhillipo declares that he has no conflict of interest.

Robert F. LaPrade reports grants and personal fees from Arthrex, Inc, grants from Linvatec, grants and personal fees from Ossur, grants and personal fees from Smith & Nephew, outside the submitted work.

References

1. Hauser RA, Dolan EE, Phillips HJ, Newlin AC, Moore RE, Woldin BA. Ligament injury and healing: a review of current clinical diagnostics and therapeutics. Open Rehabil J. 2013;6:1–20.
2. Nguyen DT, Ramwadhdoebe TH, Van Der Hart CP, Blankevoort L, Tak PP, Niek Van Dijk C. Intrinsic healing response of the human anterior cruciate ligament: an histological study of reattached ACL remnants. J Orthop Res. 2014;32:296–301.
3. Sciore P, Boykiw R, Hart DA. Semiquantitative reverse transcription-polymerase chain reaction analysis of mRNA for growth factors and growth factor receptors from normal and healing rabbit medial collateral ligament tissue. J Orthop Res. 1998;16: 429–37.

4. Chamberlain CS, Crowley E, Vanderby R. The spatio-temporal dynamics of ligament healing. Wound Repair Regen. 2009;17:206–15.
5. Maniar HH, Tawari AA, Suk M, Horwitz DS. The current role of stem cells in orthopaedic surgery. Malaysian Orthop J. 2015;9:1–7.
6. Ziegler CG, Pietrini SD, Westerhaus BD, Anderson CJ, Wijdicks CA, Johansen S, et al. Arthroscopically pertinent landmarks for tunnel positioning in single-bundle and double-bundle anterior cruciate ligament reconstructions. Am J Sports Med. 2011;39:743–52.
7. Dallo I, Chahla J, Mitchell JJ, Pascual-Garrido C, Feagin JA, LaPrade RF. Biologic approaches for the treatment of partial tears of the anterior cruciate ligament: a current concepts review. Orthop J Sports Med. 2017;5.
8. Bray RC, Leonard CA, Salo PT. Vascular physiology and long-term healing of partial ligament tears. J Orthop Res. 2002;20:984–9.
9. LaPrade RF, Engebretsen AH, Ly TV, Johansen S, Wentorf FA, Engebretsen L. The anatomy of the medial part of the knee. J Bone Jt Surg. 2007;89:2000.
10. James EW, Laprade CM, Laprade RF. Anatomy and biomechanics of the lateral side of the knee and surgical implications. Sports Med Arthrosc Rev. 2015;23:2–9.
11. LaPrade RF, Wentorf FA, Olson EJ, Carlson CS. An in vivo injury model of posterolateral knee instability. Am J Sports Med. 2006;34:1313–21.
12. Griffith CJ, Wijdicks CA, Goerke U, Michaeli S, Ellermann J, LaPrade RF. Outcomes of untreated posterolateral knee injuries: an in vivo canine model. Knee Surg Sport Traumatol Arthrosc. 2011;19:1192–7.
13. LaPrade RF, Wentorf FA, Crum JA. Assessment of healing of grade III posterolateral corner injuries: an in vivo model. J Orthop Res. 2004;22:970–5.
14. LaPrade RF, Ly TV, Wentorf FA, Engebretsen L. The posterolateral attachments of the knee. Am J Sports Med. 2003;31:854–60.
15. Zhang J, Pan T, Im H-J, Fu FH, Wang JH. Differential properties of human ACL and MCL stem cells may be responsible for their differential healing capacity. BMC Med. 2011;9:68.
16. Murray MM, Fleming BC. Biology of anterior cruciate ligament injury and repair: kappa delta ann doner vaughn award paper 2013. J Orthop Res. 2013;31:1501–6.
17. Murray MM, Spector M. The migration of cells from the ruptured human anterior cruciate ligament into collagen-glycosaminoglycan regeneration templates in vitro. Biomaterials. 2001;22:2393–402.
18. Murray MM, Spindler KP, Ballard P, Welch TP, Zurakowski D, Nanney LB. Enhanced histologic repair in a central wound in the anterior cruciate ligament with a collagen–platelet-rich plasma scaffold. J Orthop Res. 2007;25:1007–17.
19. Murray MM, Martin SD, Martin TL, Spector M. Histological changes in the human anterior cruciate ligament after rupture. J Bone Jt Surg Am. 2000;82:1387–97.
20. Dominici M, Le Blanc K, Mueller I, Slaper-Cortenbach I, Marini FC, Krause DS, et al. Minimal criteria for defining multipotent mesenchymal stromal cells. The International Society for Cellular Therapy position statement. Cytotherapy. 2006;8:315–7.
21. Riekstina U, Muceniece R, Cakstina I, Muiznieks I, Ancans J. Characterization of human skin-derived mesenchymal stem cell proliferation rate in different growth conditions. Cytotechnology. 2008;58:153–62.
22. Wagner W, Wein F, Seckinger A, Frankhauser M, Wirkner U, Krause U, et al. Comparative characteristics of mesenchymal stem cells from human bone marrow, adipose tissue, and umbilical cord blood. Exp Hematol. 2005;33:1402–16.
23. Zhang X, Yang M, Lin L, Chen P, Ma KT, Zhou CY, et al. Runx2 overexpression enhances osteoblastic differentiation and mineralization in adipose - Derived stem cells in vitro and in vivo. Calcif Tissue Int. 2006;79:169–78.
24. da Silva ML, Fontes AM, Covas DT, Caplan AI. Mechanisms involved in the therapeutic properties of mesenchymal stem cells. Cytokine Growth Factor Rev. 2009;20:419–27.
25. Saether EE, Chamberlain CS, Aktas E, Leiferman EM, Brickson SL, Vanderby R. Primed mesenchymal stem cells alter and improve rat medial collateral ligament healing. Stem Cell Rev Reports. 2016;12:42–53.
26. Narbona-Carceles J, Vaquero J, Suárez-Sancho SBS, Forriol F, Fernández-Santos ME. Bone marrow mesenchymal stem cell aspirates from alternative sources: is the knee as good as the iliac crest? Injury. 2014;45:S42–7.
27. Zakrzewski W, Dobrzyński M, Szymonowicz M, Rybak Z. Stem cells: past, present, and future. Stem Cell Res Ther. 2019;10:68.
28. Li WJ, Jiao H, Walczak BE. Emerging opportunities for induced pluripotent stem cells in orthopaedics. J Orthop Transl. 2019;17:73–81.
29. Mishra A, Tummala P, King A, Lee B, Kraus M, Tse V, Jacobs CR. Buffered platelet-rich plasma enhances mesenchymal stem cell proliferation and chondrogenic differentiation. Tissue Eng Part C Methods. 2009;15(3):431–5.
30. Feagin JA, Curl WW. Isolated tear of the anterior cruciate ligament: 5-year follow-up study. Am J Sports Med. 1976;4:95–100.
31. Geiger MH, Green MH, Monosov A, Akeson WH, Amiel D. An in vitro assay of anterior cruciate ligament (ACL) and medial collateral ligament (MCL) cell migration. Connect Tissue Res. 1994;30:215–24.
32. Cooper JA, Bailey LAO, Carter JN, Castiglioni CE, Kofron MD, Ko FK, et al. Evaluation of the anterior cruciate ligament, medial collateral ligament, achilles tendon and patellar tendon as cell sources for tissue-engineered ligament. Biomaterials. 2006;27:2747–54.
33. Amiel D, Nagineni CN, Choi SH, Lee J. Intrinsic properties of acl and mcl cells and their responses to growth factors. Med Sci Sports Exerc. 1995;27:844–51.
34. Spindler KP, Clark SW, Nanney LB, Davidson JM. Expression of collagen and matrix

metalloproteinases in ruptured human anterior cruciate ligament: an in situ hybridization study. J Orthop Res. 1996;14:857–61.

35. Nagineni CN, Amiel D, Green MH, Berchuck M, Akeson WH. Characterization of the intrinsic properties of the anterior cruciate and medial collateral ligament cells: an in vitro cell culture study. J Orthop Res. 1992;10:465–75.

36. Frank C, Amiel D, Woo SLY, Akeson W. Normal ligament properties and ligament healing. Clin Orthop Relat Res. 1985;196:15–25.

37. Samitier G, Marcano AI, Alentorn-Geli E, Cugat R, Farmer KW, Moser MW. Failure of anterior cruciate ligament reconstruction. Arch Bone Jt Surg. 2015;3:220–40.

38. Wang S-H, Chien W-C, Chung C-H, Wang Y-C, Lin L-C, Pan R-Y. Long-term results of posterior cruciate ligament tear with or without reconstruction: a nationwide, population-based cohort study. PLoS One. 2018;13:e0205118.

39. Shelbourne KD, Clark M, Gray T. Minimum 10-year follow-up of patients after an acute, isolated posterior cruciate ligament injury treated nonoperatively. Am J Sports Med. 2013;41:1526–33.

40. Yoon KH, Bae DK, Song SJ, Cho HJ, Lee JH. A prospective randomized study comparing arthroscopic single-bundle and double-bundle posterior cruciate ligament reconstructions preserving remnant fibers. Am J Sports Med. 2011;39:474–80.

41. Li Y, Li J, Wang J, Gao S, Zhang Y. Comparison of single-bundle and double-bundle isolated posterior cruciate ligament reconstruction with allograft: a prospective, randomized study. Art Ther. 2014;30:695–700.

42. LaPrade RF, Cinque ME, Dornan GJ, DePhillipo NN, Geeslin AG, Moatshe G, et al. Double-bundle posterior cruciate ligament reconstruction in 100 patients at a mean 3 years' follow-up: outcomes were comparable to anterior cruciate ligament reconstructions. Am J Sports Med. 2018;46:1809–18.

43. Hanhan S, Ejzenberg A, Goren K, Saba F, Suki Y, Sharon S, et al. Skeletal ligament healing using the recombinant human amelogenin protein. J Cell Mol Med. 2016;20:815–24.

44. Niyibizi C, Kavalkovich K, Yamaji T, Woo SL-Y. Type V collagen is increased during rabbit medial collateral ligament healing. Knee Surg Sport Traumatol Arthrosc. 2000;8:281–5.

45. Woo SL, Gomez MA, Sites TJ, Newton PO, Orlando CA, Akeson WH. The biomechanical and morphological changes in the medial collateral ligament of the rabbit after immobilization and remobilization. J Bone Joint Surg Am. 1987;69:1200–11.

46. Thornton GM, Leask GP, Shrive NG, Frank CB. Early medial collateral ligament scars have inferior creep behaviour. J Orthop Res. 2000;18:238–46.

47. Plaas AHK, Wong-Palms S, Koob T, Hernandez D, Marchuk L, Frank CB. Proteoglycan metabolism during repair of the ruptured medial collateral ligament in skeletally mature rabbits. Arch Biochem Biophys. 2000;374:35–41.

48. Indelicato PA, Hermansdorfer J, Huegel M. Nonoperative management of complete tears of the medial collateral ligament of the knee in intercollegiate football players. Clin Orthop Relat Res. 1990:174–7.

49. Lind M, Jakobsen BW, Lund B, Hansen MS, Abdallah O, Christiansen SE. Anatomical reconstruction of the medial collateral ligament and posteromedial corner of the knee in patients with chronic medial collateral ligament instability. Am J Sports Med. 2009;37:1116–22.

50. Laprade RF, Wijdicks CA. The management of injuries to the medial side of the knee. J Orthop Sport Phys Ther. 2012;42:221–33.

51. Kannus P. Nonoperative treatment of Grade II and III sprains of the lateral ligament compartment of the knee. Am J Sports Med. 1989;17:83–8.

52. Krukhaug Y, Mølster A, Rodt A, Strand T. Lateral ligament injuries of the knee. Knee Surg Sport Traumatol Arthrosc. 1998;6:21–5.

53. Haddad MA, Budich JM, Eckenrode BJ. Conservative management of an isolated grade iii lateral collateral ligament injury in an adolescent multi-sport athlete: a case report. Int J Sports Phys Ther. 2016;11: 596–606.

54. LaPrade RF, Spiridonov SI, Coobs BR, Ruckert PR, Griffith CJ. Fibular collateral ligament anatomical reconstructions. Am J Sports Med. 2010;38:2005–11.

55. Moulton SG, Matheny LM, James EW, LaPrade RF. Outcomes following anatomic fibular (lateral) collateral ligament reconstruction. Knee Surg Sport Traumatol Arthrosc. 2015;23:2960–6.

56. Jang KM, Lim HC, Jung WY, Moon SW, Wang JH. Efficacy and safety of human umbilical cord blood-derived mesenchymal stem cells in anterior cruciate ligament reconstruction of a rabbit model: new strategy to enhance tendon graft healing. Art Ther. 2015;31:1530–9.

57. Lui PPY, Wong OT, Lee YW. Application of tendon-derived stem cell sheet for the promotion of graft healing in anterior cruciate ligament reconstruction. Am J Sports Med. 2014;42:681–9.

58. Lim JK. Enhancement of tendon graft osteointegration using mesenchymal stem cells in a rabbit model of anterior cruciate ligament reconstruction. Art Ther. 2004;20:899–910.

59. Guo R, Gao L, Xu B. Current evidence of adult stem cells to enhance anterior cruciate ligament treatment: a systematic review of animal trials. Art Ther. 2018;34:331–340.e2.

60. Takayama K, Kawakami Y, Mifune Y, Matsumoto T, Tang Y, Cummins JH, et al. The effect of blocking angiogenesis on anterior cruciate ligament healing following stem cell transplantation. Biomaterials. 2015;60:9–19.

61. Mifune Y, Matsumoto T, Ota S, Nishimori M, Usas A, Kopf S, et al. Therapeutic potential of anterior cruciate ligament-derived stem cells for anterior cruciate ligament reconstruction. Cell Transplant. 2012;21:1651–65.
62. Matsumoto T, Kubo S, Sasaki K, Kawakami Y, Oka S, Sasaki H, et al. Acceleration of tendon-bone healing of anterior cruciate ligament graft using autologous ruptured tissue. Am J Sports Med. 2012;40:1296–302.
63. Oe K, Kushida T, Okamoto N, Umeda M, Nakamura T, Ikehara S, et al. New strategies for anterior cruciate ligament partial rupture using bone marrow transplantation in rats. Stem Cells Dev. 2011;20:671–9.
64. Kanaya A, Deie M, Adachi N, Nishimori M, Yanada S, Ochi M. Intra-articular injection of mesenchymal stromal cells in partially torn anterior cruciate ligaments in a rat model. Arthroscopy. 2007;23:610–7.
65. Kanazawa T, Soejima T, Noguchi K, Tabuchi K, Noyama M, Nakamura KI, et al. Tendon-to-bone healing using autologous bone marrow-derived mesenchymal stem cells in ACL reconstruction without a tibial bone tunnel—a histological study. Muscles Ligaments Tendons J. 2014;4:201–6.
66. Chen CH, Whu SW, Chang CH, Su CI. Gene and protein expressions of bone marrow mesenchymal stem cells in a bone tunnel for tendon-bone healing. Formos J Musculoskelet Disord. 2011;2:85–93.
67. Soon MYH, Hassan A, Hui JHP, Goh JCH, Lee EH. An analysis of soft tissue allograft anterior cruciate ligament reconstruction in a rabbit model: a short-term study of the use of mesenchymal stem cells to enhance tendon osteointegration. Am J Sports Med. 2007;35:962–71.
68. Van Eijk F, Saris DBF, Riesle J, Willems WJ, Van Blitterswijk CA, Verbout AJ, et al. Tissue engineering of ligaments: a comparison of bone marrow stromal cells, anterior cruciate ligament, and skin fibroblasts as cell source. Tissue Eng. 2004;10:893–903, Ann Liebert, Inc.
69. Ge Z, Goh JCH, Lee EH. The effects of bone marrow-derived mesenchymal stem cells and fascia wrap application to anterior cruciate ligament tissue engineering. Cell Transplant. 2005;14:763–73.
70. Proffen BL, Haslauer CM, Harris CE, Murray MM. Mesenchymal stem cells from the retropatellar fat pad and peripheral blood stimulate ACL fibroblast migration, proliferation, and collagen gene expression. Connect Tissue Res. 2013;54:14–21.
71. Eagan MJ, Zuk PA, Zhao KW, Bluth BE, Brinkmann EJ, Wu BM, et al. The suitability of human adipose-derived stem cells for the engineering of ligament tissue. J Tissue Eng Regen Med. 2012;6:702–9.
72. Su Y, Denbeigh JM, Camilleri ET, Riester SM, Parry JA, Wagner ER, et al. Extracellular matrix protein production in human adipose-derived mesenchymal stem cells on three-dimensional polycaprolactone (PCL) scaffolds responds to GDF5 or FGF2. Gene Reports. 2018;10:149–56.
73. Ter Huurne M, Schelbergen R, Blattes R, Blom A, De Munter W, Grevers LC, et al. Antiinflammatory and chondroprotective effects of intraarticular injection of adipose-derived stem cells in experimental osteoarthritis. Arthritis Rheum. 2012;64:3604–13.
74. Prager P, Kunz M, Ebert R, Klein-Hitpass L, Sieker J, Barthel T, et al. Mesenchymal stem cells isolated from the anterior cruciate ligament: characterization and comparison of cells from young and old donors. Knee Surg Relat Res. 2018;30:193–205.
75. Lange-Consiglio A, Tassan S, Corradetti B. Investigating the efficacy of amnion-derived compared with bone marrow-derived mesenchymal stromal cells in equine tendon and ligament injuries. Cytotherapy. 2013;15(8):1011–20.
76. Centeno CJ, Pitts J, Al-Sayegh H, Freeman MD. Anterior cruciate ligament tears treated with percutaneous injection of autologous bone marrow nucleated cells: a case series. J Pain Res. 2015;8: 437–47.
77. Silva A, Sampaio R, Fernandes R, Pinto E. Is there a role for adult non-cultivated bone marrow stem cells in ACL reconstruction? Knee Surg Sports Traumatol Arthrosc. 2014;22:66–71.
78. Wang Y, Shimmin A, Ghosh P, Marks P, Linklater J, Connell D, et al. Safety, tolerability, clinical, and joint structural outcomes of a single intra-articular injection of allogeneic mesenchymal precursor cells in patients following anterior cruciate ligament reconstruction: a controlled double-blind randomised trial. Arthritis Res Ther. 2017;19:180.
79. Youn GM, Remigio Van Gogh AM, Alvarez A, Shin Yin SS, Chakrabarti MO, McGahan PJ, et al. Stem cell–infused anterior cruciate ligament reconstruction. Arthrosc Tech. 2019;8:e1313–7.
80. Saether EE, Chamberlain CS, Leiferman EM, Kondratko-Mittnacht JR, Li WJ, Brickson SL, et al. Enhanced medial collateral ligament healing using mesenchymal stem cells: dosage effects on cellular response and cytokine profile. Stem Cell Rev Reports. 2014;10:86–96.
81. Jiang D, Yang S, Gao P, Zhang Y, Guo T, Lin H, et al. Combined effect of ligament stem cells and umbilical-cord-blood-derived CD34+ cells on ligament healing. Cell Tissue Res. 2015;362:587–95.
82. Tei K, Matsumoto T, Mifune Y, Ishida K, Sasaki K, Shoji T, et al. Administrations of peripheral blood CD34-positive cells contribute to medial collateral ligament healing via vasculogenesis. Stem Cells. 2008;26:819–30.
83. Nishimori M, Matsumoto T, Ota S, Kopf S, Mifune Y, Harner C, et al. Role of angiogenesis after muscle derived stem cell transplantation in injured medial collateral ligament. J Orthop Res. 2012;30:627–33.
84. Hoffman JK, Protzman NM, Malhotra AD. Biologic augmentation of the ulnar collateral ligament in the elbow of a professional baseball pitcher. Case Rep Orthop. 2015;2015:130157.
85. Chambers MC, Fu FH. Editorial commentary: adult stem cell potential to enhance healing of the anterior cruciate ligament. Art Ther. 2018;34:341–2.

David Figueroa, Rodrigo Guiloff,
and Francisco Figueroa

21.1 Introduction

Surgical biologic augmentation techniques, known as ortho-biologics, have been progressively increasing in the past decade [1]. Surgeons expect these biologics to enhance tissue healing in order to restore a native or near-native tissue and aid in symptoms management while reducing risks for treatment failure [1]. The most popular ortho-biologic currently used as an adjuvant for conservative and surgical approaches in the treatment of ligament injuries is platelet-rich plasma (PRP).

PRP is commonly defined as a blood sample with a platelet concentration above baseline values [2]. It derives from the centrifugation process of autologous blood [3]. This centrifugate generates a platelet concentrate mixed with growth factors and interleukins, which are believed to enhance tissue recovery [4, 5]. PRP started to be used in orthopedic surgery in the 1990s, reaching recognition in the twenty-first century because of its use in sports medicine [6]. One of the main applications studied is knee surgery, especially ligamentous injuries, with promising results in anterior cruciate ligament reconstruction (ACL-R), while conservative injective applications have been performed to address medial collateral ligament (MCL) lesions as well as ankle sprains.

However, "there is no clear algorithm for the indications, processing methodology, application, and reporting, which has led to inconsistencies in clinical and basic science results" [1]. This chapter aims to review the latest available evidence and discuss the rationale for the use of PRP in ligamentous injuries.

21.2 PRP Use in Anterior Cruciate Ligament Injuries

This book focuses on injectable biologic therapies. However, since anterior cruciate ligament (ACL) injuries have been traditionally operated, PRP is mainly associated with surgical application. Therefore, ACL-R results will be primarily discussed, followed by a description of the recent evidence regarding injective applications in ACL treatment, mostly on partial ACL injuries.

ACL-R is the gold standard for restoration of stability and knee function after an ACL rupture [7]. Restoration of the normal joint kinematics is believed to be necessary to prevent secondary knee injuries such as cartilage, menisci, and other core ligaments of the knee [8]. When done correctly, it is a surgical procedure with successful outcomes and self-reported satisfaction [8–10]. However, besides advancements in technology, rehabilitation, and a better understanding of the knee anatomy and biomechanics, there is still a 0.7–20% failure rate reported [10, 11]. In terms of returning to sports activity, a rate of 82% has

D. Figueroa (✉) · R. Guiloff · F. Figueroa
Facultad Medicina Clínica Alemana, Universidad del Desarrollo, Santiago, Chile

© ISAKOS 2022
G. Filardo et al. (eds.), *Orthobiologics*, https://doi.org/10.1007/978-3-030-84744-9_21

been reported, with only 63% reaching their pre-injury sport level [12]. The rate is even lower in elite athletes, with less than 50% returning to level I sports [13]. Moreover, the timing for return to sports has progressively increased to avoid failures [14], and few advances have been made to reduce the healing time of ACL graft and subsequent development of degenerative joint disease [14]. Summed to the increasing patients' expectations for returning to their pre-injury sport activity as soon as possible [15, 16], different biological agents, such as PRP, have been tested in ACL-R to improve functional outcomes and accelerate the rehabilitation process while diminishing the graft failure rates.

PRP contains interleukins and growth factors that have been shown to promote cell proliferation, migration, differentiation, and angiogenesis, all properties that may improve ligament, bone, and wound healing [7]. In ACL procedures, experimental preclinical evidence has highlighted overall encouraging results. Andriolo and colleagues [17] reviewed the most significant investigations, indicating that in vitro studies have shown that the use of PRP increased the expression of procollagen gene and collagen protein and also contributed to reduce apoptosis and stimulate fibroblast metabolic activity in ACL grafts. In animal models, PRP augmentation promotes superior biomechanical properties such as a higher tensile load and linear stiffness of ACL grafts [17]. Clinical data suggest beneficial effects in ACL injuries as well. There is evidence showing favorable PRP biological properties in graft maturation (intra-articular ligamentization and osteointegration of the graft-tunnel interface) and donor site morbidity, specifically for bone tendon bone (BTB) grafts [8]. Therefore, it is expected that these advantages could lead to faster and secure rehabilitation while diminishing failure rates.

A systemic review of the literature on the use of PRP and ACL-R found only 11 clinical articles (516 patients) for inclusion [6]. The review revealed that during the first decade of the present century, different clinical studies showed an enhanced effect over the ligamentization (remodeling) of the intra-articular component of the ACL-R graft [18–21]. One study showed improved integration [22]. However, most of the investigations failed to have validated methods and scores for measuring graft maturation, and different volumes and concentrations of PRP were used. Most importantly, the clinical implications of improved graft maturation remain unknown. Of the 11 analyzed articles in the mentioned systematic review, 5 investigated clinical outcomes. Of these, only one demonstrated a positive correlation with clinical evaluation, showing that patients treated with PRP had significantly better anteroposterior knee stability than patients without PRP [22]. Nevertheless, even though the authors included Tegner and Lysholm scores in their materials and methods, no functional outcomes were reported. Another recent systematic review of 34 studies concluded that MRI-based graft maturity is challenging at present times and cannot predict clinical and functional outcomes in patients at both early and long-term follow-up [23]. Further reviews agree with these observations [8, 24]. Davey and colleagues [7] reported this year (2020) a systematic review that included only randomized controlled trials. The authors concluded that current level I evidence does not support the use of PRP to improve graft healing [7].

Regarding donor site morbidity, it can be a significant source of post-operative pain in ACL-R [8]. There is clinical data with promising early results, reporting a significant reduction in the immediate post-operative pain scores (visual analogue scale (VAS), 0.6–3.8 vs. 2.6–5.1) in patients with BTB tendon harvest site treated with PRP injections into the bone gaps (patellar and tibial) immediately after skin closure [25] and through a PRP gel added directly to fill the patellar tendon gap, before suturing the peritenon [26, 27]. However, Seijas and colleagues showed that this difference in VAS scores becomes insignificant at 2 months post-op [25], while De Almeida and colleagues showed no difference in isokinetic testing and Lysholm, IKDC, Kujala, and Tegner scores at 6 months post-op [26]. Interestingly, Cervellin and colleagues [27] did not find a significant difference at 12 months post-op in VAS scores but found that the PRP group showed a significantly higher Victorian Institute of Sport Assessment (VISA) score. This observation may suggest the importance of using more than one scale when subjective data is being tested. Healing of the patellar tendon gapping has also been evaluated with controversial results. While De Almeida and colleagues showed significantly less gapping on MRI at 6 months following surgery [26], Walters and colleagues found no differences in healing indices on MRI at the same time when using PRP mixed with autologous cancellous bone chips and placed into the patellar donor site [28]. Walters et al. also did not find differences in kneeling pain, pain with daily living activities, and IKDC scores at 3, 6, 12, and 24 months of follow-up. In the already mentioned systematic review from this year, the authors concluded that current level I evidence does not support the use of PRP to improve donor site morbidity and reduce post-operative pain levels [7].

In terms of the latest clinical results, there is insufficient evidence to recommend the use of PRP in ACL-R for improving graft maturation and donor site morbidity or improving functional outcomes, no matter the method of application. Moreover, even though there is some controversial evidence to support the use of PRP for improving graft ligamentization, there is no clinical evidence that shows a correlation between these findings and faster and more secure rehabilitations with lower failure rates. A detailed analysis of the published evidence shows systematical weaknesses in the current literature. In essence, there is a scarce of high-level evidence investigation, with considerable variability of PRP preparations and methods of applications, impeding comparison between them [29]. For the exposed reasons, doubts about the use of PRP in ACL-R will remain, at least until new studies with a better scientific quality advise something different.

As already mentioned, ACL-R remains the gold standard for the treatment of ACL injuries, especially for complete ruptures. However, even though scarce, there is evidence supporting alternative treatments such as ortho-biologic injections alone and ortho-biologic augmented repairs for partial ACL lesions [30]. These alternatives can preserve the native insertion site of the remaining fibers and, therefore, its proprioceptive function, which may lead to more physiological knee biomechanics [30]. Although animal studies using PRP alone to treat ACL partial injuries have failed to restore native knee stability, few clinical studies show promising results. Seijas et al. reported a high return to sports in 19 professional soccer players with a partial ACL tear treated with intraligamentous PRP injections into the intact bundle under direct arthroscopic visualization and the articular space after emptying the knee at the end of the procedure. No complications were noted, and at 1-year follow-up, MRI evaluations showed complete ligamentization and satisfactory anatomic arrangement of all ACL remnant bundles [31]. Nevertheless, the study is limited, lacking a control group and assessment of functional parameters. Koch et al. used a similar technique, adding a trephination of the ACL origin, and followed 38 patients for 33 months (±17.4 months). The authors reported a failure rate of 9.5%. On the remaining subjects, good to excellent results in personal reported outcomes were shown. Clinical examination revealed a stable Lachman test, negative pivot shift phenomenon, and a significant reduction in anteroposterior laxity compared to preoperative status in all patients. Functional performance testing showed no significant differences between the injured and healthy side. Return to sports was achieved after a mean of 5.8 months (±3.6) in 71.1% of the included patients [32]. However, once again, the study is limited due to the absence of a control group. The retrospective design of the investigation and the exclusion of patients with complete ruptures of one of the ACL bundles must also be noted. Interestingly, Murray et al. reported the use of a collagen scaffold soaked with whole blood to deliver platelets in combination with a novel bioenhanced primary repair technique using a suture stent, named bridge-enhanced ACL repair (BEAR technique), and reported that it resulted in biomechanical properties of the repaired ACL equivalent to an ACLR after 3, 6, and 12 months of healing in an animal model [30]. Further studies with higher clinical evidence and longer follow-up are necessary to support PRP in the treatment of partial ACL tears.

21.3 PRP Use in Medial Collateral Ligament Injuries

As opposed to PRP use in ACL lesions, its use in medial collateral ligament (MCL) injuries has focused mainly on enhancing non-operative management with a focus on promoting healing and obtaining a faster rehabilitation period, with most of the studies being animal models and only a few case reports and one case series of its use in humans.

LaPrade et al., in a controlled laboratory study using New Zealand white rabbits, showed that low doses of PRP did not have any impact on healing in acute grade 3 MCL tears (mid-body injury) [33]. Contrarily, higher doses of PRP decreased the quality of reparative tissue in this study, suggesting that there is no substantial role for PRP in this specific injury and that more in vivo studies are needed to recommend its use in humans. Amar et al. published a controlled study in rats, obtaining similar results, thus not recommending PRP use in MCL injuries [34].

On the other hand, da Costa et al. [35] and Yoshioka et al. [36] published controlled laboratory studies in animal models (rabbits) reporting accelerated ligament healing and better structural properties after PRP application for MCL injuries. However, despite the positive results, both studies state that the clinical relevance of these findings is uncertain and that more studies, especially in vivo, are needed to assess the usefulness of this therapy in MCL injuries.

Regarding the clinical use of PRP in MCL lesions, the evidence is restricted mostly to case reports (1–3 patients), being Zou and colleagues [37] the only group that reported a case series study. In their article, they treated 52 patients with chronic pain (≥3-month history) after low-grade MCL injuries (6.5 ± 1.11 months) with 3 PRP intra-articular injections (1 injection weekly for 3 weeks). They obtained significantly better International Knee Documentation Committee (IKDC) and VAS scores between pre-treatment and post-treatment; however, there were no significant differences among the various post-treatment time points (1 month, 3 months, 6 months). Complete healing of proximal ligaments was identified on magnetic resonance images. Regarding the other reports [38–40], the clinical relevance to recommend its use is limited due to their small samples.

With the evidence available at this time, mostly in animal studies, and with controversial results, the authors of the present chapter conclude that the use of PRP for the treatment of MCL injuries cannot be recommended. The only human case series shows promising results for chronic pain after low-grade MCL injuries. However, more studies, especially in vivo and human studies, are needed to obtain clinical relevance and assess its efficacy in treating MCL lesions.

21.4 PRP Use in Ankle Sprains

There are limited data available on PRP in acute ankle sprains [41]; however, interest is growing regarding the use of PRP in acute syndesmotic injuries in high-level athletes. A small cohort of rugby players with anterior inferior tibiofibular ligament (AITFL) tears [42] showed a shorter time to return to play, higher agility, and better vertical jumps when treated with a single autologous PRP injection into the AITFL. There was also a lower level of fear avoidance associated with rugby in the intervention group. Another study followed 16 elite athletes receiving ultrasound-guided injections of PRP into the injured AITFL [43]. The PRP group showed a shorter time to return to play, re-stabilization of the syndesmosis joint, and less long-term residual pain.

Contrarily, regarding other varieties of ankle sprains, evidence has not shown benefits on injecting PRP into the lesion. In 2015, a double-blind, randomized control trial examined patients presenting to the emergency department with unclassified ankle sprains [44]. It failed to find a significant difference in pain or function within the first month comparing a PRP injection with a placebo injection. Another recent study in 2020 showed no differences in PRP-injected patients when the two groups were treated with rigid immobilization [45].

In spite of the promising evidence, further investigation with higher clinical evidence and specific injury classification is needed to determine if there is a role for the routine use of PRP injections in ankle ligamentous injuries.

21.5 Conclusion

In terms of the latest clinical evidence, the use of PRP for ligament injuries remains controversial. There is insufficient evidence to recommend the use of PRP in ACL-R for improving graft maturation and donor site morbidity or improving functional outcomes. The use of PRP injections for the treatment of partial ACL injuries, MCL injuries, and ankle sprains cannot be recommended. Finally, there is no reported clinical evidence of the use of PRP in other knee ligament injuries such as posterior cruciate ligament, lateral collateral ligament, and posterolateral corner.

Take-Home Messages
- The most popular and evidence-based ortho-biologic currently used as an adjuvant for conservative and surgical approaches in the treatment of ligament injuries is PRP.
- In terms of the latest clinical results, there is insufficient evidence to recommend the use of PRP in ACL reconstruction for improving graft maturation and donor site morbidity or improving functional outcomes, no matter the method of application.
- Even though there is evidence supporting ortho-biologic injections alone and ortho-biologic augmented repairs for partial ACL lesions, further studies with higher clinical evidence and longer fol-

low-up are necessary to support PRP in the treatment of partial ACL tears.
- The available evidence for the use of PRP in the treatment of MCL injuries derives mostly from animal studies and has controversial results.
- There is promising evidence regarding the use of PRP in the treatment of ankle ligamentous injuries; however, further investigation with higher clinical evidence and specific injury classification is needed to determine if there is a role for the routine use of PRP injections in ankle sprains.
- The routine use of PRP for ligament injuries remains controversial.

21.6 Summary Table

	Evidence	Authors' recommendation
PRP in ACL injuries		
ACL-R	– Controversial clinical evidence regarding the use of PRP in ACL-R for improving graft maturation and donor site morbidity or improving functional outcomes, no matter the method of application – Even though there is some controversial clinical evidence to support the use of PRP for improving graft ligamentization, there is no clinical evidence that shows a correlation between these findings and faster and more secure rehabilitations with lower failure rates [6, 7]	Insufficient evidence for recommendation – There is a scarce of high-level evidence investigation and with considerable variability of PRP preparations and methods of applications, impeding comparison between them

	Evidence	Authors' recommendation
Partial ACL injuries	– Animal studies have failed to restore native knee stability – Two clinical studies have shown promising results with PRP intraligamentous injections; however, both lack a control group [31, 32]	Not recommended – Further studies with higher clinical evidence and longer follow-up are necessary to support PRP in the treatment of partial ACL tears
PRP in MCL injuries	– Mostly in animal studies, with controversial results – The only human case series shows promising results for chronic pain after low-grade MCL injuries [37]	Not recommended – More studies, especially in vivo and human studies, are needed to obtain clinical relevance and assess its efficacy in treating MCL lesions
PRP in ankle sprains	– Promising clinical evidence for AITFL sprains – No difference within unclassified ankle sprains	Not recommended – Further studies with higher clinical evidence and specific injury classification are necessary to support PRP in the treatment of partial ankle sprains

References

1. Chahla J, Kennedy MI, Aman ZS, RF LP. Orthobiologics for ligament repair and reconstruction. Clin Sports Med. 2019:97–107.
2. Hall MP, Band PA, Meislin RT, Jazrawi LM, Cardone DA. Platelet-rich plasma: current concepts and application in sports medicine. J Am Acad Orthop Surg. 2009;17:602–8.
3. Figueroa PD, Figueroa BF, Ximena AP, Calvo RR, Vaisman BA. Uso del plasma rico en plaquetas en cirugía ligamentosa de rodilla. Rev Med Chil. 2013;141:1315–20.
4. Moraes VY, Lenza M, Tamaoki MJ, Faloppa F, Belloti JC. Platelet-rich therapies for musculoskeletal soft tissue injuries. Cochrane Database Syst Rev. 2014;2014.
5. Zhang JY, Fabricant PD, Ishmael CR, Wang JC, Petrigliano FA, Jones KJ. Utilization of platelet-rich plasma for musculoskeletal injuries: an analysis of current treatment trends in the United States. Orthop J Sports Med. 2016;4:24–6.
6. Figueroa D, Figueroa F, Calvo R, Vaisman A, Ahumada X, Arellano S. Platelet-rich plasma use in anterior cruciate ligament surgery: Systematic review of the literature. Art Ther. 2015;31:981–8.
7. Davey MS, Hurley ET, Withers D, Moran R, Moran CJ. Anterior cruciate ligament reconstruction with platelet-rich plasma: a systematic review of randomized control trials. Art Ther. 2020:1–7.
8. Riediger MD, Stride D, Coke SE, Kurz AZ, Duong A, Ayeni OR. ACL reconstruction with augmentation: a scoping review. Curr Rev Musculoskelet Med. 2019;12:166–72.
9. Nwachukwu BU, Voleti PB, Berkanish P, Chang B, Cohn MR, Williams RJ, et al. Return to play and patient satisfaction after ACL reconstruction. J Bone Jt Surg Am Vol. 2017;99:720–5.
10. Samitier G, Marcano AI, Alentorn-Geli E, Cugat R, Farmer KW, Moser MW. Failure of anterior cruciate ligament reconstruction. Arch Bone Jt Surg. 2015;3:220–40.
11. Di Benedetto P, Di Benedetto E, Fiocchi A, Beltrame A, Causero A. Causes of failure of anterior cruciate ligament reconstruction and revision surgical strategies. Knee Surg Relat Res. 2016;28:319–24.
12. Ardern CL, Webster KE, Taylor NF, Feller JA. Return to sport following anterior cruciate ligament reconstruction surgery: a systematic review and meta-analysis of the state of play. Br J Sports Med. 2011;45:596–606.
13. Webster KE, Feller JA. Return to level I sports after anterior cruciate ligament reconstruction: evaluation of age, sex, and readiness to return criteria. Orthop J Sports Med. 2018;6:1–6.
14. Chahla J, Cinque ME, Mandelbaum BR. Biologically augmented quadriceps tendon autograft with platelet-rich plasma for anterior cruciate ligament reconstruction. Arthrosc Tech. 2018;7:e1063–9.
15. Webster KE, Feller JA. Expectations for return to pre-injury sport before and after anterior cruciate ligament reconstruction. Am J Sports Med. 2019;47:578–83.
16. Feucht MJ, Cotic M, Saier T, Minzlaff P, Plath JE, Imhoff AB, et al. Patient expectations of primary and revision anterior cruciate ligament reconstruction. Knee Surg Sports Traumatol Arthrosc. 2016;24:201–7.
17. Andriolo L, Di Matteo B, Kon E, Filardo G, Venieri G, Marcacci M. PRP augmentation for ACL reconstruction. Biomed Res Int. 2015;2015.
18. Ventura A, Terzaghi C, Borgo E, Verdoia C, Gallazzi M, Failoni S. Use of growth factors in ACL surgery: preliminary study. J Orthop Traumatol. 2005;6:76–9.
19. Orrego M, Larrain C, Rosales J, Valenzuela L, Matas J, Durruty J, et al. Effects of platelet concentrate and a bone plug on the healing of hamstring tendons in a bone tunnel. Arthroscopy. 2008;24:1373–80.
20. Radice F, Yánez R, Gutiérrez V, Rosales J, Pinedo M, Coda S. Comparison of magnetic resonance imaging findings in anterior cruciate ligament grafts with and without autologous platelet-derived growth factors. Art Ther. 2010;26:50–7.

21. Sánchez M, Anitua E, Azofra J, Prado R, Muruzabal F, Andia I. Ligamentization of tendon grafts treated with an endogenous preparation rich in growth factors: gross morphology and histology. Art Ther. 2010;26:470–80.

22. Vogrin M, Rupreht M, Crnjac A, Dinevski D, Krajnc Z, Rečnik G. The effect of platelet-derived growth factors on knee stability after anterior cruciate ligament reconstruction: a prospective randomized clinical study. Wien Klin Wochenschr. 2010;122:91–5.

23. Van Dyck P, Zazulia K, Smekens C, Heusdens CHW, Janssens T, Sijbers J. Assessment of anterior cruciate ligament graft maturity with conventional magnetic resonance imaging: a systematic literature review. Orthop J Sports Med. 2019;7:1–9.

24. Le ADK, Enweze L, DeBaun MR, Dragoo JL. Platelet-rich plasma. Clin Sports Med. 2019;38:17–44.

25. Seijas R, Cuscó X, Sallent A, Serra I, Ares O, Cugat R. Pain in donor site after BTB-ACL reconstruction with PRGF: a randomized trial. Arch Orthop Trauma Surg. 2016;136:829–35.

26. De Almeida AM, Demange MK, Sobrado MF, Rodrigues MB, Pedrinelli A, Hernandez AJ. Patellar tendon healing with platelet-rich plasma: a prospective randomized controlled trial. Am J Sports Med. 2012;40:1282–8.

27. Cervellin M, de Girolamo L, Bait C, Denti M, Volpi P. Autologous platelet-rich plasma gel to reduce donor-site morbidity after patellar tendon graft harvesting for anterior cruciate ligament reconstruction: a randomized, controlled clinical study. Knee Surg Sport Traumatol Arthrosc. 2012;20:114–20.

28. Walters BL, Porter DA, Hobart SJ, Bedford BB, Hogan DE, McHugh MM, et al. Effect of intraoperative platelet-rich plasma treatment on post-operative donor site knee pain in patellar tendon autograft anterior cruciate ligament reconstruction: a double-blind randomized controlled trial. Am J Sports Med. 2018;46:1827–35.

29. Figueroa D. Editorial commentary: use of platelet-rich plasma in anterior cruciate ligament reconstruction: should we abandon? Art Ther. 2020;36:1211–2.

30. Dallo I, Chahla J, Mitchell JJ, Pascual-Garrido C, Feagin JA, LaPrade RF. Biologic approaches for the treatment of partial tears of the anterior cruciate ligament: a current concepts review. Orthop J Sports Med. 2017;5:1–9.

31. Seijas R, Ares O, Cuscó X, Álvarez P, Steinbacher G, Cugat R. Partial anterior cruciate ligament tears treated with intraligamentary plasma rich in growth factors. World J Orthop. 2014;5:373–8.

32. Koch M, Mayr F, Achenbach L, Krutsch W, Lang S, Hilber F, et al. Partial anterior cruciate ligament ruptures: advantages by intraligament autologous conditioned plasma injection and healing response technique—midterm outcome evaluation. Biomed Res Int. 2018;2018:3204869.

33. LaPrade RF, Goodrich LR, Phillips J, Dornan GJ, Turnbull TL, Hawes ML, et al. Use of platelet-rich plasma immediately after an injury did not improve ligament healing, and increasing platelet concentrations was detrimental in an in vivo animal model. Am J Sports Med. 2018;46:702–12.

34. Amar E, Snir N, Sher O, Brosh T, Khashan M, Salai M, et al. Platelet-rich plasma did not improve early healing of medial collateral ligament in rats. Arch Orthop Trauma Surg. 2015;135:1571–7.

35. da Costa EL, Teixeira LEM, Pádua BJ, de Araújo ID, de Souza Vasconcellos L, Dias LSB. Biomechanical study of the effect of platelet rich plasma on the treatment of medial collateral ligament lesion in rabbits. Acta Cir Bras. 2017;32:827–35.

36. Yoshioka T, Kanamori A, Washio T, Aoto K, Uemura K, Sakane M, et al. The effects of plasma rich in growth factors (PRGF-Endoret) on healing of medial collateral ligament of the knee. Knee Surg Sport Traumatol Arthrosc. 2013;21:1763–9.

37. Zou G, Zheng M, Chen W, He X, Cang D. Autologous platelet-rich plasma therapy for refractory pain after low-grade medial collateral ligament injury. J Int Med Res. 2020;48:1–7.

38. Yoshida M, Marumo K. An autologous leukocyte-reduced platelet-rich plasma therapy for chronic injury of the medial collateral ligament in the knee: a report of 3 successful cases. Clin J Sport Med. 2019;29:E4–6.

39. Eirale C, Mauri E, Hamilton B. Use of platelet rich plasma in an isolated complete medial collateral ligament lesion in a professional football (soccer) player: a case report. Asian J Sports Med. 2013;4:158–62.

40. Bagwell MS, Wilk KE, Colberg RE, Dugas JR. The use of serial platelet rich plasma injections with early rehabilitation to expedite grade III medial collateral ligament injury in a professional athlete: a case report. Int J Sports Phys Ther. 2018;13:520–5.

41. West TA, Williams ML. Orthobiologics. Clin Podiatr Med Surg. 2019;36(4):609–26.

42. Samra DJ, Sman AD, Rae K, Linklater J, Refshauge KM, Hiller CE. Effectiveness of a single platelet-rich plasma injection to promote recovery in rugby players with ankle syndesmosis injury. BMJ Open Sport Exerc Med. 2015;1(1):e000033.

43. Laver L, Carmont MR, McConkey MO, Palmanovich E, Yaacobi E, Mann G, Nyska M, Kots E, Mei-Dan O. Plasma rich in growth factors (PRGF) as a treatment for high ankle sprain in elite athletes: a randomized control trial. Knee Surg Sports Traumatol Arthrosc. 2015;23(11):3383–92.

44. Rowden A, Dominici P, D'Orazio J, Manur R, Deitch K, Simpson S, Kowalski MJ, Salzman M, Ngu D. Double-blind, randomized, placebo-controlled study evaluating the use of platelet-rich plasma therapy (PRP) for acute ankle sprains in the emergency department. J Emerg Med. 2015;49(4):546–51.

45. Blanco-Rivera J, Elizondo-Rodríguez J, Simental-Mendía M, Vilchez-Cavazos F, Peña-Martínez VM, Acosta-Olivo C. Treatment of lateral ankle sprain with platelet-rich plasma: a randomized clinical study. Foot Ankle Surg. 2020;26(7):750–4.

Kazunori Shimomura, David A. Hart, and Norimasa Nakamura

22.1 Introduction

The meniscus plays important roles in the knee joint, including force transmission, congruency of the joint, lubrication, and provision of joint stability [1–3]. Meniscal tears are the most common injury of the knee regardless of age, and the mean annual incidence of meniscal lesions per 10,000 population has been reported to be 9.0 for males and 4.2 for females [4]. Importantly, effective treatments for such injuries remain challenging, particularly for young, active patients. Part of this challenge is due to the meniscus having limited healing potential owing to its hypocellularity and hypovascularity, as well as its complex structure [5]. The damaged meniscus is recog-

nized to lose function in the absence of adequate treatment, and such knees are at high risk for the development of osteoarthritis (OA) [6, 7]. Therefore, orthopedic surgeons should always consider preserving the important meniscal functions as extensively as possible.

Based on meniscal anatomy and vascularity, the meniscus has limited healing capacities especially in the central two-thirds avascular zone, while meniscal tears in the peripheral vascular zone should be reparable [2]. As many studies have addressed, meniscal tears in the vascular zone of the peripheral area, such as vertical longitudinal tears, have good indications for effective meniscal repair [8, 9]. On the other hand, meniscal tears that include lesions in the avascular zone, such as radial tears which account for the majority of the tissue injuries, are not expected to spontaneously heal. Thus, such tears have been mostly treated by partial meniscectomy [10]. A partial meniscectomy increases the subsequent risk for OA at least 10- to 20-fold [11]. Therefore, the development of novel therapeutic methods for meniscal repair is both timely and necessary to address these issues.

Recently, there has been strong interest in approaches to enhance repair and healing of the injured meniscus to restore function. The currently available restorative procedures are mainly transplantation of a meniscal allograft transplant [12, 13] or implantation of a scaffold [14, 15]. Regarding a meniscal allograft implantation, a recent systematic review concluded that meniscal

K. Shimomura (✉)
Department of Orthopaedic Surgery, Osaka University Graduate School of Medicine, Osaka, Japan
e-mail: kazunori-shimomura@umin.net

D. A. Hart
McCaig Institute for Bone & Joint Health, University of Calgary, Calgary, AB, Canada

N. Nakamura
Department of Orthopaedic Surgery, Osaka University Graduate School of Medicine, Osaka, Japan

Institute for Medical Science in Sports, Osaka Health Science University, Osaka, Japan

Global Center for Medical Engineering and Informatics, Osaka University, Osaka, Japan
e-mail: norimasa.nakamura@ohsu.ac.jp

© ISAKOS 2022
G. Filardo et al. (eds.), *Orthobiologics*, https://doi.org/10.1007/978-3-030-84744-9_22

allograft transplantation appears to provide good clinical results over short-term and mid-term follow-up, with improvement in knee function [13]. On the other hand, some drawbacks of allografts were reported to include immunological reaction to the implanted tissue, potential disease transmission, graft size mismatching, and limited donor availability [13, 16]. Regarding a scaffold implantation, there have been several implants that have been made available for clinical practice, including a collagen meniscus implant (CMI®) [17] and a polyurethane polymeric implant (Actifit®) [15]. These two implants provided significant pain relief and functional improvement with safety. On the other hand, several studies reported negative outcomes based on MRI results with these two implants, despite the general observation of improved clinical scores, showing these implants were partially or totally resorbed during follow-up, accompanied by extrusion of the implanted materials and subchondral bone edema [15, 18, 19]. Thus, there is likely still room for improvement regarding the development of such implants serving as meniscal substitutes.

Alternatively, several biologics have been studied as potential therapies for OA or repair of meniscus and/or cartilage injuries. These include blood products [e.g., platelet-rich plasma (PRP)] [20], native tissue-derived cells, cells isolated and expanded in vitro [e.g., mesenchymal stromal cells (MSC)] [21], secretory products of MSCs (e.g., MSC-derived extracellular vesicles or growth factors) [22], and disease-modifying drugs [23]. Intra-articular injections of cells have the potential to provide beneficial effects for the prevention of knee OA and are seen as a potential therapy for meniscal regeneration [21, 24]. This chapter will discuss the feasibility of using injectable biologics, with a special focus on cells, for the repair of meniscal lesions.

22.2 Cell Selection

The most tissue-relevant cell source for meniscal repair may be autogenous biopsy specimens taken directly from the patients. Outgrowth of progenitor cells from meniscal tissue or cartilage tissue can be used to generate fibrochondrocytes and chondrocytes in vitro. However, a tissue biopsy sacrifices undamaged meniscus and cartilage within the same joint. Also, the number of cells obtained is usually limited. Moreover, the heterogeneous population of tissue-resident connective tissue progenitors (CTPs) in meniscus and cartilage alter their phenotype in association with their in vitro expansion [25, 26]. Therefore, the use of such cells is usually limited to experimental studies [27, 28].

Culture-expanded mesenchymal stromal cells (MSCs) represent an alternative strategy for cell-based therapy, with a number of advantages [29]. MSCs are known to secrete various cytokines and growth factors that can exhibit immunomodulatory, angiogenic, anti-inflammatory, and anti-apoptotic effects. While details regarding their mechanism(s) of action have not been fully understood, such activities could potentially enable allogenic transplantation [30, 31]. Culture-expanded MSCs, by definition, have the capability to differentiate into a variety of connective tissue cells including bone, cartilage, tendon, muscle, and adipose tissue [32]. In addition, MSC populations can be isolated from various tissues such as bone marrow, adipose tissue, skeletal muscle, synovial membrane, synovial fluid, and umbilical cord blood [33, 34].

Culture-expanded MSCs have unique characteristics that can differ from batch to batch and change based on the tissue from which cells are obtained. Bone marrow-derived MSCs (BMSCs) have been widely used in clinical practice, and bone marrow has been considered the main site to obtain MSCs for a long time [31]. However, MSC populations have been grown out from various other sources, including synovial membrane and adipose tissue, and both may serve as alternatives to bone marrow. Adipose tissue-derived MSCs (sometimes called ASCs) can also be readily generated. Adipose tissue contains a higher prevalence of colony founding CTPs than bone marrow or synovial membrane as a source [31]. In comparative studies of MSC characteristics, synovium-derived cells have been reported to exhibit the greatest chondrogenic potential [34, 35].

Yet another recent option is the generation of cellular therapies starting from a population of induced pluripotent stem (iPS) cells. Such cells showed some feasibility for cartilage repair [36]. However, iPS cells or iPS-derived cell populations are not currently approved for use in any musculoskeletal application. To date, there is no data related to the use of iPS-derived cells for meniscal repair, but iPS cells might be applied to treat meniscal lesions in the near future.

22.3 Mechanism(S) of Action of MSCs

The mechanism(s) involved related to the use of culture-expanded MSCs for tissue repair has been discussed for many years. It remains controversial as to whether implanted MSCs directly contribute to cartilage and meniscal repair and are retained within the tissue that develops post-implantation. In fact, the regeneration of meniscal tissue by transplanted cells and durable retention of the progeny of transplanted cells in local tissues have not been shown. Instead, it appears that any positive effect of transplantation of culture-expanded MSC populations appears to be through the release of trophic mediators and/or anti-inflammatory factors that enhance the involvement of endogenous cells. Recently, Caplan has proposed that MSCs home to sites of injury and/or disease and then secrete bioactive factors that are immunomodulatory and trophic regenerative mediators, indicating that these cells make forms of what could be called therapeutic drugs in situ [37]. Indeed, the patient's own local site-specific and tissue-specific CTPs are the mediators of new tissue formation, stimulated by the bioactive factors secreted by the exogenously supplied cells [38, 39]. Therefore, the secretome of MSCs is garnering much research attention, and some of those factors are being extensively studied (see Chap. 11).

Recent work has shown that MSCs were able to promote the regeneration of joint components through two of their secretory activities, those that are anti-inflammatory factors and those that are trophic factors [40]. Anti-inflammatory fac-

tors secreted by MSCs may downregulate inflammatory signals in osteoarthritic cartilage and menisci induced by interleukin (IL)-1β, IL-6, IL-8, matrix metalloproteinase (MMP)-1, and MMP-13 [41, 42]. Trophic factors are molecules that give rise to enhanced cell proliferation, decreased formation of scar tissue, and can trigger the repair of endogenous cartilage and meniscus. Examples of such factors are epithelial growth factor (EGF), insulin-like growth factor (IGF)-1, basic fibroblast growth factor (bFGF), transforming growth factor (TGF)-β, and vascular endothelial growth factor (VEGF) [41, 43]. Thus, it may be possible to utilize such secretomes in the treatment of cartilage and meniscus injuries, considering that proinflammatory cytokines such as IL-1 and TNF-α are produced in response to injury by chondrocytes, meniscal cells, synoviocytes, and macrophages, contributing to the joint destruction [44–46]. It may also follow that inhibition of IL-1 and TNF-α can lead to improvement in meniscus healing in experimental studies [46–48].

Two additional points are relevant to the future in vivo use of MSCs for meniscal tears or repair of massive meniscal defects. First, scaffold-free tissue-engineered constructs have been generated from synovial MSCs for cartilage repair in both preclinical models [49] and human studies [50]. Implantation of a construct with undifferentiated MSCs led to apparent differentiation of the cells in vivo. Therefore, cells in a three-dimensional environment established in vitro may be able to yield good results in vivo if the required biological and biomechanical cues to generate a hyaline cartilage-like tissue are available.

Second, the normal meniscus is quite complex biologically, with avascular and vascular parts and aneural and innervated parts. Furthermore, the inner part of the menisci is more cartilage-like, while the outer part is more ligament-like with regard to collagen types, so it is also heterogeneous at the composition level. At the biomechanical level, a meniscus is also heterogeneous with the inner part subjected to compressive forces, the outer part subjected to hoop stresses, and the surface subjected to shear forces. Finally, recent studies have elaborated details of the complex structure of animal and human menisci [51, 52]. Menisci have a very complex structure that may be difficult to mimic without a template to effectively repair a large meniscal defect sufficient for long-term function. Thus, repair with MSCs via the appropriate mechanism may yield satisfactory long-term results for tears, but repair/regeneration of large meniscal defects may be challenging unless new understanding arises as to how to address the above complexities.

22.4 Preclinical Studies

To date, many studies related to intra-articular injection of cells for repair of meniscal defects have been performed, and some have indicated biological repair of the damaged menisci. Small animals such as mice and rats have become widely available and allow for elucidation of a better understanding of the healing mechanism of meniscal injuries, and they confirm the role of critical anabolic and/or catabolic molecules associated with cell-based therapy treatment [53, 54]. On the other hand, unlike in humans, the natural healing responses in rodents may not be as clinically relevant as studies in a large animal model, based on host tissue reactions to such treatments in rodents as well as the physical size and volume of new tissue that must be generated to accomplish repair. In consideration of clinical relevance, it is likely preferable to utilize a large animal model such as pigs and sheep, as both the physiology and size of meniscal tissues in such animals are similar to that of humans [55]. Cost is a limiting factor, however. In Table 22.1, we outline the latest animal studies and provide some more detailed examples below.

Horie et al. [56] studied the feasibility of injecting culture-expanded rat or human cells for meniscal regeneration using a rat hemi-meniscectomy model and showed that MSCs enhanced meniscal regeneration and reduced OA progression, with increased type II collagen synthesis and activated Indian hedgehog (Ihh), parathyroid hormone-like hormone (PTHLH), and bone morphogenetic protein (BMP)-2 genes in regenerated menisci. Hatsushika et al. [57] investigated the effects of a single intra-articular injection of culture-expanded synovial cells on meniscal regeneration in a rabbit massive meniscal defect model. They showed that transplanted cells adhered around the meniscal defect and promoted meniscal regeneration. Also, articular cartilage and subchondral bone were better preserved in the cell-injected group. Caminal et al. [58] evaluated the safety and efficacy of an autologous cell-based treatment for OA using MSCs expanded from bone marrow aspirates that were administered intra-articularly in a sheep model. At 12 months of follow-up, after injection of $1.1–1.2 \times 10^7$ cells, they showed that evidence for regeneration of articular cartilage and meniscus was case-dependent, but a statistically significant improvement was found in specific macroscopic and histological parameters. Nakagawa et al. [59] examined whether direct transplantation of 2×10^7 culture-expanded synovial cells to the meniscal lesion promoted healing in a micro-minipig model with an extended longitudinal tear in the avascular area of the tissue. At 12 weeks of follow-up, meniscal healing was enhanced based on the results of macroscopic, histological, T1rho mapping analysis and biomechanical testing for tensile strength in the MSC transplanted group.

22.5 Clinical Studies

Clinical studies evaluating the effects of cellular injections in the knee joint are still limited. Available data suggests encouraging outcomes [60], but systematic reviews of published work show a paucity of rigor in experimental design and inconsistent overall impact [61, 62]. There have generally not been safety concerns identified or side effects reported in the clinical use of MSC injections [63–65]. The latest clinical studies are outlined in Table 22.2, and some examples are detailed below.

Vangsness et al. [66] investigated the effects of allogenic marrow-derived MSCs on osteoarthritic changes in the knee and gave single-cell injections for patients who had undergone a partial medial meniscectomy. Patients were randomized to one of three treatment groups: (1) an

Table 22.1 Summary of studies on intra-articular injections of MSCs for meniscal regeneration in animal models

Author (year)	Animal	Experimental model	Cell source	Cell number	Follow-up period	Outcomes
Agung et al. (2006) [72]	Rat	ACL, MM, MFC cartilage injured model	Rat bone marrow (allogenic)	1×10^6 1×10^7	4 weeks	MSCs were observed in injured sites of ACL, MM, and cartilage and contributed to tissue regeneration
Horie et al. (2009) [73]	Rat	Anterior half of MM resection	Rat synovium and bone marrow	5×10^6	2, 4, 8, 12 weeks	MSCs adhered to the lesion, differentiated into meniscal cells directly, and promoted meniscal regeneration without mobilization to distant organs
Horie et al. (2012) [56]	Rat	Hemi-meniscectomy model	Rat and human bone marrow	2×10^6	2, 4, 8 weeks	Enhanced meniscal regeneration and prevented OA progression. Increased col II synthesis and activated Ihh, PTHLH, BMP2 genes
Al Faqeh et al. (2012) [74]	Sheep	ACL and MM resection	Sheep bone marrow (autologous)	1×10^7 (cultured w/ CM or BM)	6 weeks	Retardation of cartilage destruction with CM as well as BM group and no significant ICRS scoring between the two groups with cells. Meniscus repair was observed in the knee joint treated with CM group
Duygulu et al. (2012) [75]	Sheep	Full-thickness longitudinal tear in MM	Autologous bone marrow aspirate	Unknown cell number (5 mL injected into the torn site)	16 weeks	Enhanced a bridging reparation tissue and adhesion of the torn site, neovascularization, and formation of cartilage plaques
Hatsushika et al. (2013) [57]	Rabbit	Anterior half of MM resection	Rabbit synovium	1×10^7	4, 12, 16, 24 weeks	MSCs adhered around the meniscal defect and promoted meniscal regeneration. Articular cartilage and subchondral bone were better preserved in the MSC group
Shen et al. (2013) [76]	Rabbit	Anterior half of MM resection	Rabbit meniscus (allogenic)	6×10^6 (twice)	4, 8, 12 weeks	Promoted meniscus regeneration, protected joint surface cartilage, and maintained joint space width
Shen et al. (2014) [77]	Rat	Anterior half of MM resection	Human meniscus	6×10^6 (twice)	4, 12 weeks	Enhanced meniscus regeneration through the SDF-1/CXCR4 axis. Protected articular cartilage by reduced expression of OA markers but increased expression of col II

(continued)

Table 22.1 (continued)

Author (year)	Animal	Experimental model	Cell source	Cell number	Follow-up period	Outcomes
Hatsushika et al. (2014) [78]	Pig	Anterior half of MM resection	Porcine synovium (allogenic)	5×10^7 (three times)	2, 4, 8, 12, 16 weeks	Resected meniscus regenerated based on histological and MRI analyses. Articular cartilage and subchondral bone at MFC were preserved
Caminal et al. (2014) [58]	Sheep	Full-thickness cartilage defect in MFC and meniscal tear in the anterior horn of MM	Ovine bone marrow (autologous)	1.1×10^7 1.2×10^7	6, 12 months	Evidence of regeneration of articular cartilage and meniscus was case-dependent, but statistically significant improvement was found in specific macroscopic and histological parameters
Saulnier et al. (2015) [79]	Rabbit	Complete radial transection of the MM at mid-body	Equine umbilical cord	3.5×10^6	Day 15, 56	Lower cartilage fibrillation in early treatment than in delayed one. Cellular infiltration and a reduced expression of MMP-1, MMP-3, MMP-13 in the synovium. (Not addressed in meniscal status)
Nakagawa et al. (2015) [59]	Micro-minipig	Longitudinal tear in MM	Porcine synovium (allogenic)	2×10^7 (direct transplantation into the lesion)	2, 4, 12 weeks	Enhanced meniscal healing based on macroscopically, histologically, and T1rho mapping analysis. Higher tensile strength in MSC group
Qi et al. (2016) [80]	Rabbit	Anterior half of MM resection	Rabbit subcutaneous fat	2×10^6 (iron oxide (SPIO)-labeled or unlabeled)	6, 12 weeks	SPIO-MSCs promoted meniscus regeneration while providing protective effects from OA damage. Implanted MSCs adhered to the injured meniscus and differentiated specifically into meniscal cells
Lv et al. (2018) [81]	Sheep	ACL and MM resection	Autologous SVF or ASC	1×10^7 SVF w/ HA 1 or 5×10^7 ASC w/ HA	2, 18 weeks	Low-dose ASC improved MOCART and ICRS scores, compared with SVF. High-dose ASC increased cartilage thickness of MFC, while low-dose ASC showed no improvement. (Not addressed in meniscal repair)
McKinney et al. (2019) [82]	Rat	Full-thickness MM tear	Human bone marrow	5×10^5 (encapsuled in sodium alginate microspheres)	3 weeks	Exerted a chondroprotective therapeutic role but augmented the compensatory increases in osteophyte formation. (Not addressed in meniscal status)

Table 22.1 (continued)

Author (year)	Animal	Experimental model	Cell source	Cell number	Follow-up period	Outcomes
Mahmoud et al. (2019) [69]	Rabbit	ACL resection	Rabbit bone marrow (allogenic)	1×10^6 (single or multiple injections)	2 months	Multiple injections attenuated degenerative changes including loss of matrix staining, cell clustering, osteophyte formation, meniscal remodeling, and synovitis. A single injection demonstrated more modest results

ACL Anterior cruciate ligament, *MM* Medial meniscus, *MFC* Medial femoral condyle
OA Osteoarthritis, *Col II* Type II collagen, *Ihh* Indian hedgehog, *PTHLH* Parathyroid hormone-like hormone, *BMP2* Bone morphogenetic protein 2
CM Chondrogenic media, *BM* Basal media, *ICRS* International Cartilage Repair Society
MMP Matrix metalloproteinase
SVF Stromal vascular fraction, *ASC* Adipose mesenchymal stem cell, *HA* Hyaluronic acid, *MOCART* Magnetic resonance observation of cartilage repair tissue

injection with 50×10^6 cells (low dose), (2) injection with 150×10^6 cells (high dose), and (3) injection with sodium hyaluronate as a control group. At 2 years follow-up, they showed that osteoarthritic changes were reduced in both MSC groups, meniscal volume was increased by 24% for the high-dose MSC group and 6% in the low-dose MSC group as assessed by MRI, and no clinically important safety issues were identified.

Khalifeh Soltani et al. [67] investigated the safety and efficacy of the intra-articular injection of allogenic placental MSCs to promote knee OA healing. The patients with symptomatic knee OA were randomly divided into two groups to receive intra-articular injection of either 0.5–0.6×10^8 placental MSCs or normal saline. At 24 weeks follow-up, MSC injection improved clinical scores and cartilage thickness in 3 out of 28 measurement sites on magnetic resonance arthrography in the knee joint, while no significant healing of the meniscus was noted.

Sekiya et al. [68] studied the additional use of autologous synovial MSCs following surgical repair of complex degenerative medial meniscus tears in five male patients (34–57 years of age). The patients underwent an initial arthroscopy to confirm the location and type of lesions, the lesions was repaired with sutures and then a synovial tissue biopsy was arthroscopically per-

formed. The synovial tissue was cultured and expanded in vitro for 14 days to generate an MSC population. Then a repeat arthroscopy was performed to transplant 3.2–7.0×10^7 synovial MSC cell suspension directly to the site of the repair through a 18-gauge needle attached to 1 mL syringe. All patients reported a significant clinical score improvement at 2 years, and 3D MRI assessments indicated no evidence for a residual tear at the repair sites.

22.6 Future Directions

The studies reviewed in this chapter are highly heterogeneous with regard to animal modelling, methodology, surgical indications, evaluation time points, and objective outcomes evaluated. Efficacy is strongly suggested, as is the safely intra-articular transplantation or injection of culture-expanded MSCs in a variety of animal studies and early human trials. However, the "true" mechanisms of MSC action are still unclear.

Injection of culture-expanded MSCs could be a viable therapeutic option for the repair of meniscal lesions because it could be a simple, less-invasive procedure with lower cost, particularly when compared to more elaborate tissue engineering approaches combining cells with a biomaterial scaffold.

Table 22.2 Summary of intra-articular injections of MSCs for meniscal regeneration in clinical studies

Author (year)	Type of study	Patient number	Type of injury	Cell source	Cell number	Follow-up period	Outcomes
Vangsness et al. (2014) [66]	RCT	55 (19 HA vs 18 high dose vs 18 low dose)	Post-op partial medial meniscectomy (>50%)	Allogenic bone marrow	5×10^7 Or 1.5×10^8	2 years	Osteoarthritic changes were reduced in MSC group. MRI showed meniscal volume was increased in 24% of high-dose and 6% of low-dose group. Clinically important safety issues were identified
Khalifeh Soltani et al. (2019) [67]	RCT	20 (10 MSC vs 10 saline)	Symptomatic OA (KL-II ~ IV)	Allogenic placenta	$0.5–0.6 \times 10^8$	24 weeks	Improved clinical scores and cartilage thickness in about 10% of the total knee joint area. No significant healing in MM and LM
Onoi et al. (2019) [83]	Case report	2 (3 knees)	Symptomatic OA (KL-I ~ III), 1 MM deg. and 2 deg. tear on MMPH	SVF	$5.5–6 \times 10^6$	6 months	In second-look arthroscopy, cartilage defect was covered by regenerated cartilage, some cartilage fibrillation was reduced, and meniscus tear was repaired
Sekiya et al. (2019) [68]	Case report	5	Deg. tear of MM	Autologous synovium	$3.2–7 \times 10^7$ (direct transplantation on repaired site)	2 years	Clinical scores including Lysholm, KOOS, NRS were significantly improved, and 3D MRI showed meniscus tears were indistinguishable at 2 years. No serious adverse events occurred

RCT Randomized controlled trial, *HA* Hyaluronic acid
KL Kellgren Lawrence grade, *MM* Medial meniscus, *LM* Lateral meniscus
MM Medial meniscus, *MMPH* Medial meniscus posterior horn, *deg.* Degeneration/degenerative, *SVF* Stromal vascular fraction
KOOS Knee Injury and Osteoarthritis Outcome Scale, *NRS* Numerical Rating Scale

On the other hand, the limited current clinical evidence leaves the question of the efficacy of intra-articular injections of MSCs for patients with meniscal lesions unproven and uncertain. Thus, to repair and/or regenerate injured menisci and prevent the progression of OA in the clinical practice, additional studies are needed, and careful follow-up will be necessary to draw conclusions regarding the viability of this approach.

As for the choice of cells for such applications, MSCs derived from bone marrow, fat, and synovium are currently favored, due to ease of harvest and their chondrogenic capacities, but whether a bulk expansion and selection among CTP-derived clones through in vitro competition or a more strategic selection of CTP clones based on performance will provide the ideal type of cell for meniscal repair remains to be clarified. Therefore, a comparison of MSC preparation strategies and detailed analysis of population attributes, beyond traditional MSC surface markers and tri-lineage differentiation, must be performed to ultimately determine the ideal cell/cell population.

Allogenic MSCs have been recently investigated for joint tissue repair in animal studies [49, 59, 69], as well as clinical studies [70, 71], not only showing no detectable abnormal immunological reactions but also yielding comparable results to those with autologous cells (see Chap. 5). Considering tissue harvest for cell culture and cost effectiveness, an allogenic approach might become a reasonable strategy for cell-based therapy in the future, although safety issues have not yet been completely resolved presently.

22.7　Conclusion

Current literature indicates that the intra-articular injection of culture-expanded MSCs from a variety of tissue sources for the treatment of meniscal lesions has shown improvement in clinical outcomes and some potential for meniscal regeneration. However, the evidence is currently limited. Further studies are necessary to elucidate the healing mechanism(s) of MSCs for meniscal injuries and to rigorously define the clinical impact and durability of outcomes.

Take-Home Messages
- Intra-articular injection of culture-expanded MSCs for the treatment of meniscal lesions has shown improvement in clinical outcomes and some potential for meniscal regeneration, although the evidence is currently limited.
- There have generally not been safety concerns identified or side effects reported in the clinical use of MSC injections.
- As mechanism of action, recent works have shown that MSCs promoted the regeneration of joint components through their secretory activities of anti-inflammatory factors and trophic factors.
- As for the choice of cells, MSCs derived from bone marrow, fat, and synovium are currently favored, due to ease of harvest and their chondrogenic capacities.
- To repair and/or regenerate injured menisci and prevent the progression of OA, additional studies are needed, and careful follow-up will be necessary to draw conclusions regarding the viability of this approach.

References

1. Makris EA, Hadidi P, Athanasiou KA. The knee meniscus: structure-function, pathophysiology, current repair techniques, and prospects for regeneration. Biomaterials. 2011;32(30):7411–31.
2. Woodmass JM, LaPrade RF, Sgaglione NA, Nakamura N, Krych AJ. Meniscal repair: reconsidering indications, techniques, and biologic augmentation. J Bone Joint Surg Am. 2017;99(14):1222–31.
3. Shimomura K, Hamamoto S, Hart DA, Yoshikawa H, Nakamura N. Meniscal repair and regeneration: current strategies and future perspectives. J Clin Orthop Trauma. 2018;9(3):247–53.
4. Hede A, Jensen DB, Blyme P, Sonne-Holm S. Epidemiology of meniscal lesions in the knee. 1,215 open operations in Copenhagen 1982-84. Acta Orthop Scand. 1990;61(5):435–7.
5. Andrews SHJ, Adesida AB, Abusara Z, Shrive NG. Current concepts on structure-function relationships in the menisci. Connect Tissue Res. 2017;58(3–4):271–81.
6. Fairbank TJ. Knee joint changes after meniscectomy. J Bone Joint Surg Br. 1948;30B(4):664–70.
7. Arnoczky SP, Warren RF. Microvasculature of the human meniscus. Am J Sports Med. 1982;10(2):90–5.
8. Grant JA, Wilde J, Miller BS, Bedi A. Comparison of inside-out and all-inside techniques for the repair of isolated meniscal tears: a systematic review. Am J Sports Med. 2012;40(2):459–68.
9. Fillingham YA, Riboh JC, Erickson BJ, Bach BR Jr, Yanke AB. Inside-out versus all-inside repair of isolated meniscal tears: an updated systematic review. Am J Sports Med. 2017;45(1):234–42.
10. Foad A. Self-limited healing of a radial tear of the lateral meniscus. Knee Surg Sports Traumatol Arthrosc. 2012;20(5):933–6.
11. Roos H, Lauren M, Adalberth T, Roos EM, Jonsson K, Lohmander LS. Knee osteoarthritis after meniscectomy: prevalence of radiographic changes after twenty-one years, compared with matched controls. Arthritis Rheum. 1998;41(4):687–93.
12. Vundelinckx B, Vanlauwe J, Bellemans J. Long-term subjective, clinical, and radiographic outcome evalua-

tion of meniscal allograft transplantation in the knee. Am J Sports Med. 2014;42(7):1592–9.

13. Rosso F, Bisicchia S, Bonasia DE, Amendola A. Meniscal allograft transplantation: a systematic review. Am J Sports Med. 2015;43(4):998–1007.

14. Grassi A, Zaffagnini S, Marcheggiani Muccioli GM, Benzi A, Marcacci M. Clinical outcomes and complications of a collagen meniscus implant: a systematic review. Int Orthop. 2014;38(9):1945–53.

15. Leroy A, Beaufils P, Faivre B, Steltzlen C, Boisrenoult P, Pujol N. Actifit(R) polyurethane meniscal scaffold: MRI and functional outcomes after a minimum follow-up of 5 years. Orthop Traumatol Surg Res. 2017;103(4):609–14.

16. Rodeo SA, Seneviratne A, Suzuki K, Felker K, Wickiewicz TL, Warren RF. Histological analysis of human meniscal allografts. A preliminary report. J Bone Joint Surg Am. 2000;82-a(8):1071–82.

17. Monllau JC, Gelber PE, Abat F, Pelfort X, Abad R, Hinarejos P, et al. Outcome after partial medial meniscus substitution with the collagen meniscal implant at a minimum of 10 years' follow-up. Arthroscopy. 2011;27(7):933–43.

18. Hirschmann MT, Keller L, Hirschmann A, Schenk L, Berbig R, Luthi U, et al. One-year clinical and MR imaging outcome after partial meniscal replacement in stabilized knees using a collagen meniscus implant. Knee Surg Sports Traumatol Arthrosc. 2013;21(3):740–7.

19. Schuttler KF, Haberhauer F, Gesslein M, Heyse TJ, Figiel J, Lorbach O, et al. Midterm follow-up after implantation of a polyurethane meniscal scaffold for segmental medial meniscus loss: maintenance of good clinical and MRI outcome. Knee Surg Sports Traumatol Arthrosc. 2016;24(5):1478–84.

20. Hutchinson ID, Rodeo SA, Perrone GS, Murray MM. Can platelet-rich plasma enhance anterior cruciate ligament and meniscal repair? J Knee Surg. 2015;28(1):19–28.

21. Kraeutler MJ, Mitchell JJ, Chahla J, McCarty EC, Pascual-Garrido C. Intra-articular implantation of mesenchymal stem cells, part 2: a review of the literature for meniscal regeneration. Orthop J Sports Med. 2017;5(1):2325967116680814.

22. Wang R, Jiang W, Zhang L, Xie S, Zhang S, Yuan S, et al. Intra-articular delivery of extracellular vesicles secreted by chondrogenic progenitor cells from MRL/MpJ superhealer mice enhances articular cartilage repair in a mouse injury model. Stem Cell Res Ther. 2020;11(1):93.

23. Lietman C, Wu B, Lechner S, Shinar A, Sehgal M, Rossomacha E, et al. Inhibition of Wnt/beta-catenin signaling ameliorates osteoarthritis in a murine model of experimental osteoarthritis. JCI Insight. 2018;3(3)

24. Kraeutler MJ, Mitchell JJ, Chahla J, McCarty EC, Pascual-Garrido C. Intra-articular implantation of mesenchymal stem cells, part 1: a review of the literature for prevention of postmeniscectomy osteoarthritis. Orthop J Sports Med. 2017;5(1):2325967116680815.

25. Thompson CL, Plant JC, Wann AK, Bishop CL, Novak P, Mitchison HM, et al. Chondrocyte expansion is associated with loss of primary cilia and disrupted hedgehog signalling. Eur Cell Mater. 2017;34:128–41.

26. Tan GK, Dinnes DL, Myers PT, Cooper-White JJ. Effects of biomimetic surfaces and oxygen tension on redifferentiation of passaged human fibrochondrocytes in 2D and 3D cultures. Biomaterials. 2011;32(24):5600–14.

27. Kon E, Filardo G, Tschon M, Fini M, Giavaresi G, Marchesini Reggiani L, et al. Tissue engineering for total meniscal substitution: animal study in sheep model—results at 12 months. Tissue Eng Part A. 2012;18(15–16):1573–82.

28. Kang SW, Son SM, Lee JS, Lee ES, Lee KY, Park SG, et al. Regeneration of whole meniscus using meniscal cells and polymer scaffolds in a rabbit total meniscectomy model. J Biomed Mater Res A. 2006;78(3):659–71.

29. Koizumi K, Ebina K, Hart DA, Hirao M, Noguchi T, Sugita N, et al. Synovial mesenchymal stem cells from osteo- or rheumatoid arthritis joints exhibit good potential for cartilage repair using a scaffold-free tissue engineering approach. Osteoarthr Cartil. 2016;24(8):1413–22.

30. Weiss ARR, Dahlke MH. Immunomodulation by mesenchymal stem cells (MSCs): mechanisms of action of living, apoptotic, and dead MSCs. Front Immunol. 2019;10:1191.

31. Strioga M, Viswanathan S, Darinskas A, Slaby O, Michalek J. Same or not the same? Comparison of adipose tissue-derived versus bone marrow-derived mesenchymal stem and stromal cells. Stem Cells Dev. 2012;21(14):2724–52.

32. Pittenger MF, Mackay AM, Beck SC, Jaiswal RK, Douglas R, Mosca JD, et al. Multilineage potential of adult human mesenchymal stem cells. Science. 1999;284(5411):143–7.

33. Shimomura K, Ando W, Moriguchi Y, Sugita N, Yasui Y, Koizumi K, et al. Next generation mesenchymal stem cell (MSC)–based cartilage repair using scaffold-free tissue engineered constructs generated with synovial mesenchymal stem cells. Cartilage. 2015;6(Suppl 2):13S–29S.

34. Mantripragada VP, Piuzzi NS, Bova WA, Boehm C, Obuchowski NA, Lefebvre V, et al. Donor-matched comparison of chondrogenic progenitors resident in human infrapatellar fat pad, synovium, and periosteum—implications for cartilage repair. Connect Tissue Res. 2019;60(6):597–610.

35. Sakaguchi Y, Sekiya I, Yagishita K, Muneta T. Comparison of human stem cells derived from various mesenchymal tissues: superiority of synovium as a cell source. Arthritis Rheum. 2005;52(8):2521–9.

36. Yamashita A, Morioka M, Yahara Y, Okada M, Kobayashi T, Kuriyama S, et al. Generation of scaffoldless hyaline cartilaginous tissue from human iPSCs. Stem Cell Reports. 2015;4(3):404–18.

37. Caplan AI. Mesenchymal stem cells: time to change the name! Stem Cells Transl Med. 2017;6(6):1445–51.
38. Le Blanc K, Mougiakakos D. Multipotent mesenchymal stromal cells and the innate immune system. Nat Rev Immunol. 2012;12(5):383–96.
39. Caplan AI. Adult mesenchymal stem cells: when, where, and how. Stem Cells Int. 2015;2015:628767.
40. Mianehsaz E, Mirzaei HR, Mahjoubin-Tehran M, Rezaee A, Sahebnasagh R, Pourhanifeh MH, et al. Mesenchymal stem cell-derived exosomes: a new therapeutic approach to osteoarthritis? Stem Cell Res Ther. 2019;10(1):340.
41. Ruiz M, Cosenza S, Maumus M, Jorgensen C, Noël D. Therapeutic application of mesenchymal stem cells in osteoarthritis. Expert Opin Biol Ther. 2016;16(1):33–42.
42. Richards MM, Maxwell JS, Weng L, Angelos MG, Golzarian J. Intra-articular treatment of knee osteoarthritis: from anti-inflammatories to products of regenerative medicine. Physician Sports Med. 2016;44(2):101–8.
43. Murphy MB, Moncivais K, Caplan AI. Mesenchymal stem cells: environmentally responsive therapeutics for regenerative medicine. Exp Mol Med. 2013;45(11):e54.
44. Goldring MB. The role of the chondrocyte in osteoarthritis. Arthritis Rheum. 2000;43(9):1916–26.
45. Goldring MB. Update on the biology of the chondrocyte and new approaches to treating cartilage diseases. Best Pract Res Clin Rheumatol. 2006;20(5):1003–25.
46. Lawrence JT, Birmingham J, Toth AP. Emerging ideas: prevention of posttraumatic arthritis through interleukin-1 and tumor necrosis factor-alpha inhibition. Clin Orthop Relat Res. 2011;469(12):3522–6.
47. McNulty AL, Moutos FT, Weinberg JB, Guilak F. Enhanced integrative repair of the porcine meniscus in vitro by inhibition of interleukin-1 or tumor necrosis factor alpha. Arthritis Rheum. 2007;56(9):3033–42.
48. McNulty AL, Weinberg JB, Guilak F. Inhibition of matrix metalloproteinases enhances in vitro repair of the meniscus. Clin Orthop Relat Res. 2009;467(6):1557–67.
49. Shimomura K, Ando W, Tateishi K, Nansai R, Fujie H, Hart DA, et al. The influence of skeletal maturity on allogenic synovial mesenchymal stem cell-based repair of cartilage in a large animal model. Biomaterials. 2010;31(31):8004–11.
50. Shimomura K, Yasui Y, Koizumi K, Chijimatsu R, Hart DA, Yonetani Y, et al. First-in-human pilot study of implantation of a scaffold-free tissue-engineered construct generated from autologous synovial mesenchymal stem cells for repair of knee chondral lesions. Am J Sports Med. 2018;46(10):2384–93.
51. Andrews SH, Rattner JB, Abusara Z, Adesida A, Shrive NG, Ronsky JL. Tie-fibre structure and organization in the knee menisci. J Anat. 2014;224(5):531–7.
52. Vetri V, Dragnevski K, Tkaczyk M, Zingales M, Marchiori G, Lopomo NF, et al. Advanced microscopy analysis of the micro-nanoscale architecture of human menisci. Sci Rep. 2019;9(1):18,732.
53. Poole R, Blake S, Buschmann M, Goldring S, Laverty S, Lockwood S, et al. Recommendations for the use of preclinical models in the study and treatment of osteoarthritis. Osteoarthr Cartil. 2010;18(Suppl 3):S10–6.
54. Gerwin N, Bendele AM, Glasson S, Carlson CS. The OARSI histopathology initiative—recommendations for histological assessments of osteoarthritis in the rat. Osteoarthr Cartil. 2010;18(Suppl 3):S24–34.
55. Vodicka P, Smetana K Jr, Dvorankova B, Emerick T, Xu YZ, Ourednik J, et al. The miniature pig as an animal model in biomedical research. Ann N Y Acad Sci. 2005;1049:161–71.
56. Horie M, Choi H, Lee RH, Reger RL, Ylostalo J, Muneta T, et al. Intra-articular injection of human mesenchymal stem cells (MSCs) promote rat meniscal regeneration by being activated to express Indian hedgehog that enhances expression of type II collagen. Osteoarthr Cartil. 2012;20(10):1197–207.
57. Hatsushika D, Muneta T, Horie M, Koga H, Tsuji K, Sekiya I. Intraarticular injection of synovial stem cells promotes meniscal regeneration in a rabbit massive meniscal defect model. J Orthop Res. 2013;31(9):1354–9.
58. Caminal M, Fonseca C, Peris D, Moll X, Rabanal RM, Barrachina J, et al. Use of a chronic model of articular cartilage and meniscal injury for the assessment of long-term effects after autologous mesenchymal stromal cell treatment in sheep. New Biotechnol. 2014;31(5):492–8.
59. Nakagawa Y, Muneta T, Kondo S, Mizuno M, Takakuda K, Ichinose S, et al. Synovial mesenchymal stem cells promote healing after meniscal repair in microminipigs. Osteoarthritis Cartilage. 2015;23:1007–17.
60. Jacob G, Shimomura K, Krych AJ, Nakamura N. The meniscus tear: a review of stem cell therapies. Cells. 2019;9(1):32.
61. Chahla J, Piuzzi NS, Mitchell JJ, Dean CS, Pascual-Garrido C, LaPrade RF, et al. Intra-articular cellular therapy for osteoarthritis and focal cartilage defects of the knee: a systematic review of the literature and study quality analysis. J Bone Joint Surg Am. 2016;98(18):1511–21.
62. Pas HI, Winters M, Haisma HJ, Koenis MJ, Tol JL, Moen MH. Stem cell injections in knee osteoarthritis: a systematic review of the literature. Br J Sports Med. 2017;51(15):1125–33.
63. Ha CW, Park YB, Kim SH, Lee HJ. Intra-articular mesenchymal stem cells in osteoarthritis of the knee: a systematic review of clinical outcomes and evidence of cartilage repair. Arthroscopy. 2019;35(1):277–88.e2.
64. Kim SH, Ha CW, Park YB, Nam E, Lee JE, Lee HJ. Intra-articular injection of mesenchymal stem cells for clinical outcomes and cartilage repair in osteoarthritis of the knee: a meta-analysis of randomized controlled trials. Arch Orthop Trauma Surg. 2019;139(7):971–80.
65. McIntyre JA, Jones IA, Han B, Vangsness CT Jr. Intra-articular mesenchymal stem cell therapy for the

human joint: a systematic review. Am J Sports Med. 2018;46(14):3550–63.

66. Vangsness CT Jr, Farr J 2nd, Boyd J, Dellaero DT, Mills CR, LeRoux-Williams M. Adult human mesenchymal stem cells delivered via intra-articular injection to the knee following partial medial meniscectomy: a randomized, double-blind, controlled study. J Bone Joint Surg Am. 2014;96(2):90–8.

67. Khalifeh Soltani S, Forogh B, Ahmadbeigi N, Hadizadeh Kharazi H, Fallahzadeh K, Kashani L, et al. Safety and efficacy of allogenic placental mesenchymal stem cells for treating knee osteoarthritis: a pilot study. Cytotherapy. 2019;21(1):54–63.

68. Sekiya I, Koga H, Otabe K, Nakagawa Y, Katano H, Ozeki N, et al. Additional use of synovial mesenchymal stem cell transplantation following surgical repair of a complex degenerative tear of the medial meniscus of the knee: a case report. Cell Transplant. 2019;28(11):1445–54.

69. Mahmoud EE, Adachi N, Mawas AS, Deie M, Ochi M. Multiple intra-articular injections of allogeneic bone marrow-derived stem cells potentially improve knee lesions resulting from surgically induced osteoarthritis: an animal study. Bone Joint J. 2019;101-b(7):824–31.

70. de Windt TS, Vonk LA, Slaper-Cortenbach IC, van den Broek MP, Nizak R, van Rijen MH, et al. Allogeneic mesenchymal stem cells stimulate cartilage regeneration and are safe for single-stage cartilage repair in humans upon mixture with recycled autologous chondrons. Stem Cells. 2017;35(1):256–64.

71. Song JS, Hong KT, Kim NM, Jung JY, Park HS, Lee SH, et al. Implantation of allogenic umbilical cord blood-derived mesenchymal stem cells improves knee osteoarthritis outcomes: two-year follow-up. Regen Ther. 2020;14:32–9.

72. Agung M, Ochi M, Yanada S, Adachi N, Izuta Y, Yamasaki T, et al. Mobilization of bone marrow-derived mesenchymal stem cells into the injured tissues after intraarticular injection and their contribution to tissue regeneration. Knee Surg Sports Traumatol Arthrosc. 2006;14(12):1307–14.

73. Horie M, Sekiya I, Muneta T, Ichinose S, Matsumoto K, Saito H, et al. Intra-articular injected synovial stem cells differentiate into meniscal cells directly and promote meniscal regeneration without mobilization to distant organs in rat massive meniscal defect. Stem Cells. 2009;27(4):878–87.

74. Al Faqeh H, Nor Hamdan BM, Chen HC, Aminuddin BS, Ruszymah BH. The potential of intra-articular injection of chondrogenic-induced bone marrow stem cells to retard the progression of osteoarthritis in a sheep model. Exp Gerontol. 2012;47(6):458–64.

75. Duygulu F, Demirel M, Atalan G, Kaymaz FF, Kocabey Y, Dülgeroğlu TC, et al. Effects of intra-articular administration of autologous bone marrow aspirate on healing of full-thickness meniscal tear: an experimental study on sheep. Acta Orthop Traumatol Turc. 2012;46(1):61–7.

76. Shen W, Chen J, Zhu T, Yin Z, Chen X, Chen L, et al. Osteoarthritis prevention through meniscal regeneration induced by intra-articular injection of meniscus stem cells. Stem Cells Dev. 2013;22(14):2071–82.

77. Shen W, Chen J, Zhu T, Chen L, Zhang W, Fang Z, et al. Intra-articular injection of human meniscus stem/progenitor cells promotes meniscus regeneration and ameliorates osteoarthritis through stromal cell-derived factor-1/CXCR4-mediated homing. Stem Cells Transl Med. 2014;3(3):387–94.

78. Hatsushika D, Muneta T, Nakamura T, Horie M, Koga H, Nakagawa Y, et al. Repetitive allogeneic intraarticular injections of synovial mesenchymal stem cells promote meniscus regeneration in a porcine massive meniscus defect model. Osteoarthr Cartil. 2014;22(7):941–50.

79. Saulnier N, Viguier E, Perrier-Groult E, Chenu C, Pillet E, Roger T, et al. Intra-articular administration of xenogeneic neonatal mesenchymal stromal cells early after meniscal injury down-regulates metalloproteinase gene expression in synovium and prevents cartilage degradation in a rabbit model of osteoarthritis. Osteoarthr Cartil. 2015;23(1):122–33.

80. Qi Y, Yang Z, Ding Q, Zhao T, Huang Z, Feng G. Targeted transplantation of iron oxide-labeled, adipose-derived mesenchymal stem cells in promoting meniscus regeneration following a rabbit massive meniscal defect. Exp Ther Med. 2016;11(2):458–66.

81. Lv X, He J, Zhang X, Luo X, He N, Sun Z, et al. Comparative efficacy of autologous stromal vascular fraction and autologous adipose-derived mesenchymal stem cells combined with hyaluronic acid for the treatment of sheep osteoarthritis. Cell Transplant. 2018;27(7):1111–25.

82. McKinney JM, Doan TN, Wang L, Deppen J, Reece DS, Pucha KA, et al. Therapeutic efficacy of intra-articular delivery of encapsulated human mesenchymal stem cells on early stage osteoarthritis. Eur Cell Mater. 2019;37:42–59.

83. Onoi Y, Hiranaka T, Nishida R, Takase K, Fujita M, Hida Y, et al. Second-look arthroscopic findings of cartilage and meniscus repair after injection of adipose-derived regenerative cells in knee osteoarthritis: report of two cases. Regen Ther. 2019;11:212–6.

Meniscal Lesions: Biologics

Stefano Zaffagnini, Alberto Poggi, Luca Andriolo, Angelo Boffa, and Giuseppe Filardo

23.1 Introduction

The meniscus is an important knee structure that improves weight-bearing distribution and shock absorption and increases joint congruency contributing to articular stability, overall playing a fundamental function for knee health [1]. Meniscal lesions represent one of the most frequent orthopedic injuries, with a mean annual prevalence of meniscal tears resulting in meniscectomy estimated around 61 per 100,000 inhabitants [2]. The loss of meniscal tissue, both in injured and in post-meniscectomy (partial or total) knees, can alter the joint environment jeopardizing the long-term articular homeostasis [3] with the risk to develop early osteoarthritis (OA) and a consequent impairment of the quality of life [4]. For this reason, especially in young patients, attempts to preserve as much meniscal tissue as possible are paramount [5]. Several injury types can affect the meniscal tissue, and different classification systems have been proposed based on lesion etiology and anatomic location and pattern. Among them, the ISAKOS classification is largely applied to assess tear depth, length, location, pattern, as well as tissue quality and percentage of the meniscus excised, with sufficient inter-observer reliability [6]. The proper classification of meniscal tears is fundamental for the correct management of meniscal injuries. Moreover, different factors directly related to the meniscal lesion such as etiology, tissue quality, location, pattern, and any associated lesions, as well as those related to patient age, activity level, general health status, and expectations [7], can guide the choice of the specific conservative or surgical treatment strategy.

23.2 The Management of Meniscal Lesions

The primary distinction in the management of meniscal injuries is represented by the choice between conservative and surgical treatment. In detail, non-surgical management represents the first-line treatment in degenerative lesions or in acute knee trauma, and it should be carried out at least 3–6 months, when mechanical symptoms such as catching or locking do not dominate the clinical picture [8]. On the other hand, a surgical approach should be considered in case of symptoms persisting after 3–6 months of conservative treatment or when meniscal injury determines mechanical symptoms. Partial or total meniscec-

S. Zaffagnini · A. Poggi (✉) · L. Andriolo · A. Boffa
Clinica Ortopedica e Traumatologica II, IRCCS
Istituto Ortopedico Rizzoli, Bologna, Italy
e-mail: stefano.zaffagnini@unibo.it

G. Filardo
Rizzoli Orthopaedic Institute
Bologna, Italy

Università della Svizzera Italiana, Ente Ospedaliero
Cantonale, Lugano, Switzerland
e-mail: ortho@gfilardo.com

© ISAKOS 2022
G. Filardo et al. (eds.), *Orthobiologics*, https://doi.org/10.1007/978-3-030-84744-9_23

tomy, meniscal repair, and meniscal reconstruction are the current available surgical options to address the different meniscal tears. Meniscectomy has been, and still is, a procedure largely adopted, because of its relative easiness, low post-operative morbidity, and good short-term results. Nevertheless, nowadays there is an overall evidence-based agreement to limit the use of this procedure only to selected patients with persistent mechanical symptoms derived from non-repairable meniscal tears. The reason lies with the degenerative effects of meniscal tissue loss on knee homeostasis with the consequent risk to develop early OA [9]. In this light, meniscal repair procedures are increasingly performed with the rationale to preserve as much meniscal tissue as possible, showing better long-term results in terms of clinical outcomes and cost-effectiveness compared to meniscectomy [10, 11]. This procedure can be performed through an open-access or arthroscopically with different techniques. Traumatic meniscal tears with a longitudinal pattern located in the peripheral red-red zone represent one of the main indications to perform a meniscal repair because of their high healing potential, but the indications may also be extended to horizontal lesions in young patients; radial, ramp, and root lesions; and tears located in the red-white zone [12–14]. Finally, meniscus reconstruction procedures with scaffolds or meniscal allograft transplantation (MAT) represent a surgical rescue option in relatively young patients affected by important symptomatic loss of meniscal tissue [7].

23.3 The Rationale of Orthobiologics Injections

The results of both conservative and surgical meniscal treatments are not always satisfactory. The reasons can be traced to several aspects, represented by demographic factors (age, sex), interventional factors (technique, time to surgery), and the specific features of the meniscal tear [15]. Moreover, two biological factors have to be taken into account. The consequent alterations in knee homeostasis following a meniscus lesion may lead to a clinical problem per se. Furthermore, meniscal tissue presents a lim-

ited intrinsic healing potential in terms of chemotaxis, cell proliferation, and matrix production, determined by the poor vascularity especially in the white-white zone or by degenerative processes, which hinder the healing potential [16–18].

In order to improve the outcome following conservative and surgical treatment, the injections of biologically active substances like corticosteroids, hyaluronic acid (HA), and platelet-rich plasma (PRP) have been proposed as a simple and minimally invasive treatment strategy. They may act directly onto the meniscal tissue aiming at improving its healing or indirectly by restoring or improving the joint homeostasis. The biologic injective approach entails different ways to deliver biologic substances to improve either conservative or surgical meniscal treatment. Specifically, a simple intra-articular injection or an intra-meniscal ultrasound−/fluoroscopy-guided injection directly in the lesion site (Fig. 23.1) have been proposed to enhance a non-surgical approach. On the other hand, injections have also been used combined with surgical procedures such as meniscal suture or meniscal reconstruction during the same surgical time, to obtain a direct delivery of biological substances in the specific meniscal site or following surgery to provide a more favorable healing environment.

Corticosteroids have been the first intra-articular injective option introduced in the clinical practice. They showed an anti-inflammatory short-term effect on the knee joint through a complex multiplicity of actions [19, 20], with a minimal risk of systemic effects, and thus may be used to address the homeostatic changes started by meniscal lesions. Unfortunately, long-term

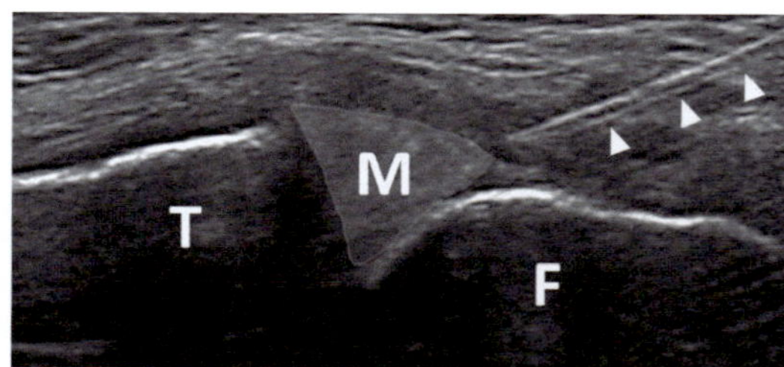

Fig. 23.1 Ultrasound image illustrating an intra-meniscal PRP injection. *T* Tibia, *F* Femur, *M* Meniscal tissue artificially highlighted, arrowheads: needle

benefits have not been confirmed, and risks of tissue atrophy, joint destruction, and cartilage degeneration have been also reported [21]. Hyaluronic acid (HA) is another injectable option, a glycosaminoglycan normally present in the synovial fluid in the adult knee with a concentration ranging from 2.5 to 4.0 mg/mL that provides joint lubrication and shock absorption and acts as the backbone for the proteoglycans of the extracellular matrix [20]. HA can be isolated from different sources, such as rooster comb or bacterial fermentation (streptococci) with or without biochemical modifications, and its intra-articular delivery showed promising results in several studies in terms of pain reduction combined to anti-inflammatory and chondro-protective effects in degenerated knees [22, 23].

More recently, research efforts have been invested in the field of hemoderivatives, aiming at developing, other than the anti-inflammatory and homeostatic effects already provided by corticosteroids and HA, a solution also presenting anabolic/regenerative effects. Platelet-rich plasma (PRP) represents a biological derivative of blood largely applied in many clinical fields, including orthopedic pathologies [24], being a simple way to exploit the potential of a "cocktail" of different growth factors like PDGF, TGF-β1, VEGF, and IFG-1. The importance of its composition lies on the evidence that the combination of different growth factors showed to improve the meniscal cell activity promoting meniscal tissue repair [25, 26], contrary to the administration of single growth factors, which offered instead poor results [27, 28]. Different preclinical studies confirmed significant positive effects of PRP growth factors on matrix production, cell activity, and proliferation [25, 26, 29], increasing the interest of orthopedic surgeons on PRP applications also to treat meniscal injuries.

According to their rationale, orthobiologics injections could represent an interesting minimally invasive treatment for meniscal injuries. In the following paragraphs, the biologic injective options available in the clinical practice are described, reporting their indications as well as the evidence about their results in the management of meniscal lesions.

23.4 Corticosteroid Injections

Corticosteroid injections have been performed for almost 60 years in the management of knee pain and especially in knee OA, due to their anti-inflammatory effects. Nevertheless, few studies have been reported about the corticosteroid injective treatment of meniscal injuries, prevalently limited to degenerative tears. This pattern of meniscal lesions represents a challenge in orthopedics, since usually they are not isolated lesions but are combined with chondral damage and a progressively altered joint environment. In these conditions, a surgical approach is not indicated, while a conservative management based on anti-inflammatory drugs combined with physical therapy represents the first line and should be tried for at least 3–6 months [7, 30]. In case of failure or aiming at improving the efficacy of a conservative treatment, intra-articular corticosteroid injections represent a further option. On this regard, Vermesan et al. [31] conducted a study including 120 consecutive cases of non-traumatic symptomatic knees affected by degenerative lesions of the medial compartment (cartilage and meniscus) assessed on MRI. Patients were randomized to receive either an intra-articular steroid injection or an arthroscopic debridement, obtaining a significant improvement of the scores for all the examined cases. After 1 month of follow-up, symptoms reappeared in 12 patients in the steroid group and 7 in the arthroscopy group, respectively. Accordingly, the authors concluded that degenerative medial meniscal tears in the presence of OA can only marginally benefit from arthroscopic debridement over intra-articular steroid injections at short-term follow-up. Six years later, Wilderman et al. performed a retrospective study on 135 patients treated with a meniscal-targeted injection of the corticosteroid triamcinolone for meniscal tears or degenerative fraying. The results obtained suggested the efficacy of ultrasound-guided meniscal injections of corticosteroids in alleviating chronic knee pain associated with OA, meniscal tears, and meniscal degeneration for a mean of 5 weeks [32].

Besides the mere conservative management, some authors explored the use of intra-articular injections of corticosteroids after partial or total meniscectomy. In detail, Rasmussen et al. [33] published a double-blind randomized study in which 60 patients undergoing arthroscopic meniscectomy were treated at the end of the procedure with either intra-articular saline as control group, intra-articular bupivacaine and morphine in the first treatment group, or with the same dose of bupivacaine and morphine, plus intra-articular methylprednisolone in the second treatment group. During the 10-day follow-up, pain related to movement and walking, leg muscle force, joint effusion, and the use of crutches were assessed, showing a significantly reduction in pain, time of immobilization, and duration of convalescence obtained with bupivacaine and morphine, with a further reduction of pain and functional impairment (including the duration of convalescence) by the addition of the intra-articular injection of methylprednisolone acetate. No data are available in the literature regarding the injection of corticosteroids after meniscal repair or after meniscal reconstruction procedures.

Overall, although literature evidence suggests an anti-inflammatory effect of steroid injections to improve the conservative treatment of degenerative lesions or the results of partial meniscectomy, it should be noted that their efficacy was demonstrated at short−/very short-term follow-up, not supporting a long-lasting effect and neither changes in the meniscal tissue. Moreover, the potential deleterious long-term effects of intra-articular steroid injections on biologic tissues further discourage the use of such treatments for meniscal lesions [21]. Thus, until stronger evidence can support safety and efficacy of this approach, indications for intra-articular injections should be limited to the more acute inflammatory phases not responsive to less invasive treatments, while more evidence is warranted for intra-meniscus injections and multiple injections should be avoided.

23.5 Hyaluronic Acid Injections

The term "viscosupplementation" was introduced in the 1970s [34] referring to an intra-articular injection with HA to reduce pain and improve joint mobility in patients affected by knee OA [35, 36]. Regardless of the potential rationale for using HA intra-articular injections in joints affected by meniscal lesions, few studies in the literature described its application in this context. In 2016, Zorzi et al. [37] investigated the efficacy of the intra-articular administration of a recently developed hydrogel formulation obtained from a HA derivative, to treat degenerative meniscal tears. Fifty patients were recruited and randomized in two groups: patients in the control group underwent conservative treatment for 2 weeks, while patients in the treatment group received also two intra-articular injections of HA 2 weeks apart. The treatment group showed a significant VAS pain reduction and a significant reduction in length and depth of the degenerative meniscal lesions assessed with MRI.

HA intra-articular injections have been applied also after meniscectomy. In 2008, Huskin et al. [38] published a prospective multicenter study on 62 patients after arthroscopic meniscectomy. Patients received three intra-articular HA injections in the target knee with 1-week intervals. Even if the absence of a control group did not allow to draw conclusions on the real effect of HA, the authors underlined that the clinical scores improved significantly, and the symptomatic efficacy was highest at 12 weeks and maintained at 26 and 52 weeks. In 2010, Thein et al. [39] performed a randomized, single-blind, controlled study evaluating the clinical outcome after one HA injection in patients who underwent arthroscopic meniscectomy. Fifty-six patients were divided into two groups: HA vs saline injected immediately post-arthroscopy. Controls reported more pain at week 1 with respect to patients in the treatment group, and at 4 weeks postoperatively, none of the patients in the treatment group had consumed analgesics, while 9 patients out of 28 in the control group reported acetaminophen intake. More recently, Filardo et al. [40] performed a larger double-blind randomized controlled trial which did not confirm the previous promising results. Ninety patients undergoing meniscectomy were randomized into two groups: a single injection of HA at the end of the arthroscopic procedure versus surgery alone. The results showed that the administration of a single HA injection at the end of the surgical procedure did not prove to be a successful strategy in providing either faster functional recovery or symptomatic improvement after meniscectomy.

The clinical application of HA to improve meniscal repair or meniscal reconstruction procedures has not yet been investigated in the clinical setting, although some preclinical evidence is available. In particular, Sonoda et al. [41] performed an animal study to assess the effect of HA on meniscus injury repair: 35 mature New Zealand White rabbits underwent bilateral meniscus injury and repair with a longitudinal tear created in the medial meniscus and repaired with horizontally placed nylon sutures. The left knee joints received five weekly intra-articular injections of HA, while the right knees were injected with phosphate-buffered saline (the carrier vehicle of the hyaluronan). Twelve weeks after repair, no significant differences between the two groups were found in the peripheral region, although a greater level of collagen remodeling was underlined in the HA group. In the inner region, poor healing response was observed in both treatment groups although the water content in the HA-treated menisci was significantly lower than that in the control group, indicating a lower level of swelling.

Considering the available literature, the data to support the use of intra-articular HA injections to address meniscal lesions are still limited (Table 23.1), with a low-level evidence supporting a possible indication for the conservative intra-articular treatment of knees affected by degenerative meniscal lesions, while direct effects on meniscal tissue or on post-meniscectomy knees remain to be proven.

23.6 Platelet-Rich Plasma Injections

PRP application has largely spread in different orthopedic fields, particularly for the treatment of knee OA [42]. The possibility to positively mod-

ulate different biological mechanisms in the joint could offer potential benefits also in the treatment of meniscal lesions [29, 43]. Accordingly, PRP injections could represent a promising and simple option for meniscal treatment, performed isolated or combined to other standard available procedures, with some encouraging preliminary findings.

No study is currently available documenting the results of intra-articular PRP injections as conservative treatment for meniscus lesions. On the other hand, there are various studies analyzing PRP injections to treat knee OA [42], a degenerative joint disease which also affects the meniscal tissue; however, no study specifically analyzed the results of intra-articular PRP specifically applied for meniscus lesions. Conversely, some studies have been performed on PRP injected with a percutaneous approach directly into the meniscal tissue for the treatment of degenerative intrasubstance meniscal tears, showing promising results. In 2015, Blanke et al. [44] performed a retrospective analysis of ten recreational athletes with intrasubstance meniscal lesions assessed on MRI who underwent three sequential weekly injections of PRP in the area of meniscal lesion under fluoroscopy guide. The results showed a significant pain reduction 6 months after treatment. Six out of ten patients reported an increase of sports activity and returned to the previous sports activity level. At the same follow-up, four out of ten patients showed an improvement of the meniscal lesion at MRI, while two patients showed meniscal lesion progression. The technique applied is limited by the uncertainty on the injection site, as disclosed by the authors themselves, while the use of ultrasound guidance may be more precise. To this regard, in 2019 Guenoun et al. [45] published a retrospective study on ten patients with degenerative meniscal tears treated with a single ultrasound-guided PRP injection in the meniscal wall and in the peri-meniscal space. All patients were assessed clinically at 6 months of follow-up and showed a significant improvement of KOOS total score, not confirmed by VAS for pain evaluation. All six patients who practiced sports regularly before the onset of symptoms were able to resume

Table 23.1 Clinical literature evidence on orthobiologics injectable therapies for meniscus lesions

	Orthobiologics injections	Clinical literature evidence
HA injections	HA injection alone	**Zorzi et al. (2015)**: Randomized controlled study: good clinical evidence
	Meniscectomy + HA injection	**Huskin et al. (2008)**: Prospective multicenter study (no control group): good clinical improvement **Thein et al. (2010)**: Randomized, single-blind, controlled study: no significant differences in knee functionality between the two groups **Filardo et al. (2016)**: Double-blind randomized controlled trial: no significant differences in the clinical improvement between the two groups
	Meniscal repair + HA injection	No studies available in the literature
	Meniscal reconstruction + HA injection	No studies available in the literature
PRP injections	PRP injection alone	**Blanke et al. (2015)**: Retrospective study on athletes who underwent 3 percutaneous PRP injections under fluoroscopy guide. Good clinical and instrumental results **Guenoun et al. (2019)**: Retrospective study on a single ultrasound-guided injection. Good clinical and instrumental results **Kaminski et al. (2019)**: Double-blind randomized controlled trial with treatment group treated by ultrasound trephination + ultrasound-guided PRP injection. Significant clinical improvement in the treatment group
	Meniscectomy + PRP injection	No studies available in the literature
	Meniscal repair + PRP injection	**Pujol et al. (2015)**: Case-control study, 34 consecutive young patients underwent an open meniscal repair: 17 pts. arthrotomic meniscal repair + PRP injection vs 17 pts. isolated meniscal repair. PRP slightly improved the clinical outcome **Griffin et al. (2015)**: Retrospective comparative study, 15 pts. treated with arthroscopic meniscal repair + PRP injection vs 20 pts. treated with isolated meniscal repair: no clinical differences **Kaminski et al. (2018)**: Randomized double-blind controlled study, 20 menisci treated with arthroscopic repair + PRP injection vs 17 menisci with isolated repair: superior healing rate of meniscus lesions in PRP group **Kemmochi et al. (2018)**: Prospective comparative study, 17 pts. underwent arthroscopic meniscus repair + PRF and PRP injection vs 5 pts. treated by isolated meniscal repair: no significant clinical differences **Dai et al. (2019)**: 29 pts. with discoid lateral meniscal tears arthroscopically treated with meniscal suture: 14 with PRP, 15 without PRP augmentation, with similar effects in pain relief and functional improvement at mid-term follow-up **Everhart et al. (2019)**: Prospective comparative study, 45 pts. underwent arthroscopic meniscal repair + PRP injection vs 106 pts. treated by isolated meniscal repair: PRP augmentation improved survival of isolated meniscal repairs
	Meniscal reconstruction + PRP injection	**Zhang et al. (2018)**: 31 pts. underwent lateral MAT combined with intra-articular PRP injection: improvements in all functions and pain scores at short-term follow-up

HA Hyaluronic acid, *PRP* Platelet-rich plasma, *PRF* Platelet-rich fibrin, *MAT* Meniscal allograft transplantation, *pts* patients

competition or training. At 6 months, the MRI performed in seven patients showed the stability of the meniscal tears, with the healing of a para-meniscal cyst in one case. In the same year, Kaminski et al. [46] performed a double-blind randomized controlled trial enrolling 72 patients affected by chronic meniscal tears: 30 patients in the control group were subjected to ultrasound-guided trephination, while 42 patients in the treatment group underwent the same treatment combined with ultrasound-guided PRP injections directly in the lesion site. At 12 months, a higher percentage of patients in the treatment group improved clinically. Moreover, the failure rate was significantly higher in the control group than in the treatment group (70% vs 48%).

With regard to PRP injections and surgical procedures, the rationale of modulating the whole joint homeostasis to reduce post-operative inflammation could justify the possibility to test it after meniscectomy, but to date, no clinical study has been published reporting the results of this procedure. Regarding meniscal augmentation repair with PRP, recent clinical data available in the literature showed controversial results: three comparative studies showed improved clinical results with PRP augmentation compared to isolated meniscal suture, while other three studies showed no significant clinical differences between the two groups (Table 23.1) [47–52]. More recently, Zaffagnini et al. [53] performed a meta-analysis of all these studies which showed overall a reduced failure risk when meniscal repair was combined to PRP augmentation versus isolated repair.

Finally, Zhang et al. [54] explored the combination of PRP injections and meniscal reconstruction. In detail, they treated 31 patients with a single PRP injection performed under arthroscopy around the meniscal tissue after lateral meniscal allograft transplantation. A retrospective evaluation at a mean follow-up of 37 months showed a clinical and functional improvement. There were no significant differences in the grade of chondral damage between the pre-operative and 2-year follow-up periods, and only three patients (9.7%) showed no improvements or had lower evaluation scores. However, the lack of a control group hindered the possibility to ascertain the specific contribution of PRP injections to the final outcome.

Considering the available clinical studies, PRP injections seem to represent a promising procedure in the case of meniscal degenerative tears approached with a non-surgical treatment, as well as a potential augmentation strategy for meniscal tear repair. However, these data are supported by a low number of clinical studies with controversial findings, and further high-level evidence is required to confirm the poten-

tial of PRP injections for the treatment of meniscal lesions.

23.7 Conclusions

The role of orthobiologics injections for the management of meniscal lesions has gained a growing interest, thanks to their minimal invasiveness, the possibility to combine them with other procedures, and the not always satisfactory results of other available treatments. To this regard, corticosteroids, HA, and PRP represent interesting treatment options with a different specific rationale and possibly different indications based on the type of meniscal tears and joint status. At the moment, corticosteroids and HA injections find their main application as intra-articular injection in osteoarthritic knees with degenerative meniscal lesions, although results are still controversial. On the other hand, PRP presents a wider spectrum of action than other biologic substances, including anabolic/regenerative properties, and showed a potential role also in the ultrasound-guided treatment of degenerative meniscal lesion, as well as to improve the success rate of meniscal repair procedures. However, literature findings are still preliminary and inconclusive, and biologic injective meniscus treatments require further higher-level studies to better understand their real efficacy and the more appropriate indications for the management of meniscal lesions.

Take-Home Messages
- The potential of orthobiologics injections for meniscus lesions is currently explored in terms of intra-articular or direct intra-meniscus delivery, as well as augmentation during and after meniscal surgical procedures.
- Intra-articular corticosteroid injections are closely limited to treat acute knee inflammatory phases, with a fast but short-term effectiveness. More clinical evidence is needed to support their effi-

cacy and safety for the treatment of meniscal lesions.

- Evidence on HA injections for meniscal lesions is limited, with low-level studies supporting a possible indication in degenerative meniscal lesions as conservative treatment.
- Intra-articular PRP injections are a promising conservative approach for meniscal degenerative tears, also representing a potential augmentation technique for meniscal surgical repair.

References

1. Fox AJ, Wanivenhaus F, Burge AJ, Warren RF, Rodeo SA. The human meniscus: a review of anatomy, function, injury, and advances in treatment. Clin Anat. 2015;28(2):269–87.
2. Ridley TJ, McCarthy MA, Bollier MJ, Wolf BR, Amendola A. Age differences in the prevalence of isolated medial and lateral meniscal tears in surgically treated patients. Iowa Orthop J. 2017;37: 91–4.
3. Longo UG, Ciuffreda M, Candela V, Rizzello G, D'Andrea V, Mannering N, et al. Knee osteoarthritis after arthroscopic partial meniscectomy: prevalence and progression of radiographic changes after 5 to 12 years compared with contralateral knee. J Knee Surg. 2019;32(5):407–13.
4. Pengas IP, Assiotis A, Nash W, Hatcher J, Banks J, McNicholas MJ. Total meniscectomy in adolescents: a 40-year follow-up. J Bone Joint Surg. 2012;94(12):1649–54.
5. Filardo G, Andriolo L, Kon E, de Caro F, Marcacci M. Meniscal scaffolds: results and indications. A systematic literature review. Int Orthop. 2015;39(1):35–46.
6. Anderson AF, Irrgang JJ, Dunn W, Beaufils P, Cohen M, Cole BJ, et al. Interobserver reliability of the International Society of Arthroscopy, Knee Surgery and Orthopaedic Sports Medicine (ISAKOS) classification of meniscal tears. Am J Sports Med. 2011;39(5):926–32.
7. Doral MN, Bilge O, Huri G, Turhan E, Verdonk R. Modern treatment of meniscal tears. EFORT Open Reviews. 2018;3(5):260–8.
8. Mordecai SC, Al-Hadithy N, Ware HE, Gupte CM. Treatment of meniscal tears: an evidence based approach. World J Orthop. 2014;5(3):233–41.
9. Pengas I, Nash W, Assiotis A, To K, Khan W, McNicholas M. The effects of knee meniscectomy on the development of osteoarthritis in the patellofemoral joint 40 years following meniscectomy. Eur J Orthop Surg Traumatol. 2019;29(8):1705–8.
10. Xu C, Zhao J. A meta-analysis comparing meniscal repair with meniscectomy in the treatment of meniscal tears: the more meniscus, the better outcome? Knee Surg Sports Traumatol Arthrosc. 2015;23(1):164–70.
11. Feeley BT, Liu S, Garner AM, Zhang AL, Pietzsch JB. The cost-effectiveness of meniscal repair versus partial meniscectomy: a model-based projection for the United States. Knee. 2016;23(4):674–80.
12. Barber-Westin SD, Noyes FR. Clinical healing rates of meniscus repairs of tears in the central-third (red-white) zone. Arthroscopy. 2014;30(1):134–46.
13. Thaunat M, Fayard JM, Guimaraes TM, Jan N, Murphy CG, Sonnery-Cottet B. Classification and surgical repair of ramp lesions of the medial meniscus. Arthrosc Tech. 2016;5(4):e871–e5.
14. Bhatia S, LaPrade CM, Ellman MB, LaPrade RF. Meniscal root tears: significance, diagnosis, and treatment. Am J Sports Med. 2014;42(12):3016–30.
15. Yeo DYT, Suhaimi F, Parker DA. Factors predicting failure rates and patient-reported outcome measures after arthroscopic meniscal repair. Arthroscopy. 2019;35(11):3146–64.e2.
16. Petersen W, Tillmann B. Age-related blood and lymph supply of the knee menisci. A cadaver study. Acta Orthop Scand. 1995;66(4):308–12.
17. Arnoczky SP, Warren RF. Microvasculature of the human meniscus. Am J Sports Med. 1982;10(2):90–5.
18. Longo UG, Campi S, Romeo G, Spiezia F, Maffulli N, Denaro V. Biological strategies to enhance healing of the avascular area of the meniscus. Stem Cells Int. 2012;2012:528359.
19. Filardo G, Kon E, Longo UG, Madry H, Marchettini P, Marmotti A, et al. Non-surgical treatments for the management of early osteoarthritis. Knee Surg Sports Traumatol Arthrosc. 2016;24(6):1775–85.
20. Kon E, Filardo G, Drobnic M, Madry H, Jelic M, van Dijk N, et al. Non-surgical management of early knee osteoarthritis. Knee Surg Sports Traumatol Arthrosc. 2012;20(3):436–49.
21. Wijn SRW, Rovers MM, van Tienen TG, Hannink G. Intra-articular corticosteroid injections increase the risk of requiring knee arthroplasty. Bone Joint J. 2020;102-B(5):586–92.
22. Moreland LW. Intra-articular hyaluronan (hyaluronic acid) and hylans for the treatment of osteoarthritis: mechanisms of action. Arthritis Res Ther. 2003;5(2):54–67.
23. Waddell DD, Bert JM. The use of hyaluronan after arthroscopic surgery of the knee. Arthroscopy. 2010;26(1):105–11.
24. Foster TE, Puskas BL, Mandelbaum BR, Gerhardt MB, Rodeo SA. Platelet-rich plasma: from basic science to clinical applications. Am J Sports Med. 2009;37(11):2259–72.
25. Bhargava MM, Hidaka C, Hannafin JA, Doty S, Warren RF. Effects of hepatocyte growth factor and platelet-derived growth factor on the repair of meniscal defects in vitro. In Vitro Cell Dev Biol Anim. 2005;41(8–9):305–10.

26. Cook JL, Smith PA, Bozynski CC, Kuroki K, Cook CR, Stoker AM, et al. Multiple injections of leukoreduced platelet rich plasma reduce pain and functional impairment in a canine model of ACL and meniscal deficiency. J Orthop Res. 2016;34(4):607–15.

27. Kopf S, Birkenfeld F, Becker R, Petersen W, Starke C, Wruck CJ, et al. Local treatment of meniscal lesions with vascular endothelial growth factor. J Bone Joint Surg Am. 2010;92(16):2682–91.

28. Petersen W, Pufe T, Starke C, Fuchs T, Kopf S, Raschke M, et al. Locally applied angiogenic factors—a new therapeutic tool for meniscal repair. Ann Anat. 2005;187(5–6):509–19.

29. Ishida K, Kuroda R, Miwa M, Tabata Y, Hokugo A, Kawamoto T, et al. The regenerative effects of platelet-rich plasma on meniscal cells in vitro and its in vivo application with biodegradable gelatin hydrogel. Tissue Eng. 2007;13(5):1103–12.

30. Giuffrida A, Di Bari A, Falzone E, Iacono F, Kon E, Marcacci M, et al. Conservative vs. surgical approach for degenerative meniscal injuries: a systematic review of clinical evidence. Eur Rev Med Pharmacol Sci. 2020;24(6):2874–85.

31. Vermesan D, Prejbeanu R, Laitin S, Damian G, Deleanu B, Abbinante A, et al. Arthroscopic debridement compared to intra-articular steroids in treating degenerative medial meniscal tears. Eur Rev Med Pharmacol Sci. 2013;17(23):3192–6.

32. Wilderman I, Berkovich R, Meaney C, Kleiner O, Perelman V. Meniscus-targeted injections for chronic knee pain due to meniscal tears or degenerative fraying: a retrospective study. J Ultrasound Med. 2019;38(11):2853–9.

33. Rasmussen S, Larsen AS, Thomsen ST, Kehlet H. Intra-articular glucocorticoid, bupivacaine and morphine reduces pain, inflammatory response and convalescence after arthroscopic meniscectomy. Pain. 1998;78(2):131–4.

34. Rydell N, Balazs EA. Effect of intra-articular injection of hyaluronic acid on the clinical symptoms of osteoarthritis and on granulation tissue formation. Clin Orthop Relat Res. 1971;80:25–32.

35. Strauss EJ, Hart JA, Miller MD, Altman RD, Rosen JE. Hyaluronic acid viscosupplementation and osteoarthritis: current uses and future directions. Am J Sports Med. 2009;37(8):1636–44.

36. Nelson F, Billinghurst RC, Pidoux I, Reiner A, Langworthy M, McDermott M, et al. Early post-traumatic osteoarthritis-like changes in human articular cartilage following rupture of the anterior cruciate ligament. Osteoarthr Cartil. 2006;14(2):114–9.

37. Zorzi C, Rigotti S, Screpis D, Giordan N, Piovan G. A new hydrogel for the conservative treatment of meniscal lesions: a randomized controlled study. Joints. 2015;3(3):136–45.

38. Huskin JP, Vandekerckhove B, Delince P, Verdonk R, Dubuc JE, Willems S, et al. Multicentre, prospective, open study to evaluate the safety and efficacy of hylan G-F 20 in knee osteoarthritis subjects presenting with pain following arthroscopic meniscectomy. Knee Surg Sports Traumatol Arthrosc. 2008;16(8):747–52.

39. Thein R, Haviv B, Kidron A, Bronak S. Intra-articular injection of hyaluronic acid following arthroscopic partial meniscectomy of the knee. Orthopedics. 2010;33(10):724.

40. Filardo G, Di Matteo B, Tentoni F, Cavicchioli A, Di Martino A, Lo Presti M, et al. No effects of early viscosupplementation after arthroscopic partial meniscectomy: a randomized controlled trial. Am J Sports Med. 2016;44(12):3119–25.

41. Sonoda M, Harwood FL, Amiel ME, Moriya H, Temple M, Chang DG, et al. The effects of hyaluronan on tissue healing after meniscus injury and repair in a rabbit model. Am J Sports Med. 2000;28(1):90–7.

42. Jubert NJ, Rodríguez L, Reverté-Vinaixa MM, Navarro A. Platelet-rich plasma injections for advanced knee osteoarthritis: a prospective, randomized, double-blinded clinical trial. Orthop J Sports Med. 2017;5(2):2325967116689386.

43. Gonzales VK, de Mulder EL, de Boer T, Hannink G, van Tienen TG, van Heerde WL, et al. Platelet-rich plasma can replace fetal bovine serum in human meniscus cell cultures. Tissue Eng Part C Methods. 2013;19(11):892–9.

44. Blanke F, Vavken P, Haenle M, von Wehren L, Pagenstert G, Majewski M. Percutaneous injections of Platelet rich plasma for treatment of intrasubstance meniscal lesions. Muscles, Ligaments Tendons J. 2015;5(3):162–6.

45. Guenoun D, Magalon J, de Torquemada I, Vandeville C, Sabatier F, Champsaur P, et al. Treatment of degenerative meniscal tear with intrameniscal injection of platelets rich plasma. Diagn Interv Imaging. 2020;101(3):169–76.

46. Kaminski R, Maksymowicz-Wleklik M, Kulinski K, Kozar-Kaminska K, Dabrowska-Thing A, Pomianowski S. Short-term outcomes of percutaneous trephination with a platelet rich plasma intrameniscal injection for the repair of degenerative meniscal lesions. a prospective, randomized, double-blind, parallel-group, placebo-controlled study. Int J Mol Sci. 2019;20(4).

47. Pujol N, Salle De Chou E, Boisrenoult P, Beaufils P. Platelet-rich plasma for open meniscal repair in young patients: any benefit? Knee Surg Sports Traumatol Arthrosc. 2015;23(1):51–8.

48. Griffin JW, Hadeed MM, Werner BC, Diduch DR, Carson EW, Miller MD. Platelet-rich plasma in meniscal repair: does augmentation improve surgical outcomes? Clin Orthop Relat Res. 2015;473(5):1665–72.

49. Kaminski R, Kulinski K, Kozar-Kaminska K, et al. A prospective, randomized, double-blind, parallel-group, placebo-controlled study evaluating meniscal healing, clinical outcomes, and safety in patients undergoing meniscal repair of unstable, complete vertical meniscal tears (bucket handle) augmented with platelet-rich plasma. Biomed Res Int. 2018;2018:9315815.

50. Kemmochi M, Sasaki S, Takahashi M, Nishimura T, Aizawa C, Kikuchi J. The use of platelet-rich fibrin with platelet-rich plasma support meniscal repair surgery. J Orthop. 2018;15(2):711–20.
51. Dai WL, Zhang H, Lin ZM, Shi ZJ, Wang J. Efficacy of platelet-rich plasma in arthroscopic repair for discoid lateral meniscus tears. BMC Musculoskelet Disord. 2019;20(1):113.
52. Everhart JS, Cavendish PA, Eikenberry A, Magnussen RA, Kaeding CC, Flanigan DC. Platelet-rich plasma reduces failure risk for isolated meniscal repairs but provides no benefit for meniscal repairs with anterior cruciate ligament reconstruction. Am J Sports Med. 2019;47(8):1789–96.
53. Zaffagnini S, Poggi A, Reale D, Andriolo L, Flanigan DC, Filardo G. Biologic augmentation reduces failure rate of meniscal repair: a systematic review and meta-analysis. Orthopaedic J Sports Med. 2021;9(2):2325967120981627.
54. Zhang H, Chen S, Qiu M, Zhou A, Yan W, Zhang J. Lateral meniscus allograft transplantation with platelet-rich plasma injections: a minimum two-year follow-up study. Knee. 2018;25(4):568–76.

Orthobiologics for the Treatment of Muscle Lesions

Alberto Grassi, Giacomo Dal Fabbro, and Stefano Zaffagnini

24.1 Introduction

The musculoskeletal system is one of the major organ systems of the body and represents a lever compound that allows body movements and normal articular kinematics. During sports activity, muscles may be exposed to an exaggerated and abnormal activity due to overuse conditions and invoking close to maximal muscular strength during practice or competition. The muscle injuries lead to considerable absence from training and competition, which may compromise team performance and athlete's career. Consequently, the medical staff faces pressure to return the athlete to training and matches as soon as possible. In order to achieve a quick yet full recovery, there is a continuous search for innovative treatments to improve and accelerate muscle healing. This chapter provides an overview of main features of muscle injuries, followed by an analysis of scientific evidence about the clinical results of different emerging orthobiologic approaches, with a particular focus on platelet-rich plasma (PRP), in the management of this kind of injuries.

24.2 Muscle Injuries

24.2.1 Epidemiology and Mechanism of Injury

Muscle injuries are common among athletes and sports people [1] and represent a challenge for patients, team and medical staff. This kind of lesions represents the first cause of injury in sprinting sports such as soccer, football and hockey. In an epidemiology study among high school girls and collegiate women field hockey athletes, muscle and tendon strains represented 31.8% of all injuries recorded during the 2008–2009 through 2013–2014 academic years [2]. In the latter study, authors reported also that the most commonly injured body parts were the hip, the thigh and the upper leg (23.6%) [2]. Among professional soccer players, muscle injuries represented more than 31% of all injuries [3]. In particular, the hamstring muscle strain injury resulted the most common injury in sports involving high speed running, such as American football, Australian football, rugby and track and field [4]. Overall, the most commonly involved muscles and muscle groups are the hamstrings, the rectus femoris and the medial head of the gastrocnemius [5].

A. Grassi (✉) · G. Dal Fabbro
Clinica Ortopedica e Traumatologica II,
IRCCS - Istituto Ortopedico Rizzoli, Bologna, Italy
e-mail: alberto.grassi@ior.it;
giacomo.dalfabbro@studio.unibo.it

S. Zaffagnini
Clinica Ortopedica e Traumatologica II,
IRCCS - Istituto Ortopedico Rizzoli, Bologna, Italy

University of Bologna, Bologna, Italy
e-mail: Stefano.zaffagnini@unibo.it

G. Filardo et al. (eds.), *Orthobiologics*, https://doi.org/10.1007/978-3-030-84744-9_24

Muscles tears can occur with different mechanisms, depending also on the age of the patient, which is associated with a different weak link in the muscle-tendon-bone chain changes [6]: in young adults usually the mechanical failure occurs at the muscle-tendon interface, in older adults coexistent tendinopathy and overload of the musculo-tendinous unit may contribute to the tearing process, and in children the weakness of apophyseal growth plates may lead to apophyseal avulsion when excessive tension is applied on the muscle-tendon-bone chain.

The trauma that leads to muscle injuries can be direct or indirect; when the trauma to the muscle is direct, the lesion occurs at the site of the impact, while in indirect trauma the tear usually occurs at the end of the belly [5]. Most of the muscle tears are due to indirect trauma, and an eccentric contraction is a major cause of injury, probably as a consequence of the forces produced by eccentric contractions compared with the forces produced by isometric or concentric contractions [7, 8]. Overall, tears occur mostly at the musculotendinous junction (MTJ), the weakest link within the muscle-tendon unit, where the tendon arises from the muscle belly [9].

Sprinting, kicking and changes of direction have been described as kinematics patterns associated with muscle injuries. In particular, the predominant injury mechanism consists in elongation or heaving eccentric load of the biarticular muscles such as hamstrings, biceps femori and quadriceps femori [10]. In a three-dimensional kinematics analysis [11] among female college students, the muscle strain of each three biarticulated hamstrings muscles reached a peak during the late swing phase of sprinting. Moreover, authors of the study found that the peak hamstring strains were negatively correlated with hamstring flexibility. These findings may suggest that a potential risk for hamstring injury exists during the late phase of sprinting.

24.2.2 Muscle Lesions: Classification

Several classification systems of muscle tears have been described. In the classic and most common system, the acute muscle injuries are classified as strains (grade I), partial tears (grade II) and complete tears (grade III) [12, 13]. Later an updated version has been proposed with the purpose to take into account the extent, the size and also the exact location of the acute muscle injury (Table 24.1). Other classification systems described in the literature are the Munich consensus statement, the ISMuLT classification and the British athletics muscle injury classification [14–16].

According to the mechanism of trauma, muscle injuries can be classified as direct and indirect. Tears due to a direct trauma could de distinguished in contusion and lacerations. The contusions could be divided in mild, moderate and severe according to the functional disability they produce [5]. The tears due to an indirect trauma are classified as nonstructural (functional) and structural, according to the presence of an anatomically evident lesion. The structural injuries are divided in three subgroups according to the severity of the lesion within the muscle fibres: grade 1 (few muscle fibres involved), grade 2 (some continuity of the fibres is maintained),

Table 24.1 Classic classification system

Imaging findings and function impairment		
Injury grade	Features	
Grade I (strain)	MRI/US findings	Few muscle fibres involved
	Function	Minimal impairment of strength and function
Grade II (partial)	MRI/US findings	Some continuity of fibres is maintained
	Function	Partial impairment of function
Grade III (complete)	MRI/US findings	Complete discontinuity of fibres
	Function	Muscle function is lost
Site of lesion		
Proximal musculotendinous junction		
Muscle belly	Proximal	Intramuscular Myofascial
	Middle	Myofascial/perifascial Myotendinous
	Distal	Combined Myotendinous
Distal musculotendinous junction		

Data from Maffulli et al. [5]

grade 3 (complete discontinuity of the fibres) (Table 24.1). Moreover, the structural injuries may be distinguished according to the location among the muscle structure in proximal, middle and distal [5].

24.2.3 Treatment Strategies and Burdens

The main purposes of the treatment are to reduce the re-injury recurrence rate and to achieve a fast return to sport. Since most muscle injuries respond well to conservative treatment [17] and surgery is indicated in few specific cases, most of muscle injuries can be managed conservatively, as skeletal muscles have an endogenous capacity of healing.

The PRICE (protection, rest, ice, compression and elevation) approach is a commonly accepted treatment approach to control the early inflammatory process [18, 19]. POLICE is a new acronym which represents protection, optimal loading, ice, compression and elevation and introduces the concept of a safe and effective loading in acute soft tissue injury management. Overall, the conservative approach consists in restriction of activity and reduced mobility, ice, nonsteroidal anti-inflammatory drugs (NSAIDs), physical therapy, specific exercises, electrotherapeutic modalities, hyperbaric oxygen therapy and photothermal therapy. Satisfactory results using conservative treatments were reported, mainly in the non-professional population; however, these treatments have no firm scientific basis, and clinical evidence to support them is sparse, so they are mainly applied as empirical medicine due to the lack of indications from high-level trials [20, 21].

The surgical management of the muscle injuries is reserved for larger tears or is indicated in specific cases such as tendon avulsion or complete lesion of the muscle belly, with the aim to avoid strength deficit and inability to return to sports practice at the pre-injury level [22].

Despite the recent efforts in the area of prevention and management, the re-injury rates remain high [23]: between 16% and 18% of recurrence was reported among footballers and track and field athletes, with 8–73 days of layoff based on severity and intra-tendinous location [3, 24]. Advances in scientific understanding of tendinopathy and muscle injuries have prompted the need for alternative therapies with the aim to improve the functional results, return to field timing and re-injury rates.

24.3 Healing Process of Muscle Injuries

The increasing knowledge of basic science related to muscle healing set the stage for the introduction of novel biological approaches to accelerate the healing of injured muscles.

Skeletal muscle tissue normally has a low turnover rate. However, following injury, it starts a rapid process, consisting of sequential overlapping phases: degeneration and necrosis, inflammation and cellular response, regeneration, repair and, finally, the remodelling and fibrosis phase [25, 26]. In particular, two competing processes are involved during the muscle healing process: the production of connective tissue scar and the regeneration of disrupted muscle functional fibres.

Degeneration and necrosis. The injured fibres undergo a necrosis process, triggered by the disruption of the plasma membrane, followed by alterations of cell permeability and calcium transfer. The tissue gap created by the injury is filled with a hematoma. Late elimination and resorption of the hematoma are known to delay skeletal muscle regeneration, to improve fibrosis and to reduce biomechanical properties of the healing muscle with a lower functional recovery in athletes.

Inflammation and cellular response. The injured tissue activates platelets and endothelial cells, which release factors that recruit resident and circulating inflammatory cells (leukocytes). First, neutrophils arrive in the damaged area, followed by monocytes and macrophages [27].

Neutrophils secrete a large number of pro-inflammatory molecules such as cytokines, chemokines and growth factors, in order to create a chemo-attractive microenvironment for other inflammatory cells such as monocytes and macrophages [28, 29]. Macrophages play a key role during the healing process and are able to participate in both the muscle regeneration process and fibrosis production [30]. At first, macrophages produce pro-inflammatory cytokines, contributing to the degeneration and inflammation process [31]. Finally, T lymphocytes infiltrate the damaged tissue playing an important role in the local vascularization through adhesion molecule secretion and the production of growth factors and cytokines.

Regeneration and repair. The regeneration phase starts during the first 4–5 days after injury, peaks at 2 weeks and then diminishes 3–4 weeks after injury. The muscle fibres are post-mitotic cells, which do not have the capacity to divide. The satellite cells (SC) are adult muscle stem and progenitor cells located between the plasma membrane of myofibre and the basal lamina, whose regenerative capability is essential to repair skeletal muscle after injury [32, 33]. After a first pro-inflammatory phase, macrophages adapt and change the microenvironment, achieving a different activation state. In this phase macrophages produce anti-inflammatory cytokines and pro-fibrotic factors such as transforming growth factor β (TGF-β) that activate fibroblast [31]. In this way, the early inflammatory phase is followed by repair processes, where the SC activate, proliferate and differentiate to restore the muscle structure, in response to several growth factors [34]. In particular, the proliferation phase is characterized by the production of an extracellular matrix associated with granulation, contraction and epithelialization.

Remodelling and fibrosis. Finally, the remodelling phase is defined by the maturation of regenerated myofibres and collagen remodelling, leading to a recovery of muscle functional capacity. On the other hand, this last phase is characterized also by fibrosis and scar tissue formation [35]. The fibrotic response activated by pro-fibrotic cytokines such as TGF-β1 leads to an excessive fibroblast and myofibroblast proliferation and to an increase in type I and III collagen, laminin and fibronectin production [36]. At first, the fibrotic response is beneficial, stabilizing the tissue and acting as a scaffold for myofibre regeneration. However, an excessive fibrosis process and scar tissue formation could lead to a loss of muscle functional capacity. A fine balance between the different phases of the healing process and their mechanisms is essential for a full recovery of the contractile muscle function, and, even though the phases of muscle healing are almost the same in all muscle injuries, the functional recovery changes. Usually, the healing process leads to muscle regeneration with scar tissue which differs from normal muscle tissue. While a minor muscle injury could regenerate completely and spontaneously, after severe injuries muscle healing is incomplete, often resulting in the formation of fibrotic tissue that compromises muscle function. Furthermore, in the rare case of major muscle injuries, some complications like myositis ossificans, cystic degeneration and heterotopic ossifications may occur [17].

24.4 Platelet-Rich Plasma (PRP) in Muscle Injury

24.4.1 PRP: Biological Rationale and Formulations

The physiologic progression through the muscle healing process phases is led by growth factors and cytokines. Many of these bioactive molecules are stored in the platelet alpha granules (Table 24.2), such as insulin-like growth factor (IGF-1), hepatocyte growth factor (HGF), fibroblast growth factor (FGF-2), vascular endothelial growth factor (VEGF) and TGF- β1, which may be the key regulators of muscle regeneration and myogenesis [37]. Moreover, platelets are the first blood component arriving at the site of tissue injury, and they have an active role in haemostasis and, releasing the growth factors, promote tissue repair and influence the reactivity of vascular and other blood cells in angiogenesis and inflammation processes. Therefore, PRP, which is blood derivative with a higher platelet concentrate than whole blood [38], gained increasing interest to deliver a high concentration of autologous growth factors and bioactive molecules in physiologic

Table 24.2 Bioactive molecules stored in the platelet α-granules

Bioactive molecules	Effects	Role
TGF-β1	Production of extra-cellular proteins	Promote fibrosis
	Conversion of myoblast in myofibroblast	
IGF-1	Myoblast proliferation	Promote regeneration and hypertrophy
	Muscle growth	
VEGF	Angiogenetic process	Promote angiogenesis
	Satellite cell migration and proliferation	
HGF	Myoblast proliferation	Satellite cell activation
	Inhibition of myoblast differentiation	
FGF-6/2	Satellite cell proliferation	Active fibroblast proliferation
	Inhibition of myogenic differentiation	
PDGF-AA/AB	Active myoblast proliferation	Promote regeneration and angiogenesis
	Stimulate angiogenesis	

Data from Cianforlini et al. and Zanon et al. [17, 35]

proportions, with low costs and in a minimally invasive way [39].

Under the acronym PRP, several procedures have been proposed, leading to heterogeneous concentrates in terms of absolute number of platelet activation methods, presence or absence of white blood cells and other factors that might influence the effect on the tissue target. In order to ensure that the platelets are suspended and not form a clot, PRP must be made from anticoagulated blood, but its preparation has significant variability [40]. In all kinds of preparation techniques, blood is collected from the patient with anticoagulant and immediately centrifuged within the hour. Through centrifugation and various other steps, the red blood cells and the platelet-poor plasma layer are discarded, and the platelet concentrate remains [41].

The use of PRP represents a possible alternative approach based on the ability of autologous growth factors to improve skeletal muscle regeneration [42, 43]. Hammond et al. showed the capability of PRP to promote and accelerate myogenesis in an experimental study investigating the biomechanical and biochemical effects in rat muscle injuries [43]. These results are in line with findings reported in other laboratory studies in which a significant acceleration of muscle healing process in animal treated with autologous PRP or related products was provided. A side effect of the use of PRP and related products may be the occurrence of fibrosis. In an in vitro analysis [44], the platelet-rich fibrin matrix (PRFM) was found to have a significantly higher concentration of TGF-β1 compared with whole blood concentrate of similar volume; TGF-β1 has the ability to significantly increase connective cell proliferation over time, thus generating fibrotic tissue. On the other side, no increase in fibrotic tissue formation was observed after PRFM treatment in comparison with controls during an in vivo investigation [45]. In this study the authors reported that the PRFM-treated muscle tears on the longissimus dorsi muscle of Wistar rats exhibited an improved muscular regeneration, an increase in neovascularization and a slight reduction of fibrosis compared with controls. To closely simulate a clinical approach, Cianforlini et al. [46] injected in the longissimus dorsi muscle of Wistar rats two different concentrations of PRP intramuscularly 24 hours after a surgical trauma and evaluated, with histological and immunohistochemical analyses, the dose-dependent effects. The histological results confirmed the effectiveness of PRP in muscle healing and showed that the increase in PRP concentrations in damaged muscle tissue accelerated the tissue regeneration process.

24.4.2 PRP: Clinical Evidence

Preclinical evaluations suggested that the local delivery of PRP might reduce recovery time [37]. Thus, the application in the clinical practice of autologous PRP injections increased rapidly in patients with sport-related injuries [47]. However, after early reports of positive preliminary experience, higher-quality studies have recently questioned the real benefit provided by PRP injections to promote the healing process and return to sport.

Only few comparative and randomized studies, which differ from each other significantly, were performed about the application of PRP in the treatment of muscle injuries. Most of the randomized clinical trials (RCTs) investigate the effect of PRP in hamstring lesions. The presence and amount of hematoma and its evacuation have been performed only in some studies and represent factors that may jeopardize and influence the healing effect [48]. A systematic review and meta-analysis about clinical outcomes of PRP treatment in acute muscle injuries provided that the procedures analysed were usually intra-lesional, with the injection performed in the location of the tear. Moreover, in the majority of studies included, the PRP injection was performed under ultrasound guidance.

Among six studies included, four reported using PRP with leukocytes, and one reported using an activating agent [38]. These different PRP protocols used in these studies make it difficult to generalize results from clinical trials using different commercial PRP preparations in heterogeneous pools of muscle injuries, in terms of type, grade and location of the tear.

Return to sport: In a recent systematic review, the outcome of return to sport in patient with muscle injuries treated with PRP was analysed in all the six studies included [38]. While in three of these studies a significantly shorter return to sport in PRP group was reported, in the two double-blind studies included, the mean difference between the PRP and control group was not significant [48, 49]. A similar finding was also provided in the random-effect meta-analysis of the three included studies evaluating only hamstring injuries. On the other hand, a significant mean difference in support of PRP was reported when the authors considered the four single-blind studies or the three studies included with heterogeneous muscle involvement [38]. In a retrospec-

tive observational study, a significantly reduced time for reaching a complete functional recovery and for returning to practice the previous sport activity was reported in patients who underwent ultrasound-guided PRP treatment for gastrocnemius strains compared with the standard treatment group [50]. No significant differences, on the contrary, were found in a case control study among National Football League players with an acute hamstring injury in time to return to play between treatment with PRP and routine rehabilitation [51] (Fig. 24.1).

Re-injuries: Nonsignificant risk differences between the PRP groups and the control groups were reported in the studies in which a clear definition of re-injury was reported [38, 48, 49, 52, 53]. Complications due to the treatment, such as discomfort at the injection site, hematoma or hyperesthesia in the posterior thigh, were reported in two studies [48, 49]; however, no significant complications risk differences between the PRP groups and the control groups were found in the latter studies [38] (Fig. 24.2).

Patient-reported outcomes: Regarding the pain evaluation during rehabilitation, the results are controversial. In a RCT, a lower pain at rest and lower resisted motion were reported in the PRP group compared with the control group [53]. On the other way, no differences were provided in a multicentre, randomized, double-blind placebo-controlled trial [48]. In a randomized trial, no significant differences were reported in a subjective score of muscle function (hamstring outcome score and satisfaction) between patients treated with PRP or isotonic saline injection [49].

Objective evaluation: In two studies the effect of treatment with PRP on range of motion or flexibility was investigated. While in a RCT no differences compared with the control group were reported [49], in a comparative study among professional athletes, higher ROM in patients treated with PRP was found [54]. In the latter study, a higher strength in patients treated with PRP was provided, but only during the first 2 weeks. However, no differences between PRP and the control group at final follow-up were reported in all studies in which muscle strength was evaluated [38].

RETURN TO SPORT: DOUBLE-BLIND STUDIES ONLY

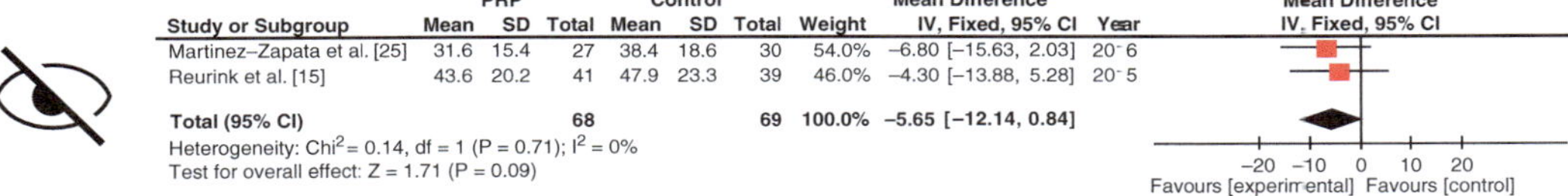

RETURN TO SPORT: HAMSTRING INJURIES ONLY

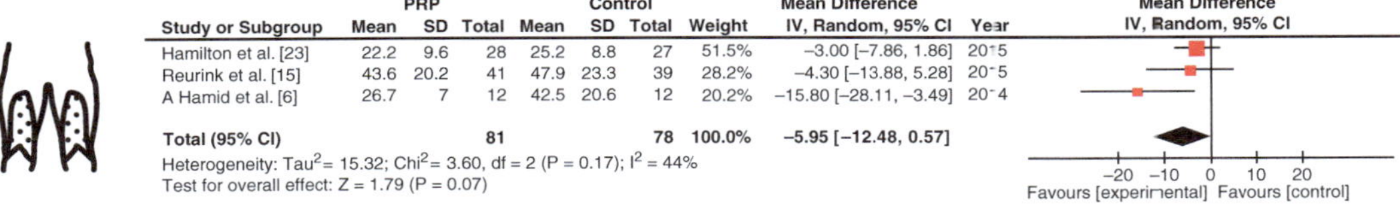

Fig. 24.1 The mean difference for return to sport for PRP and control based on double-blind studies only and studies including only hamstring injuries. Data from Grassi et al. [38]

RISK OF RE-INJURY

RISK OF COMPLICATIONS

Fig. 24.2 Pooled risk difference for re-injury (a) and complications (b) for PRP and control interventions. Data from Grassi et al. [38]

Regarding the effectiveness of PRP in the treatment of acute muscle injuries, a direct correlation between low meta-analysis quality and positive results has been demonstrated in the literature [55]; in a recent systematic review and meta-analysis [38], when all the randomized clinical trials were considered, return to sport was significantly faster in patients who underwent PRP treatment compared with control groups, but this result was not confirmed when the analysis excluded unblinded studies.

In summary, the results of the available studies in the literature suggest that PRP could be an effective tool in acute muscle tear treatment, in particular with regard to the time to return to sport, which resulted significantly shorter after PRP application [38]. However, the evidence from the available studies with the highest quality

[48, 49, 52] does not support the hypothesis that the use of a PRP injection promotes muscle healing and return to sport. These findings do not currently justify the use of PRP in acute muscle tears, but they should be interpreted as a demonstration of the need for a larger number of high-quality trials, with the purpose to gain more insight into the potential for the PRP strategy.

24.5 Other Biological Approaches

24.5.1 Culture-Expanded Stromal Cell Therapy

With respect to muscular disorders and tears, the concept of transplanting cells with a high regenerative capacity as a treatment seems logical, and it has been explored in the literature.

Mesenchymal stromal cells (MSCs) are an example of such cells. Derived through the isolation and in vitro culture of native stem and progenitor cell populations in connective tissues, characterized MSCs share unique attributes that are not present in their native counterparts yet can maintain attributes uniquely associated with their tissue of origin. Like all MSCs, muscle-derived stromal cells (MDSCs) possess an inherent ability for regeneration with in vitro attributes of long-term proliferation suggestive of self-renewal and multipotent differentiation. Furthermore, MDSCs can be genetically modified to express proteins of interest, such as growth factors or anti-fibrotic molecules [56]. MDSCs may participate in muscle regeneration processes with both new muscular fibre formation and neurovascular supply [57]. Direct intramuscular transplantation of MDSCs resulted in a decrease in the fibrotic area. However, a considerable amount of fibrosis was still noted in murine muscular contusion model study [58].

Bone marrow-derived mesenchymal stromal cells (BMMSCs) can differentiate and blend with myoblast in vitro and may contribute to the muscle healing process [59]. Therefore, the application of BMMSCs to facilitate the regeneration of defective tissues may be an effective tool for treating orthopaedic conditions such as trauma, muscle contusion and tears. In a recent murine study, increased number of regenerating myofibres and improved fast-twitch and tetanus muscle strength were reported in treated muscles after BMMSC treatment. However, in the latter study, the authors showed a rapid decay of transplanted BMMSCs, suggesting a paracrine effect of this action [60].

The tissue engineering approach based on association between stromal cells, matrix scaffold and signalling molecules, which has been developed for articular cartilage regeneration and healing of bony defects, may have a role in treatment of severe muscle lesions [61]. The stromal cell approach may have a considerable clinical potential to treat muscle injuries and represents an exciting concept that warrants further development and evaluation. However, despite promising results, the use of stromal cell therapy in muscle injuries is still limited, because of lack of evidence about its efficacy and safety in humans at long-term follow-up.

24.5.2 Glycosaminoglycans

The matrix-based therapy aims at improving extracellular matrix remodelling to support tissue repair. Glycosaminoglycans, in particular chondroitin sulphate (CS) and glucosamine (GlcN), have been widely investigated as potential treatment of musculoskeletal diseases and have demonstrated beneficial effects for the treatment of osteoarthritis both in preclinical studies and in clinical trials. The potential therapeutic effect of CS and GlcN has been related to anti-inflammatory and anti-apoptotic effects, preventing the transcription of some cytokines [62, 63]. In a preclinical study, the combination of CS and GlcN on muscle healing has been associated with an increase of intramuscular deposition of CS in the injured area, improvement in muscle force and growth of regenerating muscle fibres [64].

An analysis of the properties of a single injection of the clinical grade glycosaminoglycan mimetics RGTA® (ReGeneraTing Agents) resulted in an increase of the number of myonuclei in regenerating myofibres and an increase of capillarization of the new myofibres. These results suggested that the glycosaminoglycan therapy may have a non-neglectable potential in treatment of muscle injuries [65].

24.5.3 Anti-Fibrotic Therapy

The formation of scar tissue can be a mechanical barrier against muscle fibre regeneration and reinnervation at the site of injury, where the tear is localized, not allowing the complete recovery of the injured muscle tissue and function. Therefore, treatments which counteract fibrotic processes have the potential to enhance muscle healing. In particular, since TGF-β1 has been identified as trigger factor in the development of scar tissue by activating fibrotic cascades, anti-fibrotic strategies have mainly targeted the TGF-β1 pathway [57].

The effect of different anti-fibrotic agents was investigated in the literature. However, since no clinical trial in human exists, the future step is to investigate efficacy and safety of the anti-fibrotic therapies in humans, with the purpose, subsequently, to use this treatment in the clinical practice.

Decorin is a human proteoglycan which has been reported to block TGF-β1 action on its receptor by binding to TGF-β1 [66]. A significant decrease of the amount of fibrosis and enhanced muscle healing were reported after direct injection with decorin compared with a direct injection with saline [67]. However, it was a murine analysis, and large quantities of decorin were required in a very small muscle.

Losartan, an antagonist of angiotensin II receptor, is thought to reduce fibrosis by inactivating small mother against decapentaplegic-receptor (R-SMAD) through the upregulation of SMAD7 [68]. Moreover, losartan appeared to be associated with increased local expression of follistatin, an angiogenesis stimulator which may be able to neutralize several members of the TGF family [68]. Oral administration of losartan in mice resulted in a decrease of fibrosis and in an increase of structural and functional muscle regeneration [69].

Suramin competitively binds the TGF-β1 receptor, with an inhibitory effect on the TGF-β1 pathway. In in vitro and murine studies analysing suramin effects, it provided an antiproliferative consequence on fibroblasts, reduced scar formation and has positive effects on muscle regeneration [70, 71].

Interferon-γ has an inhibitory effect on the TGF-β1 pathway by inducing SMAD7 expression. Intramuscular injection in murine muscle tear model resulted in reduced fibrosis and an improved muscle healing process [72].

24.5.4 Actovegin

Actovegin is a biological drug with a 60-year history of controversial use as an injection therapy for sports muscle injuries. It is a calf blood deproteinized haemodialysate, which contains physiological components, electrolytes and essential trace elements, with 30% organic components including amino acids, nucleosides, intermediary products of carbohydrates and fat metabolites. Since it is ultrafiltered to 6000 Da, it does not contain growth factors or hormone-like substances [73]. Conflicting opinion have been raised about the effect of actovegin, due to a weak scientific base and the pressure to deliver cutting-edge treatment in the field of sports medicine. In vitro studies have suggested that the active ingredients of actovegin are involved in glucoplastic energy metabolism and the repair metabolism of injured muscle tissue, improving the healing of muscular cells during post-ischemic metabolic events [74].

In an in vitro cell injury model, authors showed that actovegin improved intrinsic mitochondrial respiratory capacity in injured human skeletal muscle fibres; the group concluded that their findings supported and explained the reported ergogenic properties [75]. However, results of a previous clinical trial have shown that actovegin has no effect on peak aerobic capacity in humans in vivo [76].

The results of a recent review of the past 10 years literature showed that actovegin represents a safe injectable therapy that has demonstrated some efficacy in treating muscular sports injury, in particular in reducing return-to-play time, and is unlikely to be ergogenic [74]. Despite there have been improvements in the scientific evidence base surrounding actovegin use, further expansion and research are warranted to fully understand the role of this injective option in the treatment of muscle injuries.

24.6 Conclusions

Muscle tears are extremely frequent sports injuries, and their optimal treatment is not well defined. The healing process of the muscle requires the presence of different cell populations, regulations and participation of multiple growth factors. PRP and other several emerging approaches have been proposed to enhance and accelerate muscle healing after injury, with promising results in preclinical studies. Among the emerging biological approaches for muscle injuries, only PRP has a demonstrated safety practice and improved outcomes reported in RCTs. However, clear evidence of clinical efficacy of these innovative biological approaches is still lacking.

Take-Home Messages
- In order to improve the muscle healing process and to achieve a quick yet full recovery, several orthobiologic approaches have been introduced among the management options of muscle injuries.
- Platelet-rich plasma (PRP) seems to be a safe tool in acute muscle tear treatment, with some potential in particular with regard to the time to return to sport, although results are still controversial.
- Promising preclinical results are available in the literature about other injective treatments such as mesenchymal stromal cells, glycosaminoglycans and anti-fibrotic agents.
- Clear evidence of the clinical efficacy of the innovative biological approaches in muscle tears treatment is still lacking.

References

1. Orchard JW, Seward H, Orchard JJ. Results of 2 decades of injury surveillance and public release of data in the Australian football league. Am J Sports Med. 2013;41(4):734–41.
2. Lynall RC, Gardner EC, Paolucci J, Currie DW, Knowles SB, Pierpoint LA, et al. The first decade of web-based sports injury surveillance: descriptive epidemiology of injuries in US high school girls' field hockey (2008-2009 through 2013-2014) and National Collegiate Athletic Association Women's field hockey (2004-2005 through 2013-2014). J Athl Train. 2018;53(10):938–49.
3. Ekstrand J, Healy JC, Waldén M, Lee JC, English B, Hägglund M. Hamstring muscle injuries in professional football: the correlation of MRI findings with return to play. Br J Sports Med. 2012;46(2):112–7.
4. Yu B, Li L. Research in prevention and rehabilitation of hamstring muscle strain injury. J Sport Health Sci. 2017;6(3):253–4.
5. Maffulli N, Aicale R, Tarantino D. Classification of muscle lesions. In: Canata GL, d'Hooghe P, Hunt KJ, editors. Muscle and tendon injuries. Berlin: Springer Nature; 2017. p. 95–102.
6. Boutin RD, Fritz RC, Steinbach LS. Imaging of sports-related muscle injuries. Radiol Clin N Am. 2002;40(2):333–62, vii.
7. Garrett WE Jr. Muscle strain injuries: clinical and basic aspects. Med Sci Sports Exerc. 1990;22(4):436–43.
8. Järvinen TA, Järvinen TL, Kääriäinen M, Kalimo H, Järvinen M. Muscle injuries: biology and treatment. Am J Sports Med. 2005;33(5):745–64.
9. Koh ES, McNally EG. Ultrasound of skeletal muscle injury. Semin Musculoskelet Radiol. 2007;11(2):162–73.
10. Schuermans J, Van Tiggelen D, Danneels L, Witvrouw E. Biceps femoris and semitendinosus— teammates or competitors? New insights into hamstring injury mechanisms in male football players: a muscle functional MRI study. Br J Sports Med. 2014;48(22):1599–606.
11. Wan X, Qu F, Garrett WE, Liu H, Yu B. The effect of hamstring flexibility on peak hamstring muscle strain in sprinting. J Sport Health Sci. 2017;6(3):283–9.
12. Garrett WE Jr, Safran MR, Seaber AV, Glisson RR, Ribbeck BM. Biomechanical comparison of stimulated and nonstimulated skeletal muscle pulled to failure. Am J Sports Med. 1987;15(5):448–54.
13. Palmer WE, Kuong SJ, Elmadbouh HM. MR imaging of myotendinous strain. AJR Am J Roentgenol. 1999;173(3):703–9.
14. Pollock N, James SL, Lee JC, Chakraverty R. British athletics muscle injury classification: a new grading system. Br J Sports Med. 2014;48(18):1347–51.
15. Maffulli N, Oliva F, Frizziero A, Nanni G, Barazzuol M, Via AG, et al. ISMuLT Guidelines for muscle injuries. Muscles Ligaments Tendons J. 2013;3(4):241–9.

16. Mueller-Wohlfahrt HW, Haensel L, Mithoefer K, Ekstrand J, English B, McNally S, et al. Terminology and classification of muscle injuries in sport: the Munich consensus statement. Br J Sports Med. 2013;47(6):342–50.
17. Cianforlini M, Coppa V, Grassi M, Gigante A. New strategies for muscular repair and regeneration. In: Canata GL, d'Hooghe P, Hunt KJ, editors. Muscle and tendon injuries. Berlin: Springer Nature; 2017. p. 145–56.
18. Ueblacker P, Haensel L, Mueller-Wohlfahrt HW. Treatment of muscle injuries in football. J Sports Sci. 2016;34(24):2329–37.
19. Barnett AJ, Negus JJ, Barton T, Wood DG. Reattachment of the proximal hamstring origin: outcome in patients with partial and complete tears. Knee Surg Sports Traumatol Arthrosc. 2015;23(7):2130–5.
20. Assis L, Moretti AI, Abrahão TB, de Souza HP, Hamblin MR, Parizotto NA. Low-level laser therapy (808 nm) contributes to muscle regeneration and prevents fibrosis in rat tibialis anterior muscle after cryolesion. Lasers Med Sci. 2013;28(3):947–55.
21. Hamid MSA, Mohamed Ali MR, Yusof A, George J, Lee LP. Platelet-rich plasma injections for the treatment of hamstring injuries: a randomized controlled trial. Am J Sports Med. 2014;42(10):2410–8.
22. Hofmann KJ, Paggi A, Connors D, Miller SL. Complete avulsion of the proximal hamstring insertion: functional outcomes after nonsurgical treatment. J Bone Joint Surg Am. 2014;96(12):1022–5.
23. Alonso JM, Edouard P, Fischetto G, Adams B, Depiesse F, Mountjoy M. Determination of future prevention strategies in elite track and field: analysis of Daegu 2011 IAAF championships injuries and illnesses surveillance. Br J Sports Med. 2012;46(7):505–14.
24. Pollock N, Patel A, Chakraverty J, Suokas A, James SL, Chakraverty R. Time to return to full training is delayed and recurrence rate is higher in intratendinous ('c') acute hamstring injury in elite track and field athletes: clinical application of the British athletics muscle injury classification. Br J Sports Med. 2016;50(5):305–10.
25. Ciciliot S, Schiaffino S. Regeneration of mammalian skeletal muscle. Basic mechanisms and clinical implications. Curr Pharm Des. 2010;16(8):906–14.
26. Laumonier T, Menetrey J. Muscle injuries and strategies for improving their repair. J Exp Orthop. 2016;3(1):15.
27. Chargé SB, Rudnicki MA. Cellular and molecular regulation of muscle regeneration. Physiol Rev. 2004;84(1):209–38.
28. Tidball JG. Inflammatory cell response to acute muscle injury. Med Sci Sports Exerc. 1995;27(7):1022–32.
29. Toumi H, Best TM. The inflammatory response: friend or enemy for muscle injury? Br J Sports Med. 2003;37(4):284–6.
30. Zhao W, Lu H, Wang X, Ransohoff RM, Zhou L. CX3CR1 deficiency delays acute skeletal muscle injury repair by impairing macrophage functions. FASEB J. 2016;30(1):380–93.
31. Chazaud B, Brigitte M, Yacoub-Youssef H, Arnold L, Gherardi R, Sonnet C, et al. Dual and beneficial roles of macrophages during skeletal muscle regeneration. Exerc Sport Sci Rev. 2009;37(1) 18–22.
32. Relaix F, Zammit PS. Satellite cells are essential for skeletal muscle regeneration: the cell on the edge returns centre stage. Development. 2012;139(16):2845–56.
33. Sambasivan R, Yao R, Kissenpfennig A, Van Wittenberghe L, Paldi A, Gayraud-Morel B, et al. Pax7-expressing satellite cells are indispensable for adult skeletal muscle regeneration. Development. 2011;138(17):3647–56.
34. Creaney L, Hamilton B. Growth factor delivery methods in the management of sports injuries: the state of play. Br J Sports Med. 2008;42(5):314–20.
35. Zanon G, Combi A, Benazzo F Bargagliotti M. The use of PRP in athletes with muscular lesions or classification of PRP preparations. In: Gobbi A, Espregueira-Mendes J, Lane JG, Karahan M, editors. Bio-orthopaedics, a new approach. Berlin: Springer Nature; 2017. p. 239–45.
36. Lehto M, Sims TJ, Bailey AJ. Skeletal muscle injury—molecular changes in the collagen during healing. Res Exp Med (Berl). 1985;185(2):95–106.
37. Cole BJ, Seroyer ST, Filardo G, Bajaj S, Fortier LA. Platelet-rich plasma: where are we now and where are we going? Sports Health. 2010;2(3):203–10.
38. Grassi A, Napoli F, Romandini I, Samuelsson K, Zaffagnini S, Candrian C, et al. Is platelet-rich plasma (PRP) effective in the treatment of acute muscle injuries? A systematic review and meta-analysis. Sports Med. 2018;48(4):971–89.
39. Filardo G, Kon E, Roffi A, Di Matteo B, Merli ML, Marcacci M. Platelet-rich plasma: why intra-articular? A systematic review of preclinical studies and clinical evidence on PRP for joint degeneration. Knee Surg Sports Traumatol Arthrosc. 2015;23(9):2459–74.
40. Mishra A, Harmon K, Woodall J, Vieira A. Sports medicine applications of platelet rich plasma. Curr Pharm Biotechnol. 2012;13(7):1185–95.
41. Dohan Ehrenfest DM, Bielecki T, Mishra A, Borzini P, Inchingolo F, Sammartino G, et al. In search of a consensus terminology in the field of platelet concentrates for surgical use: platelet-rich plasma (PRP), platelet-rich fibrin (PRF), fibrin gel polymerization and leukocytes. Curr Pharm Biotechnol. 2012;13(7):1131–7.
42. Hamid MS, Yusof A, Mohamed Ali MR. Platelet-rich plasma (PRP) for acute muscle injury: a systematic review. PLoS One. 2014;9(2):e90538.
43. Hammond JW, Hinton RY, Curl LA, Muriel JM, Lovering RM. Use of autologous platelet-rich plasma to treat muscle strain injuries. Am J Sports Med. 2009;37(6):1135–42.
44. Visser LC, Arnoczky SP, Caballero O, Egerbacher M. Platelet-rich fibrin constructs elute higher concentrations of transforming growth factor-β1 and increase tendon cell proliferation over time when compared to

blood clots: a comparative in vitro analysis. Vet Surg. 2010;39(7):811–7.

45. Gigante A, Del Torto M, Manzotti S, Cianforlini M, Busilacchi A, Davidson PA, et al. Platelet rich fibrin matrix effects on skeletal muscle lesions: an experimental study. J Biol Regul Homeost Agents. 2012;26(3):475–84.

46. Cianforlini M, Mattioli-Belmonte M, Manzotti S, Chiurazzi E, Piani M, Orlando F, et al. Effect of platelet rich plasma concentration on skeletal muscle regeneration: an experimental study. J Biol Regul Homeost Agents. 2015;29(4 Suppl):47–55.

47. Engebretsen L, Steffen K, Alsousou J, Anitua E, Bachl N, Devilee R, et al. IOC consensus paper on the use of platelet-rich plasma in sports medicine. Br J Sports Med. 2010;44(15):1072–81.

48. Martinez-Zapata MJ, Orozco L, Balius R, Soler R, Bosch A, Rodas G, et al. Efficacy of autologous platelet-rich plasma for the treatment of muscle rupture with haematoma: a multicentre, randomised, double-blind, placebo-controlled clinical trial. Blood Transfus. 2016;14(2):245–54.

49. Reurink G, Goudswaard GJ, Moen MH, Weir A, Verhaar JA, Bierma-Zeinstra SM, et al. Rationale, secondary outcome scores and 1-year follow-up of a randomised trial of platelet-rich plasma injections in acute hamstring muscle injury: the Dutch hamstring injection therapy study. Br J Sports Med. 2015;49(18):1206–12.

50. Borrione P, Fossati C, Pereira MT, Giannini S, Davico M, Minganti C, et al. The use of platelet-rich plasma (PRP) in the treatment of gastrocnemius strains: a retrospective observational study. Platelets. 2018;29(6):596–601.

51. Rettig AC, Meyer S, Bhadra AK. Platelet-rich plasma in addition to rehabilitation for acute hamstring injuries in NFL players: clinical effects and time to return to play. Orthop J Sports Med. 2013;1(1):2325967113494354.

52. Hamilton B, Tol JL, Almusa E, Boukarroum S, Eirale C, Farooq A, et al. Platelet-rich plasma does not enhance return to play in hamstring injuries: a randomised controlled trial. Br J Sports Med. 2015;49(14):943–50.

53. Rossi LA, Molina Rómoli AR, Bertona Altieri BA, Burgos Flor JA, Scordo WE, Elizondo CM. Does platelet-rich plasma decrease time to return to sports in acute muscle tear? A randomized controlled trial. Knee Surg Sports Traumatol Arthrosc. 2017;25(10):3319–25.

54. Bubnov R, Yevseenko V, Semeniv I. Ultrasound guided injections of platelets rich plasma for muscle injury in professional athletes. Comparative study. Med Ultrason. 2013;15(2):101–5.

55. Serner A, van Eijck CH, Beumer BR, Hölmich P, Weir A, de Vos RJ. Study quality on groin injury management remains low: a systematic review on treatment of groin pain in athletes. Br J Sports Med. 2015;49(12):813.

56. Gates CB, Karthikeyan T, Fu F, Huard J. Regenerative medicine for the musculoskeletal system based on muscle-derived stem cells. J Am Acad Orthop Surg. 2008;16(2):68–76.

57. van der Made A, Reurink G, Tol J, Marotta M, Rodas G, Kerkhoffs G. Emerging biological approaches to muscle injuries. In: Gobbi A, Espregueira-Mendes J, Lane JG, Karahan M, editors. Bio-orthopaedics, a new approach. Berlin: Springer Nature; 2017. p. 227–38.

58. Ota S, Uehara K, Nozaki M, Kobayashi T, Terada S, Tobita K, et al. Intramuscular transplantation of muscle-derived stem cells accelerates skeletal muscle healing after contusion injury via enhancement of angiogenesis. Am J Sports Med. 2011;39(9):1912–22.

59. Peçanha R, Bagno LL, Ribeiro MB, Robottom Ferreira AB, Moraes MO, Zapata-Sudo G, et al. Adipose-derived stem-cell treatment of skeletal muscle injury. J Bone Joint Surg Am. 2012;94(7):609–17.

60. Chiu C-H, Chang T-H, Chang S-S, Chang G-J, Chen AC-Y, Cheng C-Y, et al. Application of bone marrow–derived mesenchymal stem cells for muscle healing after contusion injury in mice. Am J Sports Med. 2020;48(5):1226–35.

61. McCullagh KJ, Perlingeiro RC. Coaxing stem cells for skeletal muscle repair. Adv Drug Deliv Rev. 2015;84:198–207.

62. du Souich P, García AG, Vergés J, Montell E. Immunomodulatory and anti-inflammatory effects of chondroitin sulphate. J Cell Mol Med. 2009;13(8a):1451–63.

63. Campo GM, Avenoso A, Campo S, D'Ascola A, Traina P, Samà D, et al. Purified human plasma glycosaminoglycans reduced NF-kappaB activation, pro-inflammatory cytokine production and apoptosis in LPS-treated chondrocytes. Innate Immun. 2008;14(4):233–46.

64. Contreras-Muñoz P, Fernández-Martín A, Torrella R, Serres X, De la Varga M, Viscor G, et al. A new surgical model of skeletal muscle injuries in rats reproduces human sports lesions. Int J Sports Med. 2016;37(3):183–90.

65. Bouvière J, Trignol A, Hoang DH, Del Carmine P, Goriot ME, Ben Larbi S, et al. Heparan sulfate mimetics accelerate postinjury skeletal muscle regeneration. Tissue Eng Part A. 2019;25(23–24):1667–76.

66. Li Y, Foster W, Deasy BM, Chan Y, Prisk V, Tang Y, et al. Transforming growth factor-beta1 induces the differentiation of myogenic cells into fibrotic cells in injured skeletal muscle: a key event in muscle fibrogenesis. Am J Pathol. 2004;164(3):1007–19.

67. Fukushima K, Badlani N, Usas A, Riano F, Fu F, Huard J. The use of an antifibrosis agent to improve muscle recovery after laceration. Am J Sports Med. 2001;29(4):394–402.

68. Kobayashi M, Ota S, Terada S, Kawakami Y, Otsuka T, Fu FH, et al. The combined use of losartan and muscle-derived stem cells significantly improves the functional recovery of muscle in a young mouse

model of contusion injuries. Am J Sports Med. 2016;44(12):3252–61.

69. Kobayashi T, Uehara K, Ota S, Tobita K, Ambrosio F, Cummins JH, et al. The timing of administration of a clinically relevant dose of losartan influences the healing process after contusion induced muscle injury. J Appl Physiol1985. 2013;114(2):262–73.

70. Chan YS, Li Y, Foster W, Fu FH, Huard J. The use of suramin, an antifibrotic agent, to improve muscle recovery after strain injury. Am J Sports Med. 2005;33(1):43–51.

71. Nozaki M, Ota S, Terada S, Li Y, Uehara K, Gharaibeh B, et al. Timing of the administration of suramin treatment after muscle injury. Muscle Nerve. 2012;46(1):70–9.

72. Foster W, Li Y, Usas A, Somogyi G, Huard J. Gamma interferon as an antifibrosis agent in skeletal muscle. J Orthop Res. 2003;21(5):798–804.

73. Müller-Wohlfart H-W, Hänsel L. Ueblacker P, Binder A. Conservative treatment of muscle injuries. In: Müller-Wohlfart H-W, Ueblacker P, Haensel L, Garret EW, editors. Muscle injuries in sports. Stuttgart: Georg Thieme Verlag; 2013. p. 268–95.

74. Brock J, Golding D, Smith PM, Nokes L, Kwan A, Lee PYF. Update on the role of Actovegin in musculoskeletal medicine: a review of the past 10 years. Clin J Sport Med. 2020;30(1):83–90.

75. Søndergård SD, Dela F, Helge JW, Larsen S. Actovegin, a non-prohibited drug increases oxidative capacity in human skeletal muscle. Eur J Sport Sci. 2016;16(7):801–7.

76. Lee P, Nokes L, Smith PM. No effect of intravenous Actovegin® on peak aerobic capacity. Int J Sports Med. 2012;33(4):305–9.

Cartilage Lesions and Osteoarthritis: Cell Therapy

Tiago Lazzaretti Fernandes, Kazunori Shimomura,
David A. Hart, Angelo Boffa,
and Norimasa Nakamura

25.1 Introduction

Articular cartilage is a unique avascular and aneural tissue that does not readily regenerate once damaged [1]. Chondral lesions are common knee injuries, showing a high prevalence, reaching as high as 63% in the general population and 36% among athletes [2–4]. Despite the numerous techniques available today, including regenera-

T. L. Fernandes
Sports Medicine Division, Department of
Orthopaedic Surgery, University of São Paulo
Medical School, São Paulo, Brazil
e-mail: tiago.lazzaretti@hc.fm.usp.br

K. Shimomura
Department of Orthopaedic Surgery, Osaka
University Graduate School of Medicine,
Osaka, Japan

D. A. Hart
McCaig Institute for Bone & Joint Health, University
of Calgary, Calgary, AB, Canada
e-mail: hartd@ucalgary.ca

A. Boffa
Clinica Ortopedica e Traumatologica 2, IRCCS
Istituto Ortopedico Rizzoli, Bologna, Italy

N. Nakamura (✉)
Department of Orthopaedic Surgery, Osaka
University Graduate School of Medicine,
Osaka, Japan

Institute for Medical Science in Sports, Osaka Health
Science University, 1-9-27, Tenma, Kita-ku,
Osaka, Japan

Global Centre for Medical Engineering and
Informatics, Osaka University, Osaka, Japan
e-mail: norimasa.nakamura@ohsu.ac.jp

tive techniques, complete healing of damaged or degenerated cartilage with the consistent reproduction of normal hyaline cartilage is not yet possible. If left untreated, defects in articular cartilage may enlarge and result in further lesions in the underlying subchondral bone. Such changes can lead to alteration in joint biomechanics, as well as homeostasis disturbances in the knee as a whole [5, 6]. Thus, considering the knee joint as an organ system [7, 8], failure to address articular cartilage lesions can affect the functioning of the joint. As a consequence, initial cartilage injuries can result in chronic disability, leading to development of overt osteoarthritis (OA) with significant physical limitations and decreased quality of life [1, 9, 10].

25.2 Osteoarthritis (OA)

OA is defined as a "serious condition" associated with substantial and progressive morbidity. It emerges as a condition with urgent needs for clinical treatment [11]. Chu et al. [11] stated that the twenty-first Century Cures Act dictates provisions to accelerate the development and translation of promising new therapies into clinical evaluation for conditions or diseases with substantial impact on day-to-day functioning and diseases with no satisfactory treatment available. Therefore, the main therapeutic purposes of addressing needs for OA are to reduce or eliminate pain and to improve joint function [12]. OA

© ISAKOS 2022
G. Filardo et al. (eds.), *Orthobiologics*, https://doi.org/10.1007/978-3-030-84744-9_25

is estimated to occur in up to 22.7% of the US population [13, 14], and it will continue to be a major cause of morbidity and physical limitation among individuals older than 40 years, and demand for osteotomies and total knee replacement is expected to substantially increase year to year [14, 15].

Current conservative therapy for OA is directed toward non-pharmacological treatments as physical activity through mechanical stimulation and symptomatic treatment, focusing on pain management. However, these approaches are not able to promote regeneration of degenerated cartilage or to attenuate joint inflammation in most patients [16]. Pharmacological therapies include analgesics and non-steroid and steroid anti-inflammatory drugs that provide temporary benefit for some but not all patients [17–19]. Additionally, conventional conservative therapies for OA include physiotherapy, hyaluronic acid, platelet-rich plasma, or corticosteroid-based intra-articular injections, as well as surgical interventions such as knee arthroscopic surgery or microfracture procedures [20, 21]. Effective treatment options remain limited for many patients, and progression of such degenerative conditions may occur due to failure of conventional nonsurgical therapies, leading to chronic pain and disability for daily living activities [11].

It is now well established that OA is a disease which affects all joint tissues and is characterized by progressive degeneration of the articular cartilage, vascular invasion in the deep layer of articular cartilage, subchondral bone remodeling, osteophyte formation, and synovial membrane inflammation (synovitis) [22, 23]. Dai et al. [24] also stated that the articular inflammatory environment is a key factor for the initiation and aggravation of cartilage lesions. Accumulating evidence suggests that synovial inflammation is correlated with the pathogenesis and progression of OA [22].

The development of OA involves multiple pathological changes, including apoptosis and hypertrophy of chondrocytes and activation of innate immune cells (mainly macrophages) from the synovial membrane [24]. These changes progressively alter the articular microenvironment, which is crucial for the repair of damaged cartilage tissues. To effectively treat OA, it is critical to develop a multifunctional agent with immunomodulatory effects that can diminish the catabolic inflammatory microenvironment and shift it toward a more anabolic pro-chondrogenic atmosphere [24]. Endogenous stem and progenitor cells in the synovial fluid can be increased by inflammation as a response to tissue injury, but the quantity of available bioactive cells to repair cartilage is likely not sufficient to facilitate effective repair [15].

25.3 Cell Therapies for Repair of Cartilage Lesions and OA Damage

Among the various cell therapeutic solutions and regenerative techniques, there are two main examples: autologous chondrocyte implantation (ACI) and mesenchymal stromal cell (MSC)-based therapy. ACI consists of a two-step procedure involving the arthroscopic harvest of healthy cartilage followed by expanded chondrocyte cell culture and, in a second step, the implantation of the expanded chondrocytes into the defect [25, 26]. Despite the versatility of second- and third-generation ACI, these techniques sacrifice healthy cartilage tissue and use dedifferentiated chondrocytes arising from in vitro cell culture [2, 9, 27–30]. Moreover, isolating chondrocytes from healthy donor cartilage might also elevate risk for infection and morbidity [31]. Also, the complete and reliable healing of damaged cartilage with consistent reproduction of normal hyaline cartilage has not yet been achieved with this approach [6, 8, 28, 32]. Therefore, continuous drug therapies and secondary surgeries, such as joint replacement, are common, elevating the clinical relevance for new therapeutic approaches to facilitate better repair of articular cartilage lesions [6, 33]. For these reasons, MSC-based approaches have received considerable research attention.

Cell populations meeting the criteria for MSCs are readily generated from a variety of tissue sources through competitive expansion of the mixed tissue-specific connective tissue progenitor (CTP) population in native tissues [34–36], resulting in the isolation of cells in vitro that are selected based on the most rapid and durable proliferation under in vitro culture conditions. Culture-expanded MSC populations can be characterized with respect to established surface markers and shown to exhibit in vitro differentiation into bone, cartilage, and adipose tissues (Fig. 25.1). MSC populations are relatively easy to handle; however, batches of MSCs often differ significantly between patient donors and tissues [36, 37]. They also tend to lose differentiation potential, after repeated passages in vitro. MSCs could provide a feasible cell therapy option in the clinical practice [1, 6, 19], but ongoing work is required to provide autograft or allograft MSC populations with highly repeatable and reproducible attributes and biological potential.

fabrication, and transplantation are being explored [11]. For many indications, laboratory manipulation and culture expansion are needed to isolate and adequately enrich stem and progenitor cell populations, which is not possible with minimally manipulated cell preparations [11]. Minimally manipulated autogenous cell products designed to increase the concentration and prevalence of colony-founding CTPs are already in clinical use for musculoskeletal indications and can be considered a "lower-risk" treatment. Culture-expanded populations, such as MSCs without genetic modifications, carry greater risk but can be made available with appropriate fabrication strategies and quality controls. However, translation of the use of native cell populations of cells fabricated using culture expansion requires investments in developing rigorous clinical indications and well-documented evidence of clinical efficacy [11].

25.4 Mesenchymal Stromal Cells (MSCs)

Culture-expanded MSCs may be derived from various tissues, such as bone marrow, adipose tissue, dental pulp, and synovial membrane [6, 8, 30]. The tissue-resident colony-founding CTP population that serves as a source material for MSC fabrication is a heterogeneous population of cells from a diversity of niches and differentiation states, but pericytes embedded in the basement membrane of capillaries are considered to be a common source and can be released from this tissue niche by enzymatic digestion [11].

Culture-expanded MSC populations meeting ISCT criteria have shown to have the capacity to modulate immunologic and inflammatory response and to stimulate adhesion, migration, and proliferation by paracrine effects acting on other cells [30, 38]. Like some native CTP populations, injected MSCs are reported to home to damaged tissues and contribute to their repair by secretion of cytokines, chemokines, and extracellular matrix proteins [39].

Because of the urgent need for OA treatments, practical strategies for a diversity of cell harvest,

25.5 MSCs and the Immune System

There is evidence that the therapeutic efficacy of MSCs may be attributed in part to the paracrine effect of their secreted factors such as growth factors, cytokines, and extracellular vesicles [40]. It has been proposed that culture-expanded human MSCs exhibit immuno-tolerance, a paracrine capacity, and the ability to repair cartilage lesions. Although the number of reported studies is limited, positive results have been reported in translational large animal models and preliminary clinical findings [8, 41, 42].

During culture expansion, MSCs express low levels of human leukocyte antigen (HLA) major histocompatibility complex (MHC) class I antigens and do not express co-stimulatory molecules (B7–1, B7–2, CD40, or CD40L), so that they would not normally activate alloreactive T cells. However, they can be induced to express MHC class II and Fas ligand. MSCs are also reported to inhibit dendritic cell (DC) maturation and B and T cell proliferation and differentiation, as well as attenuate natural killer (NK) cell killing, and they also support suppressive T regulatory cells

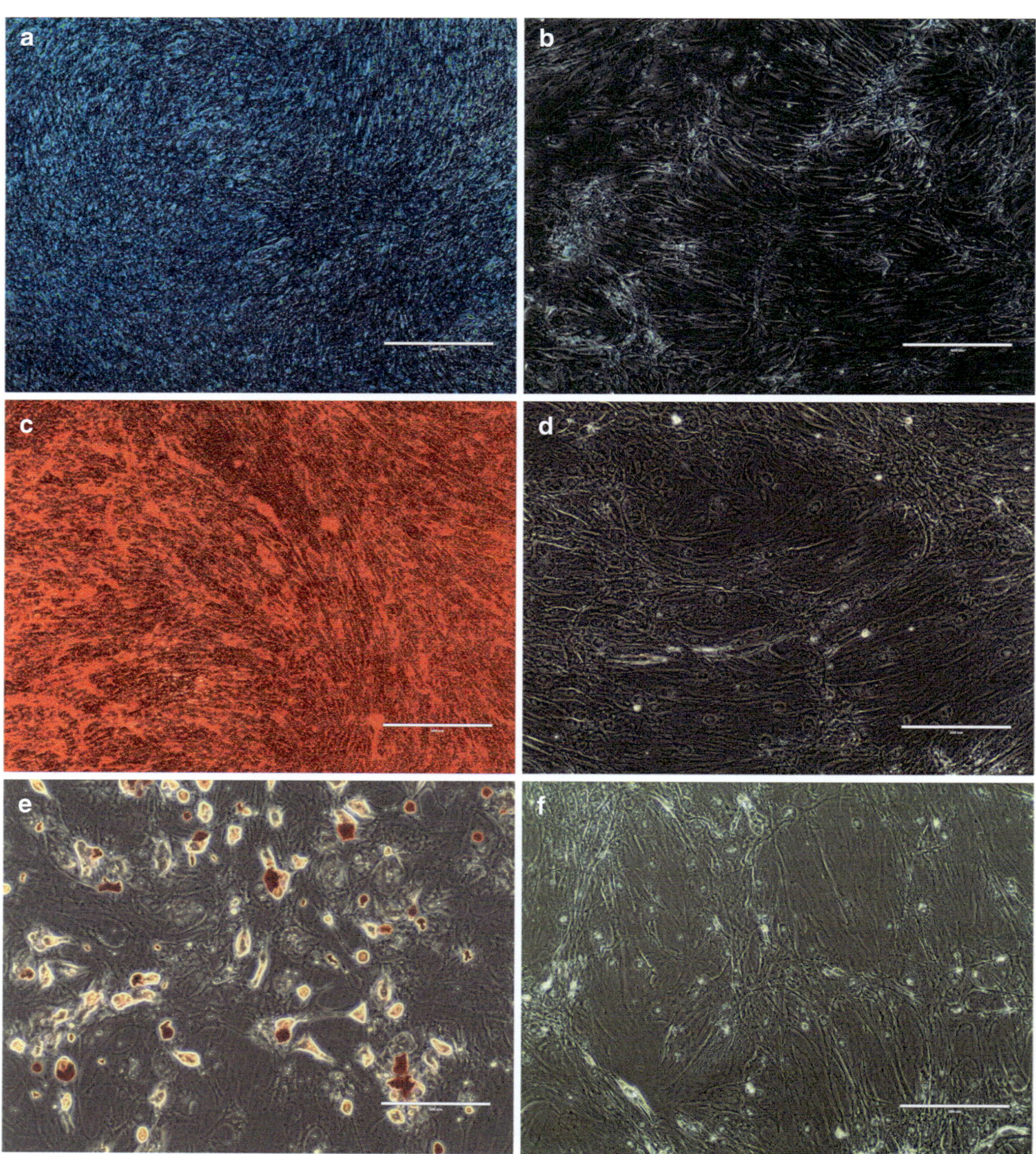

Fig. 25.1 All three MSC strains were induced to undergo chondrogenic (Alcian blue staining) (**a**), osteogenic (Alizarin Red S staining) (**c**), and adipogenic (Oil Red O staining) (**e**) differentiation, which showed that these strains had mesenchymal origins and maintained multipotentiality, and controls (**b**, **d**, and **f**), respectively

(Tregs) [39, 43]. For instance, Ankrum et al. [44] have shown that culture-expanded MSCs express low levels of MHC class I and are negative for MHC class II [44]. Several factors in the secretory profile of MSCs have been associated with these immune-modulating properties, including transforming growth factor beta (TGFb), hepatocyte growth factor (HGF), HLA-G, prostaglandin (PGE2), IL-10, indoleamine 2,3-dioxygenase (IDO), and interferon-gamma (IFNg). More recently a role for the notch family member Jagged1 in immune modulation was specifically attributed to downstream TLR signaling by MSCs [39, 45].

MSCs can release anti-inflammatory factors, potentially reducing synovial inflammation [12]. Macrophages are one of key cells in synovial inflammation that are present in the synovial lining of joints and have been shown to play a prominent role in the progression of OA [22]. After initial stimulation, undifferentiated M0 macrophages acquire a phenotype, ranging from pro-inflammatory (M1) to anti-inflammatory (M2) [46]. M1 are the "classically activated" macrophages associated with high production of pro-inflammatory mediators such as nitric oxide (NO), TNF-α, IL-6, and IL-12 [43]. TNF-α, Oncostatin M (OSM), IL-1β, and IL-6 have been identified as some of the key players involved in synovial inflammation, with synovial macrophages considered to play a prominent role in the production of these mediators [47]. The "alternatively activated" macrophages, designated M2, are known as wound-healing macrophages and participate in tissue repair; the change of M0 macrophages toward the M2 phenotype in the injury lesion may contribute to the repair of the damaged articular cartilage [24]. The presence of M2 macrophages leads to increases in TGF-β levels, a mediator that is a well-known pro-chondrogenic cytokine for MSC differentiation and expression of cartilage components. In a juxtacrine (cell-to-cell contact) and a paracrine manner (through the production of soluble factors), MSCs can inhibit the activation of inflammatory M1 macrophages and promote their conversion to the anti-inflammatory M2 phenotype [16]. MSCs are also reported to induce conversion of TNF-α and IL-1 producing inflammatory M1 macrophages into immunosuppressive IL-10 producing M2 cells that can attenuate joint inflammation and promote cartilage regeneration [16]. Recently, it has been shown that intra-articular injection of adipose-derived stromal cells (ADSCs) exerts anti-inflammatory and chondroprotective effects in preclinical models, as well as in humans [48, 49].

Overall, these data confirm that MSCs can potentially contribute to a more favorable, less inflamed local environment via impact on immunoregulation in the OA or injured joints, and these effects could also improve the potential for chondrogenesis. However, these immunomodulatory effects can vary from batch to batch of MSC fabrication, even from the same patient. Given the heterogeneity of the CTP founding cell population in individual tissues, greater attention has been called to early-stage characterization and selection among the potential founding clones from which MSCs may be derived. MSC attributes are likely influenced by the tissue or origin, by the processing to select cells of a particular niche, as well as by expansion techniques, relative purity after expansion, and other measurable characteristics of surface markers or functional performance [50, 51].

25.6   Clinical Studies of MSC Injection

As mentioned above, MSCs are considered to have "immunoprivileged" capacities [52]. Therefore, they can be safely injected into either autologous or allogenous hosts as a potential therapy for OA due to their lack of host immune reactivity [52]. Many orthopedic surgeons feel that intra-articular (IA) injections of MSCs can be considered to be an efficient and minimally invasive therapy, concentrating and retaining cells in the confined space [53].

Satué et al. performed IA injections of MSCs in the knee and reported that MSCs remained in the synovial cavity, engrafted within the cartilage lesion, and were detectable up to 1 month post-injection; and no adverse effects after the injection were observed [53]. Other authors reported that MSCs were detectable in the knee joint at 3 months [54], 8 weeks [55], and 14 days after implantation in animal studies [56].

Clinical trials are the ultimate research tool with real patients that can confirm or refute the value of these new therapies [38]. Recently, some clinical trials demonstrated that MSC injections yield beneficial outcomes for patients with OA [38]. Moreover, results of preclinical and clinical studies have provided preliminary evidence for the safety and efficacy of MSCs in the treatment of OA. Nevertheless, there is still uncertainty regarding the magnitude, reliability, and reproducibility of the clinical effectiveness of cell therapy injections, including culture-expanded MSCs [12, 57].

The latest available meta-analysis from 2020 published by Huang et al. [12] included nine randomized clinical trials (RCTs) involving 399 patients, 203 patients in the stromal cell therapy group and 196 patients in the control group [12]. Changes in VAS and IKDC scores from baseline to 24 months were superior in the cell-treated group. Huang et al. [12] showed that in three studies good clinical results were reported with the injection of approximately 20×10^6 culture-expanded cells. The changes in IKDC score at 24 months were higher in the cell-treated group than that in the control group. The authors concluded that treatment with MSCs is effective in preventing or limiting the progression of OA at an early stage. Huang et al. also pointed out that patients with knee OA at the moderate and late stages may not benefit from MSC injection therapy, and tissue engineering methods using a scaffold may be needed to repair large cartilage defects associated with overt knee OA [12]. High-quality, large-scale RCTs will be required to verify this hypothesis. Available reviews also highlighted the wide variation in methodology and the fact that cell therapy approaches and methods for MSC fabrication and characterization were quite heterogeneous. There is a need for standardization, and future studies will need to make far greater investment in rigorous characterization and documentation of the cell populations being injected.

Kim et al. [15] included in their meta-analysis a total of five RCTs (220 patients). Four of five studies were performed with bone marrow-derived MSCs, and one study was performed with adipose-derived stromal vascular fraction (SVF) containing unselected and uncharacterized adipose-derived cells. Two studies reported that MSCs, delivered in a concentration of 1.4–110×10^7 cells, led to significantly improved pain scores, including VAS scores and cumulative assessment of pain scores, as well as significant improvement in functional outcomes (Lysholm knee scale) and in cumulative period WOMAC function score in the short term of 12 or 24 months

in patients with OA. Conversely, there was no significant difference in cartilage repair assessed by MRI. Therefore, this meta-analysis demonstrated that intra-articular MSCs have a limited evidence in pain relief and functional improvement in knee OA, but it does not support the use of intra-articular MSCs for improving cartilage repair in knee OA.

Yubo et al. [20] included 11 eligible trials in their meta-analysis on 582 patients with knee OA. Studies were eligible for inclusion if (1) they were published RCTs in humans of MSC transplantation therapy for patients with knee OA, (2) the patient's detailed information was reported both prior to and after therapy, and (3) the study enrolled ten or more patients. They concluded that MSC treatment significantly decreased VAS and increased IKDC scores after a 24-month follow-up in comparison with the control group. MSC treatment led to significant decreases in WOMAC and Lequesne scores after 12 months. Analysis of Lysholm (24-month) and Tegner (12- and 24-month) scores also demonstrated favorable results for MSC treatment. These authors concluded that MSC transplantation was safe and had therapeutic potential for patients with knee OA.

Chahla et al. [57] included six eligible trials in their meta-analysis, consisting of 300 knees. The analysis revealed that IA cellular therapy injections for OA and focal cartilage defects suggested positive results with respect to clinical improvement and safety. However, a placebo effect cannot be disregarded. The methodological quality was fair, even among level II and III studies. Effective clinical assessment and optimization of injection therapies will demand greater attention to study methodology, including blinding, standardized quantitative methods for cell harvesting, processing, characterization and delivery, and standardized reporting of clinical and structural outcomes.

Finally, Xia et al. [19] evaluated 7 randomized and controlled clinical trials, studying a total of 314 participants with a diagnosis of knee OA. Results from two high-quality trials (94 patients) showed a positive effect of MSC injection on pain scores, and the authors also con-

Table 25.1 Meta-analyses of randomized and non-randomized comparative studies using MSCs for osteoarthritis and cartilage lesions in humans

Author	Year	Journal	Level of evidence	Studies	N (knees)	Conclusion
Huang et al. [12]	2020	Medicine	Not mentioned (only RCTs)	9	339	Stem cell therapy is superior to traditional treatments in the conservative treatment of knee OA; reduces pain with no obvious additional side effects
Kim et al. [15]	2019	Archives of Orthopaedic and Trauma Surgery	II	5	220	Intra-articular MSCs have a limited evidence in pain relief and functional improvement in knee osteoarthritis
Ha et al. [58]	2018	Arthroscopy	III	17	499	Intra-articular MSCs provide improvements in pain and function in knee osteoarthritis at short-term follow-up (<28 months) in many cases. Some efficacy has been shown of MSCs for cartilage repair in osteoarthritis
Yubo et al. [20]	2017	PLoS one	Not mentioned (non-randomized studies included)	11	528	MSC transplantation treatment was shown to be safe and has great potential as an efficacious clinical therapy for patients with knee OA
Chahla et al. [57]	2016	Journal of Bone and Joint Surgery	Level III	6	300	Cell injection leads to positive results with respect to clinical improvement and safety. However, the improvement was modest, and a placebo effect cannot be disregarded. The overall quality of the literature was poor, and the methodological quality was fair, even among level II and III studies
Xu et al. [59]	2015	Cytotherapy	Not mentioned (randomized and non-randomized comparative studies)	11	3558	Assessment of the comprehensive evaluation index indicated that there were no significant differences after stem cell treatment. However, assessment of clinical symptoms and cartilage morphology showed significant improvement after stromal cell treatment
Xia et al. [19]	2015	International Orthopaedics	Not mentioned (randomized and non-randomized comparative studies)	7	314	MSC injection could be potentially efficacious for decreasing pain and may improve physical function in patients with knee OA

RCT Randomized clinical trials, *OA* Osteoarthritis, *MSCs* Mesenchymal stromal cells

cluded that MSCs may improve physical function in patients with knee OA post-injection at 3, 12, and 24 months (Table 25.1).

25.7 Lessons from Preclinical Animal Studies

Muhammad et al. performed a meta-analysis of 59 animal model studies that utilized stem/stromal cells, showing a significant improvement of 266% for cartilage formation in the MSC groups in comparison to control values. MSCs were capable of producing cartilage-like tissue with a matrix under appropriate culture conditions. Macroscopic, magnetic resonance imaging, and histological evaluations of aggrecan and collagen type 2 were the main outcome measures used to investigate cartilage regeneration. Nevertheless, they observed a significant regression of the stem/stromal cell effect on articular cartilage restoration at long-term follow-up, and a risk of bias due to poor reporting for most studies was noted. The authors suggested that a checklist such as the

Animal Research: Reporting of In Vivo Experiments (ARRIVE) guideline should be strongly recommended as the gold standard for future studies in this area [60].

25.8 One-Step Approaches

25.8.1 Bone Marrow Aspirate Concentrate (BMAC)

Due to the limited evidence and complexity of cultured cell-based strategies, more attention has been recently paid toward the use of cell concentrates. In fact, the "minimal manipulation" methods are becoming a popular strategy to exploit the potential of MSCs directly on-site in a one-step treatment, reducing the duration and costs of these approaches. Two main treatment modalities have emerged as "minimally manipulated" strategies, based on bone marrow aspirate concentrate (BMAC) and adipose-derived products [61, 62].

BMAC is commonly obtained through density gradient centrifugation of bone marrow aspirate, usually collected from the posterior or anterior iliac crest. This technically easy process allows to obtain directly in the operating room a product containing entire bone marrow niches with MSCs, hematopoietic precursors, monocytes, and endothelial cells, as well as a great array of soluble factors [63, 64]. Thanks to the proprieties offered by the high amount of cells and bioactive proteins, BMAC may present the potential to alter the disease course and not just to decrease pain [65]. Despite the still limited preclinical evidence, BMAC use is growing exponentially in the clinical practice for several orthopedic procedures, including the injective treatment of OA [61]. The available clinical studies evaluating the use of BMAC to address OA investigated mainly the knee injective treatment, while only few studies focused on shoulder, hip, or ankle OA. Several studies confirmed the safety and effectiveness of BMAC for OA symptom management, although with an overall low quality of evidence [61]. The few comparative studies in the literature were not able to prove the superiority of BMAC over the other IA options, and in the only available pla-

cebo controlled blinded RCT, BMAC did not show superiority over saline at 12 months of follow-up [66]. On the other hand, IA BMAC injections combined with platelet products demonstrated better results than exercise therapy in a comparative study on knee OA patients at 24 months of follow-up [67].

A new application has been recently proposed to further exploit the potential of BMAC by targeting the subchondral bone, which is commonly involved in the OA processes [68]. The subchondral application of BMAC provided promising clinical results in several preliminary studies on knee OA. Subchondral BMAC injections in 30 young patients with knee OA secondary to osteonecrosis showed similar results compared with total knee arthroplasty (TKA), with a lower complication rate and a quicker recovery [68]. In a similar study, 140 patients with a mean age of 75 years and who planned to undergo staged-bilateral TKA for medial knee OA were treated with subchondral BMAC injections on one side and with TKA on the other side [69]. The authors reported that subchondral BMAC provided a sufficient effect on pain to post-pone or avoid TKA up to 15 years of follow-up, with only 25 patients requesting TKA in joints treated with BMAC. In a recent RCT, Hernigou et al. demonstrated the superiority of subchondral BMAC injections over IA BMAC injections in 60 patients with bilateral knee OA, showing in the subchondral group higher clinical and MRI improvements at 24 months of follow-up and a lower yearly arthroplasty incidence [70].

Several authors investigated the factors that can potentially influence BMAC efficacy, such as harvest site, patient age, and injection schedule. The harvest site may play an important role for the obtained product, with a major number of MSCs isolated from the pelvis rather than the knee (distal femur or proximal tibia), even though the two sites did not demonstrate a significant difference in the phenotype of the isolated cells [71]. Moreover, the bone marrow aspirate harvested from the posterior iliac crest showed a high number of colony-founding connective tissue progenitors compared to the anterior iliac crest [72]. Patient age also seems to influence the quality and quantity of MSCs contained in BMAC, with some authors reporting an age-related reduction in the absolute number of MSCs and a decrease of their proliferative and differentiative capacity [73, 74]. Regarding the injection schedule, multiple injections provided better results than a single injection for the treatment of patients with hip OA, reporting additional benefit with each subsequent treatment [75]. Moreover, there are several commercial systems suitable for obtaining BMAC, with some differences such as the starting bone marrow aspirate volume, the centrifuge devices, and other methodological differences, leading to products not always similar in terms of progenitor cell number, platelet count, or growth factors and cytokine concentrations [76, 77]. Due to the lack of standardization of the procedure and the limited preclinical and clinical evidence, many controversies remain on the real effectiveness of BMAC, and further high-level studies are needed to better understand the potential of this product for the treatment of cartilage degenerative lesions and OA.

25.8.2 Stromal Vascular Fraction (SVF) and Micro-Fragment Adipose Tissue (MFAT)

Adipose-derived products have been recently proposed as a promising alternative for the treatment of OA, thanks to the advantages provided by adipose tissue over other MSC sources. In fact, adipose tissue is abundant, easily accessible, and obtainable with a mini-invasive procedure that makes available a high number of cells and pericytes (precursors to MSCs) [78, 79]. Adipose tissue can be processed at the point of care into cell suspensions, producing the SFV, or as micro-fragments, producing the micro-fragmented adipose tissue (MFAT).

SVF is generally obtained with an enzymatic method, which consists of digesting the lipoaspirate with collagenase to break down the matrix and release the MSCs and other cells. Subsequently, the collagenase is removed by dilution and washing, followed by centrifugation [80]. Preclinical studies showed the safety, feasibility, and effectiveness of SVF application for the treatment of degenerative cartilage lesions and OA joints, with better improvement in the quality of cartilage with respect to control groups [81]. Clinical studies focusing on intra-articular SVF injection reported improvement in pain and functional outcome scores, with a low rate of adverse events, in patients with knee OA, although the majority of studies are case series without a comparative arm [80] A recent double-blind placebo-controlled RCT on 39 patients with knee OA showed a statistically significant improvement in the SVF group compared to the saline group, although magnetic resonance images did not reveal changes in cartilage thickness after treatment [82]. Moreover, IA SVF demonstrated better clinical and imaging results at 12 months compared to hyaluronic acid in an RCT on 32 OA knee [83]. Nevertheless, these studies investigated only small sample sizes, and thus further studies are needed to elucidate the real therapeutic potential of SVF.

On the other side, MFAT approach gained increasing interest in clinical practice, since it is obtained through a simple, minimal mechanical manipulation that determines a progressive reduction in the size of adipose tissue clusters with the elimination of oil and blood residue [84]. MFAT has the advantages in providing a high amount of cells and growth factors, without expansion or enzymatic treatment, and preserving the integrity of cell and tissue microarchitecture [85]. In particular, MFAT well preserves the stem cell "niche," maintaining the biologically intact structure of the junctions between cells and saving the basal extra-cellular matrix proteins, which also showed to counteract inflammation [86]. In vitro studies showed the benefits of this approach, demonstrating a higher qualitatively and quantitatively secretion of growth factors and cytokines involved in tissue repair compared to the enzymatic methods [87, 88]. Moreover, MFAT contains a significantly higher concentration of exosomes secreted by MSCs compared to the enzymatic method, suggesting a better preservation of the paracrine potential of adipose MSCs and thus their efficacy [89]. A recent in vivo preclinical study revealed more promising results in terms of protection of the articular surface from the joint degenerative OA processes in rabbits treated with MFAT compared to those treated with SVF or expanded ADSCs [90]. Promising preclinical results were also confirmed by clinical studies focused on patients with knee OA, where IA MFAT produced a low number of adverse events and a significant improvement in pain, function, and quality of life [91, 92]. Despite the growing number of clinical studies focusing on IA MFAT injections for knee OA, high-level studies comparing the MFAT effectiveness with other injectable products are still limited. Therefore, further studies are needed to justify MFAT use in the clinical practice to treat OA joints.

25.9 Conclusions

To effectively address OA, it is of critical importance to not only control articular inflammation but also improve the local microenvironment to support chondrogenesis and prevent chondrocytes from OA-associated apoptosis and hypertrophy. Thus, it is important to diminish the catabolic environment and promote an anabolic environment in the joint in order to enhance the effectiveness of additional interventions. Injectable cell therapies with culture-expanded MSCs or minimally manipulated concentrates are therapeutic approaches that may influence the immunologic system and exert a regulatory effect on the local inflammatory responses to promote a pro-chondrogenic and anti-inflammatory polarization of innate immune cells. This may be useful for the treatment of cartilage defects and OA.

Take-Home Messages
- To effectively treat OA, it is critical to develop a multifunctional agent with immunomodulatory effects.
- Cell-based strategies can diminish the catabolic inflammatory microenvironment and shift it toward a more anabolic pro-chondrogenic environment by paracrine effect of their secreted factors such as growth factors, cytokines, and extracellular vesicles.
- Results of preclinical studies provided preliminary evidence on the safety of autologous or allogenous MSCs, either expanded or concentrated, injected for the treatment of OA.
- Preliminary clinical evidence supports the beneficial outcome of MSC IA injections regarding pain control and function for patients with OA from baseline up to 24 months, preventing or limiting progression of OA at an early stage.
- Future studies are needed to invest on the standardization with rigorous characterization and documentation of the cell populations being injected.

References

1. Zainal Ariffin SH, Kermani S, Megat Abdul Wahab R, Senafi S, Zainal Ariffin Z, Abdul Razak M. In vitro chondrogenesis transformation study of mouse dental pulp stem cells. Scientific World J. 2012;2012:827149.
2. Fernandes TL, de SantAnna JP C, Frisene I, Gazarini JP, Gomes Pinheiro CC, Gomoll AH, et al. Systematic review of human dental pulp stem cells for cartilage regeneration. Tissue Eng Part B Rev. 2020;26(1):1–12.
3. Flanigan DC, Harris JD, Trinh TQ, Siston RA, Brophy RH. Prevalence of chondral defects in athletes' knees: a systematic review. Med Sci Sports Exerc. 2010;42(10):1795–801.
4. Perera JR, Gikas PD, Bentley G. The present state of treatments for articular cartilage defects in the knee. Ann R Coll Surg Engl. 2012;94(6):381–7.
5. Gomoll AH, Madry H, Knutsen G, van Dijk N, Seil R, Brittberg M, et al. The subchondral bone in articular cartilage repair: current problems in the surgical management. Knee Surg Sports Traumatol Arthrosc. 2010;18(4):434–47.
6. Fernandes TL, Shimomura K, Asperti A, Pinheiro CCG, Caetano HVA, Oliveira C, et al. Development of a novel large animal model to evaluate human dental pulp stem cells for articular cartilage treatment. Stem Cell Rev Rep. 2018;14(5):734–43.
7. Frank CB, Shrive NG, Boorman RS, Lo IK, Hart DA. New perspectives on bioengineering of joint tissues: joint adaptation creates a moving target for engineering replacement tissues. Ann Biomed Eng. 2004;32(3):458–65.

8. Loeser RF, Goldring SR, Scanzello CR, Goldring MB. Osteoarthritis: a disease of the joint as an organ. Arthritis Rheum. 2012;64(6):1697–707.

9. Kubosch EJ, Lang G, Furst D, Kubosch D, Izadpanah K, Rolauffs B, et al. The potential for synovium-derived stem cells in cartilage repair. Curr Stem Cell Res Ther. 2018;13(3):174–84.

10. Hunziker EB. Articular cartilage repair: basic science and clinical progress. A review of the current status and prospects. Osteoarthr Cartil. 2002;10(6):432–63.

11. Chu CR, Rodeo S, Bhutani N, Goodrich LR, Huard J, Irrgang J, et al. Optimizing clinical use of biologics in orthopaedic surgery: consensus recommendations from the 2018 AAOS/NIH U-13 conference. J Am Acad Orthop Surg. 2019;27(2):e50–63.

12. Huang R, Li W, Zhao Y, Yang F, Xu M. Clinical efficacy and safety of stem cell therapy for knee osteoarthritis: a meta-analysis. Medicine. 2020;99(11):e19434.

13. Barbour KE, Helmick CG, Boring M, Brady TJ. Vital signs: prevalence of doctor-diagnosed arthritis and arthritis-attributable activity limitation—United States, 2013-2015. MMWR Morb Mortal Wkly Rep. 2017;66(9):246–53.

14. Shen J, Chen D. Recent progress in osteoarthritis research. J Am Acad Orthop Surg. 2014;22(7):467–8.

15. Kim SH, Ha CW, Park YB, Nam E, Lee JE, Lee HJ. Intra-articular injection of mesenchymal stem cells for clinical outcomes and cartilage repair in osteoarthritis of the knee: a meta-analysis of randomized controlled trials. Arch Orthop Trauma Surg. 2019;139(7):971–80.

16. Harrell CR, Markovic BS, Fellabaum C, Arsenijevic A, Volarevic V. Mesenchymal stem cell-based therapy of osteoarthritis: Current knowledge and future perspectives. Biomed Pharmacother. 2019;109:2318–26.

17. Filardo G, Kon E, Longo UG, Madry H, Marchettini P, Marmotti A, et al. Non-surgical treatments for the management of early osteoarthritis. Knee Surg Sports Traumatol Arthrosc. 2016;24(6):1775–85.

18. Bannuru RR, Osani MC, Vaysbrot EE, Arden NK, Bennell K, Bierma-Zeinstra SMA, et al. OARSI guidelines for the non-surgical management of knee, hip, and polyarticular osteoarthritis. Osteoarthr Cartil. 2019;27(11):1578–89.

19. Xia P, Wang X, Lin Q, Li X. Efficacy of mesenchymal stem cells injection for the management of knee osteoarthritis: a systematic review and meta-analysis. Int Orthop. 2015;39(12):2363–72.

20. Yubo M, Yanyan L, Li L, Tao S, Bo L, Lin C. Clinical efficacy and safety of mesenchymal stem cell transplantation for osteoarthritis treatment: a meta-analysis. PLoS One. 2017;12(4):e0175449.

21. Erggelet C, Vavken P. Microfracture for the treatment of cartilage defects in the knee joint—a golden standard? J Clin Orthop Trauma. 2016;7(3):145–52.

22. Zhang H, Lin C, Zeng C, Wang Z, Wang H, Lu J, et al. Synovial macrophage M1 polarisation exacerbates experimental osteoarthritis partially through R-spondin-2. Ann Rheum Dis. 2018;77(10):1524–34.

23. Suri S, Gill SE, Massena de Camin S, Wilson D, McWilliams DF, Walsh DA. Neurovascular invasion at the osteochondral junction and in osteophytes in osteoarthritis. Ann Rheum Dis. 2007;66(11):1423–8.

24. Dai M, Sui B, Xue Y, Liu X, Sun J. Cartilage repair in degenerative osteoarthritis mediated by squid type II collagen via immunomodulating activation of M2 macrophages, inhibiting apoptosis and hypertrophy of chondrocytes. Biomaterials. 2018;180:91–103.

25. Marcacci M, Filardo G, Kon E. Treatment of cartilage lesions: what works and why? Injury. 2013;44(Suppl 1):S11–5.

26. Kon E, Filardo G, Di Martino A, Marcacci M. ACI and MACI. J Knee Surg. 2012;25(1):17–22.

27. Niemeyer P, Salzmann G, Feucht M, Pestka J, Porichis S, Ogon P, et al. First-generation versus second-generation autologous chondrocyte implantation for treatment of cartilage defects of the knee: a matched-pair analysis on long-term clinical outcome. Int Orthop. 2014;38(10):2065–70.

28. Hickery MS, Bayliss MT, Dudhia J, Lewthwaite JC, Edwards JC, Pitsillides AA. Age-related changes in the response of human articular cartilage to IL-1alpha and transforming growth factor-beta (TGF-beta): chondrocytes exhibit a diminished sensitivity to TGF-beta. J Biol Chem. 2003;278(52):53063–71.

29. Shimomura K, Ando W, Moriguchi Y, Sugita N, Yasui Y, Koizumi K, et al. Next generation mesenchymal stem cell (MSC)-based cartilage repair using scaffold-free tissue engineered constructs generated with synovial mesenchymal stem cells. Cartilage. 2015;6(2 Suppl):13S–29S.

30. Fernandes TL, Kimura HA, Pinheiro CCG, Shimomura K, Nakamura N, Ferreira JR, et al. Human synovial mesenchymal stem cells good manufacturing practices for articular cartilage regeneration. Tissue Eng Part C Methods. 2018;24(12):709–16.

31. Rizk A, Rabie AB. Human dental pulp stem cells expressing transforming growth factor beta3 transgene for cartilage-like tissue engineering. Cytotherapy. 2013;15(6):712–25.

32. Kon E, Di Martino A, Filardo G, Tetta C, Busacca M, Iacono F, et al. Second-generation autologous chondrocyte transplantation: MRI findings and clinical correlations at a minimum 5-year follow-up. Eur J Radiol. 2011;79(3):382–8.

33. Tuan RS, Chen AF, Klatt BA. Cartilage regeneration. J Am Acad Orthop Surg. 2013;21(5):303–11.

34. Muschler GF, Midura RJ, Nakamoto C. Practical modeling concepts for connective tissue stem cell and progenitor compartment kinetics. J Biomed Biotechnol. 2003;2003(3):170–93.

35. Muschler GF, Midura RJ. Connective tissue progenitors: practical concepts for clinical applications. Clin Orthop Relat Res. 2002;395:66–80.

36. Mantripragada VP, Piuzzi NS, Bova WA, Boehm C, Obuchowski NA, Lefebvre V, et al. Donor-matched comparison of chondrogenic progenitors resident in human infrapatellar fat pad, synovium, and peri-

osteum - implications for cartilage repair. Connect Tissue Res. 2019;60(6):597–610.

37. Qadan MA, Piuzzi NS, Boehm C, Bova W, Moos M Jr, Midura RJ, et al. Variation in primary and culture-expanded cells derived from connective tissue progenitors in human bone marrow space, bone trabecular surface and adipose tissue. Cytotherapy. 2018;20(3):343–60.

38. Robinson PG, Murray IR, West CC, Goudie EB, Yong LY, White TO, et al. Reporting of mesenchymal stem cell preparation protocols and composition: a systematic review of the clinical orthopaedic literature. Am J Sports Med. 2019;47(4):991–1000.

39. Waterman RS, Tomchuck SL, Henkle SL, Betancourt AM. A new mesenchymal stem cell (MSC) paradigm: polarization into a pro-inflammatory MSC1 or an immunosuppressive MSC2 phenotype. PLoS One. 2010;5(4):e10088.

40. Toh WS, Lai RC, Hui JHP, Lim SK. MSC exosome as a cell-free MSC therapy for cartilage regeneration: implications for osteoarthritis treatment. Semin Cell Dev Biol. 2017;67:56–64.

41. Park YB, Ha CW, Lee CH, Yoon YC, Park YG. Cartilage regeneration in osteoarthritic patients by a composite of allogeneic umbilical cord blood-derived mesenchymal stem cells and hyaluronate hydrogel: results from a clinical trial for safety and proof-of-concept with 7 years of extended follow-up. Stem Cells Transl Med. 2017;6(2):613–21.

42. Shimomura K, Ando W, Tateishi K, Nansai R, Fujie H, Hart DA, et al. The influence of skeletal maturity on allogenic synovial mesenchymal stem cell-based repair of cartilage in a large animal model. Biomaterials. 2010;31(31):8004–11.

43. Fibbe WE, Nauta AJ, Roelofs H. Modulation of immune responses by mesenchymal stem cells. Ann N Y Acad Sci. 2007;1106:272–8.

44. Ankrum JA, Ong JF, Karp JM. Mesenchymal stem cells: immune evasive, not immune privileged. Nat Biotechnol. 2014;32(3):252–60.

45. Liotta F, Angeli R, Cosmi L, Fili L, Manuelli C, Frosali F, et al. Toll-like receptors 3 and 4 are expressed by human bone marrow-derived mesenchymal stem cells and can inhibit their T-cell modulatory activity by impairing notch signaling. Stem Cells. 2008;26(1):279–89.

46. Lepage SIM, Robson N, Gilmore H, Davis O, Hooper A, St John S, et al. Beyond cartilage repair: the role of the osteochondral unit in joint health and disease. Tissue Eng Part B Rev. 2019;25(2):114–25.

47. Fahy N, de Vries-van Melle ML, Lehmann J, Wei W, Grotenhuis N, Farrell E, et al. Human osteoarthritic synovium impacts chondrogenic differentiation of mesenchymal stem cells via macrophage polarisation state. Osteoarthr Cartil. 2014;22(8):1167–75.

48. Manferdini C, Paolella F, Gabusi E, Gambari L, Piacentini A, Filardo G, et al. Adipose stromal cells mediated switching of the pro-inflammatory profile of M1-like macrophages is facilitated by PGE2: in vitro evaluation. Osteoarthr Cartil. 2017;25(7):1161–71.

49. Khatab S, van Osch GJ, Kops N, Bastiaansen-Jenniskens YM, Bos PK, Verhaar JA, et al. Mesenchymal stem cell secretome reduces pain and prevents cartilage damage in a murine osteoarthritis model. Eur Cell Mater. 2018;36:218–30.

50. Affan A, Al-Jezani N, Railton P, Powell JN, Krawetz RJ. Multiple mesenchymal progenitor cell subtypes with distinct functional potential are present within the intimal layer of the hip synovium. BMC Musculoskelet Disord. 2019;20(1):125.

51. Kwee E, Saidel G, Powell K, Heylman C, Boehm C, Muschler G. Quantifying proliferative and surface marker heterogeneity in colony-founding connective tissue progenitors and their progeny using time-lapse microscopy. J Tissue Eng Regen Med. 2019;13(2):203–16.

52. Prockop DJ. Repair of tissues by adult stem/progenitor cells (MSCs): controversies, myths, and changing paradigms. Mol Ther. 2009;17(6):939–46.

53. Satue M, Schuler C, Ginner N, Erben RG. Intra-articularly injected mesenchymal stem cells promote cartilage regeneration, but do not permanently engraft in distant organs. Sci Rep. 2019;9(1):10153.

54. Lee KB, Hui JH, Song IC, Ardany L, Lee EH. Injectable mesenchymal stem cell therapy for large cartilage defects—a porcine model. Stem Cells. 2007;25(11):2964–71.

55. Park YB, Ha CW, Kim JA, Han WJ, Rhim JH, Lee HJ, et al. Single-stage cell-based cartilage repair in a rabbit model: cell tracking and in vivo chondrogenesis of human umbilical cord blood-derived mesenchymal stem cells and hyaluronic acid hydrogel composite. Osteoarthr Cartil. 2017;25(4):570–80.

56. Xia H, Liang C, Luo P, Huang J, He J, Wang Z, et al. Pericellular collagen I coating for enhanced homing and chondrogenic differentiation of mesenchymal stem cells in direct intra-articular injection. Stem Cell Res Ther. 2018;9(1):174.

57. Chahla J, Piuzzi NS, Mitchell JJ, Dean CS, Pascual-Garrido C, LaPrade RF, et al. Intra-articular cellular therapy for osteoarthritis and focal cartilage defects of the knee: a systematic review of the literature and study quality analysis. J Bone Joint Surg Am. 2016;98(18):1511–21.

58. Ha CW, Park YB, Kim SH, Lee HJ. Intra-articular mesenchymal stem cells in osteoarthritis of the knee: a systematic review of clinical outcomes and evidence of cartilage repair. Arthroscopy. 2019;35(1):277–88. e2.

59. Xu S, Liu H, Xie Y, Sang L, Liu J, Chen B. Effect of mesenchymal stromal cells for articular cartilage degeneration treatment: a meta-analysis. Cytotherapy. 2015;17(10):1342–52.

60. Kwee E, Herderick EE, Adams T, Dunn J, Germanowski R, Krakosh F, et al. Integrated colony imaging, analysis, and selection device for regenerative medicine. SLAS Technol. 2017;22(2):217–23.

61. Cavallo C, Boffa A, Andriolo L, Silva S, Grigolo B, Zaffagnini S, et al. Bone marrow concentrate injections for the treatment of osteoarthritis: evidence from

preclinical findings to the clinical application. Int Orthop. 2021;45(2):525–38.

62. Roffi A, Nakamura N, Sanchez M, Cucchiarini M, Filardo G. Injectable systems for intra-articular delivery of mesenchymal stromal cells for cartilage treatment: a systematic review of preclinical and clinical evidence. Int J Mol Sci. 2018;19(11)

63. Fortier LA, Potter HG, Rickey EJ, Schnabel LV, Foo LF, Chong LR, et al. Concentrated bone marrow aspirate improves full-thickness cartilage repair compared with microfracture in the equine model. J Bone Joint Surg Am. 2010;92(10):1927–37.

64. Filardo G, Madry H, Jelic M, Roffi A, Cucchiarini M, Kon E. Mesenchymal stem cells for the treatment of cartilage lesions: from preclinical findings to clinical application in orthopaedics. Knee Surg Sports Traumatol Arthrosc. 2013;21(8):1717–29.

65. Fortier LA, Strauss EJ, Shepard DO, Becktell L, Kennedy JG. Biological effects of bone marrow concentrate in knee pathologies. J Knee Surg. 2019;32(1):2–8.

66. Shapiro SA, Arthurs JR, Heckman MG, Bestic JM, Kazmerchak SE, Diehl NN, et al. Quantitative T2 MRI mapping and 12-month follow-up in a randomized, blinded, placebo controlled trial of bone marrow aspiration and concentration for osteoarthritis of the knees. Cartilage. 2019;10(4):432–43.

67. Centeno C, Sheinkop M, Dodson E, Stemper I, Williams C, Hyzy M, et al. A specific protocol of autologous bone marrow concentrate and platelet products versus exercise therapy for symptomatic knee osteoarthritis: a randomized controlled trial with 2 year follow-up. J Transl Med. 2018;16(1):355.

68. Hernigou P, Auregan JC, Dubory A, Flouzat-Lachaniette CH, Chevallier N, Rouard H. Subchondral stem cell therapy versus contralateral total knee arthroplasty for osteoarthritis following secondary osteonecrosis of the knee. Int Orthop. 2018;42(11):2563–71.

69. Hernigou P, Delambre J, Quiennec S, Poignard A. Human bone marrow mesenchymal stem cell injection in subchondral lesions of knee osteoarthritis: a prospective randomized study versus contralateral arthroplasty at a mean fifteen year follow-up. Int Orthop. 2021;45(2):365–73.

70. Hernigou P, Bouthors C, Bastard C, Flouzat Lachaniette CH, Rouard H, Dubory A. Subchondral bone or intra-articular injection of bone marrow concentrate mesenchymal stem cells in bilateral knee osteoarthritis: what better postpone knee arthroplasty at fifteen years? A randomized study. Int Orthop. 2021;45(2):391–9.

71. Davies BM, Snelling SJB, Quek L, Hakimi O, Ye H, Carr A, et al. Identifying the optimum source of mesenchymal stem cells for use in knee surgery. J Orthop Res. 2017;35(9):1868–75.

72. Pierini M, Di Bella C, Dozza B, Frisoni T, Martella E, Bellotti C, et al. The posterior iliac crest outperforms the anterior iliac crest when obtaining mesenchymal stem cells from bone marrow. J Bone Joint Surg Am. 2013;95(12):1101–7.

73. Stolzing A, Jones E, McGonagle D, Scutt A. Age-related changes in human bone marrow-derived mesenchymal stem cells: consequences for cell therapies. Mech Ageing Dev. 2008;129(3):163–73.

74. Baxter MA, Wynn RF, Jowitt SN, Wraith JE, Fairbairn LJ, Bellantuono I. Study of telomere length reveals rapid aging of human marrow stromal cells following in vitro expansion. Stem Cells. 2004;22(5):675–82.

75. Centeno CJ, Al-Sayegh H, Bashir J, Goodyear S, Freeman MD. A dose response analysis of a specific bone marrow concentrate treatment protocol for knee osteoarthritis. BMC Musculoskelet Disord. 2015;16:258.

76. Hegde V, Shonuga O, Ellis S, Fragomen A, Kennedy J, Kudryashov V, et al. A prospective comparison of 3 approved systems for autologous bone marrow concentration demonstrated nonequivalency in progenitor cell number and concentration. J Orthop Trauma. 2014;28(10):591–8.

77. Dragoo JL, Guzman RA. Evaluation of the consistency and composition of commercially available bone marrow aspirate concentrate systems. Orthop J Sports Med. 2020;8(1):2325967119893634.

78. Carelli S, Messaggio F, Canazza A, Hebda DM, Caremoli F, Latorre E, et al. Characteristics and properties of mesenchymal stem cells derived from microfragmented adipose tissue. Cell Transplant. 2015;24(7):1233–52.

79. Aust L, Devlin B, Foster SJ, Halvorsen YD, Hicok K, du Laney T, et al. Yield of human adipose-derived adult stem cells from liposuction aspirates. Cytotherapy. 2004;6(1):7–14.

80. Shanmugasundaram S, Vaish A, Chavada V, Murrell WD, Vaishya R. Assessment of safety and efficacy of intra-articular injection of stromal vascular fraction for the treatment of knee osteoarthritis-a systematic review. Int Orthop. 2021;45:615–25.

81. Perdisa F, Gostynska N, Roffi A, Filardo G, Marcacci M, Kon E. Adipose-derived mesenchymal stem cells for the treatment of articular cartilage: a systematic review on preclinical and clinical evidence. Stem Cells Int. 2015;2015:597652.

82. Garza JR, Campbell RE, Tjoumakaris FP, Freedman KB, Miller LS, Santa Maria D, et al. Clinical efficacy of intra-articular mesenchymal stromal cells for the treatment of knee osteoarthritis: a double-blinded prospective randomized controlled clinical trial. Am J Sports Med. 2020;48(3):588–98.

83. Hong Z, Chen J, Zhang S, Zhao C, Bi M, Chen X, et al. Intra-articular injection of autologous adipose-derived stromal vascular fractions for knee osteoarthritis: a double-blind randomized self-controlled trial. Int Orthop. 2019;43(5):1123–34.

84. Tremolada C, Colombo V, Ventura C. Adipose tissue and mesenchymal stem cells: state of the art and Lipogems(R) technology development. Curr Stem Cell Rep. 2016;2:304–12.

85. Bianchi F, Maioli M, Leonardi E, Olivi E, Pasquinelli G, Valente S, et al. A new nonenzymatic method and device to obtain a fat tissue derivative highly

enriched in pericyte-like elements by mild mechanical forces from human lipoaspirates. Cell Transplant. 2013;22(11):2063–77.

86. Hamdi H, Planat-Benard V, Bel A, Puymirat E, Geha R, Pidial L, et al. Epicardial adipose stem cell sheets results in greater post-infarction survival than intramyocardial injections. Cardiovasc Res. 2011;91(3):483–91.

87. Vezzani B, Gomez-Salazar M, Casamitjana J, Tremolada C, Peault B. Human adipose tissue Microfragmentation for cell phenotyping and Secretome characterization. J Vis Exp. 2019;152

88. Vezzani B, Shaw I, Lesme H, Yong L, Khan N, Tremolada C, et al. Higher pericyte content and secretory activity of microfragmented human adipose tissue compared to enzymatically derived stromal vascular fraction. Stem Cells Transl Med. 2018;7(12):876–86.

89. D'Arrigo D, Roffi A, Cucchiarini M, Moretti M, Candrian C, Filardo G. Secretome and extracellular vesicles as new biological therapies for knee osteoarthritis: a systematic review. J Clin Med. 2019;4, 8(11)

90. Filardo G, Tschon M, Perdisa F, Brogini S, Cavallo C, Desando G, et al. Micro-fragmentation is a valid alternative to cell expansion and enzymatic digestion of adipose tissue for the treatment of knee osteoarthritis: a comparative preclinical study. Knee Surg Sports Traumatol Arthrosc. 2021;19

91. Heidari N, Noorani A, Slevin M, Cullen A, Stark L, Olgiati S, et al. Patient-centered outcomes of microfragmented adipose tissue treatments of knee osteoarthritis: an observational, intention-to-treat study at twelve months. Stem Cells Int. 2020;2020:8881405.

92. Hudetz D, Boric I, Rod E, Jelec Z, Kunovac B, Polasek O, et al. Early results of intra-articular microfragmented lipoaspirate treatment in patients with late stages knee osteoarthritis: a prospective study. Croat Med J. 2019;60(3):227–36.

Cartilage Lesions and Osteoarthritis of the Knee: Biologics

26

Giuseppe Filardo, Angelo Boffa, Luca Andriolo, Alberto Poggi, and Alessandro Di Martino

26.1 Introduction

Degenerative cartilage lesions are a debilitating disease, often resulting in fibrillation and subsequent degradation of the surrounding articular surface, possibly involving the subchondral bone as well and, in the end, leading to the development of osteoarthritis (OA) [1]. OA affects more than 10% of the world population aged 60 years or older and represents one of the major causes of disability worldwide, with a massive impact on society both in terms of quality of life for the individuals and high costs for the healthcare system [2]. Knee, hip, fingers, and the lower spine region are frequently affected by the occurrence of OA, inducing chronic pain, inflammation, and stiffness. Although commonly referred to as a "wear and tear" disease, OA may be initiated by various mechanisms of onset or conditions, involving complex interactions between genetic, metabolic, biochemical, and biomechanical factors, all favoring the disease progression [3].

Regardless from its etiology, these processes result in a common endpoint evolving toward symptomatic and advanced OA. Despite the possibility of a timely diagnosis of early knee OA, the conservative treatments currently used, including physiotherapy, anti-inflammatory, and anti-pain medications, have modest and short-lasting efficacy, and are not able to delay or interrupt its evolution [1]. For this reason, the demand for knee replacement is high and continuously growing. However, whereas the surgical approach can provide a high success rate and satisfaction for older patients, high functional request and longer life expectancy of young patients are an issue for joint arthroplasty.

In this context, biological approaches recently emerged as a promising option to treat articular degenerative defects and early knee OA stages, aimed at reducing symptoms, restoring a satisfactory knee function, and possibly preventing OA progression and delaying the need for metal resurfacing of the damaged articular surface. Numerous products developed for intra-articular treatment are currently applied in the clinical practice, ranging from corticosteroids and visco-supplementation to the new orthobiologic solutions including platelet concentrates.

Intra-articular injections of corticosteroids, first described in 1951, are still considered to be a cost-effective strategy among non-invasive OA treatments [1]. The rationale for using intra-articular corticosteroids in knee OA lays on their anti-inflammatory and immunosuppressive

G. Filardo
Rizzoli Orthopaedic Institute
Bologna, Italy

Università della Svizzera Italiana, Ente Ospedaliero
Cantonale, Lugano, Switzerland

A. Boffa (✉) · L. Andriolo · A. Poggi · A. Di Martino
Clinica Ortopedica e Traumatologica 2, IRCCS
Istituto Ortopedico Rizzoli, Bologna, Italy

G. Filardo et al. (eds.), *Orthobiologics*, https://doi.org/10.1007/978-3-030-84744-9_26

effects through a complex multiplicity of actions. In fact, corticosteroids act directly on nuclear steroid receptors and interrupt the inflammatory and immune cascade at several levels: the mechanisms involve the reduction of pro-inflammatory and pain mediators, including bradykinins, histamine, leukotrienes, prostaglandins, and the inhibition of mononuclear cell adhesion. Corticosteroids also attenuate the effects of IL-1 and TNF-α and inhibit metalloproteases and immunoglobulin synthesis [3]. The prolonged concentration of corticosteroids in the synovial fluid confers the maximum anti-inflammatory effect locally while minimizing the risk of systemic exposure and potential adverse side effects [1]. The inflammatory nature of OA is known even in early OA stages and is linked to symptoms and structural progression of the disease, thus supporting the role of corticosteroids. However, corticosteroids also showed anti-anabolic effects on healthy cartilage, due to the upregulation of aggrecanases, collagenases, and metalloproteinases and the reduction of lubricin, raising questions about their potential to damage the cartilage joint surface, especially in early OA degrees [3].

A completely different rationale characterizes viscosupplementation, based on the physiologic importance of hyaluronic acid (HA) in the synovial fluid. In fact, HA is a glycosaminoglycan commonly present in the joint, secreted by type B synoviocytes, chondrocytes, and fibroblasts, with a natural protective function for the articular cartilage due to its viscoelastic properties [4]. Synovial fluid with normal HA concentration acts as a viscous lubricant during slow joint movements and as an elastic shock absorber during rapid joint movements [5]. HA concentration ranges from 2.5 to 4.0 mg/mL in the normal adult knee. However, the amount of HA is reduced in an osteoarthritic knee joint, and its molecular weight has been found to decrease by as much of 33–50% [6]. For this reason, the intra-articular HA administration in OA joints was clinically introduced in 1974 [7], with the aim to restore the natural protective functions of HA by increasing synovial fluid elasticity and viscosity. Moreover, HA seems to be characterized by further mechanisms able to reduce pain and improve function in OA patients, including the inhibition of tissue nociceptors, the stimulation of endogenous HA, as well as chondroprotective effects through the inhibition of metalloproteinases activity and direct anti-inflammatory effects through the suppression of the production of pro-inflammatory mediators [8]. Nevertheless, there is still no evidence clearly supporting that viscosupplementation influences the natural progression of knee OA.

In order to address and reverse the underlying disease processes of knee OA, many research efforts have been made in the field of blood derivatives. In particular, platelet-rich plasma (PRP) has gained increasing attention due to the high concentrations of cytokines and growth factors (GFs) stored in platelet α-granules, which showed to take part in the homeostasis of articular cartilage being involved in both healing process and immunoregulation. These biologically active proteins seem to be able to influence and promote a favorable joint environment, favoring the restoration of a homeostatic balance in degenerative joints [9]. In vitro studies investigating the effect of PRP on chondrocytes showed various and heterogeneous mechanisms of action, including the increase of the chondrocyte proliferation rate, the matrix production stimulation, and the inflammation modulation [10]. Moreover, an analgesic effect of PRP was also shown, possibly by modulating cannabinoid receptors in chondrocytes [10]. PRP may also significantly enhance synoviocyte HA secretion and switch synovial angiogenesis to a more balanced status [11]. Besides the biological rationale, current preclinical evidence further supports the role of PRP in modulating the intra-articular environment by counteracting inflammation in degenerative joint diseases. After an initial pro-inflammatory action, with the stimulation of synoviocytes to release metalloproteinases and cytokines, a following phase of modulation and reduction of the inflammatory response has been demonstrated, contrasting the chemotaxis of monocyte-like cells and decreasing pro-inflammatory cytokines [12]. Based on this evidence, it is important to underline that the main PRP intra-articular effect may not be a direct promotion of tissue regeneration, but it may rather act through its different bioactive molecules by affecting tissue homeostasis, slowing down the inflammatory, catabolic, and degenerative processes, thus offering benefits in terms of symptom relief and functional improvement.

26.2   Clinical Evidence

26.2.1   Corticosteroids

Corticosteroids are the most common product used for the intra-articular treatment of joint degeneration/inflammatory processes. There are different formulations of corticosteroids that can be administered into the joint space, including triamcinolone, betamethasone, and methylprednisolone [13]. A systematic review on the comparative efficacy of various intra-articular corticosteroids found a limited number of studies focused on knee OA, without a clear superiority of one type over the other corticosteroids [14]. Corticosteroid injections can be further combined with local anesthetics, such as lidocaine and bupivacaine. The common use of this intra-articular approach in the clinical practice is favored by the ease of use, low cost, and overall safety with a low complication rate. To this regard, a post-injection flare-up of pain can occur in 2–25% of patients and last for a few days [15]. Joint infection is a possible rare complication; a non-negligible 1 in 3000 incidence of arthroplasty infections after the use of steroid injections has also been suggested [16]. This may be linked to the steroid-induced decrease in inflammatory response and subsequent local immunosuppression. In this light, contraindications to steroid injections are the presence of soft tissue infection, injured skin at the injection site, or the suspect of joint infection.

Systemic side effects of intra-articular corticosteroids are rare and include elevated blood pressure, hyperglycemia, and alterations in mood and energy. Therefore, there should be caution for steroid administration in diabetic people and patients with hypertension [17].

Corticosteroids may be indicated after failing of conservative measures, non-steroidal anti-inflammatory drugs (NSAIDs) and acetaminophen, especially in knee OA with a significant synovitis component [18]. In these cases, the anti-inflammatory role of corticosteroids may provide moderate improvement in pain and function. However, the duration of the effects of this injective treatment is limited, with a quick onset of action, typically within 24 h, but a benefit wearing off generally within 1–24 weeks [19]. Several studies suggested some prognostic factors for the response to corticosteroid injections, including the presence of knee effusion and less radiographic evidence of OA. Conversely, obesity, chronic medical problems, and a sedentary lifestyle are negative predictors of response to intra-articular corticosteroids [20]. Considering the short-term benefits, clinicians often repeat steroid injection after a few weeks. Nevertheless, repeated use remains controversial, and some researchers suggest to use them no more than once every 3 months [18], since intra-articular steroids might facilitate tissue atrophy, joint destruction, and cartilage degeneration. Therefore, their use should be used with particular caution in the early stages of OA, where it should be indicated especially for cases with evidence of synovitis. Long-term benefits have not been confirmed, and studies dealing with the long-term complications of intra-articular injections have reported contradictory findings, often in association with an increased risk of knee OA progression [21]. In this regard, in a recent saline-controlled, double-blind RCT in 140 knee OA patients with ultrasound-documented synovitis receiving an intra-articular injection of 40 mg of triamcinolone every 3 months, MRI revealed a significantly greater cartilage volume loss in patients treated with steroid injections [22]. On the other hand, intra-articular corticosteroid injections in knee OA patients showed a significantly larger short-term pain reduction compared to placebo in a meta-analysis of RCTs, especially in patients with severe knee pain at baseline [23]. Considering these controversial findings, recommendations for corticosteroids in knee OA vary along different organizations. The American Academy of Orthopaedic Surgeons (AAOS) is unable to recommend for or against the use of intra-articular corticosteroids for patients with symptomatic knee OA [24]. Conversely, the Osteoarthritis Research Society International Guidelines (OARSI) recommended intra-articular corticosteroids for short-term analgesia as second line after maximizing physiotherapy, but physicians should consider other treatments for long-term management of OA [25], while the European Society for Clinical and Economic Aspects of Osteoporosis, Osteoarthritis and Musculoskeletal Diseases (ESCEO) suggests the intra-articular use of corticosteroids only in knee OA patients with an acute inflammatory flare [26]. Finally, the American College of Rheumatology (ACR) recommended intra-articular corticosteroid injections for patients with knee OA while underlining insufficient data to judge the choice of preparations or doses [27].

26.2.2 Hyaluronic Acid

The intra-articular administration of HA in the management of knee OA is widely used, with the aim to improve the biomechanical function of the osteoarthritic knee by replacing the reduced HA and to promote pain management thanks to its physicochemical properties. There are more than 80 marketed intra-articular HA preparations worldwide, differing for many characteristics such as origin (natural or bacterial fermentation), mean molecular weight (MW) (500–6000 kDa) and MW distribution, molecular structure (linear, cross-linked, and a mix of both), method of cross-linking, sterilization process (heat or ultra-filtration), concentration (0.8–30 mg/mL), volume of injection (0.5–6.0 mL), and posology [28]. Moreover, some of the preparations include different concentrations of additives, such as mannitol, sorbitol, or chondroitin sulfate [28]. While each of these parameters may theoretically have an impact on the effect of the intra-articular HA treatment, research mostly focused on the potential differences resulting from the MW of HA. In fact, HAs available for intra-articular injections are grouped into three MW categories: low (500–800 kDa), intermediate (800–2000 kDa), and high (2000–6000 kDa), the latter including cross-linked formulations of HA [28]. Preclinical studies suggested distinct mechanisms of action among HA products of different MW. In particular, higher MW HAs could provide superior chondroprotective, proteoglycan/glycosaminoglycan synthesis, anti-inflammatory, mechanical, and analgesic effects [29]. On the other side, some researchers suggested a greater cellular effect for lower MW HA, which could have an enhanced penetration through the extracellular matrix of the synovium, thus maximizing its concentration and facilitating its interaction with target synovial cells [30]. Despite preclinical evidence, the current literature does not provide consensus regarding differences in terms of clinical efficacy between low and high MW HA [31, 32], although some comparative studies and systematic reviews suggested that high MW HA might provide overall greater therapeutic benefit than low MW HA in the treatment of knee OA [24, 33]. These results are probably due to the mechanical properties of high MW HA, especially the cross-linked types, which have superior stability and lower degradation compared to low MW HA, remaining longer in the joint to produce the desired therapeutic effect [34]. Nevertheless, cross-linked HA showed more severe immunological responses, causing most frequently an inflammatory soft tissue reaction [35]. Regarding the HA origin, HAs derived from biological fermentation demonstrated to provide an advantageous safety profile over HAs produced through the extraction of avian-derived molecules, which reported injection site flare-ups driven by avian-derived proteins [33], while it is not clear if the different product origin may lead to a different efficacy [36].

Many formulations of HA are available, ranging from a single injection to a series of up to five injections per treatment, with 1-week intervals [37]. The different efficacies between single and multiple HA injection formulations remain controversial. A recent systematic review evaluated 11 studies comparing single- with multiple-injection formulations of HA for the treatment of knee OA, reporting no consistent difference in patient-reported outcomes between different treatments. Furthermore, five-injection formulations were not superior to three-injection formulations [38]. Conversely, a previous meta-analysis showed that 2–4 and ≥5 injection regimens provided pain relief over intra-articular saline, while the single injection did not [37]. Regardless of these controversial findings, single-injection schedules are more attracting and show a growing interest, including the new reticulated HA derivatives which should have longer joint residence. Reducing the number of administrations per treatment cycle can reduce the risks related to this injective treatment, which is particularly important also considering that, due to the short-term efficacy, patients often undergo repeated injection cycles after 6 months if they are satisfied with the previous injection course. Still, overall the use of intra-articular HA injection in patients with symptomatic knee OA is considered well tolerated. In a recent meta-analysis involving more than 8000 knee OA patients, no differences were demonstrated in the risk of serious adverse events for HA compared with intra-articular saline, although patients treated with HA reported an increased risk of non-serious and transient local reactions [39]. The most common local reactions reported in this large population were injection site pain, arthralgia, joint swelling, and effusion, which subsided within 2–3 days in most instances.

Evidence supporting HA efficacy showed a 1–4 weeks later onset than corticosteroids, but the benefit was maintained for up to 6 or even

12 months [40]. A recent meta-analysis evaluated 12 RCTs comparing the efficacy of intra-articular HA versus corticosteroids in knee OA patients and demonstrated that clinical benefits were better in the corticosteroid group at 1 month, similar between the two treatments at 3 months, while HA showed more effectiveness at 6 months [41]. Differently from corticosteroids, in acute inflammation with severe effusion, viscosupplementation is not indicated. Synovitis impairs the efficacy of HA, by dilution of HA in the effusion fluid and even more due to enzymes and oxidants (hyaluronidases, free radicals) degrading the HA chains. In this light, acute episodes should be treated before with either NSAIDs or corticosteroids, postponing HA treatment to a less inflammatory phase [40]. Femoro-tibial OA seems the ideal indication, while in femoro-patellar OA viscosupplementation appeared less effective, with a response rate of around 50% [42]. Moreover, current literature demonstrated that intra-articular HA is more effective in the earlier stages of knee OA, as opposed to being employed as a later stage treatment [43]. A recent systematic review and meta-analysis demonstrated that intra-articular HA provided statistically significant pain relief compared to saline injections only for patients with early-moderate knee OA, with no increase in the risk of treatment-related adverse effects, up to 6 months post-injection. Conversely, intra-articular HA demonstrated no benefit over controls in the late OA subgroup and was associated with significantly greater treatment-related adverse events [44]. This aspect may explain the controversial results of the literature on the superiority of HA over saline [45, 46], as some studies included a considerable proportion of patients with end-stage knee disease, which might have impaired the potential benefit of HA. In this regard, in a recent cost analysis, intra-articular HA injections, especially the high MW formulations, demonstrated cost-effectiveness when compared to conservative treatment options (oral NSAIDs, braces and orthosis, physical therapy) in patients with early/mid-stage knee OA, while the cost-effectiveness of high MW HA in patients

with later-stage knee OA was not as apparent, due to the reduction in response to intra-articular HA treatment in these particular patients [47]. Despite the different effectiveness in patient subpopulations, an overall benefit has been suggested by some researchers, underlying a possible role of HA in delaying knee arthroplasty in patients with knee OA, as demonstrated in large retrospective studies where large percentages of the patients who underwent multiple courses of HA injection presented considerably longer times to knee replacement [48–50].

Regardless of the large body of literature and despite being approved by the Food and Drug Administration (FDA) in 1997, the use of intra-articular HA for the treatment of knee OA remains controversial because of the conflicting data regarding its efficacy of different meta-analyses. This is reflected by the varying recommendations reached by each society for OA management. In fact, current guidelines such as AAOS and ACR are conflicting and overall not supporting the use of HA [24, 27]. Nevertheless, intra-articular HA remains a widely used option worldwide in the management of knee OA, being often recommended as second-line treatment despite the controversial data. Moreover, recommendations have not been consistent over time, and recently OARSI included intra-articular HA among the recommended treatments of level 1B in patients with knee OA, being able to provide beneficial effects on pain at and beyond 12 weeks

of treatment and a more favorable long-term safety profile than repeated intra-articular steroids [25]. Moreover, the ESCEO treatment algorithm recommends intra-articular HA in patients who remain severely symptomatic despite the use of NSAIDs [26]. The ESCEO task force also encourages the use of repeated cycles of intra-articular HA in knee OA patients who responded to the first injection, starting a new treatment cycle as soon as the first symptoms appear. Nonetheless, the relative effectiveness of the long-term use of intra-articular HA through repeat courses of treatment remains to be determined, and the overall clinical impact of the various HA products remains unclear, which recently fostered research efforts into new injective solutions to more effectively address knee OA.

26.2.3 Platelet-Rich Plasma

PRP represents an attractive biological approach to improve the healing of tissues with a low healing potential, such as cartilage. This led to the wide use of PRP, which shows promising results as a minimally invasive injective treatment of knee cartilage degeneration and OA, both in preclinical and clinical studies [51–53]. PRP can be produced by centrifugation or filtration of the whole blood to concentrate or isolate platelets to a level higher (generally considered 3–5 times more) than normal plasma levels (Fig. 26.1) [54].

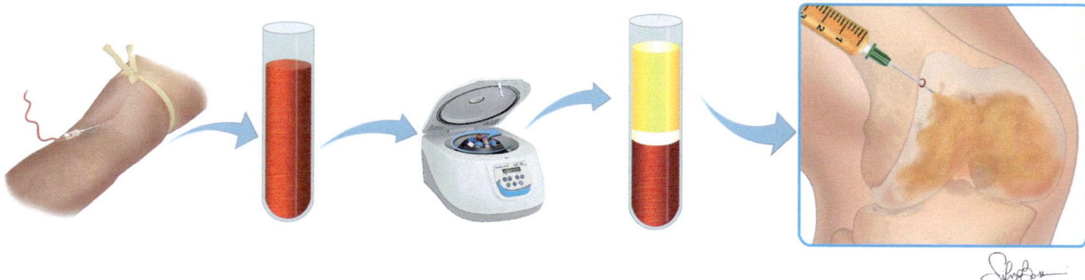

Fig. 26.1 The procedure entails a peripheral venous blood sample harvested from the patient arm. Blood is then processed using a centrifuge to separate the blood components according to their density, obtaining three layers: plasma (55% of whole blood), buffy coat with leu-kocytes and platelets (<1% of whole blood), and erythrocytes (45% of whole blood). Erythrocytes and platelet-poor plasma are discharged to obtain the layer concentrated in platelets (with or without leukocytes)—PRP, which is injected into the knee

Different preparation methods for PRP can yield products with different compositions and characteristics, with different features in terms of platelet and leukocyte content, volume of whole blood harvested, storage procedures, exogenous activation of platelets, and formation of a fibrin matrix [55]. All this makes it very difficult to compare clinical results of different studies and to gain a full understanding of the potential and limitations of PRP for the treatment of knee OA.

Although some data suggest better results with PRP formulations with leukocyte depletion, the superiority of one PRP formulation over another in terms of clinical effectiveness in knee OA has not been established [56, 57]. In this regard, a network meta-analysis on clinical trials evaluating the role of leukocyte concentration supported, through an indirect study groups comparison, a difference in favor of leukocyte-poor PRP [58], but a study directly comparing PRP with or without leukocytes documented similar clinical results at 12 months of follow-up, although patients who received a leukocyte-rich PRP were more likely to experience short-term pain and swelling after the injections [59]. Also for this treatment, it is important to underline that study results might be influenced by the study cohorts, as patients may benefit differently from PRP injections, and better results can be achieved in younger patients with a low degree of cartilage degeneration [60, 61]. The number of injections can also influence the effectiveness of PRP. In fact, after some controversial clinical reports, an in vivo preclinical study recently demonstrated that three intra-articular PRP injections provided better inflammation reduction of the synovium and more durable results than a single PRP injection [62].

While the best formulation and administration regimen remain to be defined, intra-articular PRP injections are gaining a large use in the clinical practice, due to the safety, low costs, and the simple preparation technique to obtain its biologically active content [19]. A growing number of studies support intra-articular PRP injections for the treatment of knee OA, which seems to provide significant functional improvement and reduction of pain-related symptoms up to 12 months [63–66]. Moreover, some evidence suggests that the clinical improvement provided by PRP can be perceived by some patients also beyond 24 months, with a subsequent gradual reduction over time [67].

The efficacy of intra-articular PRP was compared with the oral use of acetaminophen, an analgesic drug commonly used in the management of knee OA, showing superiority at 24 weeks after the injection [68]. As intra-articular injections present a significant placebo component, in particular when dealing with the new orthobiologics [69], PRP efficacy was also compared with saline solution in multiple randomized studies, consistently showing to be more effective than groups injected with saline up to 12 months [65, 70–72]. Accordingly, the results of several meta-analyses [73–75] converge in indicating that intra-articular PRP injections may have more benefit in terms of pain relief and functional improvement than placebo without increasing the risk of adverse events [76]. Moreover, in a recent meta-analysis performed on 34 RCTs, PRP injections showed to provide better results than other injectable options such as corticosteroids or HA. This benefit increased over time, being not significant at earlier follow-ups but becoming clinically significant after 6–12 months (Table 26.1). However, although substantial, the improvement remained partial and supported by low level of evidence [76].

The clinical effectiveness of intra-articular PRP was confirmed by the majority of the available studies; on the other hand, its ability to affect cartilage regeneration and OA progression has not been demonstrated [77]. A recent small trial, evaluating the changes of cartilage after intra-articular PRP injections with qualitative MRI, demonstrated better MRI findings in terms of patello-femoral cartilage volume and synovitis in the PRP group compared with the control

Table 26.1 RCTs on PRP efficacy

Outcome	Follow-up (months)	Trials	Patients	PRP results
PRP vs placebo				
WOMAC overall	1	6	266	🔴
	3	4	153	🔴
	6	6	266	🔴
	12	3	129	🟢
WOMAC pain	1	5	210	🔴
	3	4	153	🟢
	6	5	210	🟢
WOMAC stiffness	1	5	210	🔴
	3	4	153	🟢
	6	5	210	🟢
WOMAC pain	1	5	210	🔴
	3	4	153	🔴
	6	5	210	🔴
VAS pain	1	3	140	🟢
	6	4	238	🟢
PRP vs steroids				
VAS pain	6	4	206	🟢
KOOS symptoms	6	3	170	🔴
KOOS pain	6	3	170	🟢
KOOS ADL	6	3	170	🟢
KOOS sports/rec	6	3	170	🔴
KOOS QoL	6	3	170	🟢
PRP vs hyaluronic acid				
WOMAC overall	1	5	338	🟡
	3	5	356	🟡
	6	10	790	🟢
	12	7	553	🟢
WOMAC pain	1	5	325	🔴
	3	5	324	🔴
	6	9	702	🟡
	12	6	440	🟢
WOMAC stiffness	1	4	201	🔴
	3	4	200	🟡
	6	8	565	🟡
	12	6	445	🟢
WOMAC function	1	4	228	🟡
	3	4	228	🟡
	6	8	605	🟡
	12	6	486	🟢

Table 26.1 (continued)

Outcome	Follow-up (months)	Trials	Patients	PRP results
VAS pain	1	6	345	🔴
	3	8	481	🔴
	6	9	596	🟡
	12	6	398	🟡
IKDC	6	5	475	🔴
	12	4	324	🔴

🟢 Statistically and clinically significant
🟡 Statistically significant
🔴 No statistically significant

group at 12 months of follow-up [78], but these isolated findings need to be confirmed by larger trials investigating if PRP can actually provide structural effects beside the demonstrated clinical improvement. Thus, due to the recent introduction in the clinical practice and the lack of clear objective results, despite the large number of clinical trials on the intra-articular use of PRP, this product is not yet recommended by international societies. The AAOS work group interpreted the evidence to be inconclusive on the benefit of intra-articular PRP injection and was unable to recommend for or against the use of intra-articular PRP injection in their guideline for patients with symptomatic OA of the knee [79], and also OARSI and ESCEO did not mention intra-articular PRP injections in their guidelines for the non-surgical management of knee OA [25, 26]. Finally, the ACR strongly recommended against intra-articular PRP injections in patients with knee OA, because of the heterogeneity and lack of standardization in available preparations, as well as techniques used, making it difficult to identify exactly what is being injected [27]. Accordingly, the need for novel classification and coding systems has been recently brought to attention to improve and foster further improvements in the field [80]. A better understanding of PRP characteristics could help matching the optimal PRP product to specific patient factors, leading to improved outcomes and the elucidation of the cost-effectiveness of this treatment with respect to other injective options for knee OA.

26.3 Conclusions

Intra-articular orthobiologic injections are frequently performed in patients with knee degenerative cartilage lesions and symptomatic OA, even though indications and guidelines are not always clear. Despite the long experience and the common use in the clinical practice, the literature on injective treatments is still poor and not able to guide the clinicians in choosing the best product, indication, formulation, injection schedule, and duration of treatment. Corticosteroids may be indicated in knee OA after failing oral NSAIDs and acetaminophen, especially in patients with synovitis and a significant inflammatory component, with a fast but short-term duration of action, generally between 1 and 24 weeks. Viscosupplementation should be reserved mainly for symptomatic femorotibial OA, especially for earlier stages, while in femoro-patellar involvement and in severe OA grades, HA appeared less effective. The duration of action of HA is longer than corticosteroids, approximately 6 months, and patients are generally advised to repeat the injection schedule by 6 months if they are satisfied with the previous injection course. Intra-articular PRP injections showed promising clinical results in patients affected by early/moderate OA, reducing pain and improving knee functional status. Conversely, this biological treatment used as a "salvage procedure" in severe OA knees produced a less favorable outcome and therefore presents a limited indication for these patients. Despite the numerous evidences on the effectiveness of intra-articular PRP injection and the superiority compared with saline, corticosteroids, and HA, the recent introduction in the clinical practice and the lack of a standardized PRP protocol hindered the inclusion of this orthobiologic injective strategy in the guidelines for the treatment of knee degenerative cartilage lesions and symptomatic OA.

> **Take-Home Messages**
> - Intra-articular biological approaches can represent a suitable option to reduce symptoms and possibly delay surgery in patients with cartilage lesion and OA of the knee.
> - Corticosteroids can be indicated after failing oral NSAIDs and acetaminophen, especially in patients with synovitis and a significant inflammatory component, with a fast but short-term duration of action.
> - HA should be reserved mainly for symptomatic femoro-tibial OA, especially for earlier stages, while in femoro-patellar involvement, in acute inflammation, and in severe OA, HA appeared less effective.
> - Intra-articular PRP injections showed to provide better results than other injectable options such as saline, corticosteroids, or HA, with longer-lasting duration.
> - Several variables could influence the effects of PRP, but current literature cannot clearly identify the most suitable PRP formulation and application modality for knee OA.

References

1. Filardo G, Kon E, Longo UG, Madry H, Marchettini P, Marmotti A, et al. Non-surgical treatments for the management of early osteoarthritis. Knee Surg Sports Traumatol Arthrosc. 2016;24(6):1775–85.
2. de Girolamo L, Kon E, Filardo G, Marmotti AG, Soler F, Peretti GM, et al. Regenerative approaches for the treatment of early OA. Knee Surg Sports Traumatol Arthrosc. 2016;24(6):1826–35.

3. Jones IA, Togashi R, Wilson ML, Heckmann N, Vangsness CT Jr. Intra-articular treatment options for knee osteoarthritis. Nat Rev Rheumatol. 2019;15(2):77–90.
4. Bellamy N, Campbell J, Robinson V, Gee T, Bourne R, Wells G. Viscosupplementation for the treatment of osteoarthritis of the knee. Cochrane Database Syst Rev. 2006;2:CD005321.
5. Brockmeier SF, Shaffer BS. Viscosupplementation therapy for osteoarthritis. Sports Med Arthrosc Rev. 2006;14(3):155–62.
6. Kon E, Filardo G, Drobnic M, Madry H, Jelic M, van Dijk N, et al. Non-surgical management of early knee osteoarthritis. Knee Surg Sports Traumatol Arthrosc. 2012;20(3):436–49.
7. Peyron JG, Balazs EA. Preliminary clinical assessment of Na-hyaluronate injection into human arthritic joints. Pathol Biol (Paris). 1974;22(8):731–6.
8. Reid MC. Viscosupplementation for osteoarthritis: a primer for primary care physicians. Adv Ther. 2013;30(11):967–86.
9. Filardo G, Kon E, Roffi A, Di Matteo B, Merli ML, Marcacci M. Platelet-rich plasma: why intra-articular? A systematic review of preclinical studies and clinical evidence on PRP for joint degeneration. Knee Surg Sports Traumatol Arthrosc. 2015;23(9):2459–74.
10. Park SI, Lee HR, Kim S, Ahn MW, Do SH. Time-sequential modulation in expression of growth factors from platelet-rich plasma (PRP) on the chondrocyte cultures. Mol Cell Biochem. 2012;361(1–2):9–17.
11. Assirelli E, Filardo G, Mariani E, Kon E, Roffi A, Vaccaro F, et al. Effect of two different preparations of platelet-rich plasma on synoviocytes. Knee Surg Sports Traumatol Arthrosc. 2015;23(9):2690–703.
12. Pereira RC, Scaranari M, Benelli R, Strada P, Reis RL, Cancedda R, et al. Dual effect of platelet lysate on human articular cartilage: a maintenance of chondrogenic potential and a transient proinflammatory activity followed by an inflammation resolution. Tissue Eng Part A. 2013;19(11–12):1476–88.
13. Douglas RJ. Corticosteroid injection into the osteoarthritic knee: drug selection, dose, and injection frequency. Int J Clin Pract. 2012;66(7):699–704.
14. Garg N, Perry L, Deodhar A. Intra-articular and soft tissue injections, a systematic review of relative efficacy of various corticosteroids. Clin Rheumatol. 2014;33(12):1695–706.
15. Levy DM, Petersen KA, Scalley Vaught M, Christian DR, Cole BJ. Injections for knee osteoarthritis: corticosteroids, viscosupplementation, platelet-rich plasma, and autologous stem cells. Arthroscopy. 2018;34(5):1730–43.
16. McGarry JG, Daruwalla ZJ. The efficacy, accuracy and complications of corticosteroid injections of the knee joint. Knee Surg Sports Traumatol Arthrosc. 2011;19(10):1649–54.
17. Cook CS, Smith PA. Clinical update: why PRP should be your first choice for injection therapy in treating osteoarthritis of the knee. Curr Rev Musculoskelet Med. 2018;11(4):583–92.
18. Bert JM, Bert TM. Nonoperative treatment of unicompartmental arthritis: from bracing to injection. Clin Sports Med. 2014;33(1):1–10.
19. Ayhan E, Kesmezacar H, Akgun I. Intraarticular injections (corticosteroid, hyaluronic acid, platelet rich plasma) for the knee osteoarthritis. World J Orthop. 2014;5(3):351–61.
20. Uthman I, Raynauld JP, Haraoui B. Intra-articular therapy in osteoarthritis. Postgrad Med J. 2003;79(934):449–53.
21. Zeng C, Lane NE, Hunter DJ, Wei J, Choi HK, McAlindon TE, et al. Intra-articular corticosteroids and the risk of knee osteoarthritis progression: results from the osteoarthritis initiative. Osteoarthr Cartil. 2019;27(6):855–62.
22. McAlindon TE, LaValley MP, Harvey WF, Price LL, Driban JB, Zhang M, et al. Effect of intra-articular triamcinolone vs saline on knee cartilage volume and pain in patients with knee osteoarthritis: a randomized clinical trial. JAMA. 2017;317(19): 1967–75.
23. Van Middelkoop M, Arden NK, Atchia I, Birrell F, Chao J, Rezende MU, et al. The OA trial Bank: meta-analysis of individual patient data from knee and hip osteoarthritis trials show that patients with severe pain exhibit greater benefit from intra-articular glucocorticoids. Osteoarthr Cartil. 2016;24(7):1143–52.
24. Jevsevar DS. Treatment of osteoarthritis of the knee: evidence-based guideline, 2nd edition. J Am Acad Orthop Surg. 2013;21(9):571–6.
25. Bannuru RR, Osani MC, Vaysbrot EE, Arden NK, Bennell K, Bierma-Zeinstra SMA, et al. OARSI guidelines for the non-surgical management of knee, hip, and polyarticular osteoarthritis. Osteoarthr Cartil. 2019;27(11):1578–89.
26. Bruyere O, Honvo G, Veronese N, Arden NK, Branco J, Curtis EM, et al. An updated algorithm recommendation for the management of knee osteoarthritis from the European Society for Clinical and Economic Aspects of Osteoporosis, Osteoarthritis and Musculoskeletal Diseases (ESCEO). Semin Arthritis Rheum. 2019;49(3):337–50.
27. Kolasinski SL, Neogi T, Hochberg MC, Oatis C, Guyatt G, Block J, et al. 2019 American College of Rheumatology/Arthritis Foundation guideline for the management of osteoarthritis of the hand, hip, and knee. Arthritis Care Res. 2020;72(2):149–62.
28. Cooper C, Rannou F, Richette P, Bruyere O, Al-Daghri N, Altman RD, et al. Use of intraarticular hyaluronic acid in the management of knee osteoarthritis in clinical practice. Arthritis Care Res. 2017;69(9): 1287–96.

29. Altman RD, Manjoo A, Fierlinger A, Niazi F, Nicholls M. The mechanism of action for hyaluronic acid treatment in the osteoarthritic knee: a systematic review. BMC Musculoskelet Disord. 2015;16:321.

30. Ghosh P, Guidolin D. Potential mechanism of action of intra-articular hyaluronan therapy in osteoarthritis: are the effects molecular weight dependent? Semin Arthritis Rheum. 2002;32(1):10–37.

31. Rutjes AW, Juni P, da Costa BR, Trelle S, Nuesch E, Reichenbach S. Viscosupplementation for osteoarthritis of the knee: a systematic review and meta-analysis. Ann Intern Med. 2012;157(3): 180–91.

32. Lee PB, Kim YC, Lim YJ, Lee CJ, Sim WS, Ha CW, et al. Comparison between high and low molecular weight hyaluronates in knee osteoarthritis patients: open-label, randomized, multicentre clinical trial. J Int Med Res. 2006;34(1):77–87.

33. Altman RD, Bedi A, Karlsson J, Sancheti P, Schemitsch E. Product differences in intra-articular hyaluronic acids for osteoarthritis of the knee. Am J Sports Med. 2016;44(8):2158–65.

34. Kwasek B, Bogdal D. The use of hyaluronic acid in the treatment of osteoarthritis of knee cartilage. Technical Transactions; 2014;1-Ch(Chemistry).

35. Ottaviani RA, Wooley P, Song Z, Markel DC. Inflammatory and immunological responses to hyaluronan preparations. Study of a murine biocompatibility model. J Bone Joint Surg Am. 2007;89(1):148–57.

36. Colen S, van den Bekerom MP, Mulier M, Haverkamp D. Hyaluronic acid in the treatment of knee osteoarthritis: a systematic review and meta-analysis with emphasis on the efficacy of different products. BioDrugs. 2012;26(4):257–68.

37. Concoff A, Sancheti P, Niazi F, Shaw P, Rosen J. The efficacy of multiple versus single hyaluronic acid injections: a systematic review and meta-analysis. BMC Musculoskelet Disord. 2017;18(1):542.

38. McElheny K, Toresdahl B, Ling D, Mages K, Asif I. Comparative effectiveness of alternative dosing regimens of hyaluronic acid injections for knee osteoarthritis: a systematic review. Sports Health. 2019;11(5):461–6.

39. Miller LE, Bhattacharyya S, Parrish WR, Fredericson M, Bisson B, Altman RD. Safety of intra-articular hyaluronic acid for knee osteoarthritis: systematic review and meta-analysis of randomized trials involving more than 8,000 patients. Cartilage. 2019:1947603519888783.

40. Legre-Boyer V. Viscosupplementation: techniques, indications, results. Orthop Traumatol Surg Res. 2015;101(1 Suppl):S101–8.

41. He WW, Kuang MJ, Zhao J, Sun L, Lu B, Wang Y, et al. Efficacy and safety of intraarticular hyaluronic acid and corticosteroid for knee osteoarthritis: a meta-analysis. Int J Surg. 2017;39:95–103.

42. Clarke S, Lock V, Duddy J, Sharif M, Newman JH, Kirwan JR. Intra-articular hylan G-F 20 (Synvisc) in the management of patellofemoral osteoarthritis of the knee (POAK). Knee. 2005;12(1):57–62.

43. Altman RD, Farrokhyar F, Fierlinger A, Niazi F, Rosen J. Analysis for prognostic factors from a database for the intra-articular hyaluronic acid (euflexxa) treatment for osteoarthritis of the knee. Cartilage. 2016;7(3):229–37.

44. Nicholls M, Shaw P, Niazi F, Bhandari M, Bedi A. The impact of excluding patients with end-stage knee disease in intra-articular hyaluronic acid trials: a systematic review and meta-analysis. Adv Ther. 2019;36(1):147–61.

45. Vincent P. Intra-articular hyaluronic acid in the symptomatic treatment of knee osteoarthritis: a meta-analysis of single-injection products. Curr Ther Res Clin Exp. 2019;90:39–51.

46. Xing D, Wang B, Liu Q, Ke Y, Xu Y, Li Z, et al. Intra-articular hyaluronic acid in treating knee osteoarthritis: a PRISMA-compliant systematic review of overlapping meta-analysis. Sci Rep. 2016;6:32790.

47. Rosen J, Niazi F, Dysart S. Cost-effectiveness of treating early to moderate stage knee osteoarthritis with intra-articular hyaluronic acid compared to conservative interventions. Adv Ther. 2020;37(1): 344–52.

48. Ong KL, Runa M, Lau E, Altman R. Is intra-articular injection of synvisc associated with a delay to knee arthroplasty in patients with knee osteoarthritis? Cartilage. 2019;10(4):423–31.

49. Ong KL, Anderson AF, Niazi F, Fierlinger AL, Kurtz SM, Altman RD. Hyaluronic acid injections in medicare knee osteoarthritis patients are associated with longer time to knee arthroplasty. J Arthroplast. 2016;31(8):1667–73.

50. Altman R, Lim S, Steen RG, Dasa V. Hyaluronic acid injections are associated with delay of total knee replacement surgery in patients with knee osteoarthritis: evidence from a large U.S. health claims database. PloS One. 2015;10(12):e0145776.

51. Kon E, Filardo G, Di Matteo B, Marcacci M. PRP for the treatment of cartilage pathology. Open Orthop J. 2013;7:120–8.

52. Sen EI, Yildirim MA, Yesilyurt T, Kesiktas FN, Diracoglu D. Effects of platelet-rich plasma on the clinical outcomes and cartilage thickness in patients with knee osteoarthritis. J Back Musculoskelet Rehabil. 2020;33(4):597–605.

53. Boffa A, Salerno M, Merli G, De Girolamo L, Laver L, Magalon J, Sánchez M, Tischer T, Filardo G. Platelet-rich plasma injections induce disease-modifying effects in the treatment of osteoarthritis in animal models. Knee Surg Sports Traumatol Arthrosc. 2021.

54. Hall MP, Band PA, Meislin RJ, Jazrawi LM, Cardone DA. Platelet-rich plasma: current concepts and appli-

cation in sports medicine. J Am Acad Orthop Surg. 2009;17(10):602–8.

55. Arnoczky SP, Sheibani-Rad S. The basic science of platelet-rich plasma (PRP): what clinicians need to know. Sports Med Arthrosc Rev. 2013;21(4):180–5.

56. Mariani E, Canella V, Cattini L, Kon E, Marcacci M, Di Matteo B, et al. Leukocyte-rich platelet-rich plasma injections do not up-modulate intra-articular pro-inflammatory cytokines in the osteoarthritic knee. PLoS One. 2016;11(6):e0156137.

57. Filardo G, Kon E, BDI M, ADI M, Sessa A, Merli ML, et al. Leukocyte-poor PRP application for the treatment of knee osteoarthritis. Joints. 2013;1(3):112–20.

58. Riboh JC, Saltzman BM, Yanke AB, Fortier L, Cole BJ. Effect of leukocyte concentration on the efficacy of platelet-rich plasma in the treatment of knee osteoarthritis. Am J Sports Med. 2016;44(3):792–800.

59. Filardo G, Kon E, Pereira Ruiz MT, Vaccaro F, Guitaldi R, Di Martino A, et al. Platelet-rich plasma intra-articular injections for cartilage degeneration and osteoarthritis: single- versus double-spinning approach. Knee Surg Sports Traumatol Arthrosc. 2012;20(10):2082–91.

60. Kon E, Buda R, Filardo G, Di Martino A, Timoncini A, Cenacchi A, et al. Platelet-rich plasma: intra-articular knee injections produced favorable results on degenerative cartilage lesions. Knee Surg Sports Traumatol Arthrosc. 2010;18(4):472–9.

61. Filardo G, Kon E, Buda R, Timoncini A, Di Martino A, Cenacchi A, et al. Platelet-rich plasma intra-articular knee injections for the treatment of degenerative cartilage lesions and osteoarthritis. Knee Surg Sports Traumatol Arthrosc. 2011;19(4):528–35.

62. Chouhan DK, Dhillon MS, Patel S, Bansal T, Bhatia A, Kanwat H. Multiple platelet-rich plasma injections versus single platelet-rich plasma injection in early osteoarthritis of the knee: an experimental study in a Guinea pig model of early knee osteoarthritis. Am J Sports Med. 2019;47(10):2300–7.

63. Southworth TM, Naveen NB, Tauro TM, Leong NL, Cole BJ. The use of platelet-rich plasma in symptomatic knee osteoarthritis. J Knee Surg. 2019;32(1):37–45.

64. Chang KV, Hung CY, Aliwarga F, Wang TG, Han DS, Chen WS. Comparative effectiveness of platelet-rich plasma injections for treating knee joint cartilage degenerative pathology: a systematic review and meta-analysis. Arch Phys Med Rehabil. 2014;95(3):562–75.

65. Dai WL, Zhou AG, Zhang H, Zhang J. Efficacy of platelet-rich plasma in the treatment of knee osteoarthritis: a meta-analysis of randomized controlled trials. Arthroscopy. 2017;33(3):659–70.e1.

66. Altamura SA, Di Martino A, Andriolo L, Boffa A, Zaffagnini S, Cenacchi A, et al. Platelet-rich plasma for sport-active patients with knee osteoarthritis: limited return to sport. Biomed Res Int. 2020;2020:8243865.

67. Di Martino A, Di Matteo B, Papio T, Tentoni F, Selleri F, Cenacchi A, et al. Platelet-rich plasma versus hyaluronic acid injections for the treatment of knee osteoarthritis: results at 5 years of a double-blind, randomized controlled trial. Am J Sports Med. 2019;47(2):347–54.

68. Simental-Mendia M, Vilchez-Cavazos JF, Pena-Martinez VM, Said-Fernandez S, Lara-Arias J, Martinez-Rodriguez HG. Leukocyte-poor platelet-rich plasma is more effective than the conventional therapy with acetaminophen for the treatment of early knee osteoarthritis. Arch Orthop Trauma Surg. 2016;136(12):1723–32.

69. Previtali D, Merli G, Di Laura FG, Candrian C, Zaffagnini S, Filardo G. The long-lasting effects of "placebo injections" in knee osteoarthritis: a meta-analysis. Cartilage. 2020:1947603520906597.

70. Gormeli G, Gormeli CA, Ataoglu B, Colak C, Aslanturk O, Ertem K. Multiple PRP injections are more effective than single injections and hyaluronic acid in knees with early osteoarthritis: a randomized, double-blind, placebo-controlled trial. Knee Surg Sports Traumatol Arthrosc. 2017;25(3):958–65.

71. Smith PA. Intra-articular autologous conditioned plasma injections provide safe and efficacious treatment for knee osteoarthritis: an FDA-sanctioned, randomized, double-blind, placebo-controlled clinical trial. Am J Sports Med. 2016;44(4):884–91.

72. Lin KY, Yang CC, Hsu CJ, Yeh ML, Renn JH. Intra-articular injection of platelet-rich plasma is superior to hyaluronic acid or saline solution in the treatment of mild to moderate knee osteoarthritis: a randomized, double-blind, triple-parallel, placebo-controlled clinical trial. Arthroscopy. 2019;35(1):106–17.

73. Xu Z, Luo J, Huang X, Wang B, Zhang J, Zhou A. Efficacy of platelet-rich plasma in pain and self-report function in knee osteoarthritis: a best-evidence synthesis. Am J Phys Med Rehabil. 2017;96(11):793–800.

74. Kanchanatawan W, Arirachakaran A, Chaijenkij K, Prasathaporn N, Boonard M, Piyapittayanun P, et al. Short-term outcomes of platelet-rich plasma injection for treatment of osteoarthritis of the knee. Knee Surg Sports Traumatol Arthrosc. 2016;24(5):1665–77.

75. Shen L, Yuan T, Chen S, Xie X, Zhang C. The temporal effect of platelet-rich plasma on pain and physical function in the treatment of knee osteoarthritis: systematic review and meta-analysis of randomized controlled trials. J Orthop Surg Res. 2017;12(1):16.

76. Filardo G, Previtali D, Napoli F, Candrian C, Zaffagnini S, Grassi A. PRP injections for the treatment of knee osteoarthritis: a meta-analysis of randomized controlled trials. Cartilage. 2020:1947603520931170.

77. Buendia-Lopez D, Medina-Quiros M, Fernandez-Villacanas Marin MA. Clinical and radiographic comparison of a single LP-PRP injection, a single hyaluronic acid injection and daily NSAID administration with a 52-week follow-up: a randomized controlled trial. J Orthop Traumatol. 2018;19(1):3.

78. Raeissadat SA, Ghorbani E, Sanei Taheri M, Soleimani R, Rayegani SM, Babaee M, et al. MRI changes after platelet rich plasma injection in knee osteoarthritis (randomized clinical trial). J Pain Res. 2020;13:65–73.

79. Brown GA. AAOS clinical practice guideline: treatment of osteoarthritis of the knee: evidence-based guideline, 2nd edition. J Am Acad Orthop Surg. 2013;21(9):577–9.

80. Kon E, Di Matteo B, Delgado D, Cole BJ, Dorotei A, Dragoo JL, et al. Platelet-rich plasma for the treatment of knee osteoarthritis: an expert opinion and proposal for a novel classification and coding system. Expert Opin Biol Ther. 2020;20(12):1447–60.

Cartilage Lesions and Osteoarthritis of the Hip and Ankle: Orthobiologics

27

Francesca Vannini, Simone Ottavio Zielli, and Cesare Faldini

27.1 Introduction

This chapter summarizes current strategies and evidence for the use of orthobiologic injectable therapies for both hip and ankle cartilage lesions and osteoarthritis (OA).

The use of corticosteroids for arthritic joints dates back to the 1950s, with injections of hydrocortisone acetate. Since then, corticosteroids have always been the golden standard injection therapy for symptomatic OA in multiple joints, hip and ankle among them. Still, the effects are short-term and palliative. Major side effects have been documented such as an acceleration in the cartilage degeneration, especially with repeated injections. Thus, new therapeutic options are being developed to solve the intrinsic limitations of corticosteroid injections, to reduce symptomatology, and to improve cartilage viability, protecting joints from further damage [1, 2].

Hyaluronic acid (HA) is a viscoelastic glycosaminoglycan constituent of synovial fluid and cartilage. HA can stimulate chondrocyte metabolism and synthesis of cartilage matrix components and inhibit chondro-degenerative enzymes, reducing the inflammatory process [3]. In 1999 the use of HA was approved for OA both in the hip and ankle [4], and this topic started to be extensively studied and reported in the literature.

Platelet-rich plasma (PRP) has also gained high popularity in the last years, and its use has been investigated mostly in the knee [5, 6], but experiences exist also in the hip and ankle [7–9].

Adult cells can be obtained from the bone marrow and adipose tissue, and umbilical cord tissue has also been used with therapeutic intent [10–13]. Although there is tremendous potential for cellular therapies, there are still many questions that need to be answered, such as the best cell source(s), the best cell type(s), autologous vs allogeneic transplant, and how to optimally prime or stimulate the implanted cells. Although the use of cellular therapies in association with surgical procedures, especially in the ankle, for either osteochondral lesion or OA treatment, has gained major popularity [14–17], its use as injection therapy is still scarcely documented.

27.2 State of the Art of the Injective Therapies in the Hip

Hip OA is one of the most burdensome causes of disability in our society [18]. Despite the excellent and durable results that can be obtained by total hip replacement, there is profound need and demand for therapeutic strategies that can delay or avoid the need for this major surgery. Such "conservative" options have always been a hot topic in orthopaedic surgery. As a consequence, a large number of studies are available describing

F. Vannini (✉) · S. O. Zielli · C. Faldini
I Clinic, Rizzoli Orthopaedic Institute, Bologna University, Bologna, Italy

© ISAKOS 2022
G. Filardo et al. (eds.), *Orthobiologics*, https://doi.org/10.1007/978-3-030-84744-9_27

injection therapies (Table 27.1). Hip anatomy can make it challenging to reliably inject materials into the hip joint. Either fluoroscopic guidance or ultrasound (US) [19] guided procedures are generally used to help injective procedures in the hip [20, 21].

27.2.1 Hyaluronic Acid

The use of HA has been compared with the use of local anesthetics or corticosteroids in the arthritic hip. In a double-blinded study, Migliore et al. [22] compared an intra-articular bacterial-derived HA (Hyalubrix®) with local analgesia (mepivacaine) for OA of the hip in 42 patients. This is the only level 2 comparison to an anesthetic, to our knowledge. It demonstrated safety as well as HA superiority in symptom relief at 3 and 6 months of follow-up.

The group of studies comparing HA and corticosteroids is larger. In particular, among these Spritzer et al. in 2015 compared the efficacy and safety of intra-articular hylan G-F 20 with methylprednisolone acetate (MPA) for treating symptomatic Kellgren-Lawrence grade (KL) 2 or 3 hip OA. Response rates were higher with hylan G-F 20 in patients with more advanced disease (KL 3) and were similar between hylan G-F 20 and MPA in patients with less advanced disease (KL 2). The authors concluded that HA was an appropriate option for treating hip OA [23]. These results were not confirmed by Qvistgaard et al. in a 101-patient study with 3 groups. The authors compared HA with saline solution and with corticosteroids and found no statistically significant effect of HA on any outcome measure [24]. To date, most HA studies support the notion of an immediate rather than a long-lasting effect of injections, whatever the substance used, perhaps associated with a strong placebo effect, as commonly observed for injectable treatments [25].

The molecular weight (MW) of HA may be an important variable (Table 27.2). Most HA studies have used low molecular weight. The ideal MW to be used in OA, the most effective number injections, and the choice of therapy for different stages of OA are major topics of debate. Despite

many studies on the topic [26–29], there is still no consensus on the number of injections, the dosage per injection, and the most appropriate formulation of HA. The intrinsic features of the hip joint, being the deepest articulation of our body, and its anatomical conformation result in a more difficult approach for the injective therapy compared to the knee, with a smaller joint space requiring US guidance to be properly targeted. A large number of HA preparations are commercially available, each with different characteristics and properties [29, 30]. Whether MW differences are associated with different therapeutic effects or durability, it is still to be clarified. In this chapter we considered low MW HA those with an average MW equal or inferior to 1.5 million Da (such as Ostenil PLUS®, Adant®, Synocrom®), medium MW HA those with an average MW included between 1.5 and 3 million Da (such as Hyalubrix 60®), high MW HA those with an average molecular weight superior to 3 million Da (such as Synvisc®), and ultra-high MW HA those obtained by a complex process of cross-linking with a potentially very high MW but not quantifiable (such as Fermathron S).

Clementi et al. compared the efficacy of an ultra-high MW viscosupplement (UHMW-HA, Fermathron S) with a medium MW hyaluronan (MMW-HA, Hyalubrix 60) on 54 patients with a grade 3 KL hip OA. The authors concluded that a single dose of UHMW-HA was as effective as two doses of MMW-HA resulting in similar reductions of pain and disability. The authors concluded that the final effect is similar but UHWM-HA allowed half the number of injections to get the same results [27]. While comparing two commercially available HAs with high and medium MW, De Lucia et al. [26] reported no significant differences between the two HA formulations. This finding was also confirmed by Tikiz et al. and Bekerom et al. [29, 30]. Another important finding of the De Lucia study was that HA injections had a consistent clinical effect in terms of pain reduction, independent of the clinical and radiological severity of the disease [26]. Abate et al. combined low and high MW HA. The aim of this paper was to evaluate the efficacy of a new hybrid preparation for

Table 27.1 (a–c) Studies available regarding injective therapies in the hip

Author Journal Year	DOI/PMID	Study type	Level of evidence	Injected product	Patient	Follow-up	Injectable schedule	Outcome measures
(a) Hyaluronic acid								
Migliore et al. Orthopedic Research and Reviews 2020	10.2147/ORR.S239355	Case series	IV	HA	198	6–12 months	1 injection	VAS, Lequesne index, NSAID consumption
De Lucia et al. Frontiers in Pharmacology 2019	10.3389/fphar.2019.01007	Retrospective cohort study	III	HMW-HA vs MMW-HA vs controls	122 HMW, $n = 43$; MMW, $n = 79$; controls, $n = 20$	1, 6, 12, and 24 months	3 injections once a month	VAS score, WOMAC, number of days of NSAID/analgesic consumption, adverse events, and causes of treatment discontinuation
Brander et al. Osteoarthritis and Cartilage 2019	10.1016/j.joca.2018.08.018	Randomized, double-blind, multicenter, parallel-group study	I	HA vs saline	357 HA, $n = 182$; saline, $n = 175$	26 weeks	1 injection	"Pain on walking" (WOMAC-A1 and -A), Patient Global Self-Assessment (PTGA), WOMAC-A1 responder rate ($+\geq 2$ points on NRS), and adverse events (AEs)
Mauro et al. Journal of Biological Regulators and Homeostatic Agents 2018	30334430	Retrospective cohort study	III	HA	96 1 inj = 19; 2 inj = 24; 3 inj = 44	7 months	1/2/3 injections	VAS, WOMAC score, number of intra-articular injections, reasons for interrupting the treatment, adverse events

(continued)

Table 27.1 (continued)

Author Journal Year	DOI/PMID	Study type	Level of evidence	Injected product	Patient	Follow-up	Injectable schedule	Outcome measures
Clementi et al. European Journal of Orthopaedic Surgery and Traumatology 2018	10.1007/s00590-017-2083-9	Multicenter, independent, prospective, randomized controlled trial	I	UHMW-HA vs. MMW-HA	UHMW = 23; MMW = 27	1–3–6–12 months	1 injection	Lequesne index, VAS, and WOMAC score
Mauro et al. Clinical cases in mineral and bone metabolism : the official journal of the Italian Society of Osteoporosis, Mineral Metabolism, and Skeletal Diseases 2017	10.11138/ ccmbm/2017.14.1.146	Retrospective cohort study	IV	HA + exercise	40	10 weeks	3 injections, once every 45 days	VAS, Thomas test, Trendelenburg test, ROM
Migliore et al. European Review for Medical and Pharmacological Sciences 2017	28429341	Prospective, post-marketing, cohort study	IV	Hyalubrix (MMW)	1022	Every 6 months for 7 years	1 injection at least every 6 months	Lequesne index, pain, VAS, NSAID intake, global medical and patient assessments
Abate et al. International Journal of Immunopathology and Pharmacology 2017	10.1177/ 0394632016689275	Historically controlled comparative cohort study	IV	L+MMW vs MMW HA	40 L + H = 20; H = 20	3–6 months	1 injection	Visual analogue scale [VAS] for pain, Lequesne index, Harris Hip Score
Eymard et al. BMC Musculoskeletal Disorders 2017	10.1186/s12891-016-1359-2	Prospective, observational, multicenter, open-label, pilot study	II	HA	90	3 months	1 injection	WOMAC pain and function scores and patient global assessment (PGA)

Atchia et al. Annals of the Rheumatic Diseases 2011	10.1136/ard.2009.127183	Prospective, randomized controlled trial	I	HA vs. no injection vs. NS vs. steroid	78 HA, $n = 19$; NS, $n = 19$; corticosteroid, $n = 20$; no injection, $n = 20$	1 week 1–2 months	1 injection	Outcome Measures Numerical rating scale (NRS 0–10) "worst pain," Western Ontario and McMaster Universities Arthritis Index (WOMAC) pain/function
Migliore et al. Archives of Orthopaedic and Trauma Surgery 2011	10.1007/s00402-011-1353-y	Prospective cohort study	III	HA	120	1 inj	1 injection every 6 months	Lequesne algofunctional index, VAS, concomitant use of nonsteroidal anti-inflammatory drugs
Spitzer et al. Physician and Sportsmedicine 2010	10.3810/psm.2010.06.1781	Prospective, randomized study	I	(HMW) HA vs. methylprednisolone	305 HA = 150; MPA = 155	Weeks 4, 8, 12, 16, 20, and 26	2 injections after 2 weeks	VAS, WOMAC A, WOMAC A1, total WOMAC score, clinical observer global assessment (COGA)
Migliore et al. Arthritis Research and Therapy 2009	10.1186/ar2875	Prospective, double-blind, randomized trial	I	HA vs anesthetic	42 HA = 22; ane = 20	3–6 months	2 injections once a month	Lequesne index of the hip, VAS score, NSAID consumption, patient's global assessment, examining physician's global assessment
Migliore et al. Current Medical Research and Opinion 2008	10.1185/030079908 X291930	Prospective observational case series	III	HA	250	3–6–9–12	1 up to 4 injections	Lequesne index, VAS, physician global assessment of the target hip

(continued)

Table 27.1 (continued)

Author Journal Year	DOI/PMID	Study type	Level of evidence	Injected product	Patient	Follow-up	Injectable schedule	Outcome measures
Bekerom et al. Archives of Orthopaedic and Trauma Surgery 2007	10.1007/s00402-007-0374-z	Prospective observational case series	III	L vs L vs H MWHA	120 0.9 Mda = 91; 1.6 Mda = 20; Mda = 15	6 weeks/ 24 months	1 injection plus other optional injections	Visual analogue scale and Harris Hip Score
Qvistgaard et al. Archives of Orthopaedic and Trauma Surgery 2006	10.1007/s00402-007-0374-z	RCT	II	Saline vs CS vs LMW-HA	101 HA = 34; saline = 36; CS = 34	3 months	3 injection HA, 1 CS every 14 days	VAS, Lequesne score, the Western Ontario and McMas- ter Universities (WOMAC) total osteoarthritis index, and "patient global assessment"
Tikiz et al. Clinical Rheumatology 2005	10.1007/s10067-004-1013-5	Prospective observational cohort study	II	L vs H MWHA LMW HA) (Ostenil) with a higher molecular weight viscosupplement (hylan G-F 20, Synvisc	43 patients (56 hips), 25 (32 hips) received LMW HA and the remaining 18 patients (24 hips) received hylan G-F 20	6 months	3 weekly injections	VAS, Lequesne index, WOMAC
Migliore et al. Clinical Rheumatology 2012	10.1007/s10067-012-1994-4	Retrospective case series	IV	HA	176	24–48 months	1 injection	Pain visual analog scale, Lequesne Algofunctional Index, global patient assessment, global physician assessment, nonsteroidal anti-inflammatory drug intake

Migliore et al. Current Medical Research and Opinion 2012	10.1185/03007995.2011.645563	Retrospective case series	IV	HA	224	12–24 months	1 injection	Total hip replacement performed, VAS, Lequesne OA score. Consumption of NSAIDs and/or analgesics

(b) Platelet-rich plasma

Author Journal Year	DOI/PMID	Level of evidence	Injected product	Injectable schedule	Patient	Follow-up	Outcome measures	Injection approach
Doria et al. Joints 2017	10.1055/s-0037-1605584	Level I	HA vs PRP (Regen Kit®)	3 weekly injections	80 HA = 40; PRP = 40	6–12 months	WOMAC, VAS, and Harris Hip Score	US-guided
Di Sante et al. Medical Ultrasonography 2016	10.11152/mu-874	Level I	HA vs PRP	3 weekly injections	43 HA = 22; PRP = 21	16 weeks	VAS and WOMAC	US-guided
Dallari et al. American Journal of Sports Medicine 2016	10.1177/0363546515620383	Level I	HA (Hyalubrix) vs. PRP vs. HA+PRP	3 weekly injections	111 HA = 36; PRP = 44; PRP + HA = 31	2–6–12 months	VAS, Harris Hip Score, Western Ontario and McMaster Universities Osteoarthritis Index (WOMAC)	US-guided
Battaglia et al. Orthopedics 2013	10.3928/01477447-20131120-13	Level I	HA vs. PRP	3 injections (once every 2 weeks)	100 HA = 50; PRP = 50	1–3–6–12 months	Harris Hip Score (HHS) and visual analog scale (VAS)	US-guided

(continued)

Table 27.1 (continued)

Author Journal Year	DOI/PMID	Study type	Level of evidence	Injected product	Patient	Follow-up	Injectable schedule	Outcome measures
Villanova-Lopez et al. Revista Espanola de Cirugia Ortopedica y Traumatologia 2020	10.1016/j.recot.2019.09.008	Level I—phase III	HA vs PRP	1 injection	74	12 months	VAS score, HHS, and WOMAC	US-guided
Sanchez et al. Rheumatology 2011	10.1093/rheumatology/ker303	Level II	PRP	3 weekly injections	40	6 months	WOMAC, VAS, HHS	US-guided
Singh et al. Pain Medicine 2019	10.1093/pm/pnz041	Level IV	PRP	1 injection	36	6 months	VAS HHS WOMAC	US-guided

(c) Cell-based therapies

Author Journal Year	DOI/PMID	Level of evidence	Injected product	Number of patients	Follow-up length	Injective schedule	Outcome measures	Imaging guidance
Hauser et al. Clinical Medicine Insights: Arthritis and Musculoskeletal Disorders 2013	10.4137/CMAMD.S10951	Level IV	Unfractioned BMA	5	6 months	1	Arbitrary rating scale	Unspecified
Rodriguez-Fontan et al. PM and R 2018	10.1016/j.pmrj.2018.05.016	Level IV	BM-aspirate-concentrate	15	13.2 ± 6.3 months (range, 6–24 months)	1	WOMAC score	Radiographic or ultrasound guidance

Mardones et al. Journal of Hip Preservation Surgery 2017	10.1093/jhps/hnx011	Level IV	Ex vivo expanded autologous BM-MSC	10	7–30 months	3 weekly	Harris Hip (HHS), visual analog scale (VAS), Vail Hip and Western Ontario and McMaster Universities Osteoarthritis Index (WOMAC)	Unspecified
Emadedin et al. Archives of Iranian Medicine 2015	PMID: 26058927	Level IV	Ex vivo expanded autologous BM-MSC	5	30 months	1	VAS, WOMAC, walking distance	Fluoroscopy guide

HA hyaluronic acid, *VAS* visual analog scale, *n* number, *HMW* high molecular weight, *WOMAC* Western Ontario and McMaster Universities Osteoarthritis Index, *inj* injection, *ROM* range of motion, *MMW* medium molecular weight, *L* low, *CS* corticosteroid, *PRP* platelet rich plasma, *BMA* bone marrow aspirate, *BM-MSC* bone marrow mesenchymal stromal cells

Table 27.2 Examples of hyaluronic acid molecular weights

UHMW (ultra-high molecular weight)	>6 MDa	Fermathron S (not quantifiable)
HMW (high molecular weight)	>3 MDa	Hylan GF-20 (6 MDa)
MMW (medium molecular weight)	>1.5 MDa	Hyalubrix 60 (1.3–3.6 Mda)
LMW (low molecular weight)	<1.5 MDa	Ostenil PLUS® (1.6 MDa) Adant® (0.6–1.2 MDa) Synocrom® (1.6 MDa) HANOX-M-XL (1 Mda)

patients suffering from hip OA and to compare the results with historical data from a cohort of patients treated with high MW HA [28]. The preparation consists in a dynamic hybrid complex, by weak hydrogen linking HMW HA (1100–1400 kDa) to LMW HA (80–100 kDa), which should favor a cooperative action between the two HA preparations (chemically non-modified HA of bio-fermentative origin). Despite some limitations, this study showed that the combination of low and high MW HA was effective and safe in the management of patients suffering from hip OA and provided better therapeutic results than high MW HA.

Though the majority of the studies have employed one single HA injection, such as Migliore et al. [19], Brander et al. [31], Clementi et al. [27], and Abate et al. [28], the number and the frequency of injections have also been explored. In a retrospective study of 2018, Mauro et al. evaluated the clinical and functional outcome in patients with mild-moderate hip OA treated with a course of one, two, or three HA intra-articular injections. Ninety-six patients were included. Intra-articular injections for mild-moderate hip OA were effective in reducing pain and improving function. A full course of three injections provided the best pain control [32]. In a recent publication, De Lucia et al. performed a schedule based on three monthly injections of HA, followed by further maintenance injections administered every 6 months for 2 years. The authors stated that this method obtained analgesic effectiveness, functional recovery, and reduced joint stiffness up to 24 months. In addition, since the major improvement was found between 12 and 24 months, it is possible to infer that repeated administrations may achieve an additive effect [26].

There is scarce data available on intra-articular hyaluronan's ability to modify the progression of OA. Migliore et al. [33] in 2012 assessed the impact of treatment with hylan G-F 20 on the progression to total hip replacement (THR) in patients with symptomatic hip OA. Of the 224 selected patients, 84 patients (37.5%) progressed to THR. Two hundred six patients (92.0%) achieved a 12-month survival, 170 patients (75.9%) achieved a 24-month survival, and 69 patients (30.8%) achieved a 5-year survival. These results suggest that hylan G-F 20 could be included in the management of symptomatic hip OA before recommending for THR.

27.2.2 Platelet-Rich Plasma

Sanchez et al. in 2012 were the first to provide a preliminary non-controlled prospective study, supporting the safety, tolerability, and efficacy of PRP injections for pain relief and improved function in hip OA, although in only 40 patients [34]. Subsequently, the majority of studies compared the effects of PRP on hip OA compared to HA. Battaglia and colleagues [21] performed a non-blinded, randomized trial comparing ultrasound-guided PRP versus HA injections for hip OA in 100 consecutive patients. Patients underwent three 5 mL injections of autologous PRP, one every 2 weeks, or 2 mL HA every 2 weeks. Using the Harris Hip Score (HHS) and VAS, patients in both groups demonstrated significant improvements between 1-month and 3-month follow-ups. Although patients showed worsening of symptoms between 6-month and 12-month follow-ups, scores were still significantly improved compared with baseline. No significant differences were found between PRP and HA groups. Another randomized controlled trial by Dallari et al. [35] compared HA and PRP in 111 patients and showed that the PRP group was superior. Furthermore, the combination of PRP and HA did not lead to a significant improvement in pain symptoms. A different trend was described by Di Sante et al. [36], who found that, although intra-articular PRP had an immediate effect on pain (as measured by VAS and by WOMAC pain subscale at 4 weeks), this was not maintained at a longer-term follow-up (16 weeks). On the contrary, the effects of intra-articular HA, which were not significant at the 4-week follow-up, were evident at the 16-week follow-up both in VAS and WOMAC scores. Similar PRP outcomes were reported in a meta-analysis performed by Ye et al. in 2018 [37]. The authors concluded that PRP was associated with a significantly better reduction of VAS score at 2 months compared with HA. However, it did not show significantly better outcomes at 6 and 12 months.

A significant challenge in all studies on PRP is surely the great variability that exists in different products and patient-specific preparations [38]. These variables are likely to have significant impact on the response that individual patients have to these treatments [39]. Recently, Villanova-Lopez et al. [40] observed that the mean platelet concentration was different between responders and non-responders (at 1 month, non-responders 449 [range 438–578] $\times$ 10^3 platelets/μL versus responders 565 [range 481–666] $\times$ 10^3 platelets/μL, $p < 0.044$) following PRP hip injection. Furthermore, a high positive and significant correlation was observed between the results of the stiffness subscale of the WOMAC and the concentration of leukocytes, thus suggesting that low leukocyte concentration may contribute to better clinical outcomes with PRP injective therapy. Finally, patients with early-stage hip OA showed a statistically significative higher response rate to PRP compared with late stage disease. Due to these findings, they concluded that the cellular composition of PRP, as well as a better selection of the subjects with respect to the OA degree, may be key factors in achieving a better clinical response. The same trend was also observed by Singh et al. [41] in their retrospective analysis published in 2019. They stratified the 36 patients undergoing a single PRP hip injection according to their modified K-L classification. A significant improvement was noted in the K-L 1 and 2 subgroups, suggesting that patients with mild/moderate hip OA may experience better pain relief and functional improvement after a PRP injection than patients with later stage disease.

27.2.3 Cell-Based Therapies

To our knowledge, only four studies described the use of bone marrow as a source of cells for cell-based therapies for the treatment of hip OA. All of them have a low patient number and therefore limited power.

A simple and affordable preparation, based on unfractionated bone marrow cells, in four of seven cases combined with prolotherapy (hyperosmotic dextrose), was employed by Hauser et al. [42] in a seven-patient case series with hip, knee, or ankle OA. The preliminary findings suggested that OA treatment with whole bone marrow (WBM) injection merited further investigation. However, the potential for injury to the bone marrow-derived cells, by placing them in a hyperosmolar environment, was not discussed.

Bone marrow aspirate concentrate (BMC/BMAC) is the most common cell-based therapy reported to date. Rodriguez-Fontan et al. [43] tried to determine its efficacy, safety, and benefit in hip and knee OA. Nineteen patients (16 females and 3 males), totaling 25 joints (10 knees, 15 hips), were treated with intra-articular BMC for early OA between 2014 and 2016. All patients had autologous bone marrow aspirate harvested from the iliac crest and centrifuged to obtain BMC for intra-articular injection. Besides the low quality of the study, intra-articular injections of BMC were safe and demonstrated satisfactory results in 63.2% of patients.

Mardones et al. [44] focused their attention on investigating the safety and efficacy of the intra-articular infusion of ex vivo expanded autologous bone marrow-derived mesenchymal stromal cells (BM-MSC) in a cohort of ten patients with hip OA. The procedure consisted in the selection of mononuclear cells (BM-MNC) and the expansion, by means of cell culture procedures of the minor population of "native" MSCs present in BM-MNC. The treatment proved clinical benefit in terms of pain. No difference was found in the radiographic scores at follow-up. Similar results were found by Emadedin et al. [45] in a comparable case series of five patients treated by expanded MSCs.

27.3 State of the Art of the Injective Therapies in the Ankle

Osteochondral lesions of the talus (OLT) are an increasingly relevant pathology, due to the wider diffusion of sports activity in a larger range of ages. The ideal solution is generally surgical; nevertheless, the use of injections is still under evaluation. Moreover, although less common than hip or knee OA [7], ankle OA is a leading cause of chronic disability. Furthermore, it occurs in younger individuals, usually being post-traumatic in origin. Nonoperative treatment for acute, nondisplaced osteochondral lesions of the talus and cystic lesions has been associated with successful clinical results in about 50% of cases [46–48]. It consists of activity modification, bracing, nonsteroidal anti-inflammatory drugs (NSAIDs), physical therapy, and protected weight-bearing in a walking boot, with the aim of improving symptoms [49, 50]. Surgical options depend on the size of the lesion and the presence of OA; either regenerative or reparative procedures are to be considered in OLT [51], while the treatment of high-grade OA may require total ankle replacement or fusion [52]. Injection therapies may play a significative role in delaying or even avoiding the need for more invasive surgical procedures in the ankle (Table 27.3).

A recent systematic review and meta-analysis by Boffa et al. [53] underlined the safety of intra-articular treatment for ankle OA and OLT, even though only a very low evidence supported the efficacy of HA in terms of better results versus placebo for the treatment of ankle OA. Other conclusions were hindered by the scarcity of the available literature.

27.3.1 Hyaluronic Acid

In 2006 Salk et al. [54] set up a blinded randomized study to gather preliminary data on the efficacy and safety of five weekly intra-articular injections of Hyalgan (sodium hyaluronate, MW 500–730 kDa) vs placebo for the treatment of ankle OA. Significant improvement in the mean Ankle Osteoarthritis Score from baseline was seen at all follow-up visits from 1 to 6 months in both the sodium hyaluronate group and the saline solution group ($p < 0.0001$). The positive results both in terms of pain relief and improved function encouraged the application of HA injective therapy. On the other hand, one randomized controlled trial conducted by DeGroot et al. [55] found that HA was not superior to placebo injections. In fact, a single intra-articular injection of low-MW, non-cross-linked HA was not superior to a single intra-articular injection of saline solution for the treatment of OA of the ankle in 64 patients. However, the majority of the HA studies in the ankle have positive trends, but without significant difference due to the paucity of patients involved, possibly due to limitations in sample size and power. For example, both Sun et al. [56] and Cohen et al. [57] found positive trends comparing HA injections to controls. Cohen et al. noted these trends at week 2, week 6, and month 6, but between-group comparisons were not statistically significant.

As in the hip, the ideal MW and the injection schedule to be used are points of debate in the ankle. Both Han et al. [58] and Lucas y Hernandez et al. [59] employed a HMW HA in their studies, using a schedule of three consecutive weekly injections. No control group is available in these studies. Nevertheless, these prospective studies showed that viscosupplementation had a significant positive effect, respectively, at a follow up of 13 months for the study by Han et al. and 45 months for the study of Lucas y Hernandez et al. Based on their experience, Lucas y Hernandez et al. suggested three-injection protocol every 2 years on average but did not provide long-term data describing the outcome of this approach. A different approach was employed by

Table 27.3 Studies available regarding injective therapies in the ankle

Author Journal Year	DOI/PMID	Level of evidence	Ankle disease	Injected products	Injection schedule	Number of patients	Follow-up length (months)	Outcome measures	Imaging guidance
Bossert et al. Clinical Medicine Insights: Arthritis and Musculoskeletal Disorders 2016	10.4137/CMaMd.s40401	Level IV	OA	HMWHA+mannitol vs HMWHA (HANOX-M-XL)	1 inj	Total = 50; 35/15	19	Patient's self-evaluation of pain, satisfaction, treatment efficacy, and tolerability was collected	Fluoroscopy, ultrasonography, or landmark guidance
Cohen MM et.al Foot and Ankle International 2008	10.3113/FAI.2008.0657	Level I	OA	LMW HA (Hyalgan) vs saline	5 inj	28 (15 + 13)	6	AOS, VAS, WOMAC, SF-36	Fluo-guided
DeGroot H et al. Journal of Bone and Joint Surgery - Series A 2013	10.2106/JBJS.J.01763	Level I	OA	LMWHA vs saline	1 inj	39 (24/15)	3	VAS, AOFAS, AOS	Fluo-guided
Han et al. Yonsei Medical Journal 2014	10.3349/ymj.2014.55.4.1080	Level IV	OA	HMWHA	3 inj	40; 19/21	13	VAS	No guidance
Lucas y Hernandez et al. Foot & ankle international 2013	10.1177/1071100/14565900	Level IV	OA	HMWHA (Synvisc)	3 inj every 2 weeks	18; 13/5	45.5	AOFAS	FLUO-guided
Karatosun et al. Rheumatology International 2008	10.1007/s00296-005-0592-z	Level I – RCT not blinded	OA	LMWHA (Adant)	3 inj weekly	15; 6/9	13	AOFAS	No guidance

(continued)

Table 27.3 (continued)

Author Journal Year	DOI/PMID	Level of evidence	Ankle disease	Injected products	Injection schedule	Number of patients	Follow-up length (months)	Outcome measures	Imaging guidance
Luciani et al. La Chirurgia degli Organi di Movimento 2008	10.1007/s12306-008-0066-z	Level II – prospective cohort	OA	HMW HA (Synvisc)	3 weekly inj	21 (13/8)	18	AOS	No guidance
Mei dan et al. Foot and Ankle International 2008	10.3113/FAI.2008.1171	Level IV	OLT	MMW HA	3 weekly inj	15	6	AOFAS	No guidance
Mei dan et al. Journal of the American Podiatric Medical Association 2010	10.7547/1000093	Level IV	OA	LMW HA	5 weekly inj no guidance	15	8	AOFAS	No guidance
Murphy et al. Journal of Foot and Ankle Surgery 2017	10.1053/j.jfas.2016.09.017	Level IV	OA	LMW HA	3 inj every 2 weeks	50	12	Foot and Ankle Outcomes Score	No guidance
Salk et al. Journal of Bone and Joint Surgery - Series A 2006	10.2106/JBJS.E.00193	Level I	OA	LMW HA (Hyalgan) vs saline	5 weekly inj no guidance	17	6	VAS, WOMAC, SF-12	No guidance
Sun et al. Osteoarthritis and Cartilage 2006	10.1016/j.joca.2006.03.003	Level IV	OA	LMWHA	5 weekly inj no guidance	75	6	AOS, AOFAS, ROM	No guidance
Sun et al. Journal of Foot and Ankle Research 2014	10.1186/1757-1146-7-9	Level I	OA	Botulinum vs LMWHA (Hyalgan)	1 inj no guidance	75	6	VAS, AOS, AOFAS, SLS, TUG	No guidance

Younger et al. Journal of Foot and Ankle Surgery 2017	10.1053/j.jfas.2018.10.003	Level IV	OA	LMWHA (NASHA)	1 inj no guidance	37	6	VAS	No guidance
Fukawa et al. Foot and Ankle International 2017	10.1177/1071100717700377	Level IV	OA	PRP	3 inj	20; 5/15	6	VAS, JSSF, SAFE-Q	US-guided
Akpancar et al. Medical Science Monitor 2019	10.12659/MSM.914111	Level III	OLT	PRP vs prolotherapy	3 inj	Total = 49; PRP = 22; prolotherapy = 27	12	VAS, AOFAS, AOS	US-guided
Angthong et al. Journal of Foot and Ankle Surgery 2013	10.1053/j.jfas.2013.04.005	Level IV	OA	PRP	1 inj	5	16	VAS-FA SCORE, SF-36 SCORE	US-guided
Mei dan et al. American Journal of Sports Medicine 2012	10.1177/0363546511431238	Level II – not blinded RCT	OLT	MMW HA vs PRP	3 weekly inj	32	6	AOFAS, AHFS, VAS	No guidance
Repetto et al. Journal of Foot and Ankle Surgery 2017	10.1053/j.jfas.2016.11.015	Level IV	OA	PRP	4 weekly injec	20	17.7	VAS, FADI	No guidance
Emadedin et al. Archives of Iranian Medicine 2015	015186/AIM.003	Level IV	OA	Ex vivo expanded bone marrow derived MSC	1 inj	6	30	WOMAC, FAOS	Fluo-guided

OA osteoarthritis, *HA* hyaluronic acid, *VAS* visual analog scale, *n* number, *HMW* high molecular weight, *WOMAC* Western Ontario and McMaster Universities Osteoarthritis Index, *inj* injection, *ROM* range of motion, *MMW* medium molecular weight, *LMW* low molecular weight, *AOFAS* American Orthopaedic Foot & Ankle Society, *AOS* Ankle osteoarthritis scale, *TUG* timed up and go test, *SLS* single leg stance test, *JSSF* Japanese Society for Surgery of the Foot ankle/hindfoot scale, *SAFE-Q* Self-Administered Foot Evaluation Questionnaire, *AHFS* Ankle-Hindfoot Scale, *FAOS* foot and ankle outcomes score, *FADI* Foot & Ankle Disability Index

Bossert et al. [60], who utilized a HMW HA with a high concentration of mannitol (35 g/g), conferring a very high viscosity, with the aim to increase the HA residence time and consequently allowing a single injection regimen. The study showed the safety of HANOX-M-XL, confirming the good tolerability of the combination HA + mannitol, and the results were significantly better in patients who received injection under US guidance. In addition to the aforementioned studies, Younger et al. [61] also deployed a single-injection approach using a cross-linked HA, based on the assumption of a longer intra-articular residence time.

Consistent with prior studies in the hip, Han et al. [58] also found that early-stage OA disease and a duration of pain less than 1 year are independent predictors associated with higher satisfaction. This may explain why other studies involving patients with low (I or II) KL grade ankle OA tended to have better clinical results. In contrast Lucas y Hernandez et al. [59] found that neither etiology nor severity of OA was predictive of the response to HA injective therapy. However, this study had many limitations: low number of patients (18 patients, 26 ankles), absence of randomization, and only intermediate follow-up with an average of 45.5 months.

The demand for novel OA treatments led to testing of new orthobiologic solutions in the management of ankle OA. Sun et al. [62] compared the efficacy of intra-articular botulinum toxin type A (BoNT-A) and intra-articular hyaluronate plus rehabilitation with ankle OA. Targeted rehabilitation was performed in the HA group. It has been suggested that BoNT-A suppresses the secretion of neurotransmitters directly decreasing peripheral sensitization with a strong effect in terms of pain. No difference was found between the two methods. Both BoNT-A and HA injection plus rehabilitation exercise showed clinical improvements in terms of pain, physical function, and balance in patients with ankle OA. The inclusion of a specific exercise protocol is highly relevant. Back in 2008 Karatosun et al. already acknowledged the importance of physical exercise in the management of ankle OA and used exercise therapy as a control

group to compare the efficacy of three intra-articular HA injections. Total AOFAS Ankle-Hindfoot score improved in both groups, with no statistically significant differences between the groups. Exercise, in fact, may play a major role in early-stage OA [63]. Future studies will need to devote attention to providing adequate guidance regarding rehabilitation and exercise protocols in ankle OA and the potential of injectable treatments to further improve exercise results.

While many authors have tested HA for ankle OA, there is limited experience regarding the efficacy of HA in osteochondritis dissecans (OCD) of the ankle. Mei Dan et al. [64] reported their initial results with this treatment on 15 patients aged 18–60 years treated for OCD by three weekly injections of intra-articular HA. The treatment resulted in a decrease in terms of pain and an increase in global function over a short period of time (within 12 weeks) which then lasted for more than 6 months with minimal adverse events.

27.3.2 Platelet-Rich Plasma

PRP therapy has also been applied to ankle OA, following the design of knee and hip trials, but there is still very little evidence of efficacy. Anghtong et al. [65] reported the first study evaluating PRP in ankle OA dates back to 2013 reporting on a retrospective cohort of 12 adult patients with different diseases of the hindfoot and ankle but included 3 patients treated for ankle OA. Benefit was inconclusive. A larger PRP study was published by Repetto et al. [66] in 2017, providing mid- to long-term clinical results for PRP injections in 20 patients (20 ankles) with ankle OA. The results suggested that the use of PRP injection was a safe alternative and may postpone the need for surgery. Similar results were gained by Fukawa et al. [67] in 20 patients with ankle OA, showing that patients with late-stage OA had worse scores in all outcomes than those with early-stage OA.

PRP properties were also applied in patients with OLT by Mei Dan et al. [68]. In their randomized controlled trial, they evaluated the short-term efficacy and safety of PRP compared with HA in reducing pain and disability caused by OLTs. The authors concluded that osteochondral lesions of the ankle treated with intra-articular injections of PRP and HA resulted in a statistically significant decrease in pain scores and an increase in function for at least 6 months, with minimal adverse events. Between the two groups, a significant best performance of PRP over HA was found only for function at 28 weeks. The support for PRP treatment leading to significantly better outcomes than HA was also supported by a retrospective study by Akpancar et al. [69].

27.3.3 Cell-Based Therapy

The only available study evaluating the injection of cells into the arthritic ankle was a case series published by Emadedin et al. [45]: of 18 patients, 6 were treated for ankle OA. They employed an injection of approximately 5×105 cells/kg/bw culture-expanded MSCs in a 10 mL volume. The study concluded that the procedure was safe and had therapeutic potential, although the sample was extremely limited and uncontrolled.

27.4 Conclusions

Injection therapies have become increasingly common both in the hip and ankle in particular for the treatment of OA, and some preliminary results also showed promising findings for the injective treatment also of OLTs. The number of papers published about this topic is increasing, along with the availability of new products. Though none of the available options demonstrated to be capable of reversing OA, they may have beneficial effects on symptoms and functionality and may contribute to delay the need for major surgeries.

HA is available as an "off-the-shelf" product and due to its safety and relatively low cost. In some ways, HA is the gold standard of injection therapies with the largest number of studies reporting positive data. PRP has been one of the hottest topics of the last decade, despite challenges with standardization of preparations and unpredictable efficacy, which requires greater critical attention in future studies.

The use of bone marrow-derived cell populations or the use of culture-expanded cell populations (e.g., MSCs) may provide a valuable weapon in the armamentarium against OA. However, the efficacy of cellular therapies for OA of the ankle has not been proven in the treatment of OA of the hip or ankle.

Higher-quality studies are needed in all aspects of OA care of the hip and ankle, to optimize treatment approaches. This includes rigorous documentation of disease starting and ending states (clinical and radiographic outcomes), as well as rigorous standardization and documentation of the composition of cellular therapies such as PRP, BMAC, or culture-expanded cell populations.

It is important to note that both hip and ankle OA are strongly influenced by rehabilitation. Therefore, the future evaluation of injection therapies in both joints must include the concept of combined therapy, in which physical exercise and physical therapy play also a role.

> **Take-Home Messages**
> - Intra-articular biological approaches can represent a suitable option to reduce symptoms and possibly delay surgery in patients with either hip OA or OLT/OA of the ankle.
> - HA presents a larger number of studies showing symptom reduction at follow-up both in hip and ankle OA. However, some studies show that the efficacy is often similar to placebo, and the most effective MW and injective schedule remain controversial.
> - Intra-articular PRP injections, although providing clinical improvement, did not generally showed superiority over HA in the hip or ankle.
> - BMAC injections seem safe, both in hip and ankle, but available data are still insufficient.

References

1. Filardo G, et al. Non-surgical treatments for the management of early osteoarthritis. Knee Surg Sports Traumatol Arthrosc. 2016;24(6):1775–85.
2. Madry H, et al. Early osteoarthritis of the knee. Knee Surg Sports Traumatol Arthrosc. 2016;24(6):1753–62.
3. Suppan VKL, et al. Randomized controlled trial comparing efficacy of conventional and new single larger dose of intra-articular viscosupplementation in management of knee osteoarthritis. J Orthop Surg. 2017;25:230949901773162.
4. Wu B, Li YM, Liu YC. Efficacy of intra-articular hyaluronic acid injections in hip osteoarthritis: A meta-analysis of randomized controlled trials. Oncotarget. 2017;8:86,865–76.
5. Filardo G, et al. Platelet-rich plasma intra-articular knee injections show no superiority versus viscosupplementation: a randomized controlled trial. Am J Sports Med. 2015;43:1575–82.
6. Filardo G, et al. Leukocyte-poor PRP application for the treatment of knee osteoarthritis. Joints. 2013;1:112–20.
7. Anitua E, Andia I, Ardanza B, Nurden P, Nurden AT. Autologous platelets as a source of proteins for healing and tissue regeneration. Thromb Haemost. 2004;91:4–15.
8. O'Keefe RJ, Crabb ID, Edward Puzas J, Rosier RN. Effects of transforming growth factor-β1 and fibroblast growth factor on DNA synthesis in growth plate chondrocytes are enhanced by insulin-like growth factor-I. J Orthop Res. 1994;12: 299–310.
9. Song SU, et al. Hyaline cartilage regeneration using mixed human chondrocytes and transforming growth factor-β1-producing chondrocytes. Tissue Eng. 2005;11:1516–26.
10. de Girolamo L, et al. Regenerative approaches for the treatment of early OA. Knee Surg Sports Traumatol Arthrosc. 2016;24(6):1826–35.
11. Perdisa F, et al. Adipose-derived mesenchymal stem cells for the treatment of articular cartilage: a systematic review on preclinical and clinical evidence. Stem Cells Int. 2015;2015(597652)
12. Filardo G, Perdisa F, Roffi A, Marcacci M, Kon E. Stem cells in articular cartilage regeneration. J Orthop Surg Res. 2016;11:42.
13. Filardo G, et al. Micro-fragmentation is a valid alternative to cell expansion and enzymatic digestion of adipose tissue for the treatment of knee osteoarthritis: a comparative preclinical study. Knee Surg Sports Traumatol Arthrosc. 2021.
14. Vannini F, et al. Return to Sports After Bone Marrow–Derived Cell Transplantation for Osteochondral Lesions of the Talus. Cartilage. 2017;8:80–7.
15. Buda R, et al. "One-step" bone marrow-derived cells transplantation and joint debridement for osteochondral lesions of the talus in ankle osteoarthritis: clinical and radiological outcomes at 36 months. Arch Orthop Trauma Surg. 2016;136:107–16.
16. Murphy EP, Curtin M, McGoldrick NP, Thong G, Kearns SR. Prospective Evaluation of Intra-Articular Sodium Hyaluronate Injection in the Ankle. J Foot Ankle Surg. 2017;56:327–31.
17. Correa Bellido P, Wadhwani J, Gil Monzo E. Matrix-induced autologous chondrocyte implantation grafting in osteochondral lesions of the talus: Evaluation of cartilage repair using T2 mapping. J Orthop. 2019;16:500–3.
18. Zhang Y, Jordan JM. Epidemiology of osteoarthritis. Clin Geriatr Med. 2010;26:355–69.
19. Migliore A, et al. Efficacy of a single intra-articular HYMOVIS ONE injection for managing symptomatic hip osteoarthritis: A 12-month follow-up retrospective analysis of the ANTIAGE register data. Orthop Res Rev. 2020;12:19–26.
20. Migliore A, et al. The symptomatic effects of intra-articular administration of hylan G-F 20 on osteoarthritis of the hip: Clinical data of 6 months follow-up. Clin Rheumatol. 2006;25:389–93.
21. Battaglia M, et al. Efficacy of ultrasound-guided intra-articular injections of platelet-rich plasma versus hyaluronic acid for hip osteoarthritis. Orthopedics. 2013;36.
22. Migliore A, et al. Comparative, double-blind, controlled study of intra-articular hyaluronic acid (Hyalubrix®) injections versus local anesthetic in osteoarthritis of the hip. Arthritis Res Ther. 2009;11:R183.
23. Spitzer AI, et al. Hylan G-F 20 improves hip osteoarthritis: A prospective, randomized study. Phys Sportsmed. 2010;38:35–47.
24. Qvistgaard E, Christensen R, Torp-Pedersen S, Bliddal H. Intra-articular treatment of hip osteoarthritis: A randomized trial of hyaluronic acid, corticosteroid, and isotonic saline. Osteoarthr Cartil. 2006;14:163–70.
25. Previtali D, et al. The Long-Lasting Effects of "Placebo Injections" in Knee Osteoarthritis: A Meta-Analysis. Cartilage. 2020;18:1947603520906597.
26. De Lucia O, et al. Effectiveness and tolerability of repeated courses of viscosupplementation in symptomatic hip osteoarthritis: a retrospective observational cohort study of high molecular weight vs medium molecular weight hyaluronic acid vs no viscosupplementation. Front Pharmacol. 2019;10:1007.
27. Clementi D, et al. Efficacy of a single intra-articular injection of ultra-high molecular weight hyaluronic acid for hip osteoarthritis: a randomized controlled study. Eur J Orthop Surg Traumatol. 2018;28:915–22.
28. Abate M, Salini V. Efficacy and safety study on a new compound associating low and high molecular weight hyaluronic acid in the treatment of hip osteoarthritis. Int J Immunopathol Pharmacol. 2017;30:89–93.
29. Bekerom MPJ, Rys B, Mulier M. Viscosupplementation in the hip: Evaluation of hyaluronic acid formulations. Arch Orthop Trauma Surg. 2008;128:275–80.
30. Tikiz C, Ünlü Z, Şener A, Efe M, Tüzün Ç. Comparison of the efficacy of lower and higher molecular weight viscosupplementation in the treatment of hip osteoarthritis. Clin Rheumatol. 2005;24:244–50.

31. Brander V, et al. Evaluating the use of intra-articular injections as a treatment for painful hip osteoarthritis: a randomized, double-blind, multicenter, parallel-group study comparing a single 6-mL injection of hylan G-F 20 with saline. Osteoarthr Cartil. 2019;27:59–70.

32. Mauro GL, Scaturro D, Sanfilippo A, Benedetti MG. Intra-articular hyaluronic acid injections for hip osteoarthritis. J Biol Regul Homeost Agents. 2018;32:1303–9.

33. Migliore A, et al. The impact of treatment with hylan G-F 20 on progression to total hip arthroplasty in patients with symptomatic hip OA: a retrospective study. Curr Med Res Opin. 2012;28:755–60.

34. Sánchez M, Guadilla J, Fiz N, Andia I. Ultrasound-guided platelet-rich plasma injections for the treatment of osteoarthritis of the hip. Rheumatology (Oxford). 2012;51(144–50).

35. Dallari D, et al. Ultrasound-guided injection of platelet-rich plasma and hyaluronic acid, separately and in combination, for hip osteoarthritis. Am J Sports Med. 2016;44:664–71.

36. Di Sante L, et al. Intra-articular hyaluronic acid vs platelet-rich plasma in the treatment of hip osteoarthritis. Med Ultrason. 2016;18:463–8.

37. Ye Y, Zhou X, Mao S, Zhang J, Lin B. Platelet rich plasma versus hyaluronic acid in patients with hip osteoarthritis: a meta-analysis of randomized controlled trials. Int J Surg. 2018;53:279–87.

38. Assirelli E, et al. Effect of two different preparations of platelet-rich plasma on synoviocytes. Knee Surg Sports Traumatol Arthrosc. 2015;23(9):2690–703.

39. Chahla J, et al. Biological therapies for cartilage lesions in the hip: a new horizon. Orthopedics. 2016;39:e715–23.

40. Villanova-López MM, et al. Randomized, double-blind, controlled trial, phase III, to evaluate the use of platelet-rich plasma versus hyaluronic acid in hip coxarthrosis. Rev Esp Cir Ortop Traumatol. 2020;64:134–42.

41. Singh JR, Haffey P, Valimahomed A, Gellhorn AC. The effectiveness of autologous platelet-rich plasma for osteoarthritis of the hip: a retrospective analysis. Pain Med. 2019;20:1611–8.

42. Hauser RA, Orlofsky A. Regenerative injection therapy with whole bone marrow aspirate for degenerative joint disease: a case series. Clin Med Insights Arthritis Musculoskelet Disord. 2013;6:CMAMD.S10951.

43. Rodriguez-Fontan F, Piuzzi NS, Kraeutler MJ, Pascual-Garrido C. Early clinical outcomes of intra-articular injections of bone marrow aspirate concentrate for the treatment of early osteoarthritis of the hip and knee: a cohort study. PM R. 2018;10:1353–9.

44. Mardones R, Jofré CM, Tobar L, Minguell JJ. Mesenchymal stem cell therapy in the treatment of hip osteoarthritis. J Hip Preserv Surg. 2017;4:159–63.

45. Emadedin M, et al. Long-term follow-up of intra-articular injection of autologous mesenchymal stem cells in patients with knee, ankle, or hip osteoarthritis. Arch Iran Med. 2015;18:336–44.

46. Berndt AL, Haty M. Transchondral fractures (osteochondritis dissecans) of the talus. J Bone Joint Surg Am. 2004;86:1336.

47. Verhagen RAW, Struijs PAA, Bossuyt PMM, Van Dijk CN. Systematic review of treatment strategies for osteochondral defects of the talar dome. Foot Ankle Clin. 2003;8:233–42.

48. Dekker TJ, Dekker PK, Tainter DM, Easley ME, Adams SB. Treatment of Osteochondral lesions of the talus: A critical analysis review. JBJS Rev. 2017;5.

49. Pettine KA, Morrey BF. Osteochondral fractures of the talus. A long-term follow-up. J Bone Joint Surg Br. 1987;69:89–92.

50. Bauer M, Jonsson K, Linden B. Osteochondritis dissecans of the ankle. A 20-year follow-up study. J Bone Joint Surg Br. 1987;69:93–6.

51. Giannini S. Surgical treatment of osteochondral lesions of the talus in young active patients. J Bone Joint Surg. 2005;87:28.

52. Giannini S, et al. The treatment of severe posttraumatic arthritis of the ankle joint. J Bone Joint Surg Am. 2007;89:15–28.

53. Boffa A, et al. Evidence on ankle injections for osteochondral lesions and osteoarthritis: a systematic review and meta-analysis. Int Orthop. 2021;45(2):509–23.

54. Salk RS, Chang TJ, D'Costa WF, Soomekh DJ, Grogan KA. Sodium hyaluronate in the treatment of osteoarthritis of the ankle: a controlled, randomized, double-blind pilot study. J Bone Joint Surg Am. 2006;88:295–302.

55. DeGroot H, Uzunishvili S, Weir R, Al-omari A, Gomes B. Intra-articular injection of hyaluronic acid is not superior to saline solution injection for ankle arthritis: a randomized, double-blind, placebo-controlled study. J Bone Joint Surg Am. 2012;94:2–8.

56. Sun SF, et al. Efficacy of intra-articular hyaluronic acid in patients with osteoarthritis of the ankle: a prospective study. Osteoarthr Cartil. 2006;14:867–74.

57. Cohen MM, et al. Safety and efficacy of intra-articular sodium hyaluronate (Hyalgan®) in a randomized, double-blind study for osteoarthritis of the ankle. Foot Ankle Int. 2008;29:657–63.

58. Han SH, Park DY, Kim TH. Prognostic factors after intra-articular hyaluronic acid injection in ankle osteoarthritis. Yonsei Med J. 2014;55:1080–6.

59. Lucas y Hernandez J, Darcel V, Chauveaux D, Laffenêtre O. Viscosupplementation of the ankle: a prospective study with an average follow-up of 45.5 months. Orthop Traumatol Surg Res. 2013;99:593–9.

60. Bossert M, et al. Imaging guidance improves the results of viscosupplementation with HANOX-M-XL in patients with ankle osteoarthritis: Results of a clinical survey in 50 patients treated in daily practice. Clin Med Insights Arthritis Musculoskelet Disord. 2016;9:195–9.

61. Younger ASE, et al. Nonanimal hyaluronic acid for the treatment of ankle osteoarthritis: a prospective,

single-arm cohort study. J Foot Ankle Surg. 2019;58:514–8.

62. Sun SF, et al. Efficacy of intraarticular botulinum toxin A and intraarticular hyaluronate plus rehabilitation exercise in patients with unilateral ankle osteoarthritis: a randomized controlled trial. J Foot Ankle Res. 2014;7:9.

63. Vannini F, et al. Sport and early osteoarthritis: the role of sport in aetiology, progression and treatment of knee osteoarthritis. Knee Surg Sports Traumatol Arthrosc. 2016;24:1786–96.

64. Mei-Dan O, et al. Treatment of osteochondritis dissecans of the ankle with hyaluronic acid injections: a prospective study. Foot Ankle Int. 2008;29:1171–8.

65. Angthong C, Khadsongkram A, Angthong W. Outcomes and quality of life after platelet-rich plasma therapy in patients with recalcitrant hindfoot and ankle diseases: a preliminary report of 12 patients. J Foot Ankle Surg. 2013;52:475–80.

66. Repetto I, Biti B, Cerruti P, Trentini R, Felli L. Conservative treatment of ankle osteoarthritis: can platelet-rich plasma effectively postpone surgery? J Foot Ankle Surg. 2017;56:362–5.

67. Fukawa T, et al. Safety and efficacy of intra-articular injection of platelet-rich plasma in patients with ankle osteoarthritis. Foot Ankle Int. 2017;38:596–604.

68. Mei-Dan O, et al. Platelet-rich plasma or hyaluronate in the management of osteochondral lesions of the talus. Am J Sports Med. 2012;40:534–41.

69. Akpancar S, Gül D. Comparison of platelet rich plasma and prolotherapy in the management of osteochondral lesions of the talus: a retrospective cohort study. Med Sci Monit. 2019;25:5640–7.

Injectable Orthobiologics for the Treatment of Subchondral Insufficiency Fractures of the Knee (SIFK) and Related Pathogenic Processes

28

Kyle N. Kunze, Zaamin B. Hussain, Mikel Sánchez, and Jorge Chahla

28.1 Introduction

Osteonecrosis of the knee is a debilitating disease that can progress to end-stage osteoarthritis (OA). Spontaneous osteonecrosis of the knee (SONK or SPONK) is a focal, superficial subchondral lesion. Subchondral bone has always been present in the pathogenesis of OA, and for more than 40 years, it has been considered an effective shock absorber, suggesting a cause-and-effect connection between mechanical stress, subchondral bone sclerosis, and OA [1, 2]. SONK is composed of one of three categories used to sub-stratify osteonecrosis, the other two being secondary and post-arthroscopic. The etiology and pathogenesis of SONK have been poorly characterized; however, in a recent systematic review by Hussain et al. [3], the authors suggested SONK may not be a spontaneous pathological process, but rather the result of subchondral insufficiency fractures that lead to the development of osteonecrosis. Indeed, this group has advocated for a change in nomenclature to the term "subchondral insufficiency fractures of the knee" (SIFK). The precise etiology is unknown, and the pathological mechanisms involved remain poorly characterized, as evidenced by cases of young asymptomatic athletes with bone marrow lesions [4, 5]. Therefore, making effective management presents a considerable challenge.

One of the most challenging aspects of treating SIFK and its sequelae is the anatomical location of the lesion and limited options available to access and treat the subchondral bone in a minimally invasive manner. A more recent field of thought is to employ orthobiologic agents in order to mitigate further SIFK progression and the risk of osteonecrosis development. Currently used orthobiologics for this pathology include calcium phosphate and its derivatives, platelet-rich plasma (PRP), bone marrow aspirate concentrate (BMAC), and rheumatoid biologic agents. Despite the array of orthobiologic agents being used as methods to manage SIFK, there

K. N. Kunze
Department of Orthopaedic Surgery, Hospital for Special Surgery, New York, NY, USA

Z. B. Hussain
Harvard Graduate School of Education, Harvard University, Cambridge, MA, USA

Department of Orthopaedic Surgery, Emory University School of Medicine, Atlanta, GA, USA

M. Sánchez
UNIDAD DE CIRUGÍA ARTROSCÓPICA, Hospital Vithas San José, Gasteiz, Araba, Spain
e-mail: mikel.sanchez@ucatrauma.com

J. Chahla (✉)
Division of Sports Medicine, Department of Orthopaedic Surgery, Rush University Medical Center, Chicago, IL, USA

© ISAKOS 2022
G. Filardo et al. (eds.), *Orthobiologics*, https://doi.org/10.1007/978-3-030-84744-9_28

remains uncertainty regarding the efficacy of these treatments.

The purpose of the current chapter is to provide a comprehensive review of current injectable orthobiologic treatments for the treatment of SIFK and related lesions of the knee. This chapter will provide an update on clinical studies employing such treatments and summarize the evidence supporting or opposing their use. It is the goal of this chapter to help delineate the current state of orthobiologic treatment of SIFK and in doing so identify areas where future research is needed.

28.2 Etiology and Pathogenesis

28.2.1 Pathophysiology

The etiology and pathogenesis of SONK and post-arthroscopic osteonecrosis remain highly controversial, and many theories have been proposed. Historically, SONK has been thought to occur secondary to ischemia, ultimately resulting in necrosis. However, a more recent theory has been popularized which proposes that this process is better described as subchondral insufficiency fractures in osteopenic bone without evidence of spontaneous necrosis [6]. Histopathological studies have supported the insufficiency fracture theory for the development of SONK [7–10]. Yamamoto and Bullough [7] first described subchondral insufficiency fractures as the primary events leading to SONK based on gross and histological findings in 14 patients with SONK. Additionally, Tanaka et al. [9] characterized articular bone plate insufficiency fractures with endochondral ossification, reactive cartilage tissue formation, and proliferation of fibroneurovascular tissue, further supporting a pathological process. In addition, mesenchymal stem/stromal cells are essential during bone remodeling. Several studies have shown a high recruitment of these cells in bone marrow lesions, although their proliferation and mineralization are decreased with aberrant and senescent forms, and thus their repair effect is compromised [11]. The applica-

tion of certain treatments for articular pathologies must also be considered as possible cause of osteonecrosis, such as the prolonged use of corticosteroids [12]. The current understanding of the etiology and pathogenesis outlined in Fig. 28.1 shows that insufficiency fractures were hypothesized to lead to fluid accumulation in the bone marrow, resulting in subsequent edema with focal ischemia and eventual necrosis.

Within the physiopathology of osteonecrosis, subchondral pain is noteworthy. Subchondral bone is a highly vascularized and innervated tissue with heat receptors, chemoreceptors, and mechanoreceptors. Therefore, nociceptive stimuli coming from a harmful microenvironment, because of non-physiological mechanical load, can be a source of pain. Molecular patterns asso-

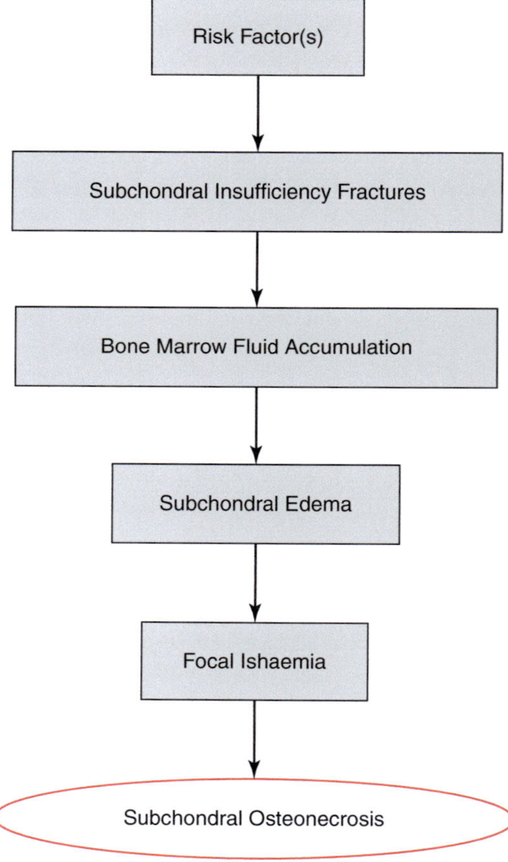

Fig. 28.1 Proposed pathogenesis of SONK/SIFK

ciated with damage could initially lead to peripheral pain and eventually both peripheral and neuropathic pains by mechanisms not yet fully identified. In addition, pro-inflammatory cytokines present in this microenvironment may contribute to pain by stimulating hyperalgesia and sensitizing nociceptors to other stimuli [13].

28.2.2 Role of Demographic Risk Factors

Several demographic risk factors are thought to contribute to the development of SONK, with the most influential variables being female sex [14–16] and increasing age [15, 16]. A recent systematic review of the SONK literature found that the range of mean ages of patients with SONK was 33–73 years of age (mean 60.4 years), with a strong predilection for older patients [3].

28.2.3 Role of Anatomic Risk Factors

Anatomic risk factors are thought to play a substantial role in the development and propagation of SONK, including load axis alterations [17], cartilage degeneration [18, 19], low bone mineral density (BMD) [18, 20, 21], and medial meniscus posterior root tears [22, 23]. Interestingly, there is variable evidence surrounding the BMD theory. Akamatsu et al. [20] reported the possible association of low BMD with SONK development; however, it was suggested this was not likely to be an etiological mechanism by Nelson et al. [24].

Many studies have suggested an association between meniscal tears and SONK, most commonly medial meniscus and posterior meniscal root tears [22–26]. The prevalence of meniscal tears in SONK patients has been found to occur in the range of 50–100% of patients, suggesting a strong association between meniscal lesions and SONK [22, 24–32].

Increased contact pressure secondary to meniscal pathology is one mechanism by which insufficiency fractures could develop as the hoop mechanism of the menisci facilitate load distribution, force absorption, lubrication, and stabilization. Given that the medial meniscus has less inherent mobility due to its robust attachment to the medial tibial plateau [33, 34] and has a role in transmitting force during weight bearing, it is more vulnerable to tearing and degeneration. However, it is notable that in patients with normal alignment, the lateral meniscus transmits significant forces as well which may also lead to insufficiency fractures. Additionally, the posterior horn of the meniscus has been shown to carry more load than the anterior horn through its strong, bony insertion [35]. Therefore, disruption of the posterior medial meniscus root can result in a loss of hoop tension, increased contact pressures, and subsequent alteration of normal knee biomechanics, bearing similarities to the biomechanics seen in total meniscectomy [36]. Additionally, medial meniscal root tears have been hypothesized to increase the peak pressure of the femoral condyle more than horizontal tears of the posterior horn [23]. The resulting increased contact pressures would provide a plausible mechanism for the insufficiency fracture theory as proposed above.

28.2.4 Role of Procedural Risk Factors

Several studies [10, 37–43] have reported the development of SONK after arthroscopic meniscectomy, thereby suggesting a possible role in the etiology and pathogenesis previously mentioned. This hypothesis is further validated by studies reporting that a knee without a medial meniscus experiences twice the peak pressures with loading as compared to a knee with an intact meniscus [44]. Therefore, the increased tibiofemoral contact pressures after meniscectomy could lead to subchondral insufficiency fractures from altered load transmission. Convincingly, Turker et al. [43] reported MRI changes suggestive of SONK in 75 patients after arthroscopic meniscectomy. Other arthroscopic procedures, including arthroscopic laser or radiofrequency treatment [45] and ACL reconstruction [46], have also been implicated in the etiology of SONK.

28.3 Injectable Orthobiologic Treatments

Currently used joint preserving treatments for SIFK and related pathological processes are those introduced and discussed in previous chapters: calcium phosphate, PRP, BMAC, and, rarely, the off-label use of biologics developed for other conditions such as rheumatoid arthritis. Notably, though the current chapter focuses specifically on the use of these adjuncts for treatment in the knee, recent investigations have demonstrated efficacy in other joints such as the hip and ankle [47–50].

28.3.1 Calcium Phosphate

Calcium phosphate is considered an agent that may modulate osteogenesis and influence the local cytokine milieu [51]. Indeed, calcium phosphate has previously demonstrated the ability to stimulate bone formation in defects induced in basic science models via VEG-F and BMP-7 [52, 53]. The majority of studies which have described the use of subchondroplasty report using calcium phosphate injections in areas of bone marrow edema as it is thought that this edema is caused by subchondral bone attrition and microfractures of the trabecular bone (Fig. 28.1). With such injections, it is theorized that autologous osteoclasts and osteoblasts remodel around the injection sites to reconstitute the structural integrity of the subchondral bone [54].

Liu et al. [55] described a technique for placement of AccuFill (Zimmer Knee Creations, Exton, PA) in patients with pain correlating to bone marrow lesions identified on MRI. The authors described correlating the location of the bone marrow lesion on sagittal MRI cuts with its position on intraoperative fluoroscopy of the knee. This process was repeated with coronal MRI cuts, and the position was correlated on these views with the position of the lesion on an anteroposterior radiograph. Horizontal and vertical lines were drawn based on the positioning coordinates above, and the intersection of these lines marked where the cannula will be inserted.

Through the cannula, a wire driver was drilled into the bone, and localization was again confirmed. Once confirmed, the authors remove the inner stylus and insert a bone substitute after which radiographs are taken to confirm complete filling of the lesion.

Despite the publication of few studies which have reported on the safety and efficacy of these injections [56, 57], in general there is a paucity of prospective and long-term data which demonstrates its success. Chua et al. [58] retrospectively studied the efficacy of subchondroplasty with calcium pyrophosphate in a small series of 12 patients with bone marrow lesions of the knee. At 1-year follow-up, they reported statistically significant improvements in the mean visual analog scale (VAS) for pain, Western Ontario and McMaster Universities Osteoarthritis Index (WOMAC), and Knee Injury and Arthritis Outcome Scores (KOOS). Astur et al. [59] conducted a literature review in which they identified a total of 164 patients with bone marrow lesions of the knee who underwent subchondroplasty. This group reported that all studies reported improvements in pain and function, although 25% of patients still reported some type of pain complaint. This group also reported that within 2 years there was a 70% reduction in conversion to total knee arthroplasty (TKA) in patients with previous surgical indication who choose calcium phosphate treatment. Multiple other studies have reported functional improvements, pain reduction, and significant reductions in TKA conversion rates in patients with bone marrow lesions treated with subchondroplasty [60, 61].

28.3.2 Bone Marrow Aspirate Concentrate (BMAC)

The administration of BMAC into areas of bone marrow edema and SIFK-related pathology has also been investigated. Hernigou et al. [62] performed a randomized controlled study to evaluate the administration of BMAC (95,000 ± 25,000 cells from the iliac crest) to extra-articular osteonecrotic areas of the femoral condyle using conventional fluoroscopy or computer navigation in

12 patients (24 knees). This group found that computer navigation allowed for a more ideal position of the trocar in terms of tip-to-subchondral distance and center position. At a minimum of 5 years follow-up, they reported that only one patient in the computer navigation group experienced collapse, while four in the conventional group experienced collapse. Furthermore, the computer navigation group experienced better overall repair volume for the lesions on average (11.4 vs. 4.2 cm^3).

Another randomized controlled trial by Hernigou et al. [63]. sought to investigate the efficacy of TKA versus BMAC for patients ($n = 30$, 60 knees) with bilateral secondary osteonecrosis of the knee related to corticosteroids. For each patient, one knee was randomized to undergo TKA and the other BMAC injection. For knees that received BMAC, an average injection of 6500 cells/mL was administered into the subchondral bone of the femur and tibia. At a mean 12 years follow-up, they found a significantly higher incidence of complications and need for subsequent surgery (6 vs. 1) in the TKA group. There were no significant differences in the mean Knee Society Score (KSS) between the groups; however, knees that had received BMAC showed improvements in bone marrow lesions at the site of subchondral injection. Specifically, after treatment with bone marrow injection, femorotibial compartment bone marrow lesion volume demonstrated regression over 24 months and 5 years (respectively, mean 3.5 cm^3, range 1.2–5.3 cm^3; mean 4.1 cm^3, range 3.2–6.2 cm^3), representing a decrease in size by an average of 40%. Among the 30 patients, 21 preferred the knee with cell therapy, while only 9 preferred the knee with TKA ($P < 0.05$).

Kasik et al. [64] reported the outcomes of 20 patients who underwent injections of BMAC with demineralized bone matrix for the treatment of symptomatic areas of bone marrow edema of the knee. At a mean 14.5-month follow-up, this group reported significant improvements in VAS pain scores, as well as the International Knee Documentation Committee (IKDC) score and SF-12 physical and mental scores. Interestingly, MRI at latest follow-up revealed that 75% of the bone marrow lesions were completely healed.

28.3.3 Platelet-Rich Plasma (PRP)

The use of PRP has also been investigated in the treatment of SIFK given the variable success in the treatment of various muscle, tendon, and cartilage pathologies. In particular, intraosseous injections may benefit these patients in localizing their effect to the pathological areas as opposed to intra-articular injections (Fig. 28.2).

Sanchez et al. [65] described combining intra-articular PRP injections with subchondral PRP injections to the medial femoral condyle and medial tibial plateau in 14 patients, allowing the action of PRP directly in this structure (Fig. 28.3). This group reported significant improvements on average in the KOOS and VAS pain scores 6 months after this regimen. The same authors in another observational study reported a significant improvement in pain and function in patients treated with this type of administration at 1-year follow-up [66]. Su et al. [67] performed a randomized controlled trial to compare the efficacy of intra-articular and subchondral PRP injections with intra-articular PRP injections alone and hyaluronic acid injections alone. This group reported that the combination of intra-articular and subchondral PRP injections resulted in significantly better VAS, WOMAC, and quality of life measures at 18 months compared to the other treatments. In addition, this technique has proved to be effective on patients with bone marrow lesions. These patients showed good quality of life improvement, significant pain reduction, and essential MRI changes at 1-year follow-up [68].

28.3.4 Other Biological Approaches

Few reports of the off-label use of biologics developed for the management of rheumatoid arthritis and other disease processes have been published. These studies have investigated the use of such biologics in treating SIFK and related lesions. It is the thought that iloprost and its analogues may help induce bone regeneration on the cellular level in local areas of injection [69]. Administration of adalimumab, a TNF-α inhibitor, may reach subchondral bone by penetrating

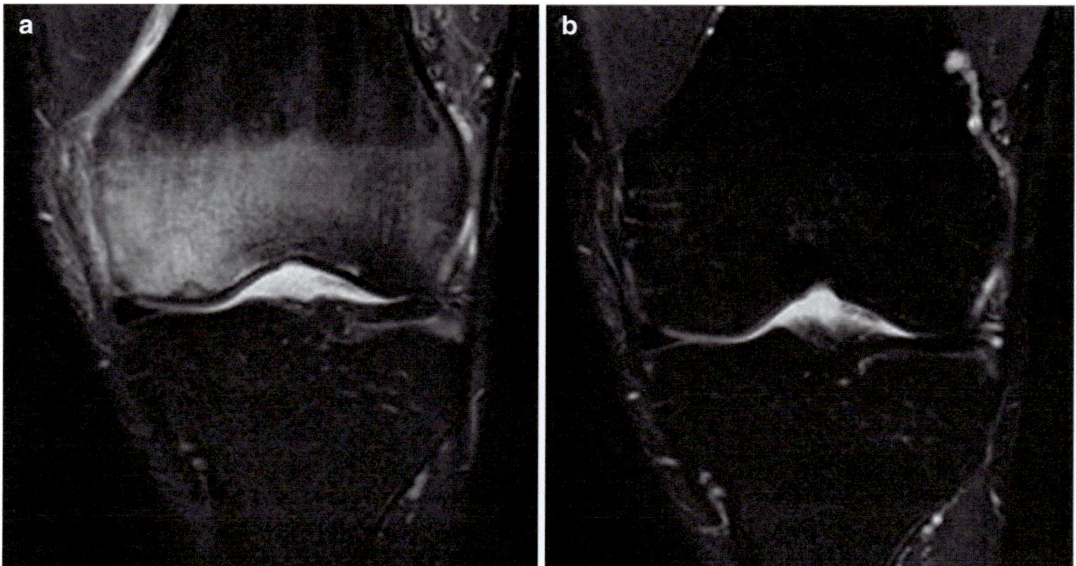

Fig. 28.2 Potential benefits of intraosseous PRP. (**a**) MRI of femoral condyle osteonecrosis. (**b**) MRI after treatment with intraosseous platelet-rich plasma

cartilage through pump mechanisms or through vascular supply after being resorbed by synovium, thereby decreasing the osteoclastic proliferation and recruitment through its inhibition of various cytokines.

Meizer et al. [70] investigated the efficacy of injectable iloprost (prostaglandin I_2) for the treatment of idiopathic and post-traumatic bone marrow edema in 104 patients. Only 50 of these patients had bone marrow edema of the knee. The therapy protocol consisted of a series of five iloprost infusions with either 20, 25, or 50 micrograms of iloprost given in normal saline over 6 h on five consecutive days. Although patients were not stratified by affected joint, at 4 months post-treatment, they found that 73% of patients reported reductions in pain and that 65% of patients had reductions of BME size of complete normalization on MRI examination. Other groups have also reported on the efficacy of this treatment for early-stage osteonecrosis or bone marrow edema, reporting significant improvements in pain, function, and size of the lesion on MRI [69].

One case report described the efficacy of 40 mg of intra-articular adalimumab for osteonecrosis of the knee in a 54-year-old patient with rheumatoid arthritis [71]. Within 1 week of treatment, the patient reported complete pain relief and resolution of knee stiffness. Two weeks following the first treatment, the patient underwent another injection. At 1 month following the first injection, MRI demonstrated almost complete resolution of the osteonecrosis.

Another novel treatment that is showing promising results in preclinical phases is the inhibition of TGF-β. Cellular populations from sclerotic subchondral bone have an elevated TGF-β expression that promotes the excessive synthesis of proangiogenic factors such as VEGF, FGF, and IL-8. The excessive presence of TGF-β and VEGF in subchondral bone could generate changes in osteoblast-osteoclast coupling, bone remodeling imbalance, NGF expression, and fibroneurovascular growth leading to cartilage degradation, pain, and an osteoarthritic joint [13]. Zhen et al. transected the ACL of mice to generate a destabilized OA model with a lower volume of subchondral bone in comparison to sham-operated controls. The authors found that, by inhibiting TGF-β signaling in a specific population of mesenchymal stem/stromal cells (MSCs) present at the subchondral bone, joint degeneration was attenuated [72].

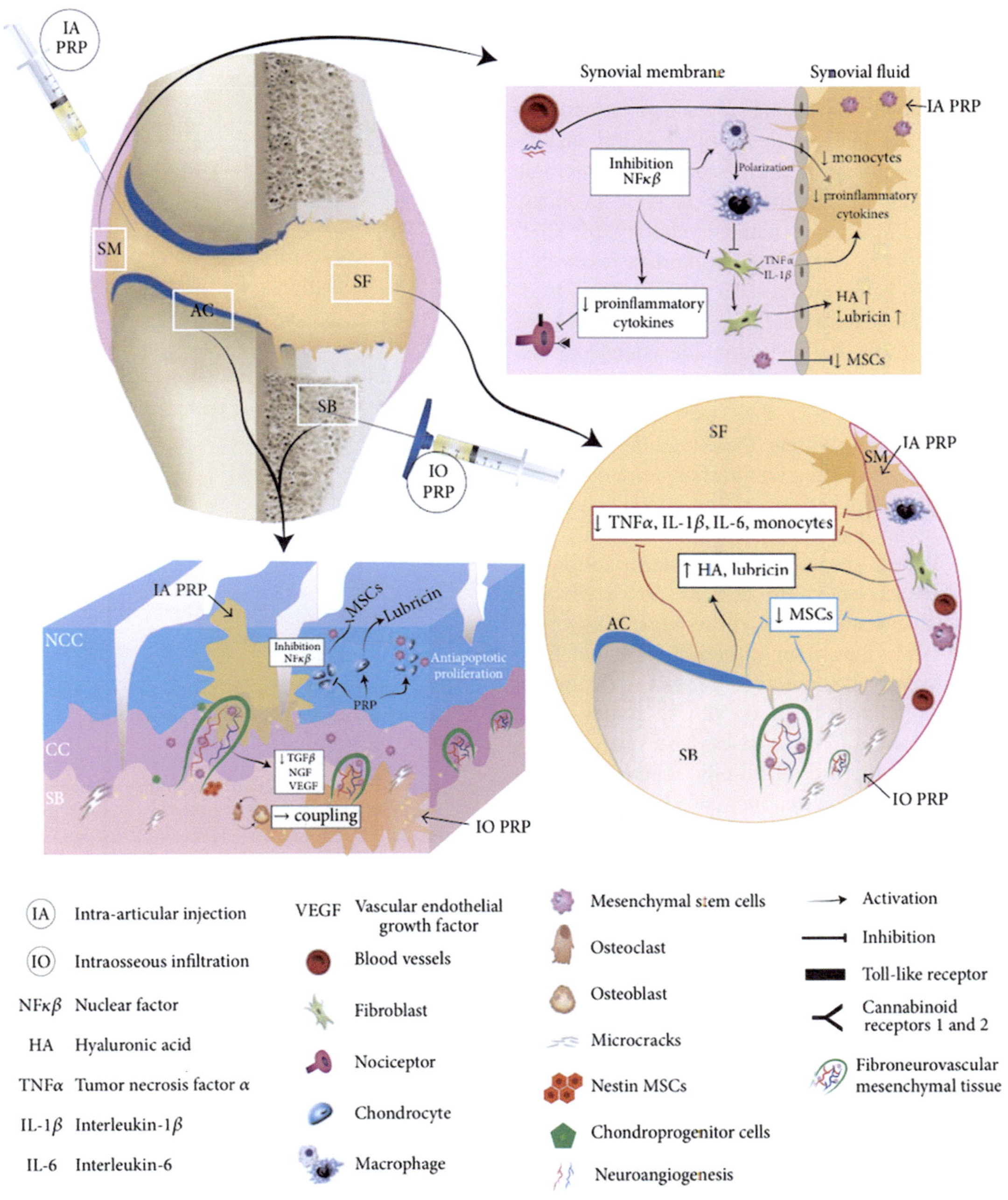

Fig. 28.3 Proposed mechanisms of intra-articular and intraosseous PRP injections. Various growth factors, cytokines, and chemokines trapped in a fibrin network of PRP might inhibit NFkB on synovial macrophages, fibroblasts, and chondrocytes. This, in turn, dampens the inflammatory response of SM and AC. Furthermore, IO may buffer excess TGF-B1 and restore HGF on osteoblasts, leading to a new homeostatic balance between TGF-B1 and HGF. The buffer effect of PRP may reduce nestin MSC concentration in SF and serve as an anti-fibrotic mechanism

SM Synovial membrane, *SF* Synovial fluid, *AC* Articular cartilage, *NCC* Noncalcified cartilage, *SB* Subchondral bone, *IA* Intra-articular, *IO* Intraosseous, *PRP* Platelet-rich plasma, *MSC* Mesenchymal stem/stromal cell, *TGF-B1* Transforming growth factor beta-1, *HGF* Hepatocyte growth factor

28.4 Conclusions

Despite an increasing focus on the etiologies and pathogenesis of avascular disease processes concerning the knee such as subchondral insufficiency fractures, research concerning biologic approaches to treatment is lacking in many domains. The exact pathophysiology of SIFK remains poorly understood, and future research is needed to better characterize the processes leading to these lesions, including questions like bone marrow lesions in asymptomatic young populations. As outlined earlier, there is a paucity of literature describing the use of injectable orthobiologics to treat these pathologies, with few studies reporting on the short-term outcomes of small patient series pertaining to subchondroplasty with calcium phosphate, BMAC, and PRP. Future studies with long-term follow-up and prospective study designs are needed to establish causation for such findings. It may also be possible that other orthobiologics that have not currently been studied for SIFK treatment, such as adipose-derived MSCs, confer some healing benefit in these cases. Like many studies concerning the use of biologic adjuncts, the literature concerning treatment of SIFK also suffers from variable reporting and lack of clarity concerning protocols for formulation and preparation. Future research and standardization of biologic treatment in these cases will be imperative to appropriately compare studies and further develop optimal formulations and standardize timing of administration.

> **Take-Home Messages**
> - Spontaneous osteonecrosis of the knee may be more appropriately described as a finding derived from the cumulative effects of subchondral insufficiency fractures.
> - More efficacious outcomes for patients with this pathology may be observed with treatments targeted at reversing or halting subchondral insufficiency fractures.

> - Current orthobiologic therapies include platelet-rich plasma (PRP), bone marrow aspirate concentrate (BMAC), calcium phosphate mixtures, and anti-rheumatologic agents.
> - Early evidence from few randomized controlled trials have demonstrated improved subjective outcomes, function, and improved appearance of bone marrow lesions.
> - Further studies with longer-term follow-up and more homogenous populations are needed to better understand the indications and efficacy of orthobiologics for subchondral insufficiency fractures and spontaneous osteonecrosis of the knee.

References

1. Radin EL, Paul IL, Rose RM. Role of mechanical factors in pathogenesis of primary osteoarthritis. Lancet. 1972;1(7749):519–22.
2. Radin EL, Rose RM. Role of subchondral bone in the initiation and progression of cartilage damage. Clin Orthop Relat Res. 1986;213:34–40.
3. Hussain ZB, Chahla J, Mandelbaum BR, Gomoll AH, LaPrade RF. The role of meniscal tears in spontaneous osteonecrosis of the knee: a systematic review of suspected etiology and a call to revisit nomenclature. Am J Sports Med. 2019;47(2):501–7.
4. Pappas GP, Vogelsong MA, Staroswiecki E, Gold GE, Safran MR. Magnetic resonance imaging of asymptomatic knees in collegiate basketball players: the effect of one season of play. Clin J Sport Med. 2016;26(6):483–9.
5. Matiotti SB, Soder RB, Becker RG, Santos FS, Baldisserotto M. MRI of the knees in asymptomatic adolescent soccer players: a case-control study. J Magn Reson Imaging. 2017;45(1):59–65.
6. Karim AR, Cherian JJ, Jauregui JJ, Pierce T, Mont MA. Osteonecrosis of the knee: review. Ann Transl Med. 2015;3(1):6.
7. Yamamoto T, Bullough PG. Spontaneous osteonecrosis of the knee: the result of subchondral insufficiency fracture. J Bone Joint Surg Am. 2000;82(6):858–66.
8. Mears SC, McCarthy EF, Jones LC, Hungerford DS, Mont MA. Characterization and pathological characteristics of spontaneous osteonecrosis of the knee. Iowa Orthop J. 2009;29:38–42.
9. Tanaka Y, Mima H, Yonetani Y, Shiozaki Y, Nakamura N, Horibe S. Histological evaluation of spontane-

ous osteonecrosis of the medial femoral condyle and short-term clinical results of osteochondral autografting: a case series. Knee. 2009;16(2):130–5.

10. Higuchi H, Kobayashi Y, Kobayashi A, Hatayama K, Kimura M. Histologic analysis of postmeniscectomy osteonecrosis. Am J Orthop (Belle Mead, NJ). 2013;42(5):220–2.

11. Delgado D, Garate A, Vincent H, Bilbao AM, Patel R, Fiz N, et al. Current concepts in intraosseous platelet-rich plasma injections for knee osteoarthritis. J Clin Orthop Trauma. 2019;10(1):36–41.

12. Weinstein RS. Glucocorticoid-induced osteonecrosis. Endocrine. 2012;41(2):183–90.

13. Sanchez M, Anitua E, Delgado D, Sanchez P, Prado R, Goiriena JJ, et al. A new strategy to tackle severe knee osteoarthritis: combination of intra-articular and intraosseous injections of platelet rich plasma. Expert Opin Biol Ther. 2016;16(5):627–43.

14. Aglietti P, Insall JN, Buzzi R, Deschamps G. Idiopathic osteonecrosis of the knee. Aetiology, prognosis and treatment. J Bone Joint Surg. 1983;65(5):588–97.

15. Ahlback S, Bauer GC, Bohne WH. Spontaneous osteonecrosis of the knee. Arthritis Rheum. 1968;11(6):705–33.

16. Akamatsu Y, Kobayashi H, Kusayama Y, Aratake M, Kumagai K, Saito T. Predictive factors for the progression of spontaneous osteonecrosis of the knee. Knee Surg Sports Traumatol Arthrosc. 2017;25(2):477–84.

17. Felson DT, McLaughlin S, Goggins J, LaValley MP, Gale ME, Totterman S, et al. Bone marrow edema and its relation to progression of knee osteoarthritis. Ann Intern Med. 2003;139(5 Pt 1):330–6.

18. Houpt JB, Pritzker KP, Alpert B, Greyson ND, Gross AE. Natural history of spontaneous osteonecrosis of the knee (SONK): a review. Semin Arthritis Rheum. 1983;13(2):212–27.

19. Zywiel MG, McGrath MS, Seyler TM, Marker DR, Bonutti PM, Mont MA. Osteonecrosis of the knee: a review of three disorders. Orthop Clin North Am. 2009;40(2):193–211.

20. Akamatsu Y, Mitsugi N, Hayashi T, Kobayashi H, Saito T. Low bone mineral density is associated with the onset of spontaneous osteonecrosis of the knee. Acta Orthop. 2012;83(3):249–55.

21. Zanetti M, Romero J, Dambacher MA, Hodler J. Osteonecrosis diagnosed on MR images of the knee. Relationship to reduced bone mineral density determined by high resolution peripheral quantitative CT. Acta Radiol. 2003;44(5):525–31.

22. Robertson DD, Armfield DR, Towers JD, Irrgang JJ, Maloney WJ, Harner CD. Meniscal root injury and spontaneous osteonecrosis of the knee: an observation. J Bone Joint Surg. 2009;91(2):190–5.

23. Sung JH, Ha JK, Lee DW, Seo WY, Kim JG. Meniscal extrusion and spontaneous osteonecrosis with root tear of medial meniscus: comparison with horizontal tear. Arthroscopy. 2013;29(4):726–32.

24. Nelson FR, Craig J, Francois H, Azuh O, Oyetakin-White P, King B. Subchondral insufficiency fractures and spontaneous osteonecrosis of the knee may not be related to osteoporosis. Arch Osteoporos. 2014;9:194.

25. Yamagami R, Taketomi S, Inui H, Tahara K, Tanaka S. The role of medial meniscus posterior root tear and proximal tibial morphology in the development of spontaneous osteonecrosis and osteoarthritis of the knee. Knee. 2017;24(2):390–5.

26. Yao L, Stanczak J, Boutin RD. Presumptive subarticular stress reactions of the knee: MRI detection and association with meniscal tear patterns. Skelet Radiol. 2004;33(5):260–4.

27. Chambers C, Craig JG, Zvirbulis R, Nelson F. Spontaneous osteonecrosis of knee after arthroscopy is not necessarily related to the procedure. American journal of orthopedics (Belle Mead, NJ). 2015;44(6):E184–9.

28. Norman A, Baker ND. Spontaneous osteonecrosis of the knee and medial meniscal tears. Radiology. 1978;129(3):653–6.

29. Plett SK, Hackney LA, Heilmeier U, Nardo L, Yu A, Zhang CA, et al. Femoral condyle insufficiency fractures: associated clinical and morphological findings and impact on outcome. Skelet Radiol. 2015;44(12):1785–94.

30. Ramnath RR, Kattapuram SV. MR appearance of SONK-like subchondral abnormalities in the adult knee: SONK redefined. Skelet Radiol. 2004;33(10):575–81.

31. Valenti Nin JR, Leyes M, Schweitzer D. Spontaneous osteonecrosis of the knee. Treatment and evolution. Knee Surg Sports Traumatol Arthrosc. 1998;6(1):12–5.

32. Yasuda T, Ota S, Fujita S, Onishi E, Iwaki K, Yamamoto H. Association between medial meniscus extrusion and spontaneous osteonecrosis of the knee. Int J Rheum Dis. 2017;

33. Bhatia S, LaPrade CM, Ellman MB, LaPrade RF. Meniscal root tears: significance, diagnosis, and treatment. Am J Sports Med. 2014;42(12):3016–30.

34. LaPrade RF, Engebretsen AH, Ly TV, Johansen S, Wentorf FA, Engebretsen L. The anatomy of the medial part of the knee. J Bone Joint Surg Am. 2007;89(9):2000–10.

35. Kan A, Oshida M, Oshida S, Imada M, Nakagawa T, Okinaga S. Anatomical significance of a posterior horn of medial meniscus: the relationship between its radial tear and cartilage degradation of joint surface. Sports Med Arthrosc Rehabil Ther Technol. 2010;2:1.

36. Allaire R, Muriuki M, Gilbertson L, Harner CD. Biomechanical consequences of a tear of the posterior root of the medial meniscus. Similar to total meniscectomy. J Bone Joint Surg Am. 2008;90(9):1922–31.

37. Muscolo DL, Costa-Paz M, Makino A, Ayerza MA. Osteonecrosis of the knee following arthroscopic meniscectomy in patients over 50-years old. Arthroscopy. 1996;12(3):273–9.

38. Prues-Latour V, Bonvin JC, Fritschy D. Nine cases of osteonecrosis in elderly patients follow-

ing arthroscopic meniscectomy. Knee Surg Sports Traumatol Arthrosc. 1998;6(3):142–7.

39. Brahme SK, Fox JM, Ferkel RD, Friedman MJ, Flannigan BD, Resnick DL. Osteonecrosis of the knee after arthroscopic surgery: diagnosis with MR imaging. Radiology. 1991;178(3):851–3.

40. Johnson TC, Evans JA, Gilley JA, DeLee JC. Osteonecrosis of the knee after arthroscopic surgery for meniscal tears and chondral lesions. Arthroscopy. 2000;16(3):254–61.

41. Kobayashi Y, Kimura M, Higuchi H, Terauchi M, Shirakura K, Takagishi K. Juxta-articular bone marrow signal changes on magnetic resonance imaging following arthroscopic meniscectomy. Arthroscopy. 2002;18(3):238–45.

42. Schmid RB, Wirz D, Gopfert B, Arnold MP, Friederich NF, Hirschmann MT. Intra-operative femoral condylar stress during arthroscopy: an in vivo biomechanical assessment. Knee Surg Sports Traumatol Arthrosc. 2011;19(5):747–52.

43. Turker M, Cetik O, Cirpar M, Durusoy S, Comert B. Postarthroscopy osteonecrosis of the knee. Knee Surg Sports Traumatol Arthrosc. 2015;23(1):246–50.

44. Song Y, Greve JM, Carter DR, Koo S, Giori NJ. Articular cartilage MR imaging and thickness mapping of a loaded knee joint before and after meniscectomy. Osteoarthr Cartil. 2006;14(8):728–37.

45. Bonutti PM, Seyler TM, Delanois RE, McMahon M, McCarthy JC, Mont MA. Osteonecrosis of the knee after laser or radiofrequency-assisted arthroscopy: treatment with minimally invasive knee arthroplasty. J Bone Joint Surg Am. 2006;88(Suppl 3):69–75.

46. Lansdown DA, Shaw J, Allen CR, Ma CB. Osteonecrosis of the knee after anterior cruciate ligament reconstruction: a report of 5 cases. Orthop J Sports Med. 2015;3(3):2325967115576120.

47. Pilge H, Bittersohl B, Schneppendahl J, Hesper T, Zilkens C, Ruppert M, et al. Bone marrow aspirate concentrate in combination with intravenous Iloprost increases bone healing in patients with avascular necrosis of the femoral head: a matched pair analysis. Orthop Rev (Pavia). 2016;8(4):6902.

48. Ghasemi SA, Zhang D, Fragomen A, Rozbruch SR. Subtalar distraction arthroplasty with bone marrow aspirate concentrate (BMAC), preliminary results of a new joint preservation technique. Foot Ankle Surg. 2020;

49. Murphy EP, McGoldrick NP, Curtin M, Kearns SR. A prospective evaluation of bone marrow aspirate concentrate and microfracture in the treatment of osteochondral lesions of the talus. Foot Ankle Surg. 2019;25(4):441–8.

50. Houdek MT, Wyles CC, Collins MS, Howe BM, Terzic A, Behfar A, et al. Stem cells combined with platelet-rich plasma effectively treat corticosteroid-induced osteonecrosis of the hip: a prospective study. Clin Orthop Relat Res. 2018;476(2):388–97.

51. Yuan W, He X, Zhang J, Chen Y, Gong T, Zhu Y. Calcium phosphate silicate and calcium silicate cements suppressing osteoclasts activity through cytokine regulation. J Nanosci Nanotechnol. 2018;18(10):6799–804.

52. Schlickewei C, Klatte TO, Wildermuth Y, Laaff G, Rueger JM, Ruesing J, et al. A bioactive nano-calcium phosphate paste for in-situ transfection of BMP-7 and VEGF-A in a rabbit critical-size bone defect: results of an in vivo study. J Mater Sci Mater Med. 2019;30(2):15.

53. Schlickewei CW, Laaff G, Andresen A, Klatte TO, Rueger JM, Ruesing J, et al. Bone augmentation using a new injectable bone graft substitute by combining calcium phosphate and bisphosphonate as composite--an animal model. J Orthop Surg Res. 2015;10:116.

54. Bajammal SS, Zlowodzki M, Lelwica A, Tornetta P 3rd, Einhorn TA, Buckley R, et al. The use of calcium phosphate bone cement in fracture treatment. A meta-analysis of randomized trials. J Bone Joint Surg Am. 2008;90(6):1186–96.

55. Liu JN, Shields TG, Gowd AK, Amin NH. Surgical treatment of insufficiency fractures of the knee. Arthrosc Tech. 2019;8(11):e1327–e32.

56. Bonadio MB, Giglio PN, Helito CP, Pecora JR, Camanho GL, Demange MK. Subchondroplasty for treating bone marrow lesions in the knee—initial experience. Rev Bras Ortop. 2017;52(3):325–30.

57. Cohen SB, Sharkey PF. Subchondroplasty for treating bone marrow lesions. J Knee Surg. 2016;29(7):555–63.

58. Chua K, Kang JYB, Ng FDJ, Pang HN, DTT L, Silva A, et al. Subchondroplasty for bone marrow lesions in the arthritic knee results in pain relief and improvement in function. J Knee Surg. 2019;

59. Astur DC, de Freitas EV, Cabral PB, Morais CC, Pavei BS, Kaleka CC, et al. Evaluation and management of subchondral calcium phosphate injection technique to treat bone marrow lesion. Cartilage. 2019;10(4):395–401.

60. Byrd JM, Akhavan S, Frank DA. Mid-term outcomes of the subchondroplasty procedure for patients with osteoarthritis and bone marrow edema. Orthopaedic Journal of Sports Medicine. 2017;5(7_suppl6)

61. Astur DC, de Freitas EV, Cabral PB, Morais CC, Pavei BS, Kaleka CC, et al. Evaluation and management of subchondral calcium phosphate injection technique to treat bone marrow lesion. Cartilage. 2018:1947603518770249.

62. Hernigou P, Gerber D, Auregan JC. Knee osteonecrosis: cell therapy with computer-assisted navigation. Surg Technol Int. 2020;36

63. Hernigou P, Auregan JC, Dubory A, Flouzat-Lachaniette CH, Chevallier N, Rouard H. Subchondral stem cell therapy versus contralateral total knee arthroplasty for osteoarthritis following secondary osteonecrosis of the knee. Int Orthop. 2018;42(11):2563–71.

64. Kasik CS, Martinkovich S, Mosier B, Akhavan S. Short-term outcomes for the biologic treatment of bone marrow edema of the knee using bone marrow aspirate concentrate and injectable demineralized bone matrix. Arthrosc Sports Med Rehabil. 2019;

65. Sanchez M, Delgado D, Sanchez P, Muinos-Lopez E, Paiva B, Granero-Molto F, et al. Combination of intra-

articular and intraosseous injections of platelet rich plasma for severe knee osteoarthritis: a pilot study. Biomed Res Int. 2016;2016:4868613.

66. Sanchez M, Delgado D, Pompei O, Perez JC, Sanchez P, Garate A, et al. Treating severe knee osteoarthritis with combination of intra-osseous and intra-articular infiltrations of platelet-rich plasma: an observational study. Cartilage. 2019;10(2):245–53.

67. Su K, Bai Y, Wang J, Zhang H, Liu H, Ma S. Comparison of hyaluronic acid and PRP intra-articular injection with combined intra-articular and intraosseous PRP injections to treat patients with knee osteoarthritis. Clin Rheumatol. 2018;37(5):1341–50.

68. Lychagin A, Lipina M, Garkavi A, Islaieh O, Timashev P, Ashmore K, et al. Intraosseous injections of platelet rich plasma for knee bone marrow lesions treatment: one year follow-up. Int Orthop. 2020;

69. Jager M, Tillmann FP, Thornhill TS, Mahmoudi M, Blondin D, Hetzel GR, et al. Rationale for prostaglandin I2 in bone marrow oedema—from theory to application. Arthritis Res Ther. 2008;10(5):R120.

70. Meizer R, Radda C, Stolz G, Kotsaris S, Petje G, Krasny C, et al. MRI-controlled analysis of 104 patients with painful bone marrow edema in different joint localizations treated with the prostacyclin analogue iloprost. Wien Klin Wochenschr. 2005;117(7–8):278–86.

71. Kobak S. Osteonecrosis and monoarticular rheumatoid arthritis treated with intra-articular adalimumab. Mod Rheumatol. 2008;18(3):290–2.

72. Zhen G, Wen C, Jia X, Li Y, Crane JL, Mears SC, et al. Inhibition of TGF-beta signaling in mesenchymal stem cells of subchondral bone attenuates osteoarthritis. Nat Med. 2013;19(6):704–12.

Injections: Orthobiologics and the Power of Placebo

Davide Previtali, Marco Cuzzolin,
Giorgio Di Laura Frattura, Christian Candrian,
and Giuseppe Filardo

29.1 Introduction: A Historical Note on the "Placebo Effect"

The term "placebo" was defined in the 1811 edition of the *Shorter Oxford Dictionary* as "a medicine given more to please than to benefit the patient" [1]. Since then, the term "placebo" assumed more and more nuanced connotations, until, in the 1950s, the medical community came to use the term "placebo" to indicate a treatment able to induce physiological or physical effects thanks to a therapeutic suggestion [1]. The "placebo effect" was thus defined as "a change in a patient's symptoms that is the result of the therapeutic intent and not the specific physicochemical nature of a medical procedure" [2]. The medical community also quickly realized the importance of taking into account the placebo effect in clinical trials [1, 3]. In 1954, Gold postulated that, to demonstrate the "pharmacological potency" of a new drug, it must prove to be more effective than an inert drug "of such physical properties as to render a distinction between the two impossible." In the same seminal paper, Gold also stated that the pharmacological nature of the two "treatments" should be unknown to both patient and doctor; Gold's guidelines later became the basis of what today is known as a "double-blind placebo-controlled trial" [4]. The placebo effect was quantified for the first time in 1955 in the seminal study carried out by Beecher, who demonstrated that more than 35% of patients enrolled in randomized controlled trials improved after having received a pharmacologically "ineffective" pill [3]. The thousands of similar trials that were performed in the years after Beecher's study led to an increased awareness of the relevance of the placebo effect.

Nonetheless, in the late 1990s, the efficacy of placebos was questioned. Kienle et al. reanalyzed all the studies evaluated in Beecher's original paper, and they concluded that reasons other than the placebo effect could have explained the improvement documented in the trials analyzed by Beecher. Accordingly, the results reported were explainable as a misleading

D. Previtali (✉) · M. Cuzzolin · G. D. L. Frattura
Orthopaedic and Traumatology Unit, Ospedale
Regionale di Lugano, EOC, Lugano, Switzerland
e-mail: davide.previtali@eoc.ch; marco.cuzzolin@eoc.ch; giorgio.dilaurafrattura@eoc.ch

C. Candrian
Orthopaedic and Traumatology Unit, Ospedale
Regionale di Lugano, EOC, Lugano, Switzerland

USI—Università della Svizzera Italiana, Faculty of
Biomedical Sciences, Lugano, Switzerland
e-mail: christian.candrian@eoc.ch

G. Filardo
Rizzoli Orthopaedic Institute
Bologna, Italy

Università della Svizzera Italiana, Ente Ospedaliero
Cantonale, Lugano, Switzerland
e-mail: ortho@gfilardo.com

© ISAKOS 2022
G. Filardo et al. (eds.), *Orthobiologics*, https://doi.org/10.1007/978-3-030-84744-9_29

interpretation of the data [5]. This finding sparked a lively debate. In 2001, Hróbjartsson et al. attempted to quantify the placebo effect by comparing the response in the placebo and in the no-treatment groups of all three-arm (i.e., active treatment-placebo-no-treatment) randomized placebo-controlled trials published to that date. They could not find a placebo effect in objectively measured outcomes or in binary subjectively measured outcomes, and they reported only a marginal effect in continuous subjectively measured outcomes and in the studies that investigated pain [6]. These findings were confirmed in 2004 by the same authors in a stronger analysis, which included 54 more trials than their previous one [7]. However, in 2005 Wampold et al. criticized both Hróbjartsson et al.'s papers arguing that, to evaluate the placebo effect, a better characterization of the studies should have been performed. Starting from data similar to those analyzed by Hróbjartsson et al., they reached different conclusions: they found that for outcomes amenable to psychological effects and evaluated in double-blind studies in which the drug administration was not surreptitious (and thus perceived by the patients), placebo is indeed effective.

The debate on the effectiveness of placebos remains open to this day [8, 9], and the evidence on this topic is continuously growing. Still, while researchers on both sides of the debate recognize the importance of placebo-controlled trials for the evaluation of the effectiveness of new drugs, the reasons for the frequently documented improvement in the placebo arm remain a point of discussion.

29.2 Mechanisms of the Placebo Effect

A modern and often referenced definition of placebo is the one from Shapiro et al.: "[A placebo is] any therapy (or that component of any therapy) that is intentionally or knowingly used for its nonspecific, psychological, or psychophysiological, therapeutic effect, or that is used for a presumed specific therapeutic effect on a patient, symptom, or illness but is without specific activity for the condition being treated" [10]. Several studies have been published to investigate the exact meaning of this "non-specific therapeutic effect," trying to describe the reactions to placebo administration and the biological mechanisms underlying the placebo effect.

It is important to make a distinction between the terms "perceived" placebo effect and "proper" placebo effect [11]. A "perceived" placebo effect occurs when the improvement of the symptoms upon administration of a placebo is not attributable to the placebo itself. For example, when a placebo is administered for a disease whose natural history implies spontaneous healing, any consequent improvements of the symptoms should not be attributed entirely to the placebo. A perceived placebo effect may also occur as a consequence of three other phenomena (or a combination of them): regression to the mean, investigator effect, and Hawthorne effect-like changes in patient behavior. In medicine, the "regression to the mean" concept refers to the fact that patients are more prone to seek medical care—and therefore being involved in a trial—after a re-activation of the disease which then may regress to a better state even without treatment [12]. The "investigator effect" refers to the effect (positive or negative) that the presence of a doctor may have on the patient [13]. The "Hawthorne effect" refers to a change in the behavior of individuals upon becoming aware of being observed [14]: during a clinical trial, patients may change their habits and refrain themselves from potentially dangerous activities or even be prone to try to please the doctor who, in their opinion, is particularly interested in obtaining positive results from the study. Scientists discrediting the placebo effect believe that the documented improvement of patients' symptoms after the administration of placebos is completely explainable as perceived placebo effect [7]; conversely, the advocates of the placebo effect attribute such improvements mostly to the efficacy of the "proper" placebo effect. Regardless of where they stand on this issue, researchers conducting studies featuring placebo trials must be aware of the existence of the perceived placebo effect, since the possibility of its presence complicates the interpretation of the results of the studies trying to identify, quantify, and explain proper placebo effect, as both types of effect may have contributed to the outcomes observed in the study.

The mechanisms underlying the proper placebo effect are still not completely understood, but several experiments demonstrated that patient conditioning, patient expectation, and meaning response can trigger biologic reactions able to influence the clinical outcome. "Classical conditioning" was first described in 1927 by Pavlov, who documented that animals used to receiving morphine anticipated its administration, starting when he was preparing the instrument for morphine injection [15]. A similar reaction was later reported in humans: the ritual of treatment administration, the place where it is administered, and the administrator himself provoke in the individual a positive response based on previous similar experiences which led to a benefit (or in Pavlovian term to a "reward") [16]. However, the significance of classical conditioning is not supported by all researchers, some of whom have advocated that it is the expectation of the patient on what the outcome will be that is the main trigger of the placebo effect in humans [17]. Other researchers suggested that conditioning could be the main trigger of unconscious responses (such as vital sign changes) to placebos, whereas expectation could be more influential on conscious responses, such as pain and symptom improvement [18].

Expectation of benefit is a powerful stimulus, able to induce a benefit by itself [19]. Albeit mostly involved in conscious responses, expectation cannot be exclusively conscious, since it is based on a deep belief that lies in the subconscious of patients [13]. Expectation can be influenced by various factors, such as positive results experienced by other subjects, the exotic nature of some therapies, or the high level of the technology used (such as the use of orthobiologics with "regenerative" potential). As long as this belief is present, the effect of a treatment will be enhanced (or completely determined, in the case of inert substances) by the placebo effect; once this belief wears out, so does the placebo effect. To this regard, the nineteenth-century French physician Armand Trousseau stated: "You should treat as many patients as possible with the new drugs while they still have the power to heal."

Expectation can be strengthened by meaning: if the patients think the treatment is meaningful, they usually experience better results [20]. A famous example of the power of expectation and meaning comes from a study by Blackwell et al., in which pink and blue pills with the same inert content were given to a class of medical students. The students who took a pink pill experienced an excitatory effect, whereas the students who took a blue pill experienced a relaxing effect; moreover, the students who took two pills of the same color had stronger effects compared to the ones who took only one pill [21]. One interpretation of these results is that most students assume pink substances to be stimulants and blue substances to be relaxing and that they assume that two pills contain more active substance than one pill. This hypothesis is further strengthened by the results of studies documenting an increased efficacy of pills with a brand name on them: the presence of a brand gives the pills an air of officiality which makes them be perceived as "powerful" [22].

Conditioning, expectation, and meaning are, then, the triggers of the proper placebo effect. These mechanisms influence the clinical outcome through neural pathways after the administration of a placebo, inducing several biological responses in the individual [23]. In fact, functional brain imaging showed that placebos activate the same areas normally activated during the administration of an active treatment [24, 25]. Moreover, besides the activation of specific brain areas, placebo administration increases the release of active molecules that are directly related to a specific clinical benefit. Among these, during placebo administration dopamine and opioids, whose activity in pain modulation is extensively documented, are augmented in several brain areas involved in pain processing [26, 27]. The increased opioid and dopamine release and activity mediate pain perception after placebo administration and could be the biological mechanism underlying the response to musculoskeletal placebo injections [28].

29.3 Placebo and Musculoskeletal Injections

The literature on the effect of placebos in musculoskeletal injections is vast. A research on PubMed with the string ((musculoskeletal OR bone OR joint OR intra-articular OR knee OR ankle OR shoulder OR hip OR elbow OR wrist OR peri-articular OR muscle OR soft tissues OR tendon) AND (injection) AND (placebo)) leads to the identification of more than 1800 randomized controlled trials. In clinical trials the placebo effect is usually considered a nuisance, a response that the "active" injective treatment should overcome to prove its efficacy. Almost all placebo-controlled trials show an improvement in the placebo injection group, and a recent meta-analysis documented that placebo injections have

a long-lasting and clinically significant benefit on pain, stiffness, and function in patients with knee osteoarthritis. These findings confirm the importance of placebo control groups in the research field of musculoskeletal injections [29] (Fig. 29.1) and even raised questions on ethical legitimacy for the use of placebo in the clinical practice [30]. In particular, the use of placebo injections is reported in the clinical practice of general physicians and orthopedics who advocate that expectancy and conditioning play an important role in improving patient symptoms while reducing the risks of collateral effects of the actual drugs [30, 31]. It is therefore evident that the placebo effect is highly relevant in both the clinical and the research fields of musculoskeletal injections. However, its exact underlying mechanisms are still unclear, and there is a need for an accurate

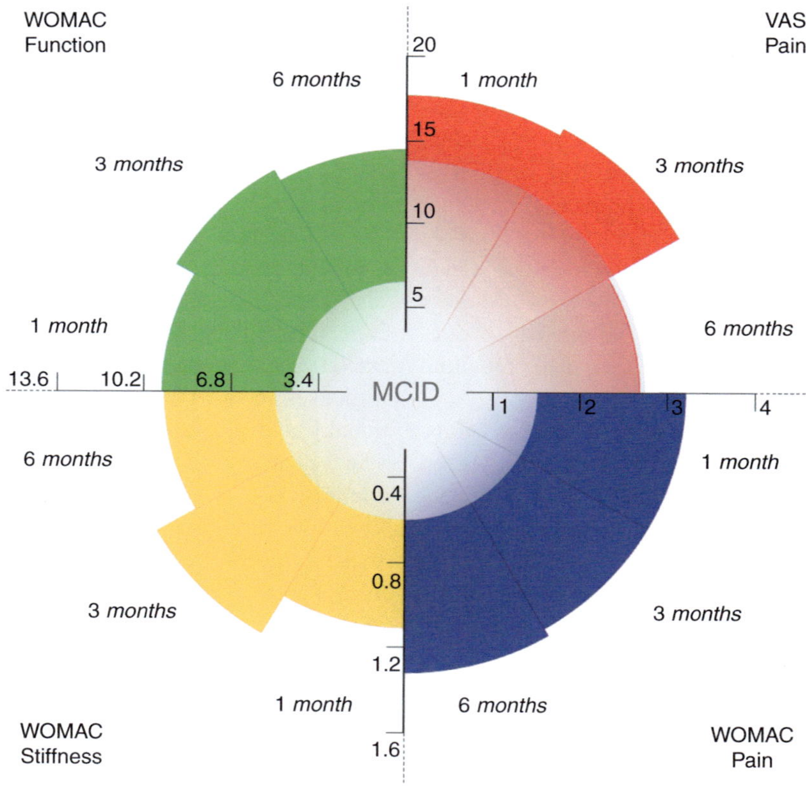

Fig. 29.1 The benefit of placebo injections in knee osteoarthritis. The results are reported in terms of VAS pain (0–100), WOMAC pain (0–20), WOMAC stiffness (0–8), and WOMAC function (0–68) at the different follow-ups (1-month, 3-month, 6-month) and are compared to the minimal clinically important difference (gray area). *MCID* Minimal clinically important difference. From *Previtali D, Merli G, Di Laura Frattura G, Candrian C, Zaffagnini S, Filardo G. The Long-Lasting Effects of "Placebo Injections" in Knee Osteoarthritis: A Meta-Analysis. Cartilage. 2020:1947603520906597* with kind permission from CARTILAGE

analysis of the possible factors that can trigger it. Indeed, when a placebo effect is observed, it may be an instance of perceived placebo effect, of proper placebo effect, or of the injected substance having an unexpected positive effect.

For trials dealing with the injection of placebos to treat musculoskeletal diseases, the perceived placebo effect could represent an important confounder in the quantification of the response. Several of the diseases that are treated with musculoskeletal injections tend to improve spontaneously, meaning that if a patient's condition improves after a placebo injection, the improvement may be due to the natural history of the disease, rather than to a proper placebo effect. Moreover, the regression to the mean phenomenon is also well documented in several chronic musculoskeletal diseases frequently treated with injections, such as osteoarthritis and tendon-related diseases [32, 33]. The investigator effect and the Hawthorne-like effect are more difficult to document, but their influence on the results of placebo injection cannot be excluded. Consequently, care needs to be taken when interpreting the results of studies on the efficacy of placebo injections in musculoskeletal diseases: the surprising benefits documented may be the results of a bias in the interpretation of data and not just a consequence of a proper placebo effect.

Nonetheless, musculoskeletal injections are characteristically prone to induce a placebo effect. In particular, musculoskeletal injections are non-surreptitious (non-surreptitiousness being a necessary condition to observe a proper placebo effect). Indeed, the ritual of the injection (i.e., preparation of the drug and of the syringe, evaluation and disinfection of the injection point) is a strong stimulus for patient conditioning. In this regard, a meta-analysis on the determinants of the placebo effect in osteoarthritis treatments/trials found that placebos administered through injections are the most effective [34]. Expectation of benefit can be a further trigger of a powerful placebo effect in musculoskeletal injections: the perceived high technological level implied in the production of novel orthobiologics, such as platelet-rich plasma or mesenchymal stem/stromal cells, further reinforces the expectations of patients. For example, the fact that famous athletes were recently successfully treated with platelet-rich plasma generated a worldwide hype around platelet-rich plasma, thus inducing regular people treated with the same orthobiologic as famous athletes to expect the same type of result—expectation which probably enhanced their response to the treatment [35, 36]. Meaning might represent another important factor in determining the placebo effect documented for musculoskeletal injections. It is likely that if patients understand the principles behind injections of hyaluronic acid, steroids, and—to an even greater extent—platelet-rich plasma (the compelling anabolic strength of platelet-derived "growth factors"…) or mesenchymal "stem cells" (with their "regenerative potential"…), may fully persuade the patient that these treatments work, thus laying the groundwork for the development of expectation able to induce a strong placebo effect. Thus, in the field of orthobiologics, the hope for tissue regeneration due to the trophic substances, the growth factors, and the powerful cells injected strongly increases the expectation of the patient on the possibility of achieving a benefit. For some applications, aspects other than the substance itself could reinforce the placebo effect even further. For example, patients may be more prone to believe in the effectiveness of a treatment because of a preconceived notion that the specific number and timing of the injections performed are based on a specific scientific rationale. It seems likely, therefore, that the improvement frequently documented in the placebo groups of randomized controlled trials that treat patients with musculoskeletal injections could be explained, at least in part, by a proper placebo effect [29, 37].

The remarkable results of placebo administration made some authors question whether the substances injected were indeed placebos or actually had an unforeseen active effect due to one or more of their components [38]. The evaluation of the presence of an active effect of some components of the substances used as injected placebos is complicated by the fact that more than 70% of the randomized placebo-controlled trials do not report enough information on the characteristics of the injected placebo [39]. Taking this limitation into account, the literature shows that saline is the most common ingredient in placebo injections; less frequently, cell culture media and inactive drug solvents are used instead [34]. The injection of saline inside the joint may dilute the concentration of the inflammatory molecules that characterize the micro-environment of the disease. Moreover, synovial fluid aspiration, which is frequently performed before saline injections, may further reduce the concentration of pro-inflammatory molecules [37]. Micro-environment modifications are known to induce a response after the injection of active drugs, and it is possible that a similar result is obtained after saline injections [40]. However, studies trying to test this hypothesis failed to find a correlation between the volume of saline injected and the strength of the clinical response; the real effectiveness of dilution due to saline injections has not been proved yet [29]. Furthermore, recent evidence from a study on knee osteoarthritis found that saline and sham injections produce similar results [41]. In light of this finding, the hypothesis of the presence of an active effect of saline seems unfounded, and the improvement documented after musculoskeletal injections should be considered entirely due to perceived and/or proper placebo effects.

The relevance of the placebo effect in the field of orthobiologics is evident and should not be underestimated: despite the absence of a specific therapeutic effect of saline injections, "active" treatment groups frequently report results that are similar to those of the control groups; indeed, the benefits of the injection of orthobiologics are often considered to be in great part due to the placebo effect. The literature provides several examples of injective treatments that only slightly overcame their inactive controls, and the placebo effect is often given as the main reason for the response to orthobiologics in fields such as muscle injuries, tendon diseases, or osteoarthritis [42–44]. Moreover, the difference between the active and the placebo groups is frequently reduced in the analyses including only double-blind studies, which are necessary to properly evaluate the influence of the placebo effect [42]. Thus, even though in most cases it remains difficult to distinguish the components and the mechanisms determining the placebo effect, it clearly plays an important role in the field of orthobiologics injections in musculoskeletal tissues.

29.4 Conclusions

Since its identification, the placebo effect has stimulated a great interest in clinical researchers, resulting in a lively debate regarding its true efficacy and in a great scientific production evaluating its underlying mechanisms. Musculoskeletal injections are unavoidably influenced by the placebo effect, with some physicians remaining concerned that many potentially promising injectable therapies are nothing more than expensive placebos. These beliefs are based on the fact that placebo injections lead to results which often equal the efficacy of tested active drugs. When dealing with placebo-controlled trials, attention should be paid to the experimental context; in particular, it is important to document and pay attention to the components of the substance used as placebo and to the potential presence of "proper" or "perceived" placebo effect. However, regardless of the specific placebo components (which may or may not have an active role in the benefit experienced by the patient), the placebo effect is undoubtedly one of the causes for the recent successes of orthobiologics injections, thus warranting high-level trials before extending their application for the treatment of musculoskeletal diseases in the clinical practice.

> **Take-Home Messages**
> - In the published clinical trials, placebo injections are seen to provide clinical benefits that often equal the efficacy of tested "active" drugs.
> - Placebo effect may be an instance of perceived placebo effect, proper placebo effect, or the injected substance having an unexpected positive effect.
> - The placebo effect is undoubtedly one of the causes for the results obtained with

orthobiologics injections, with many physicians still advocating that potentially promising injectable drugs are nothing more than expensive placebos.

- High-level trials are needed to confirm the real effectiveness of orthobiologics before extending their application for the treatment of musculoskeletal diseases in the clinical practice.

References

1. Gaddum JH. Clinical pharmacology. Proc R Soc Med. 1954;47(3):195–204.
2. Liberman R. An analysis of the placebo phenomenon. J Chronic Dis. 1962;15(8):761–83.
3. Beecher HK. The powerful placebo. J Am Med Assoc. 1955;159(17):1602–6.
4. Gold H, Grace W, Ferguson F, Jablons B, Cattell M, Reznikoff P, et al. How to evaluate a new drug. Am J Med. 1954;17(5):722–7.
5. Kienle GS, Kiene H. The powerful placebo effect: fact or fiction? J Clin Epidemiol. 1997;50(12):1311–8.
6. Hróbjartsson A, Gøtzsche PC. Is the placebo powerless? An analysis of clinical trials comparing placebo with no treatment. N Engl J Med. 2001;344(21):1594–602.
7. Hróbjartsson A, Gøtzsche PC. Is the placebo powerless? Update of a systematic review with 52 new randomized trials comparing placebo with no treatment. J Intern Med. 2004;256(2):91–100.
8. Wampold BE, Imel ZE, Minami T. The story of placebo effects in medicine: evidence in context. J Clin Psychol. 2007;63(4):379–90.
9. Hróbjartsson A, Gøtzsche PC. Powerful spin in the conclusion of Wampold et al.'s re-analysis of placebo versus no-treatment trials despite similar results as in original review. J Clin Psychol. 2007;63(4):373–7.
10. Shapiro AK, Shapiro E. The powerful placebo: from ancient priest to modern physician. Baltimore: JHU Press; 2000.
11. Ernst E, Resch KL. Concept of true and perceived placebo effects. BMJ. 1995;311(7004):551–3.
12. McDonald CJ, Mazzuca SA, McCabe GP Jr. How much of the placebo 'effect' is really statistical regression? Stat Med. 1983;2(4):417–27.
13. Mommaerts J, Devroey D. The placebo effect: how the subconscious fits in. Perspect Biol Med. 2012;55(1):43–58.
14. McCambridge J, Witton J, Elbourne DR. Systematic review of the Hawthorne effect: new concepts are needed to study research participation effects. J Clin Epidemiol. 2014;67(3):267–77.
15. Pavlov IP. Conditioned reflex. Feldsher Akush. 1927;11:6–12.
16. Wickramasekera I. A conditioned response model of the placebo effect. Biofeedback Self Regul. 1980;5(1):5–18.
17. Montgomery GH, Kirsch I. Classical conditioning and the placebo effect. Pain. 1997;72(1–2):107–13.
18. Benedetti F, Pollo A, Lopiano L, Lanotte M, Vighetti S, Rainero I. Conscious expectation and unconscious conditioning in analgesic, motor, and hormonal placebo/nocebo responses. J Neurosci. 2003;23(10):4315–23.
19. Kirsch I. Response expectancy as a determinant of experience and behavior. Am Psychol. 1985;40(11):1189.
20. Moerman DE, Jonas WB. Deconstructing the placebo effect and finding the meaning response. Ann Intern Med. 2002;136(6):471–6.
21. Blackwell B, Bloomfield S, Buncher CR. Demonstration to medical students of placebo responses and non-drug factors. Lancet. 1972;299(7763):1279–82.
22. Branthwaite A, Cooper P. Analgesic effects of branding in treatment of headaches. Br Med J (Clin Res Ed). 1981;282(6276):1576–8.
23. Goffaux P, Léonard G, Marchand S, Rainville P. Placebo analgesia. Pharmacology of pain. Seattle: IASP Press; 2010. p. 451–73.
24. Mayberg HS, Silva JA, Brannan SK, Tekell JL, Mahurin RK, McGinnis S, et al. The functional neuroanatomy of the placebo effect. Am J Psychiatr. 2002;159(5):728–37.
25. Wager TD, Rilling JK, Smith EE, Sokolik A, Casey KL, Davidson RJ, et al. Placebo-induced changes in FMRI in the anticipation and experience of pain. Science. 2004;303(5661):1162–7.
26. Scott DJ, Stohler CS, Egnatuk CM, Wang H, Koeppe RA, Zubieta J-K. Placebo and nocebo effects are defined by opposite opioid and dopaminergic responses. Arch Gen Psychiatry. 2008;65(2):220–31.
27. Zubieta J-K, Bueller JA, Jackson LR, Scott DJ, Xu Y, Koeppe RA, et al. Placebo effects mediated by endogenous opioid activity on μ-opioid receptors. J Neurosci. 2005;25(34):7754–62.
28. Wager TD, Scott DJ, Zubieta J-K. Placebo effects on human μ-opioid activity during pain. Proc Natl Acad Sci. 2007;104(26):11056–61.
29. Previtali D, Merli G, Di Laura FG, Candrian C, Zaffagnini S, Filardo G. The long-lasting effects of "placebo injections" in knee osteoarthritis: a meta-analysis. Cartilage. 2020:1947603520906597.
30. Miller FG, Colloca L. The legitimacy of placebo treatments in clinical practice: evidence and ethics. Am J Bioeth. 2009;9(12):39–47.
31. Howick J, Bishop FL, Heneghan C, Wolstenholme J, Stevens S, Hobbs FR, et al. Placebo use in the United Kingdom: results from a national survey of primary care practitioners. PLoS One. 2013;8(3).
32. Collins JE, Katz JN, Dervan EE, Losina E. Trajectories and risk profiles of pain in persons

with radiographic, symptomatic knee osteoarthritis: data from the osteoarthritis initiative. Osteoarthr Cartil. 2014;22(5):622–30.

33. Centeno CJ, Al-Sayegh H, Bashir J, Goodyear S, Freeman MD. A prospective multi-site registry study of a specific protocol of autologous bone marrow concentrate for the treatment of shoulder rotator cuff tears and osteoarthritis. J Pain Res. 2015;8:269.

34. Bannuru RR, McAlindon TE, Sullivan MC, Wong JB, Kent DM, Schmid CH. Effectiveness and implications of alternative placebo treatments: a systematic review and network meta-analysis of osteoarthritis trials. Ann Intern Med. 2015;163(5):365–72.

35. Filardo G, Kon E. PRP: product rich in placebo? Springer; 2016.

36. Rachul C, Rasko JE, Caulfield T. Implicit hype? Representations of platelet rich plasma in the news media. PLoS One. 2017;12(8).

37. Altman RD, Devji T, Bhandari M, Fierlinger A, Niazi F, Christensen R. Clinical benefit of intra-articular saline as a comparator in clinical trials of knee osteoarthritis treatments: a systematic review and meta-analysis of randomized trials. Semin Arthritis Rheum. 2016:151–9.

38. Golomb BA. Paradox of placebo effect. Nature. 1995;375(6532):530.

39. Golomb BA, Erickson LC, Koperski S, Sack D, Enkin M, Howick J. What's in placebos: who knows? Analysis of randomized, controlled trials. Ann Intern Med. 2010;153(8):532–5.

40. Andia I, Maffulli N. Platelet-rich plasma for managing pain and inflammation in osteoarthritis. Nat Rev Rheumatol. 2013;9(12):721–30.

41. Yazici Y, Tambiah J, Swearingen C, Kennedy S, Strand V, Cole B, et al. Comparison of intra-articular sham and vehicle injection from a phase 2b trial of SM04690, a small-molecule Wnt inhibitor, for knee osteoarthritis. Osteoarthr Cartil. 2019;27:S241–S2.

42. Grassi A, Napoli F, Romandini I, Samuelsson K, Zaffagnini S, Candrian C, et al. Is platelet-rich plasma (PRP) effective in the treatment of acute muscle injuries? A systematic review and meta-analysis. Sports Med. 2018;48(4):971–89.

43. C-j L, Yu K-l, J-b B, D-h T, G-l L. Platelet-rich plasma injection for the treatment of chronic Achilles tendinopathy: a meta-analysis. Medicine. 2019;98(16).

44. Filardo G, Di Matteo B, Kon E, Merli G, Marcacci M. Platelet-rich plasma in tendon-related disorders: results and indications. Knee Surg Sports Traumatol Arthrosc. 2018;26(7):1984–99.

Future Directions

Pluripotent Stem Cells: Embryonic/Fetal Stem Cells and Induced Pluripotent Stem Cells

Gun-Il Im

30.1 Introduction

Recent advances in regenerative medicine for the musculoskeletal system indicate that stem cell therapies may evolve into an established treatment for many musculoskeletal tissues. The most challenging and potentially important is the preservation of cartilage tissue, reversal of degenerative change, or regeneration of cartilage tissue in chondral defects and osteoarthritis (OA). While most of reported approaches have been based on adult stem cells that have limited potential for expansion, pluripotent stem cells also merit a consideration as potential cell sources. Embryonic stem cells (ESCs), derived from the inner cell mass of the blastocyst [1, 2], and the cells in the germinal ridge of the embryo [3] had been thought to be the only type of known pluripotent cells, until the seminal work of Yamanaka [4] showed that adult cells can be activated to revert to a pluripotent state, generating induced pluripotent stem cells (iPSCs).

ESCs can be expanded almost infinitely, and not undergoing senescence, as typically seen in adult stem cells. However, the derivation of ESCs from the early embryos raises ethical concern, and the inherent immunogenicity associated with allograft transplantation continues to pose limita-

tions for the clinical applications [1–3]. iPSCs are induced by reprogramming somatic cells with forced expression of specific transcription factors [5, 6]. iPSCs are similar to the ESCs in gene expression, surface markers, cell morphology, proliferation potential, pluripotency, and several other aspects [7, 8]. However, as iPSCs can be derived from autologous somatic cells of candidate recipients, iPSCs do not involve such religious and ethical issues as in ESCs and are free from the risk of immune rejection [9, 10].

In this chapter, iPSC generation, chondrogenic induction of iPSC, and reported in vivo application of iPSCs for cartilage regeneration and OA treatment are presented and reviewed from the author's perspectives.

30.2 Generation of iPSCs

30.2.1 Methods Used to Reprogram Somatic Cells into iPSCs

iPSCs were first generated from dermal fibroblasts in vitro by retroviral transduction and forced expression of genes of four transcription factors Oct-4, Sox-2, Klf-4, and c-Myc also called Yamanaka factors [5, 6]. These four factors were identified by testing numerous combinations of 24 factors that are plentifully expressed in ESCs. These factors reprogram the nuclei of somatic cells to make them pluripotent. These cells generated teratomas which

G.-I. Im (✉)
Integrative Research Institute for Regenerative
Biomedical Engineering, Dongguk University,
Goyang, Republic of Korea

G. Filardo et al. (eds.), *Orthobiologics*, https://doi.org/10.1007/978-3-030-84744-9_30

contained tissues from all three germ layers, when transplanted into immunodeficient mice, indicating their pluripotency [11].

Theoretically, any actively dividing somatic cell type may be reprogrammed. Therefore, iPSCs have been induced from various somatic cells [5, 6, 12–14]. Skin fibroblast has been the most frequently used cells for reprogramming due to efficiency, safety, and minimal morbidity in harvesting cells. On the other hand, the easy accessibility and even lower harvesting invasiveness of blood cells can make them further attractive cellular sources [15]. The efficiency in reprogramming for iPSC generation varies with different somatic cell sources. While reprogramming efficiency is quite important in investigating reprogramming and its mechanism in the laboratory setting, it may not be a critical variable when generating iPSCs for clinical application. In this case, selection of the optimal iPSC clones and documentation of their safety and quality become more important.

In the beginning, iPSCs were induced using either retrovirus [5] or lentivirus [6]. However, these viral vectors are associated with insertional mutagenesis and tumor formation due to the random integration of transgenes [16, 17]. Therefore, non-integrating vectors, such as adenovirus [18] and Sendai virus [19], are currently preferred for the generation of iPSCs. Adenovirus has a low efficiency and kinetics, while the Sendai virus shows fairly efficient transduction. In addition, nonviral methods to generate integration-free iPSCs have been developed, including plasmids [20], recombinant proteins [21, 22], mRNAs [23], episomal vectors [24], and piggybacks [25, 26]. Even small molecule combinations are known to induce iPSCs [27, 28].

30.2.2 Epigenetic Signature of iPSCs

Although the hiPSCs fundamentally share the properties of hESCs, there is evidence that their differentiation to a lineage related to their origin is more facilitated [64], probably due to residual epigenetic memory [7, 29].

iPSCs repeat errors in DNA methylation during the reprogramming process [30], though it has not been proven whether these epigenetic abnormalities are the result of cellular reprogramming itself or of the iPSC induction methods. While this has posed concerns about the safety and stability of the iPSCs, ESCs and iPSCs showed very few consistent differences in the gene expression profiles [31]. Also, the different epigenetic signatures observed initially in the iPSCs dissipate with prolonged passaging, suggesting that cell-specific memory may not be functionally relevant [29]. The origin of primary cells may also influence the reprogramming and differentiation thereafter. Rim et al. [32] reprogrammed hiPSCs from four different types of primary cells such as dermal fibroblasts (DF), peripheral blood mononuclear cells (PBMC), cord blood mononuclear cells (CBMC), and OA fibroblast-like synoviocytes (OAFLS). Established hiPSCs were differentiated into chondrogenic pellets. All told, the relative rank of expression of cartilage-specific markers was CBMC > DF > PBMC > FLS. On the other hand, Nasu et al. [33] generated genetically matched human iPSCs from different origins using bone marrow stromal cells and dermal fibroblasts of the same donor. Global gene expression profile, DNA methylation status, and the chondrogenic and osteogenic differentiation properties of each lineage were analyzed. After cell autonomous and induced differentiation, each iPSC clone exhibited various differentiation properties, which did not correlate with the cell of origin.

30.3 Induction of Chondrogenesis from iPSCs

Several methods to differentiate ESCs/iPSCs toward chondrocytes have been developed. These methods can be interchangeably used for both ESCs and iPSCs. ESCs and iPSCs can be expanded almost indefinitely due to their capacity for self-renewal [33, 34]. With enhanced efficiency and low cost of induction, iPSCs may become a useful cell source for regenerative medicine for chondral defects and OA, provided that safe and reliable ways of producing chondrocytes from these cells are established [12].

To employ the iPSC technology for cartilage regeneration, it is very important to understand the normal developmental processes of chondrocytes. Cartilage formation is regulated by a number of signal transduction pathways that regulate a series of events, including condensation of mesenchymal cells and nodule formation followed by chondrogenic differentiation, the hallmark of which is the expression of Sox-9. Several critical signaling molecules regulate this process, including soluble factors such as transforming growth factor beta (TGF-β), Wnt and cell adhesion molecules, bone morphogenetic proteins (BMPs), and fibroblast growth factors (FGFs). These factors activate essential targets to initiate and maintain chondrocyte phenotypes [35]. Induction of chondrogenesis from iPSCs is not yet standardized, with several different methods showing variable results. Protocols for chondrogenic differentiation of ESCs/iPSCs are grouped into four categories [36].

1. Co-culture with primary chondrocytes either in direct [37] or indirect ways [33, 38].
2. Via embryoid body (EB) formation [39, 40].
3. Through generation of intermediate induced mesenchymal stromal cells (iMSCs) and subsequent differentiation into chondrocytes [41, 42].
4. Direct chondrogenic differentiation using chondro-inductive factors [43, 44].

30.3.1 Co-Culture with Primary Chondrocytes

Co-culture takes advantage of paracrine factors secreted from the chondrocytes that can stimulate the differentiation of iPSCs into chondrocytes [34, 36]. On the other hand, co-culture conditions may increase the risk of contamination by other undesired cells [45]. The direct co-culture [38] has higher risk of contamination compared to the indirect co-culture [33, 37].

This strategy was reported by Wei et al. using human healthy chondrocytes [34] and Qu et al.

using bovine articular chondrocytes [46]. Adding BMPs or other TGF-β family molecules to the culture medium may improve the quality of chondrogenic differentiation [47].

30.3.2 Via Embryoid Body Formation

Chondrogenic differentiation of iPSCs via EB formation is the most commonly used approach to obtain hiPSC-derived chondrocytes [48]. The process involves the formation of EBs, allowing auto-induction (spontaneous differentiation) of MSC-like cells as fibroblastic outgrowths from the EBs, followed by induction of chondrogenic differentiation of the MSCs [39]. The main disadvantages of this method are the potential for unpredictable differentiation, heterogeneous cell populations, and low efficiency. EB's three-dimensional (3D) structure is similar to that in the early post-implantation embryo; thus the cells in the EB can differentiate into cells of three germ layers [49]. However, several groups employed this strategy with some success [50–53].

30.3.3 Through Intermediate iMSC

In this strategy, iPSCs are stimulated exogenously to differentiate into an MSC-like population (iMSC), followed by differentiation to chondrocytes. As this method directly generates MSCs, it can limit the spontaneous differentiation of iPSCs into undesired cell types, even though these iMSCs may be more prone to differentiate into fibro- and hypertrophic cartilage [34, 36].

Zou et al. [54] derived iMSCs from human iPSCs by culturing the iPSCs in MSC differentiation medium containing DMEM-low glucose and 10% fetal bovine serum, followed by serial trypsinization-based passaging. For chondrogenic differentiation, pellets were formed and cultured in chondrogenic medium containing TGF-β3. Similarly, other groups reported other approaches involving direct induction of hMSCs under specific cell culture conditions, followed by chondrogenic differentiation with TGF-β3 [55] or TGF-β3 and BMP2 [56]. Another strategy for generating MSCs from iPSCs utilized specific coatings during cell culture. Liu et al. reported a one-step method to derive MSC-like cells from hiPSCs using plates coated with fibrillar type I collagen. This thin layer of collagen fibrils on the plates successfully stimulated the derivation of MSC-like cells [57].

30.3.4 Direct Differentiation Using Growth Factors

This approach, also known as directed differentiation, is grounded on mimicking the events during the embryo development [17, 48]. Using different mixtures of defined factors at different developmental stages, a defined protocol to direct differentiation of the pluripotent stem cells toward the chondrocytes was reported [44]. Cheng et al. [49] successfully applied an iPSC protocol that had been developed for the direct differentiation of hESCs toward chondrocytes. This protocol involves the use of different growth factors including activin-A, Wnt3a, FGF2, BMP4, neurotrophin-4, and growth differentiation factor 5 (GDF5) in a timed sequence at specific concentrations. This protocol is also applied by Saito et al. to differentiate hiPSCs with similar results [58]. Protocol of Yamashita et al. [59] includes initially differentiating hiPSCs into mesodermal cells and then culturing them in chondrogenic medium containing TGF-β1, BMP2, GDF5, and ascorbic acid. Thereafter, chondro-induced cells are sorted according to type II collagen expression and cultured in 3D. Protocol of Borestrom et al. [14] comprises a 3D pellet pre-differentiation followed by monolayer expansion of chondrogenic progenitors. These progenitors are subsequently cultured in a second chondrogenic 3D pellet and differentiated into chondrocytes using chondrogenic medium containing growth factors.

It is not yet clear which is the best method for deriving chondrocytes from iPSCs as each reported protocol used different iPSC lines derived from different somatic cell types, has different genetic backgrounds, and employed different reprogramming methods [47, 60]. One of the unique characteristics seen in in vitro chondrogenic differentiation of iPSCs is the low expression of hypertrophic markers such as type X collagen and alkaline phosphatase [61–63]. The difference can be attributed to the heavier methylation of promotor sites of hypertrophic genes [63]. Lower expression of hypertrophic markers may mean that chondrocytes derived from iPSCs have stable phenotypes, unlike chondroid cell-derived traditional MSC populations. If true, this may represent a profound advantage for iPSCs as a cell source for the regeneration of articular cartilage.

30.4 The Use of iPSCs for Cartilage Regeneration

iPSCs can be a promising cell source for cartilage tissue engineering because plenty of accessible and autologous cells are available iPSC fabrication. iPSCs bypass ethical concerns and overcome the limited proliferation potential of adult cells such as MSCs or chondrocytes [64]. hiPSC-derived chondrocytes are more similar to juvenile chondrocytes. Cartilage from juveniles has more anabolic activity and is less antigenic than those from adults [65–67]. This reduced antigenicity may imply that cartilaginous tissue derived from a single allogeneic hiPSC or hESC clone could be used for many patients. This, in turn, could allow greater flexibility in iPSC clone selection and control the quality and lower the clinical cost of this regenerative cell therapies.

To apply iPSCs for cartilage repair, efficient and reproducible protocols to differentiate iPSCs toward chondrocytes are necessary. While a number of protocols for chondrogenic differentiation are described, so far there is no general agreement on the best approach to obtain chondrocytes from iPSCs. It is premature to state that iPSCs are better than MSCs for cartilage regeneration. On the other hand, iPSCs can resolve several issues, including cell number, accessibility, engraftment, or phenotype loss with passaging.

A number of in vivo preclinical assessments have been reported testing the use of iPSC-derived cells to treat chondral defects and OA. Zhu et al. [68] investigated the repair of cartilage defects in osteoarthritic rats with hiPSC-derived chondrocytes. 10^6 chondro-induced hiPSCs were injected into chemical OA-induced knees of rats. After 15 weeks transplantation, no immune responses were observed, micro-CT showed improvement of subchondral plate integrity, and histological examinations demonstrated articular cartilage matrix production. Rim et al. tested the repair potential of human iPSC-derived chondrocytes in a rat osteochondral defect model. hiPSC-derived chondrocytes were either implanted as pellets or injected into the joint. Both transplanted chondrogenic pellets and chondrocytes had positive effects in the osteochondral defect rat model. Detection of human proteins in the joints proved that the cells were successfully delivered and retained in the defect [62].

Kotaka et al. [69] investigated the effect of magnetically labeled iPSCs (m-iPSCs) delivered into an osteochondral defect by magnetic field on the repair of articular cartilage. The histologic grading score was significantly better in the treatment group compared to the control group. m-iPSCs maintained pluripotency, and the magnetic delivery system proved useful and safe for cartilage repair using iPSCs. Xu et al. [70] evaluated the use of MSCs derived from hiPSCs for the regeneration of cartilage defects in a rabbit model. Cartilage defects were made in the patellar grooves of New Zealand white rabbits. MSCs were generated from hiPSCs via a step of EB formation. Following flow cytological analysis, the hiPSCs-MSCs were plated onto poly(lactic-co-glycolide) and then transplanted into the cartilage defects in the experimental group. At 6 weeks, cartilage-like tissue was observed in the experimental group but not in the control or scaffold implantation groups. Chijimatsu et al. [56] investigated the feasibility of MSC-like cells originated from iPSCs via neural crest cells (NCCs) for osteochondral repair. Initially, MSC-like cells derived from iPSC-NCCs (iNCCs) were generated and characterized in vitro. When iNCC-derived tissue-engineered constructs were implanted into rat osteochondral defects, the implanted cells remained alive at the implanted site, whereas they failed to repair the defects, with only scarce development of osteochondral tissue in vivo. Our group implanted human iPSCs-derived chondrocytes into immunosuppressed rats. Cartilage was regenerated in the defects created in the articular cartilage of these rats, without any teratoma or tumor formation, suggesting that iPSC-derived chondrocytes are a promising source of cells for transplantation (Fig. 30.1) [63].

It remains to be proven that chondrogenically differentiated hiPSCs can definitely generate articular cartilage that are equal to natural hyaline cartilage in vivo. Further, in order to minimize the risk of teratoma in in vivo implantation, undifferentiated cells should not be left behind after the differentiation of iPSCs into chondrocytes. It should be also remembered that the reprogramming process in iPSCs can add another potential risk of tumor formation not present in ESCs. The efficacy and safety of such transplantation remain to be investigated in larger animal models, to provide a more accurate assessment of the repair capacity of iPSCs [71].

Of note, in vivo studies of iPSC treatment to regenerate cartilage in osteochondral defects or OA have demonstrated the survival and engraftment of implanted or injected chondro-induced iPSCs for several months. This is distinctly different from MSCs which mostly disappear from the joint within 1–3 weeks [62, 63]. While those results are very encouraging and approach the original concept of cell therapy, corroboration of this finding is necessary in large animal models of longer follow-up.

30.5 Strategy for Clinical Application: iPSC Banking

While patient-specific iPSCs are a possibility with huge advantages, individual preparation of iPSCs under good manufacturing practice (GMP) guidelines can be expensive. To tackle this issue, allogeneic clinical iPSC cell line banks should be

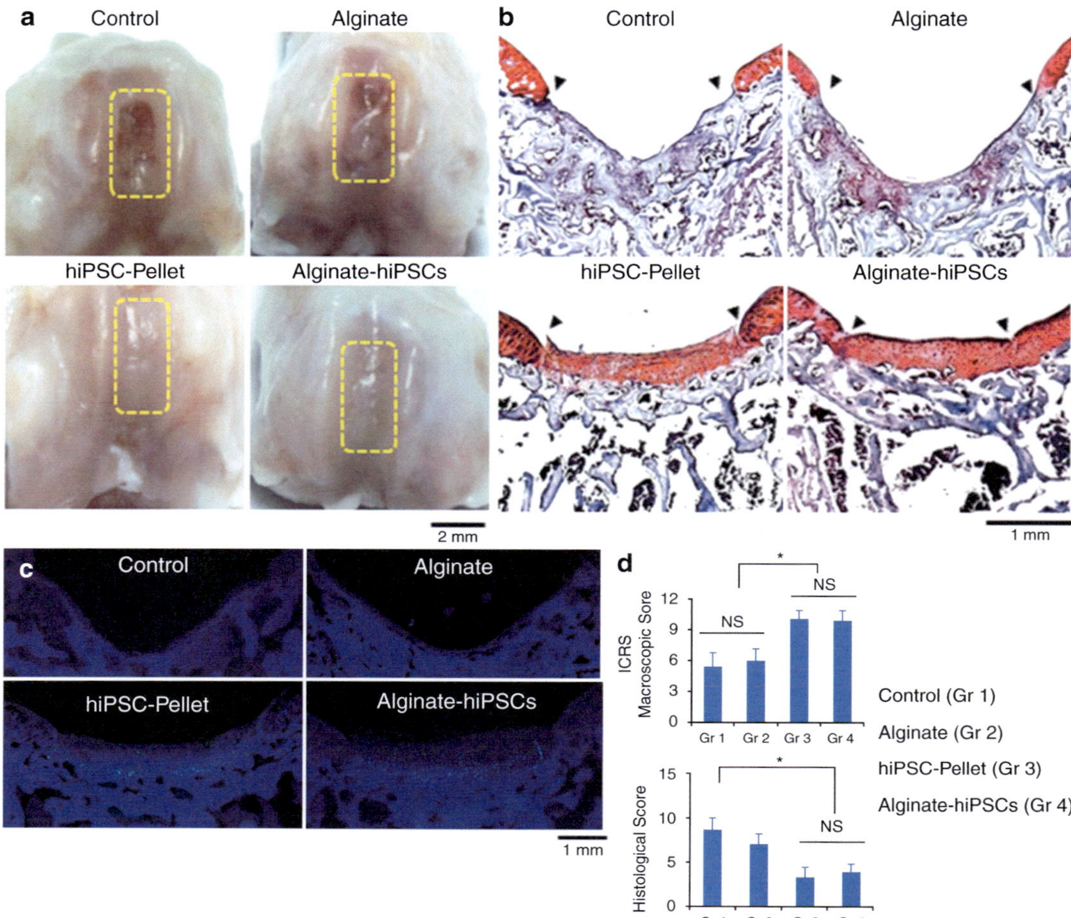

Fig. 30.1 To investigate whether chondro-induced hiPSCs promote cartilage repair, hiPSCs in either pellet state or alginate hydrogel were implanted in the osteochondral defects created on the patellar groove of immunosuppressed rats, and the status of the created defects was observed 12 weeks after implantation. The defects treated with chondro-induced hiPSC implantation were repaired with smooth, glistening, firm tissue, while the control defects showed raw surface with or without thin fibrous covering tissue (**a**). Histological appearance revealed good restoration of the articular surface albeit with reduced amount of proteoglycans compared with adjacent normal cartilage in the defects treated with hiPSC-pellets (group 3) or hiPSC-alginate hydrogel (group 4). On the other hand, the control defects (group 1) and defects treated with alginate alone (group 2) showed little evidence of cartilage regeneration (**b**). Immunohistochemistry for human nuclear antigen in groups 3 and 4 revealed that the majority of cells inside the regenerated cartilage were implanted hiPSCs and that hiPSCs successfully engrafted in the created defect (**c**). The macroscopic score was significantly better in groups 3 and 4 than that in groups 1 and 2 ($P < 0.05$). Groups 3 and 4 also had a significantly better histological score than group 1 ($P < 0.05$, **d**). Bar represents mean ± SE. $N = 6$, *$P < 0.05$ (reproduced with permission from Ko et al., Biomaterials. 2014 Apr;35(11):3571–81)

considered for lowering the cost of iPSC therapy [72, 73]. iPSC banks can be set up with samples from homozygous donors for common HLA types. Chondro-induced iPSCs with an HLA type that matches the patient's HLA types can be selected from the iPSC bank and used for transplantation. It is much less difficult to prepare homozygous HLA hiPSCs than hESCs. Individuals who bear homozygous HLA types and are willing to donate their somatic cells for iPSC generation are far more easily found than embryo donors for ESC generation [72].

The avascular nature of cartilage provides some protection against host immune response to some extent compared with other tissues. It is evidenced from reports that transplanted cartilage from unrelated donors elicits little acute or chronic immune response [64, 74]. Allogeneic cartilage has been transplanted in a large number of patients without matching for HLA types and without the administration of immunosuppressive drugs. The allogeneic transplantation of particulated juvenile articular cartilage has also shown good clinical results [75].

Thus, it may be possible to transplant iPSC bank-derived chondrocytes with less optimal HLA matching compared with other cell types. However, it remains to be seen how much of HLA matching is required for successful allogeneic transplantation of cartilage or chondrocytes [76].

30.6 Direct Conversion to Chondrocytes Without the Need for iPSCs

Somatic cells can be directly converted to another type without going through the generation of iPSCs. Cell type conversion has been demonstrated in some cell types. The transduction of fibroblasts with MyoD results in conversion into myoblasts [77]. Fibroblasts can be converted into neurons by the forced expression of Ascl1, Brn2, and Myt1l [78]. Also, cardiomyocytes can be generated from fibroblasts by forced expression of Gata4, Mef2c, and Tbx5 [79].

The Tsumaki group [80] reported the generation of induced chondrocytes (iChon) from mouse and human fibroblasts using two reprogramming factors (c-Myc and Klf -4) and one cartilage forming factor (Sox-9). The resulting iChon cells form hyaline cartilage expressing only type II collagen. As iChon cells did not express Nanog, a marker of pluripotent cells, these cells would not theoretically cause teratoma. While in vivo direct conversion for the treatment of OA has not been reported until now, the technique is likely to materialize in the near future with recent advancements in in vivo gene transfer and smart biomaterials. In vivo direct conversion might be applied in vivo to rejuvenate diseased chondrocytes that have lost the chondrocyte phenotypes or to convert synovial or connective tissue progenitors resident in the bone marrow into chondrocytes when used in combination with microfracture. In vivo direct conversion could be a relatively non-invasive method for cartilage regeneration if converting vectors can be applied into the OA lesion topical or arthroscopic placement while excluding exposure and potential reprogramming of synovial tissues.

30.7 Conclusions

Starting from the initial report by Yamanaka et al. who used retroviral transduction to reprogram somatic cells into iPSCs, numerous methods were developed to generate iPSCs. The efficiency of induction has also greatly improved, which makes patient-derived iPSCs a clinical possibility with decreased cost. An efficient nonviral induction method would greatly enhance the safety of iPSCs. While the attempts for clinical application of iPSCs started with a focus on retinal or neural disease, it has potential applications in cartilage repair and OA if lower cost and impeccable control over cartilage differentiation and safety can be achieved.

The enhanced survival and engraftment are critical for restoring cartilage form and function. As the prime purpose of cell therapy in OA is the improvement of structure in articular cartilage by regeneration, enhanced survival and engraftment would increase the chance of matrix synthesis and cartilage regeneration by the implanted cells, rather than reliance on paracrine effects targeting endogenous cells. On the other hand, unlike adult stem cells, safety issues in terms of teratoma formation from insufficiently differentiated cells pose risk to the use of iPSCs for a nonlethal disease such as OA.

There is relative lack of in vivo investigation of iPSC implantation compared with culture-expanded MSCs. While comparable or even better results were reported in small animals compared with MSCs, it is not known whether the same good results can be obtained in large animals. The efficacy and safety of such transplantation remain to be investigated in larger animal models, which would allow for a more accurate assessment of the repair capabilities of iPSCs.

> **Take-Home Messages**
> - Pluripotent stem cells merit a consideration as potential cell sources for cartilage regeneration with their advantage in terms of unlimited expansion potential.
> - Despite similar cellular characteristics, iPSCs are preferred to ESCs because iPSCs do not have religious and ethical issues and autogenous iPSCs are free from the risk of immune rejection.
> - iPSCs may become a useful cell source for regenerative therapies for chondral defects and OA, provided that safe and reliable ways of producing iPSC-derived chondrocytes are established.
> - Low expression of hypertrophic markers is seen during in vitro chondrogenic differentiation of iPSCs, suggesting a more stable phenotype in iPSC-derived chondrocytes.
> - iPSCs show enhanced survival and engraftment when implanted in vivo, which may increase the chance of articular cartilage regeneration and durable matrix synthesis.

References

1. Evans MJ, Kaufman MH. Establishment in culture of pluripotential cells from mouse embryos. Nature. 1981;292:154–6.
2. Thomson JA, Itskovitz-Eldor J, Shapiro SS, Waknitz MA, Swiergiel JJ, Marshall VS, et al. Embryonic stem cell lines derived from human blastocysts. Science. 1998;282:1145–7.
3. Nichols J, Smith A. The origin and identity of embryonic stem cells. Development. 2007;138:3–8.
4. Takahashi K, Yamanaka S. Induction of pluripotent stem cells from mouse embryonic and adult fibroblast cultures by defined factors. Cell. 2006;126:663–76.
5. Takahashi K, Tanabe K, Ohnuki M, Narita M, Ichisaka T, Tomoda K, et al. Induction of pluripotent stem cells from adult human fibroblasts by defined factors. Cell. 2007;131:861–72.
6. Yu J, Vodyanik MA, Smuga-Otto K, Antosiewicz-Bourget J, Frane JL, Tian S, et al. Induced pluripotent stem cell lines derived from human somatic cells. Science. 2007;318:1917–20.
7. Kim K, Doi A, Wen B, Ng K, Zhao R, Cahan P, et al. Epigenetic memory in induced pluripotent stem cells. Nature. 2010;467:285–90.
8. Huang C, Jiang S-W. Induced pluripotent stem cells (iPSCs): safe and efficient induction strategies. Chin J Biochem Mol Biol. 2012;28:1005–10.
9. Guha P, Morgan JW, Mostoslavsky G, Rodrigues NP, Boyd AS. Lack of immune response to differentiated cells derived from syngeneic induced pluripotent stem cells. Cell Stem Cell. 2013;12:407–12.
10. Morizane A, Doi D, Kikuchi T, Okita K, Hotta A, Kawasaki T, et al. Direct comparison of autologous and allogeneic transplantation of iPSC-derived neural cells in the brain of a nonhuman primate. Stem Cell Rep. 2013;1:283–92.
11. Okita K, Ichisaka T, Yamanaka S. Generation of germline-competent induced pluripotent stem cells. Nature. 2007;448:313–7.
12. Lee J, Kim Y, Yi H, Diecke S, Kim J, Jung H, et al. Generation of disease-specific induced pluripotent stem cells from patients with rheumatoid arthritis and osteoarthritis. Arthritis Res Ther. 2014;16:R41.
13. Aasen T, Raya A, Barrero MJ, Garreta E, Consiglio A, Gonzalez F, et al. Efficient and rapid generation of induced pluripotent stem cells from human keratinocytes. Nat Biotechnol. 2008;26:1276–84.
14. Borestrom C, Simonsson S, Enochson L, Bigdeli N, Brantsing C, Ellerstrom C, et al. Footprint-free human induced pluripotent stem cells from articular cartilage with redifferentiation capacity: a first step toward a clinical-grade cell source. Stem Cells Transl Med. 2014;3:433–47.
15. Li Y, Liu T, Van Halm-Lutterodt N, Chen J, Su Q, Hai Y. Reprogramming of blood cells into induced pluripotent stem cells as a new cell source for cartilage repair. Stem Cell Res Ther. 2016;7:31.
16. Tapia N, Schöler HR. Molecular obstacles to clinical translation of iPSCs. Cell Stem Cell. 2016;19:298–309.
17. Augustyniak E, Trzeciak T, Richter M, Kaczmarczyk J, Suchorska W. The role of growth factors in stem cell-directed chondrogenesis: a real hope for damaged cartilage regeneration. Int Orthop. 2015;39:995–1003.
18. Stadtfeld M, Nagaya M, Utikal J, Weir G, Hochedlinger K. Induced pluripotent stem cells generated without viral integration. Science. 2008;322:945–9.
19. Ban H, Nishishita N, Fusaki N, Tabata T, Saeki K, Shikamura M, et al. Efficient generation of transgene-free human induced pluripotent stem cells (iPSCs) by temperature-sensitive Sendai virus vectors. Proc Natl Acad Sci U S A. 2011;108:14234–9.

20. Okita K, Nakagawa M, Hong H, Ichisaka T, Yamanaka S. Generation of mouse induced pluripotent stem cells without viral vectors. Science. 2008;322:949–53.
21. Zhou H, Wu S, Joo JY, Zhu S, Han DW, Lin T, et al. Generation of induced pluripotent stem cells using recombinant proteins. Cell Stem Cell. 2009;4:381–4.
22. Kim D, Kim CH, Moon JI, Chung YG, Chang MY, Han BS, et al. Generation of human induced pluripotent stem cells by direct delivery of reprogramming proteins. Cell Stem Cell. 2009;4:472–6.
23. Warren L, Manos PD, Ahfeldt T, Loh YH, Li H, Lau F, et al. Highly efficient reprogramming to pluripotency and directed differentiation of human cells with synthetic modified mRNA. Cell Stem Cell. 2010;7:618–30.
24. Yu J, Hu K, Smuga-Otto K, Tian S, Stewart R, Slukvin II, et al. Human induced pluripotent stem cells free of vector and transgene sequences. Science. 2009;324:797–801.
25. Woltjen K, Michael IP, Mohseni P, Desai R, Mileikovsky M, Hämäläinen R, et al. PiggyBac transposition reprograms fibroblasts to induced pluripotent stem cells. Nature. 2009;458:766–70.
26. Kaji K, Norrby K, Paca A, Mileikovsky M, Mohseni P, Woltjen K. Virus-free induction of pluripotency and subsequent excision of reprogramming factors. Nature. 2009;458:771–5.
27. Long Y, Wang M, Gu H, Xie X. Bromodeoxyuridine promotes full chemical induction of mouse pluripotent stem cells. Cell Res. 2015;25:1171–4.
28. Hou P, Li Y, Zhang X, Liu C, Guan J, Li H, et al. Pluripotent stem cells induced from mouse somatic cells by small-molecule compounds. Science. 2013;341:651–4.
29. Polo JM, Liu S, Figueroa ME, Kulalert W, Eminli S, Tan KY, et al. Cell type of origin influences the molecular and functional properties of mouse induced pluripotent stem cells. Nat Biotechnol. 2010;28:848–56.
30. Ma H, Morey R, O'Neil RC, He Y, Daughtry B, Schultz MD, et al. Abnormalities in human pluripotent cells due to reprogramming mechanisms. Nature. 2014;511:177–83.
31. Guenther MG, Frampton GM, Soldner F, Hockemeyer D, Mitalipova M, Jaenisch R, et al. Chromatin structure and gene expression programs of human embryonic and induced pluripotent stem cells. Cell Stem Cell. 2010;7:249–57.
32. Rim YA, Nam Y, Park N, Jung H, Jang Y, Lee J, et al. Different Chondrogenic potential among human induced pluripotent stem cells from diverse origin primary cells. Stem Cells Int. 2018;2018:9432616.
33. Nasu A, Ikeya M, Yamamoto T, Watanabe A, Jin Y, Matsumoto Y, Hayakawa K, Amano N, Sato S, Osafune K, Aoyama T, Nakamura T, Kato T, Toguchida J. Genetically matched human iPS cells reveal that propensity for cartilage and bone differentiation differs with clones, not cell type of origin. PLoS One. 2013;8(1):e53771.
34. Wei Y, Zeng W, Wan R, Wang J, Zhou Q, Qiu S, et al. Chondrogenic differentiation of induced pluripotent stem cells from osteoarthritic chondrocytes in alginate matrix. Eur Cell Mater. 2012;23:1–12.
35. Matta C, Mobasheri A. Regulation of chondrogenesis by protein kinase C: emerging new roles in calcium signalling. Cell Sign. 2014;26:979–1000.
36. Tsumaki N, Okada M, Yamashita A. iPS cell technologies and cartilage regeneration. Bone. 2015;70:48–54.
37. Bigdeli N, Karlsson C, Strehl R, Concaro S, Hyllner J, Lindahl A. Coculture of human embryonic stem cells and human articular chondrocytes results in significantly altered phenotype and improved chondrogenic differentiation. Stem Cells. 2009;27:1812–21.
38. Hwang NS, Varghese S, Elisseeff J. Derivation of chondrogenically-committed cells from human embryonic cells for cartilage tissue regeneration. PLoS One. 2008;3:e2498.
39. Kawaguchi J, Mee PJ, Smith AG. Osteogenic and chondrogenic differentiation of embryonic stem cells in response to specific growth factors. Bone. 2005;36:758–69.
40. Yang Z, Sui L, Toh WS, Lee EH, Cao T. Stage-dependent effect of TGF-beta1 on chondrogenic differentiation of human embryonic stem cells. Stem Cells Dev. 2009;18:929–40.
41. Lian Q, Lye E, Suan Yeo K, Khia Way Tan E, Salto-Tellez M, Liu TM, et al. Derivation of clinically compliant MSCs from CD105þ, CD24- differentiated human ESCs. Stem Cells. 2007;25:425–36.
42. Drissi H, Gibson JD, Guzzo RM, Xu RH. Derivation and chondrogenic commitment of human embryonic stem cell-derived mesenchymal progenitors. Methods Mol Biol. 2015;1330:65–78.
43. Tanaka H, Murphy CL, Murphy C, Kimur M, Kawai S, Polak JM. Chondrogenic differentiation of murine embryonic stem cells: effects of culture conditions and dexamethasone. J Cell Biochem. 2004;93:454–62.
44. Oldershaw RA, Baxter MA, Lowe ET, Bates N, Grady LM, Soncin F, et al. Directed differentiation of human embryonic stem cells toward chondrocytes. Nat Biotech. 2010;28:1187–94
45. Lietman SA. Induced pluripotent stem cells in cartilage repair. World J Orthop. 2016;7:149–55.
46. Qu C, Puttonen KA, Lindeberg H, Ruponen M, Hovatta O, Koistinaho J, et al. Chondrogenic differentiation of human pluripotent stem cells in chondrocyte co-culture. Int J Biochem Cell Biol. 2013;45:1802–12.
47. Castro-Viñuelas R, Sanjurjo-Rodríguez C, Piñeiro-Ramil M, Hermida-Gómez T, Fuentes-Boquete IM, de Toro-Santos FJ, et al. Induced pluripotent stem cells for cartilage repair: current status and future perspectives. Eur Cell Mater. 2018;36:96–109.
48. Suchorska WM, Augustyniak E, Richter M, Trzeciak T. Comparison of four protocols to generate chondrocyte-like cells from human induced pluripotent stem cells (hiPSCs). Stem Cell Rev. 2017;13:299–308.
49. Cheng A, Kapacee Z, Peng J, Lu S, Lucas RJ, Hardingham TE, et al. Cartilage repair using human embryonic stem cell-derived chondroprogenitors. Stem Cells Transl Med. 2014;3:1287–94.

50. Craft AM, Rockel JS, Nartiss Y, Kandel RA, Alman BA, Keller GM. Generation of articular chondrocytes from human pluripotent stem cells. Nat Biotechnol. 2015;33:638–45.
51. Umeda K, Zhao J, Simmons P, Stanley E, Elefanty A, Nakayama N. Human chondrogenic paraxial mesoderm, directed specification and prospective isolation from pluripotent stem cells. Sci Rep. 2012;2:455–66.
52. Lee J, Taylor SE, Smeriglio P, Lai J, Maloney WJ, Yang F, et al. Early induction of a prechondrogenic population allows efficient generation of stable chondrocytes from human induced pluripotent stem cells. FASEB J. 2015;29:3399–410.
53. Medvedev SP, Grigor'eva EV, Shevchenko AI, Malakhova AA, Dementyeva EV, Shilov AA, et al. Human induced pluripotent stem cells derived from fetal neural stem cells successfully undergo directed differentiation into cartilage. Stem Cells Dev. 2011;20:1099–112.
54. Zou L, Luo Y, Chen M, Wang G, Ding M, Petersen CC, et al. A simple method for deriving functional MSCs and applied for osteogenesis in 3D scaffolds. Sci Rep. 2013;3:2243.
55. Nejadnik H, Diecke S, Lenkov OD, Chapelin F, Donig J, Tong X, et al. Improved approach for chondrogenic differentiation of human induced pluripotent stem cells. Stem Cell Rev. 2015;11:242–53.
56. Chijimatsu R, Ikeya M, Yasui Y, Ikeda Y, Ebina K, Moriguchi Y, et al. Characterization of mesenchymal stem cell-like cells derived from human iPSCs via neural crest development and their application for osteochondral repair. Stem Cells Int. 2017;2017:1960965.
57. Liu Y, Goldberg AJ, Dennis JE, Gronowicz GA, Kuhn LT. One-step derivation of mesenchymal stem cell (MSC)-like cells from human pluripotent stem cells on a fibrillar collagen coating. PLoS One. 2012;7:e33225.
58. Saito T, Yano F, Mori D, Kawata M, Hoshi K, Takato T, et al. Hyaline cartilage formation and tumorigenesis of implanted tissues derived from human induced pluripotent stem cells. Biomed Res. 2015;36:179–86.
59. Yamashita A, Morioka M, Yahara Y, Okada M, Kobayashi T, Kuriyama S, et al. Generation of scaffoldless hyaline cartilaginous tissue from human iPSCs. Stem Cell Reports. 2015;4:404–18.
60. Lach M, Trzeciak T, Richter M, Pawlicz J, Suchorska W. Directed differentiation of induced pluripotent stem cells into chondrogenic lineages for articular cartilage treatment. J Tissue Eng. 2014;30:5:2041731414552701.
61. Diederichs S, Klampfleuthner FAM, Moradi B, Richter W. Chondral differentiation of induced pluripotent stem cells without progression into the endochondral pathway. Front Cell Dev Biol. 2019;7:270.
62. Rim YA, Nam Y, Park N, Lee J, Park SH, Ju JH. Repair potential of nonsurgically delivered induced pluripotent stem cell-derived chondrocytes in a rat osteochondral defect model. J Tissue Eng Regen Med. 2018;12(8):1843–55.
63. Ko JY, Kim KI, Park S, Im GI. In vitro chondrogenesis and in vivo repair of osteochondral defect with human induced pluripotent stem cells. Biomaterials. 2014;35:3571–81.
64. Diekman BO, Christoforou N, Willard VP, Sun HS, Sanchez-Adams J, Leong KW, et al. Cartilage tissue engineering using differentiated and purified induced pluripotent stem cells. Proc Natl Acad Sci U S A. 2012;109:19172–7.
65. Adkisson HD, Milliman C, Zhang X, Mauch K, Maziarz RT, Streeter PR. Immune evasion by neocartilage-derived chondrocytes: implications for biologic repair of joint articular cartilage. Stem Cell Res. 2010;4(1):57–68.
66. AdkssonHD MJA, Amendola RL, Milliman C, Mauch KA, Katwal AB, Seyedin M, Amendola A, Streeter PR, Buckwalter JA. The potential of human allogeneic juvenile chondrocytes for restoration of articular cartilage. Am J Sports Med. 2010;38(7):1324–33.
67. Lee J, Smeriglio P, Chu CR, Bhutani N. Human iPSC-derived chondrocytes mimic juvenile chondrocyte function for the dual advantage of increased proliferation and resistance to IL-1beta. Stem Cell Res Ther. 2017;8(1):244.
68. Zhu Y, Wu X, Liang Y, Gu H, Song K, Zou X, Zhou G. Repair of cartilage defects in osteoarthritis rats with induced pluripotent stem cell derived chondrocytes. BMC Biotechnol. 2016;16(1):78.
69. Kotaka S, Wakitani S, Shimamoto A, Kamei N, Sawa M, Adachi N, Ochi M. Magnetic targeted delivery of induced pluripotent stem cells promotes articular cartilage repair stem. Cells Int. 2017;2017:9514719.
70. Xu X, Shi D, Liu Y, Yao Y, Dai J, Xu Z, et al. In vivo repair of full-thickness cartilage defect with human iPSC-derived mesenchymal progenitor cells in a rabbit model. Exp Ther Med. 2017;14:239–45.
71. Liu H, Yang L, Yu FF, Wang S, Wu C, Cu C, Lammi MJ, Guo X. The potential of induced pluripotent stem cells as a tool to study skeletal dysplasias and cartilage-related pathologic conditions. Osteoarthr Cartil. 2017;25(5):616–24.
72. Okita K, Matsumura Y, Sato Y, Okada A, Morizane A, Okamoto S, et al. A more efficient method to generate integration-free human iPS cells. Nat Methods. 2011;8:409–12.
73. Turner M, Leslie S, Martin NG, Peschanski M, Rao M, Taylor CJ, et al. Toward the development of a global induced pluripotent stem cell library. Cell Stem Cell. 2013;13:382–4.
74. Chesterman PJ, Smith AU. Homotransplantation of articular cartilage and isolated chondrocytes. An experimental study in rabbits. J Bone Joint Surg Br. 1968;50:184–97.
75. Farr J, Tabet SK, Margerrison E, Cole BJ. Clinical, radiographic, and histological outcomes after cartilage repair with particulated juvenile articular cartilage: a 2-year prospective study. Am J Sports Med. 2014;42(6):1417–25.
76. Yamashita A, Tamamura Y, Morioka M, Karagiannis P, Shima N, Nsumaki N. Considerations in hiPSC-

derived cartilage for articular cartilage repair. Inflamm Regen. 2018;38:17.

77. Davis RL, Weintraub H, Lassar AB. Expression of a single transfected cDNA converts fibroblasts to myoblasts. Cell. 1987;51:987–1000.

78. Vierbuchen T, Ostermeier A, Pang ZP, Kokubu Y, Sudhof TC, Wernig M. Direct conversion of fibroblasts to functional neurons by defined factors. Nature. 2010;463:1035–41.

79. Ieda M, Fu JD, Delgado-Olguin P, Vedantham V, Hayashi Y, Bruneau BG, et al. Direct reprogramming of fibroblasts into functional cardiomyocytes by defined factors. Cell. 2010;142:375–86.

80. Outani H, Okada M, Yamashita A, Nakagawa K, Yoshikawa H, Tsumaki N. Direct induction of chondrogenic cells from human dermal fibroblast culture by defined factors. PLoS One. 2013;8:e77365.

Gene Therapy

31

Magali Cucchiarini

31.1 Introduction

Despite the availability of a number of therapeutic options to manage various musculoskeletal disorders in the clinics, none has been able thus far to reliably restore both the original structure and biomechanical functions of the injured tissues in the affected population. Specifically, the treatment of articular cartilage lesions resulting from trauma (focal defects) or in osteoarthritis (OA) by either marrow-stimulating techniques (microfracture), cell/tissue transplantation (autologous chondrocytes, mesenchymal stromal cells), or replacement surgery [1–3] mostly leads to the formation of a fibrocartilaginous tissue composed of type-I collagen with poor mechanical properties instead of the native hyaline cartilage (type-II collagen, proteoglycans) [4] and may further progress to OA [5, 6]. On the other side, bone tissue has an intrinsic ability to heal, at least in part, but not when the fracture gap is too big or unstable, while treatment by either autograft or devitalized cadaveric allograft tissue remains challenging due to limited graft availability and donor site morbidity (autografts) and to a limited integration with the host tissue (allografts) [7]. Injuries affecting connective elastic tendons and ligaments are also common pathologies that represent clinical challenges due to their poor responses to therapy (suture, autografts, allografts, synthetic prostheses), leading to the formation of tissues with incomplete strength and/or restricted mobility and to adhesions, inflammation, and fibrosis [8]. Highly prevalent meniscal lesions such as tears resulting from trauma or related to age-associated degenerative changes as risk factors for OA [9, 10] do not fully regenerate, especially those in the central avascular zone, despite the availability of a variety of meniscal preservation, repair, and reconstruction procedures (arthroscopic partial meniscectomy, sutures, autologous tissues, meniscal allografts, meniscal artificial substitutes) [11–15]. Also, a number of drawbacks were reported with such techniques, including compromised biomechanical properties, lesser vascularization, changes in shape and structure of the repair tissue, immunological/infectious responses upon grafting problems of donor availability, size matching, and graft preservation, extrusion, subchondral bone edema, and in some cases negative MRI results despite improved clinical scores [10].

The gene therapy technology provides powerful tools that enable a stable delivery and expression of therapeutic editing sequences in musculoskeletal lesions over extended periods of time compared with the administration of recombinant factors with short pharmacological half-lives, allowing for a durable healing of damaged tissues [14, 16–19]. Among different approaches allowing to achieve gene transfer in vivo, the

M. Cucchiarini (✉)
Center of Experimental Orthopaedics, Saarland
University Medical Center, Homburg/Saar, Germany

G. Filardo et al. (eds.), *Orthobiologics*, https://doi.org/10.1007/978-3-030-84744-9_31

direct administration of gene-based treatments locally in sites of musculoskeletal injury via injectable procedures is a particularly attractive strategy to improve the current therapeutic options for musculoskeletal disorders. Injectable gene therapy provides convenient (non-invasive) and off-the-shelf (patient-independent) systems that might be directly applied in translational approaches in the clinics to treat affected individuals compared with the complex, indirect (multi-step) administration of genetically modified cells [14, 16–19]. The next chapters provide an overview of gene therapy technologies with a focus on treating musculoskeletal lesions via injectable protocols.

31.2 Gene Therapy: Principles

Gene therapy is based on either (i) the delivery of exogenous candidate gene sequences (transgenes) in target cells or (ii) the endogenous editing of genome sequences directly inside these cells like with the clustered regularly interspaced short palindromic repeats (CRISPR)- associated protein-9 nuclease (Cas9) technology, both requiring the use of a gene shuttle based on non-viral or viral constructs (vectors) (Table 31.1) [16–30]. Such a technology provides tools to treat affected individuals via prolonged expression of a transgene to circumvent the problem of short-term pharmacological half-life of the therapeutic product itself.

Nonviral vectors are safe, avoiding the risk to both acquire replication competence inherent to virus-based constructs and raise immune responses in the host. However, these vectors exhibit low gene transfer (transfection) efficiencies, require cell division (a feature limited to very few cell populations in adults), and allow only for a brief expression of the transgene they carry (only some days) [31, 32], making them more amenable to ex vivo (indirect) gene therapy approaches based on the implantation of genetically modified cells rather than to injection protocols.

Viral vectors that employ the natural entry pathways of viruses within target cells for gene transfer (transduction) are commonly derived from adenoviruses, retro−/lentiviruses, herpes simplex virus (HSV), and adeno-associated virus (AAV). Adenoviral vectors support very high transduction efficiencies, allowing for in vivo (direct) gene therapy approaches and injection protocols, yet they exhibit a high immunogenicity (low safety) and only short-term transgene expression (days to a 1–2 weeks) [33]. HSV-based vectors have similar features to adenoviral vectors, but they mostly target brain cells [34]. Retro−/lentiviral vectors allow for high transduction efficiencies over extended periods of time by integration in the host genome, but they show low transduction efficiencies (making them more suitable for ex vivo gene therapy approaches rather than for injection protocols), require cell division (except for lentiviral vectors that are derived from the pathogenic human immunodeficiency virus—HIV), and might induce unsafe tumor gene activation upon integration [35, 36]. Recombinant AAV (rAAV) vectors are safe due to a complete removal of viral sequences in their genome and to their maintenance as stable unintegrated forms in the target cells, allowing for very high transduction efficiencies and prolonged transgene expression (months to years), making them amenable to in vivo gene therapy approaches

Table 31.1 Gene transfer vectors for human gene therapy

Vector class	Efficacy	Safety	Expression	Approaches	Injection
Nonviral vectors	Low	High	Brief	Ex vivo	Unsuitable
Viral vectors					
Adenoviral vectors	Very high	Low	Brief	In vivo	Adapted
HSV-based vectors	High	Low	Brief	In vivo	Adapted
Retro−/lentiviral vectors	Low	Low	Prolonged	Ex vivo	Unsuitable
rAAV vectors	Very high	High	Prolonged	In vivo	Adapted

HSV Herpes simplex virus, *rAAV* Recombinant adeno-associated virus and injection protocols [37–40]. In light of such properties, the next paragraphs will therefore focus on using vectors suited for injection protocols in musculoskeletal tissues, i.e., adenoviral and rAAV vectors.

31.3 Injectable Gene Therapy for Musculoskeletal Applications

Several experimental gene transfer trials in animal models in vivo have been reported to treat disorders affecting musculoskeletal tissues (cartilage, bone, tendons, ligaments, meniscus) (Fig. 31.1) based on the injection of either adeno-

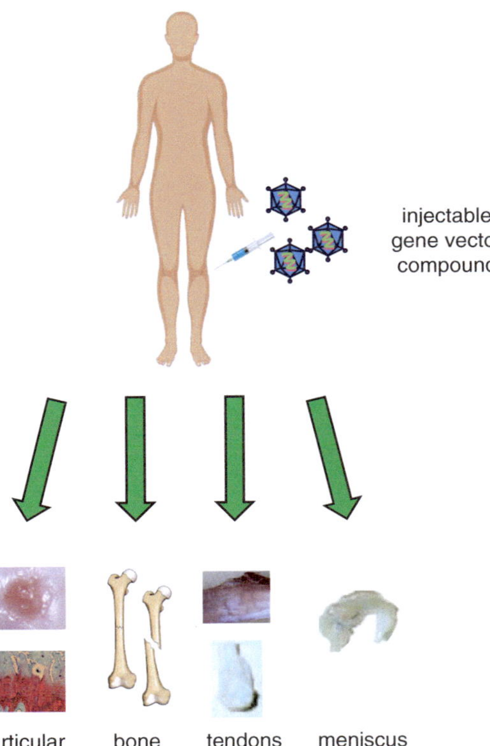

Fig. 31.1 Translational applications of injectable musculoskeletal gene therapy. Candidate gene vectors may be directly provided by non-invasive injection in patients to treat lesions or injuries located in the articular cartilage (focal defects, OA lesions), bone (bone defects, fractures), tendons and ligaments (lesions, ruptures), and meniscus (tears) in order to enhance the mechanisms of tissue healing

viral or rAAV vectors for the delivery and overexpression of a variety of therapeutic gene sequences (Table 31.2).

31.3.1 Cartilage Repair

Cartilage repair has been evidenced upon intra-articular injection of adenoviral vectors coding for:

- An interleukin-1 receptor antagonist (IL-1Ra) alone [41, 42] or combined with a soluble tumor necrosis factor-alpha receptor type I (sTNF-RI) [43] or with the insulin-like growth factor I (IGF-I) [44, 45]
- The transforming growth factor beta (TGF-β) with Smad7 (an anti-fibrotic agent) [46]
- Kallistatin (an anti-inflammatory agent) [47]
- Basic fibroblast growth factor (FGF-2) alone or combined with an IL-1Ra and/or IGF-I [45]
- Thrombospondin-1 (TSP-1, an angiogenesis inhibitor) [48]
- Bone morphogenetic proteins 2 and 6 (BMP-2, BMP-6) [49]
- Pro-opiomelanocortin (POMC, a neuropeptide) [50]
- Dickkopf-1 (Dkk-1, a Wnt antagonist) [51]
- Proteoglycan 4 (Prg4) alone [52] or combined with an IL-1Ra [53]
- Ras homolog enriched in brain (RHEB) [54]
- Histone deacetylase-4 (HDAC4) [55]
- Lysyl oxidase-like 2 (LOXL2, an amine oxidase) [56]

Such approaches were mostly tested in experimental in vivo models of OA in mice [42, 46, 51, 52, 54, 56], rats [47, 48, 50, 53, 55], rabbits [43, 45], and horses [41, 42] but also to treat focal defects in horses [44, 49], leading to reduced cartilage destruction in experimental OA for up to 3 months in mice [56], 3.5 months in rats [53], 3 weeks in rabbits [45], and 72 days in horses [42] and to focal cartilage repair for up to 52 weeks in horses [49].

rAAV vectors have been also manipulated for direct intra-articular injection in experimental OA models in vivo to deliver Dkk-1 in rats [57]

Table 31.2 Injectable gene therapy for musculoskeletal applications

Application	Vectors	Genes	Models	References
Cartilage	Adenoviral vectors	IL-1Ra	OA reduction (mouse, horse)	[41, 42]
		IL-1Ra/sTNF-RI	OA reduction (rabbit)	[43]
		IL-1Ra/IGF-I	Focal repair (horse)	[44]
			OA reduction (rabbit)	[45]
		TGF-β/Smad7	OA reduction (mouse)	[46]
		Kallistatin	OA reduction (rat)	[47]
		FGF-2, FGF-2/IL-1Ra, FGF-2/IL-1Ra/IGF-I	OA reduction (rabbit)	[45]
		TSP-1	OA reduction (rat)	[48]
		BMP-2, BMP-6	Focal repair (horse)	[49]
		POMC	OA reduction (rat)	[50]
		Dkk-1	OA reduction (mouse)	[51]
		Prg4		[52]
		Prg4/IL-1Ra	OA reduction (rat)	[53]
		RHEB	OA reduction (mouse)	[54]
		HDAC	OA reduction (rat)	[55]
		LOXL2	OA reduction (mouse)	[56]
	rAAV vectors	Dkk-1	OA reduction (rat)	[57]
		CRISPR-Cas9 editing of IL-1β or MMP-13	OA reduction (mouse)	[58]
Bone	Adenoviral vectors	BMP-2	Bone healing (rabbit)	[64, 65]
			Bone healing (rat)	[66, 68, 69]
			Bone healing (sheep)	[67]
			Bone healing (horse)	[70]
		TGF-β	Bone healing (rabbit)	[64]
		VEGF		[71]
	rAAV vectors	COX-2	Bone formation (mouse)	[72]
		siRNA (ApoE)		[73]
Tendons, ligaments	Adenoviral vectors	BMP-12	Tendon healing (chicken)	[74]
			Ligament healing (rabbit)	[75]
		GDF-5	Tendon healing (rat)	[76]
	rAAV vectors	FGF-2	Tendon healing (chicken)	[77–79]
		VEGF		[79, 80]

rAAV Recombinant adeno-associated virus, *IL-1Ra* Interleukin-1 receptor antagonist, *sTNF-RI* Soluble tumor necrosis factor-alpha receptor type I, *IGF-I* Insulin-like growth factor I, *TGF-β* Transforming growth factor beta, *FGF-2* Basic fibroblast growth factor, *TSP-1* Thrombospondin-1, *BMP* Bone morphogenetic protein, *POMC* Pro-opiomelanocortin; *Dkk-1* Dickkopf-1, *Prg4* Proteoglycan 4, *RHEB* Ras homolog enriched in brain, *HDAC* Histone deacetylase-4, *LOXL2* Lysyl oxidase-like 2, *CRISPR-Cas9* Clustered regularly interspaced short palindromic repeats, *MMP-13* Matrix metalloproteinase 13, *VEGF* Vascular endothelial growth factor, *COX-2* Cyclooxygenase-2, *siRNA* Small interfering RNA, *ApoE* Apolipoprotein E, *GDF-5* Growth and differentiation factor-5

and the CRISPR-Cas9 gene editing components to target OA-associated genes such as IL-1β and the matrix metalloproteinase 13 (MMP-13) in mice [58], leading to improved bone architecture and decreased osteophytosis for 13 weeks [57] and to reduced cartilage deterioration for 3 months [58]. Regarding focal lesions, most of the work using rAAV-mediated gene transfer to enhance cartilage repair thus far has been reported using application of these vectors via arthrotomy [59–63] and not by intra-articular injection in the joint space.

31.3.2 Bone Healing

Bone healing has been described by injection of adenoviral vectors coding for:

- BMP-2 [64–70]
- TGF-β [64]
- The vascular endothelial growth factor (VEGF) [71]

in experimental in vivo models of bone defects in rats [66, 68, 69], rabbits [64, 65, 71], sheep [67], and horses [70], leading to defect healing for up to 8 weeks in rats [66], 12 weeks in rabbits [64, 71], 8 weeks in sheep [67], and 6 weeks in horses [70].

rAAV vectors were also employed to deliver the cyclooxygenase-2 (COX-2) in a mouse fracture model [72] and a small interfering RNA (siRNA) against apolipoprotein E (ApoE) in mice [73], allowing for fracture union after 21 days [72] and for enhanced bone deposition and mechanical strength for 21 days [73].

31.3.3 Tendon and Ligament Healing

The healing of tendons and ligaments has been evidenced upon injection of adenoviral vectors coding for BMP-12 [74, 75] and the growth and differentiation factor-5 (GDF-5) [76] into lacerated and transected tendons in rats [76] and chickens [74] and in lacerated ligaments in rabbits [75], leading to tissue repair for up to 8 weeks in rats [76], 4 weeks in chickens [74], and 26 weeks in rabbits [75].

rAAV vectors were also employed to directly deliver gene sequences for FGF-2 [77–79] and VEGF [79, 80] in transected tendons in chickens [77–79], allowing for tendon healing for up to 12 weeks [77].

31.3.4 Meniscal Repair

Meniscal repair via direct injection of viral vectors in tissue lesions in experimental models in vivo has not been reported thus far, yet there is evidence in the literature that adenoviral [81] and rAAV vectors [14, 82] may provide strong tools to achieve this goal using, for instance, the delivery of therapeutic hepatocyte growth factor

(HGF, an angiogenic factor) [83], TGF-β [84, 85], FGF-2 [86], or IGF-I [87].

31.4 Perspectives

A variety of experimental, preclinical studies in relevant animal models in vivo currently support the concept of applying injectable gene-based treatments as effective, non-invasive systems to enhance the healing of lesions affecting musculoskeletal tissues like the articular cartilage, bone, tendons, ligaments, and meniscus. Although such approaches have not been tested clinically for the management of focal cartilage defects, bone defects, and fractures, in tendon/ligament lesions and ruptures, and in meniscal tears, interestingly a phase 1 clinical study (NCT02790723) is currently pending to evaluate the safety of injecting an AAV-derived construct carrying an IL-1Ra sequence in subjects with moderate knee OA [88], showing the interest in injectable gene therapy for musculoskeletal disorders.

Overall, more work is still needed to identify the optimal system(s) for safe, effective disease-specific therapy [38, 89, 90], including:

- The vector type (rAAV versus adenoviral vectors; viral serotype—at least 13 serotypes of AAV have been identified)
- The vector dose (vector genome concentrations; single versus repeated injection)
- The therapeutic gene (growth factor, transcription factor, signalling molecule, anti-inflammatory agent, etc.; single gene or gene combination)
- The levels of therapeutic gene expression (high-level expression promoter, tissue-specific promoter, disease-regulatable promoter)
- The follow-up of the outcomes (time course evaluations for durable, not suboptimal effects; safety, i.e., absence of toxicity of the treatment and of deleterious immune reactions).

Also, more laboratory studies will be critical to evaluate the potential benefits of genome editing with the CRISPR-Cas9 system versus classical (trans)gene therapy in the musculoskeletal field as only one study thus far reported that editing of IL-1β and MMP-13 via rAAV vector injection reduced cartilage degradation in experimental OA in mice in vivo [58], while most work was otherwise performed in vitro as a means to rejuvenate chondrocytes and MSCs or to allow for disease (OA) modeling [91–96].

31.5 Conclusion

Overall, the various preclinical evaluations demonstrate the potential value of injectable gene therapy to heal musculoskeletal diseases. Clinical translation in the future will require continued collaborative discussion between both clinicians, scientists and regulatory agencies to address the remaining questions and challenges to enable safe clinical trials.

Take-Home Messages
- Many current clinical interventions fail to reliably heal musculoskeletal lesions with original tissue structure and biomechanical function in patients.
- Gene therapy is a promising tool that supports either transient or prolonged expression of therapeutic and editing sequences in sites of musculoskeletal damage.
- Injectable gene therapy has the potential to provide clinically relevant, non-invasive tools to directly treat lesions affecting the articular cartilage, bones, tendons/ligaments, and meniscus.
- Musculoskeletal lesions have been successfully treated in adapted animal models in vivo following the injection of therapeutic gene vehicles.
- Clinical treatment is currently being explored to manage knee OA in patients via injection of an rAAV therapeutic gene vector coding for an IL-1Ra sequence.

References

1. Brittberg M, Lindahl A, Nilsson A, Ohlsson C, Isaksson O, Peterson L. Treatment of deep cartilage defects in the knee with autologous chondrocyte transplantation. N Engl J Med. 1994;331(14):889–95.
2. Horas U, Pelinkovic D, Herr G, Aigner T, Schnettler R. Autologous chondrocyte implantation and osteochondral cylinder transplantation in cartilage repair of the knee joint. A prospective, comparative trial. J Bone Joint Surg Am. 2003;85(2):185–92.
3. Knutsen G, Engebretsen L, Ludvigsen TC, Drogset JO, Grontvedt T, Solheim E, et al. Autologous chondrocyte implantation compared with microfracture in the knee. A randomized trial. J Bone Joint Surg Am. 2004;86(3):455–64.
4. Johnstone B, Alini M, Cucchiarini M, Dodge GR, Eglin D, Guilak F, et al. Tissue engineering for articular cartilage repair-the state of the art. Eur Cell Mater. 2013;25:248–67.
5. Goldring MB, Goldring SR. Osteoarthritis. J Cell Physiol. 2007;213(3):626–34.
6. Minas T. A primer in cartilage repair. J Bone Joint Surg Br. 2012;94(11 Suppl A):141–6.
7. Younger EM, Chapman MW. Morbidity at bone graft donor sites. J Orthop Trauma. 1989;3(3):192–5.
8. Docheva D, Muller SA, Majewski M, Evans CH. Biologics for tendon repair. Adv Drug Deliv Rev. 2015;84:222–39.
9. Makris EA, Hadidi P, Athanasiou KA. The knee meniscus: structure-function, pathophysiology, current repair techniques, and prospects for regeneration. Biomaterials. 2011;32(30):7411–31.
10. Shimomura K, Hamamoto S, Hart DA, Yoshikawa H, Nakamura N. Meniscal repair and regeneration: current strategies and future perspectives. J Clin Orthop Trauma. 2018;9(3):247–53.
11. Kurzweil PR, Lynch NM, Coleman S, Kearney B. Repair of horizontal meniscus tears: a systematic review. Arthroscopy. 2014;30(11):1513–9.
12. Filardo G, Andriolo L, Kon E, de Caro F, Marcacci M. Meniscal scaffolds: results and indications. A systematic literature review. Int Orthop. 2015;39(1):35–46.
13. Rosso F, Bisicchia S, Bonasia DE, Amendola A. Meniscal allograft transplantation: a systematic review. Am J Sports Med. 2015;43(4):998–1007.
14. Cucchiarini M, McNulty AL, Mauck RL, Setton LA, Guilak F, Madry H. Advances in combining gene therapy with cell and tissue engineering-based approaches to enhance healing of the meniscus. Osteoarthr Cartil. 2016;24(8):1330–9.
15. Moulton SG, Bhatia S, Civitarese DM, Frank RM, Dean CS, LaPrade RF. Surgical techniques and outcomes of repairing meniscal radial tears: a systematic review. Arthroscopy. 2016;32(9):1919–25.
16. Cucchiarini M, Madry H. Gene therapy for cartilage defects. J Gene Med. 2005;7(12):1495–509.

17. Madry H, Cucchiarini M. Advances and challenges in gene-based approaches for osteoarthritis. J Gene Med. 2013;15(10):343–55.
18. Evans CH, Huard J. Gene therapy approaches to regenerating the musculoskeletal system. Nat Rev Rheumatol. 2015;11(4):234–42.
19. Cucchiarini M. Human gene therapy: novel approaches to improve the current gene delivery systems. Discov Med. 2016;21(118):495–506.
20. Kotterman MA, Chalberg TW, Schaffer DV. Viral vectors for gene therapy: translational and clinical outlook. Annu Rev Biomed Eng. 2015;17:63–89.
21. Naldini L. Gene therapy returns to central stage. Nature. 2015;526(7573):351–60.
22. Adkar SS, Brunger JM, Willard VP, Wu CL, Gersbach CA, Guilak F. Genome engineering for personalized arthritis therapeutics. Trends Mol Med. 2017;23(10):917–31.
23. Almarza D, Cucchiarini M, Loughlin J. Genome editing for human osteoarthritis - a perspective. Osteoarthr Cartil. 2017;25(8):1195–8.
24. Bougioukli S, Evans CH, Alluri RK, Ghivizzani SC, Lieberman JR. Gene therapy to enhance bone and cartilage repair in orthopaedic surgery. Curr Gene Ther. 2018;18(3):154–70.
25. Dunbar CE, High KA, Joung JK, Kohn DB, Ozawa K, Sadelain M. Gene therapy comes of age. Science. 2018;359(6372):eaan4672.
26. Evans CH, Ghivizzani SC, Robbins PD. Gene delivery to joints by intra-articular injection. Hum Gene Ther. 2018;29(1):2–14.
27. Grol MW, Lee BH. Gene therapy for repair and regeneration of bone and cartilage. Curr Opin Pharmacol. 2018;40:59–66.
28. Choi YR, Collins KH, Lee JW, Kang HJ, Guilak F. Genome engineering for osteoarthritis: from designer cells to disease-modifying drugs. Tissue Eng Regen Med. 2019;16(4):335–43.
29. Cucchiarini M, Madry H. Biomaterial-guided delivery of gene vectors for targeted articular cartilage repair. Nat Rev Rheumatol. 2019;15(1):18–29.
30. High KA, Roncarolo MG. Gene therapy. N Engl J Med. 2019;381(5):455–64.
31. Hill AB, Chen M, Chen CK, Pfeifer BA, Jones CH. Overcoming gene-delivery hurdles: physiological considerations for nonviral vectors. Trends Biotechnol. 2016;34(2):91–105.
32. Schmeer M, Buchholz T, Schleef M. Plasmid DNA manufacturing for indirect and direct clinical applications. Hum Gene Ther. 2017;28(10):856–61.
33. Gao J, Mese K, Bunz O, Ehrhardt A. State-of-the-art human adenovirus vectorology for therapeutic approaches. FEBS Lett. 2019;593(24):3609–22.
34. Artusi S, Miyagawa Y, Goins WF, Cohen JB, Glorioso JC. Herpes simplex virus vectors for gene transfer to the central nervous system. Diseases. 2018;6(3):4–19.
35. Poletti V, Mavilio F. Interactions between retroviruses and the host cell genome. Mol Ther Methods Clin Dev. 2017;8:31–41.
36. Milone MC, O'Doherty U. Clinical use of lentiviral vectors. Leukemia. 2018;32(7):1529–41.
37. Kotterman MA, Schaffer DV. Engineering adeno-associated viruses for clinical gene therapy. Nat Rev Genet. 2014;15(7):445–51.
38. Rey-Rico A, Cucchiarini M. Controlled release strategies for rAAV-mediated gene delivery. Acta Biomater. 2016;29:1–10.
39. Wang D, Tai PWL, Gao G. Adeno-associated virus vector as a platform for gene therapy delivery. Nat Rev Drug Discov. 2019;18(5):358–78.
40. Li C, Samulski RJ. Engineering adeno-associated virus vectors for gene therapy. Nat Rev Genet. 2020;21(4):255–72.
41. Frisbie DD, Ghivizzani SC, Robbins PD, Evans CH, McIlwraith CW. Treatment of experimental equine osteoarthritis by in vivo delivery of equine interleukin-1 receptor antagonist gene. Gene Ther. 2002;9(1):12–20.
42. Nixon AJ, Grol MW, Lang HM, Ruan MZC, Stone A, Begum L, et al. Disease-modifying osteoarthritis treatment with interleukin-1 receptor antagonist gene therapy in small and large animal models. Arthritis Rheumatol. 2018;70(11):1757–58.
43. Wang HJ, Yu CL, Kishi H, Motoki K, Mao ZB, Muraguchi A. Suppression of experimental osteoarthritis by adenovirus-mediated double gene transfer. Chin Med J. 2006;119(16):1365–73.
44. Morisset S, Frisbie DD, Robbins PD, Nixon AJ, McIlwraith CW. IL-1Ra/IGF-1 gene therapy modulates repair of microfractured chondral defects. Clin Orthop Relat Res. 2007;462:221–8.
45. Chen B, Qin J, Wang H, Magdalou J, Chen L. Effects of adenovirus-mediated bFGF, IL-1Ra and IGF-1 gene transfer on human osteoarthritic chondrocytes and osteoarthritis in rabbits. Exp Mol Med. 2010;42(10):684–95.
46. Blaney Davidson EN, Vitters EL, van den Berg WB, van der Kraan PM. TGFbeta-induced cartilage repair is maintained but fibrosis is blocked in the presence of Smad 7. Arthritis Res Ther. 2006;8(3):R65–73.
47. Hsieh JL, Shen PC, Shiau AL, Jou IM, Lee CH, Teo ML, et al. Adenovirus-mediated kallistatin gene transfer ameliorates disease progression in a rat model of osteoarthritis induced by anterior cruciate ligament transection. Hum Gene Ther. 2009;20(2):147–58.
48. Hsieh JL, Shen PC, Shiau AL, Jou IM, Lee CH, Wang CR, et al. Intraarticular gene transfer of thrombospondin-1 suppresses the disease progression of experimental osteoarthritis. J Orthop Res. 2010;28(10):1300–6.
49. Menendez MI, Clark DJ, Carlton M, Flanigan DC, Jia G, Sammet S, et al. Direct delayed human adenoviral BMP-2 or BMP-6 gene therapy for bone and cartilage regeneration in a pony osteochondral model. Osteoarthr Cartil. 2011;19(8):1066–75.
50. Shen PC, Shiau AL, Jou IM, Lee CH, Tai MH, Juan HY, et al. Inhibition of cartilage damage by pro-opiomelanocortin prohormone overexpression in a rat

model of osteoarthritis. Exp Biol Med (Maywood). 2011;236(3):334–40.

51. Oh H, Chun CH, Chun JS. Dkk-1 expression in chondrocytes inhibits experimental osteoarthritic cartilage destruction in mice. Arthritis Rheum. 2012;64(8):2568–78.

52. Ruan MZ, Erez A, Guse K, Dawson B, Bertin T, Chen Y, et al. Proteoglycan 4 expression protects against the development of osteoarthritis. Sci Transl Med. 2013;5(176):176ra34–43.

53. Stone A, Grol MW, Ruan MZC, Dawson B, Chen Y, Jiang MM, et al. Combination of Prg4 and IL-1ra gene therapy protects against hyperalgesia and cartilage degeneration in post-traumatic osteoarthritis. Hum Gene Ther. 2019;30(2):225–35.

54. Ashraf S, Kim BJ, Park S, Park H, Lee SH. RHEB gene therapy maintains the chondrogenic characteristics and protects cartilage tissue from degenerative damage during experimental murine osteoarthritis. Osteoarthr Cartil. 2019;27(10):1508–17.

55. Gu XD, Wei L, Li PC, Che XD, Zhao RP, Han PF, et al. Adenovirus-mediated transduction with histone deacetylase 4 ameliorates disease progression in an osteoarthritis rat model. Int Immunopharmacol. 2019;75:105752–9.

56. Tashkandi M, Ali F, Alsaqer S, Alhousami T, Cano A, Martin A, et al. Lysyl oxidase-like 2 protects against progressive and aging related knee joint osteoarthritis in mice. Int J Mol Sci. 2019;20(19):4798.

57. Mason JB, Gurda BL, Hankenson KD, Harper LR, Carlson CS, Wilson JM, et al. Wnt10b and Dkk-1 gene therapy differentially influenced trabecular bone architecture, soft tissue integrity, and osteophytosis in a skeletally mature rat model of osteoarthritis. Connect Tissue Res. 2017;58(6):542–52.

58. Zhao L, Huang J, Fan Y, Li J, You T, He S, et al. Exploration of CRISPR/Cas9-based gene editing as therapy for osteoarthritis. Ann Rheum Dis. 2019;78(5):676–82.

59. Cucchiarini M, Madry H, Ma C, Thurn T, Zurakowski D, Menger MD, et al. Improved tissue repair in articular cartilage defects in vivo by rAAV-mediated overexpression of human fibroblast growth factor 2. Mol Ther. 2005;12(2):229–38.

60. Hiraide A, Yokoo N, Xin KQ, Okuda K, Mizukami H, Ozawa K, et al. Repair of articular cartilage defect by intraarticular administration of basic fibroblast growth factor gene, using adeno-associated virus vector. Hum Gene Ther. 2005;16(12):1413–21.

61. Cucchiarini M, Orth P, Madry H. Direct rAAV SOX9 administration for durable articular cartilage repair with delayed terminal differentiation and hypertrophy in vivo. J Mol Med (Berl). 2013;91(5):625–36.

62. Cucchiarini M, Madry H. Overexpression of human IGF-I via direct rAAV-mediated gene transfer improves the early repair of articular cartilage defects in vivo. Gene Ther. 2014;21(9):811–9.

63. Cucchiarini M, Asen AK, Goebel L, Venkatesan JK, Schmitt G, Zurakowski D, et al. Effects of TGF-β overexpression via rAAV gene transfer on the early

64. Baltzer AW, Lattermann C, Whalen JD, Ghivizzani S, Wooley P, Krauspe R, et al. Potential role of direct adenoviral gene transfer in enhancing fracture repair. Clin Orthop Relat Res. 2000;(379 Suppl):S120–5.

65. Southwood LL, Frisbie DD, Kawcak CE, Ghivizzani SC, Evans CH, McIlwraith CW. Evaluation of Ad-BMP-2 for enhancing fracture healing in an infected defect fracture rabbit model. J Orthop Res. 2004;22(1):66–72.

66. Betz OB, Betz VM, Nazarian A, Pilapil CG, Vrahas MS, Bouxsein ML, et al. Direct percutaneous gene delivery to enhance healing of segmental bone defects. J Bone Joint Surg Am. 2006;88(2):355–65.

67. Egermann M, Lill CA, Griesbeck K, Evans CH, Robbins PD, Schneider E, et al. Effect of BMP-2 gene transfer on bone healing in sheep. Gene Ther. 2006;13(17):1290–9.

68. Betz VM, Betz OB, Glatt V, Gerstenfeld LC, Einhorn TA, Bouxsein ML, et al. Healing of segmental bone defects by direct percutaneous gene delivery: effect of vector dose. Hum Gene Ther. 2007;18(10):907–15.

69. Betz OB, Betz VM, Nazarian A, Egermann M, Gerstenfeld LC, Einhorn TA, et al. Delayed administration of adenoviral BMP-2 vector improves the formation of bone in osseous defects. Gene Ther. 2007;14(13):1039–44.

70. Ishihara A, Zekas LJ, Weisbrode SE, Bertone AL. Comparative efficacy of dermal fibroblast-mediated and direct adenoviral bone morphogenetic protein-2 gene therapy for bone regeneration in an equine rib model. Gene Ther. 2010;17(6):733–44.

71. Liu YG, Zhou Y, Hu X, Fu JJ, Pan Y, Chu TW. Effect of vascular endothelial growth factor 121 adenovirus transduction in rabbit model of femur head necrosis. J Trauma. 2011;70(6):1519–23.

72. Lakhan R, Baylink DJ, Lau KH, Tang X, Sheng MH, Rundle CH, et al. Local administration of AAV-DJ pseudoserotype expressing COX2 provided early onset of transgene expression and promoted bone fracture healing in mice. Gene Ther. 2015;22(9):721–8.

73. Huang R, Zong X, Nadesan P, Huebner JL, Kraus VB, White JP, et al. Lowering circulating apolipoprotein E levels improves aged bone fracture healing. JCI Insight. 2019;4(18):e129144–56.

74. Lou J, Tu Y, Burns M, Silva MJ, Manske P. BMP-12 gene transfer augmentation of lacerated tendon repair. J Orthop Res. 2001;19(6):1199–202.

75. Ma Y, Zhang X, Wang J, Liu P, Zhao L, Zhou C, et al. Effect of bone morphogenetic protein-12 gene transfer on posterior cruciate ligament healing in a rabbit model. Am J Sports Med. 2009;37(3):599–609.

76. Rickert M, Wang H, Wieloch P, Lorenz H, Steck E, Sabo D, et al. Adenovirus-mediated gene transfer of growth and differentiation factor-5 into tenocytes and the healing rat Achilles tendon. Connect Tissue Res. 2005;46(4–5):175–83.

77. Tang JB, Cao Y, Zhu B, Xin KQ, Wang XT, Liu PY. Adeno-associated virus-2-mediated bFGF

gene transfer to digital flexor tendons significantly increases healing strength. An in vivo study. J Bone Joint Surg Am. 2008;90(5):1078–89.

78. Tang JB, Chen CH, Zhou YL, McKeever C, Liu PY. Regulatory effects of introduction of an exogenous FGF2 gene on other growth factor genes in a healing tendon. Wound Repair Regen. 2014;22(1):111–8.

79. Tang JB, Wu YF, Cao Y, Chen CH, Zhou YL, Avanessian B, et al. Basic FGF or VEGF gene therapy corrects insufficiency in the intrinsic healing capacity of tendons. Sci Rep. 2016;6:20643–54.

80. Mao WF, Wu YF, Yang QQ, Zhou YL, Wang XT, Liu PY, et al. Modulation of digital flexor tendon healing by vascular endothelial growth factor gene transfection in a chicken model. Gene Ther. 2017;24(4):234–40.

81. Goto H, Shuler FD, Lamsam C, Moller HD, Niyibizi C, Fu FH, et al. Transfer of lacZ marker gene to the meniscus. J Bone Joint Surg Am. 1999;81(7):918–25.

82. Madry H, Cucchiarini M, Kaul G, Kohn D, Terwilliger EF, Trippel SB. Menisci are efficiently transduced by recombinant adeno-associated virus vectors in vitro and in vivo. Am J Sports Med. 2004;32(8):1860–5.

83. Hidaka C, Ibarra C, Hannafin JA, Torzilli PA, Quitoriano M, Jen SS, et al. Formation of vascularized meniscal tissue by combining gene therapy with tissue engineering. Tissue Eng. 2002;8(1):93–105.

84. Steinert AF, Palmer GD, Capito R, Hofstaetter JG, Pilapil C, Ghivizzani SC, et al. Genetically enhanced engineering of meniscus tissue using ex vivo delivery of transforming growth factor-beta 1 complementary deoxyribonucleic acid. Tissue Eng. 2007;13(9):2227–37.

85. Cucchiarini M, Schmidt K, Frisch J, Kohn D, Madry H. Overexpression of TGF-β via rAAV-mediated gene transfer promotes the healing of human meniscal lesions ex vivo on explanted menisci. Am J Sports Med. 2015;43(5):1197–205.

86. Cucchiarini M, Schetting S, Terwilliger EF, Kohn D, Madry H. rAAV-mediated overexpression of FGF-2 promotes cell proliferation, survival, and alpha-SMA expression in human meniscal lesions. Gene Ther. 2009;16(11):1363–72.

87. Zhang H, Leng P, Zhang J. Enhanced meniscal repair by overexpression of hIGF-1 in a full-thickness model. Clin Orthop Relat Res. 2009;467(12):3165–74.

88. www.clinicaltrials.gov/ct2/show/NCT02790723?ter m=evans&cond=Osteoarthritis&draw=3&rank=1.

89. Pleticha J, Heilmann LF, Evans CH, Asokan A, Samulski RJ, Beutler AS. Preclinical toxicity evaluation of AAV for pain: evidence from human AAV studies and from the pharmacology of analgesic drugs. Mol Pain. 2014;10:54–63.

90. Mingozzi F, High KA. Overcoming the host immune response to adeno-associated virus gene delivery vectors: the race between clearance, tolerance, neutralization, and escape. Annu Rev Virol. 2017;4(1):511–34.

91. Varela-Eirín M, Varela-Vázquez A, Guitián-Caamaño A, Paíno CL, Mato V, Largo R, et al. Targeting of chondrocyte plasticity via connexin43 modulation attenuates cellular senescence and fosters a pro-regenerative environment in osteoarthritis. Cell Death Dis. 2018;9(12):1166–81.

92. D'Costa S, Rich MJ, Diekman BO. Engineered cartilage from human chondrocytes with homozygous knockout of cell cycle inhibitor p21. Tissue Eng Part A. 2020;26(7–8):441–9.

93. Hsu MN, Huang KL, Yu FJ, Lai PL, Truong AV, Lin MW, et al. Coactivation of endogenous Wnt10b and Foxc2 by CRISPR activation enhances BMSC osteogenesis and promotes calvarial bone regeneration. Mol Ther. 2020;28(2):441–51.

94. Seidl CI, Fulga TA, Murphy CL. CRISPR-Cas9 targeting of MMP13 in human chondrocytes leads to significantly reduced levels of the metalloproteinase and enhanced type II collagen accumulation. Osteoarthr Cartil. 2019;27(1):140–7.

95. Dicks A, Wu CL, Steward N, Adkar SS, Gersbach CA, Guilak F. Prospective isolation of chondroprogenitors from human iPSCs based on cell surface markers identified using a CRISPR-Cas9-generated reporter. Stem Cell Res Ther. 2020;11(1):66–79.

96. Huynh NP, Gloss CC, Lorentz J, Tang R, Brunger JM, McAlinden A, et al. Long non-coding RNA GRASLND enhances chondrogenesis via suppression of interferon type II signaling pathway. elife. 2020;9:e49558–604.

In Situ Targeting of Stem and Progenitor Cells in Native Tissues

Cierra A. Clark, Takeshi Oichi, Joshua M. Abzug, and Satoru Otsuru

32.1 Introduction

All tissues remodel, and the vast majority of them has some capacity for repair/regeneration after injury. Tissue repair or regeneration always requires the generation of new tissues. New tissues require the generation of new cells. Therefore, all effective tissue repair and regeneration are grounded in the availability of cells that can be activated to proliferate and generate progeny capable of contributing to the formation of new cells that can differentiate into the cell types needed in the new tissue that is desired.

In general, tissue repair/regeneration initiates with a series of cellular events. Homing signals result in the infiltration of inflammatory cells into the damaged tissues. Activation of local endothelial progenitor cells (EPCs) is essential for angiogenesis. Similarly, activation of tissue-specific connective tissue progenitors (CTPs) is essential for generation of new connective tissues. Tissue-specific CTPs are present in all connective tissues and are defined as tissue-resident cells that are capable of proliferating and generating progeny that can differentiate into one or more connective tissue phenotypes [1–4]. Activation signals are proliferation, migration, and differentiation of stem and progenitor cells into cells that differen-

tiate and elaborate an extracellular matrix and functional attributes specific to the new tissue that is required [5–9]. Some stem and progenitor cells can also be mobilized into systemic circulation and "home" to sites of tissue repair or regeneration by signaling mechanisms that are increasingly understood.

Cell-based therapy aims to enhance the repair/regeneration by exogenously providing culture-expanded cells (e.g., mesenchymal stromal cells (MSCs)) which have the ability to differentiate into the desired cell types [10, 11]. However, this cell-based therapy typically requires steps to isolate and expand cells. This results in treatment delays, especially in the case of acute injuries. Additionally, it has been demonstrated that ex vivo expanded cells represent a highly selected and behaviorally altered population that responds differently in vivo. Culture-expanded cells are less likely to contribute to repair/regeneration by direct cell replacement than native progenitor populations [12, 13].

To bypass these drawbacks, many approaches discussed elsewhere in this text have focused on the harvest and "transplantation" of mixed tissue-derived populations that are enriched populations of stem and progenitor cells. However, a series of alternatively novel strategies are becoming available to "target", selectively activate, and modulate local stem and progenitor populations "in situ". These strategies can enhance the repair/regeneration by recruiting more intrinsic stem and progenitor

---

C. A. Clark · T. Oichi · J. M. Abzug · S. Otsuru (✉)
Department of Orthopaedics, University of Maryland School of Medicine, Baltimore, MD, USA
e-mail: sotsuru@som.umaryland.edu

G. Filardo et al. (eds.), *Orthobiologics*, https://doi.org/10.1007/978-3-030-84744-9_32

cells rather than providing ex vivo expanded ones into the damaged tissue [7, 14, 15].

This new "homing" strategy stems from the idea that one or more populations of native CTPs can be mobilized from bone marrow or other sources into systemic circulation in blood and then inducted by chemotactic factors or adhesion molecules to home to sites of local tissue repair/regeneration and contribute to tissue formation. Indeed, cells capable of proliferation to form progeny that can differentiate into connective tissues have been isolated from the peripheral blood [16–18]. Animal studies using a parabiosis model and transgenic mouse expressing green fluorescence protein (GFP) in bone marrow cells have demonstrated that GFP-positive bone marrow-derived cells can travel through the circulation and contributed to bone formation in the other mouse [19, 20]. These findings support the possibility that tissue repair/regeneration could be enhanced by mobilizing more intrinsic stem and progenitor cells from the bone marrow and/or recruiting more of them to the damaged tissue (Fig. 32.1).

In order to implement this new strategy, it is necessary to understand the mechanisms of how CTP populations are mobilized, how they are recruited to a repair or remodeling, and how they are activated at the target site. These mechanisms likely vary from tissue to tissue, but many mechanisms may be shared. Therefore, identification of these tissue-specific mechanisms will be necessary.

In orthopedics, fracture healing is one of the most well-characterized tissue repairing processes, and this new strategy has been pre-clinically tested in fracture repair [9, 19, 21–25]. Therefore, this chapter will focus on fracture healing as an example, though the principles and mechanisms outlined here are likely to be applicable to other settings as well. In the following section, we highlight stem and progenitor cell mobilization, recruitment, and differentiation, respectively.

"Mobilization of intrinsic CTPs" induces their transition from the bone marrow to peripheral circulation. These circulating CTPs can subsequently be recruited to the damaged tissue by a separate set of chemotactic signals and adhesion molecules, where they may contribute to local tissue and tissue repair/regeneration, both by local paracrine and cell-cell interactions, as well as by proliferation, migration, and differentiation into a desired tissue.

32.2 Mobilization of Stem/Progenitor Cells

It has been demonstrated that one or more populations of CTPs, including osteoblast progenitor cells, can exist in and transit through peripheral

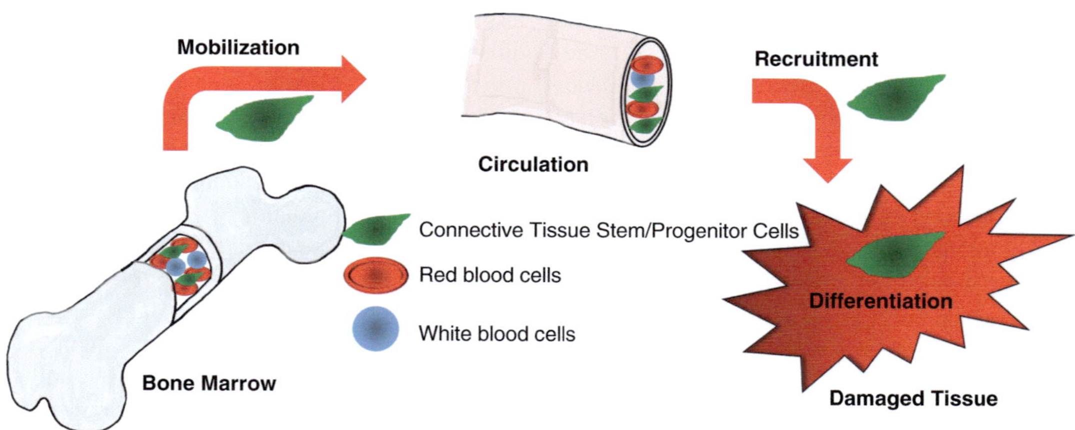

Fig. 32.1 Schematic diagram of the new strategy of tissue repair/regeneration

blood [16–18, 26–28]. Although the source or sources of these circulating progenitors have not been fully elucidated, animal studies using parabiosis models suggest that the bone marrow is a likely source of at least some of these circulating cells [19, 20]. The bone marrow contains a diversity of stem and progenitor cells, including hematopoietic stem cells, connective tissue progenitors, and endothelial progenitor cells.

Mobilization of hematopoietic stem cells from the bone marrow into systemic circulation has been extensively investigated. The cellular and molecular mechanisms associated with the release and escape of HSCs from their niche have been identified [29, 30]. Indeed, granulocyte colony-stimulating factor (G-CSF) is widely used clinically to mobilize hematopoietic stem cells into circulation, where they can be harvested and enriched for bone marrow transplantation [31].

The niches for CTPs in the native marrow are less well characterized but likely include both perivascular cells and pericytes (often CD146 positive) as well as cells on the surface of trabecular cancellous bone.

- SDF1/CXCR4: The chemokine stromal cell derived factor 1 (SDF1) and its receptor CXCR4 have been demonstrated to play an important role in retaining hematopoietic stem cells in the bone marrow. The disruption of the SDF1/CXCR4 interaction by G-CSF or AMD3100 (a CXCR4 antagonist) leads to the mobilization of hematopoietic stem cells from the bone marrow to the peripheral blood [32–35]. AMD3100 has also been shown to increase the number of circulating colony-founding CTPs in the peripheral blood, suggesting that, in addition to HSCs, AMD3100 also mobilizes one or more CTP populations into the circulation [36, 37]. When mice with femoral fractures were treated with AMD3100 for 3 days after sustaining the fracture, fracture healing was enhanced, and there was significantly higher bone mineral density [37]. Interestingly, pretreatment of animals with vascular endothelial growth factor (VEGF) or insulin-like

growth factor 1 (IGF1), but not G-CSF, prior to AMD3100 administration further enhanced mobilization of CTPs and accelerated fracture healing with significantly greater bone mass [36, 38, 39]. These results suggest that the SDF1/CXCR4 signaling pathway is also critical to keeping CTPs in the bone marrow and possibly other niches. These preclinical studies demonstrated the feasibility of the mobilization and homing strategy of stimulating intrinsic stem cells to aid in the healing of fractures.

- HMGB1: High-mobility group box 1 (HMGB1) is a nuclear protein residing in the nucleus during the steady state. Upon injury, HMGB1 is secreted by necrotic cells as well as inflammatory cells, such as macrophages and monocytes [40–44]. It has been demonstrated that the secreted HMGB1 has strong chemotactic activity to endothelial precursor cells, mesoangioblasts, and fibroblasts [41, 43, 45–47]. Of interest, HMGB1 has also been shown to mobilize platelet-derived growth factor receptor (PDGFR) α-positive cells from the bone marrow into the circulation, contributing to skin regeneration and recovery of heart function after a myocardial infarction [44, 48–50]. Moreover, it has been reported that HMGB1 not only mobilizes CTPs from marrow but also induces osteoblastic differentiation [51–56]. These findings suggest that HMGB1 has the potential to enhance bone regeneration.

- Substance P: Substance P is a neuropeptide produced by neuronal cells as well as inflammatory cells including macrophages and neutrophils. It has been shown that the serum levels of substance P are significantly increased after injury [57–59]. This injury-induced substance P has been demonstrated to mobilize CD29-positive cells from the bone marrow into the peripheral blood [57]. These mobilized CD29-positive cells in the circulation were found to be multipotent. In addition, substance P injection has accelerated wound healing in a murine burn model [57]. Moreover, the systemic injection of substance P enhanced bone repair in a murine calvarial defect model by promoting mobilization of CD29-positive cells from the bone marrow [60].

In addition to tissue-resident CTPs, endothelial progenitor cells (EPCs) also play a critical role in tissue repair/regeneration. Endothelial progenitor cells are required for angiogenesis and subsequent mass transport of nutrition, oxygen, and cells to damaged tissues [61–63]. Since providing culture-expanded endothelial progenitor cells into fracture sites promoted fracture healing [64], mobilization of endothelial progenitor cells has been attempted to enhance tissue repair/ regeneration. It should be noted that both VEGF and SDF1 have been shown to induce EPC mobilization, suggesting that tissue-resident CTPs and EPCs may share mechanisms to egress and home from the bone marrow and/or they may structurally interact with each other. Moreover, the mobilization of one cell type may affect the other one.

- VEGF: Vascular endothelial growth factor (VEGF) plays an important role in vessel formation. In addition to induction of angiogenesis, VEGF mobilized EPCs from the bone marrow to the peripheral blood [65]. Local administration of VEGF in a murine fracture model demonstrated enhanced fracture healing with increased blood vessel formation [66, 67].
- SDF1/CXCR4: The elevation of serum levels of SDF1 or disruption of SDF1/CXCR4 interaction by AMD3100 showed mobilization of EPCs from the bone marrow to the circulation [36, 68, 69]. Given that the SDF1/CXCR4 pathway is involved in mobilization of CTPs and EPCs, enhanced fracture healing by AMD3100 could be induced by both cell types [37, 39].
- G-CSF: G-CSF has been demonstrated to mobilize the CD34-positive bone marrow population which is enriched with hematopoietic stem cells (HSCs) and EPCs [70, 71]. G-CSF treatment has shown to enhance bone healing in a rat bone defect model by increasing mobilization of CD34-positive cells into the circulation [72]. The local transplantation of autologous circulating CD34-positive cells

mobilized by G-CSF administration into patients with a fracture non-union accelerated fracture healing [73–75]. In vitro expansion of mobilized CD34-positive cells also enhanced fracture repair after local transplantation. This suggests that, in addition to HSCs and EPCs, at least one of the CTP populations that may be available for mobilization may be CD34-positive.

Overall, these studies have demonstrated that mobilization of several stem and progenitor cells populations can have a positive impact on tissue repair/regeneration such as fracture healing. It remains unknown whether this approach can be applicable to hypo-vascular tissues such as cartilage. Additionally, further understanding regarding the roles of different subsets of CTPs and EPCs and if either more precise timing or more selective induction of these individual cell populations during the repair process would facilitate the development of more efficient and controlled regenerative therapies.

32.3 Homing of Stem/Progenitor Cells into Sites of Tissue Repair and Regeneration

To participate in tissue repair/regeneration, circulating stromal/progenitor cells mobilized into systemic circulation need to be recruited to and retained in the damaged tissue. This can involve the generation of gradients of soluble chemotactic agents, as well as changes in local endothelial cells so that they express surface antigens that induce sticking and rolling of circulating cells expressing matching ligands and subsequent diapedesis of those cells out of the capillary bed and into local tissues.

It has been demonstrated that damaged tissues release chemoattractant molecules to recruit cells which are required for tissue repair/regeneration; this includes inflammatory cells as well as circulating stem and progenitor cells. At the same time, endothelial cells in and around the damaged tissues coordinate an increase in the expression of adhesion molecules, such as very late antigen-4

(VLA-4), vascular cell adhesion molecule-1 (VCAM-1), and intercellular adhesion molecule-1 (ICAM-1) to catch circulating cells [76]. Some of the same molecules are also involved in mobilizing stromal/progenitor cells from the bone marrow to the peripheral blood. In particular, the SDF1/CXCR4 ligand complex, which is active in many tissues, is one of the most widely active and well-studied homing mechanisms.

- SDF1/CXCR4: It has been demonstrated that the expression of SDF1 is significantly increased in damaged tissues [20, 77–80]. Specifically, in repairing bones, SDF1 is highly expressed by endothelial cells in newly formed vessels, osteoblasts in regenerating bones, and in the periosteum [20, 79]. In vitro migration of bone marrow-derived stem and progenitor cells toward SDF1 was enhanced in a dose-dependent manner, and this migration was suppressed by blocking CXCR4 [20, 79, 81]. As mentioned above, when mice with a femoral fracture were treated with AMD3100 for a short period, the mobilization of CTPs from the bone marrow to the circulation was stimulated, and, as a result, fracture healing was accelerated [37]. However, interestingly, long-term treatment of fractured mice with AMD3100 ended with inferior fracture healing with less bone volume, suggesting that mobilized stromal/progenitor cells failed to home successfully into the fracture site, possibly due to the disruption of SDF1/CXCR4 interaction [82]. The significance of SDF1 for CTP recruitment is further supported by the findings that the addition of SDF1 enhanced bone formation both in fracture and heterotopic ossification models [83, 84].

The increased expression of SDF1 in damaged tissues seems to be a generalized and intrinsic part of the process of tissue repair/regeneration in a variety of tissues. This may explain why increased mobilization of stem/progenitor cells without enhancing recruiting signals leads to improved tissue repair/regeneration. However, in the setting of medical conditions that may inhibit the regular tissue repair process (e.g., infection or diabetes), or if the regular repair process is blunted due to a chronic milieu (e.g., fracture non-union), supplementary homing signals might be required.

32.4 Differentiation of Stem/Progenitor Cells

Once a mobilized CTP or EPC has migrated into the damaged tissue, in order to contribute to new tissue formation they must survive, proliferate, migrate, and differentiate into a mature cell phenotype that contributes to new tissue formation. While signals related to the mobilization and homing steps of stromal/progenitor cells may be common among a diversity of tissues, the process pathways and signals driving differentiation are likely to be specific to each tissue. Efficient differentiation of CTPS and EPCs in local tissues is critical to enhance the repair or regeneration process.

In the case of fracture healing, various cytokines and growth factors driving chondrogenic and osteogenic differentiation are expressed in and around the fracture site [9, 21, 23, 85]. These include variants of transforming growth factor-β (TGF-β) (mostly TGF-β1), a diversity of bone morphogenetic proteins (BMPs) (mostly BMP-2, -4, -6), fibroblast growth factors (FGFs) (mostly basic FGF, bFGF), insulin-like growth factors (IGFs), and platelet-derived growth factors (PDGFs), all of which have been therapeutic targets to promote various steps in osteoblastic and chondroblastic differentiation during fracture healing [9, 21, 23, 85]. Each of these factors is available for potential therapeutic use, if targeted and localized appropriately in time and space in a fracture site.

- BMP-2: Recombinant human BMP-2 has been applied to randomized trials for patients with open tibial fractures and showed accelerated fracture healing with a significantly reduced frequency of secondary interventions [86–88].
- FGF-2: Recombinant human FGF-2 has also been tested in a randomized double-blind

trial for patients with tibial shaft fractures. The local injection of FGF-2 into the fracture gap enhanced fracture healing and had a significantly higher cumulative percentage of patients with bone union [89].

- PTH: In addition to these osteogenic factors endogenously released in the fracture site, parathyroid hormone 1–34 (PTH 1–34), which is known as an anabolic drug for the treatment of osteoporosis, has been utilized to treat postmenopausal women who suffered from a distal radius fracture [90]. Patients treated with a daily injection of 20 µg of PTH 1–34 demonstrated significantly shorter times to healing compared to those treated with placebo, suggesting that PTH 1–34 may accelerate fracture healing. In another prospective randomized controlled study for postmenopausal patients with pelvic fractures, daily injection of 100 µg of PTH 1–84 also significantly shortened the time to healing compared to the control group which had saline injected [91].

These findings indicate that local stimulation of CTPs (and possibly EPCs) in fracture sites to increase proliferation, matrix synthesis, and osteoblastic differentiation can have therapeutic effects on fracture healing. Similar strategies could be applicable for repair/regeneration of other tissues once the biologic molecules responsible for the differentiation to the target tissue cells are identified. Given that temporal and spatial expression patterns vary among differentiation signals in damaged tissues [9, 85], it should be noted that understanding the repairing mechanisms is critical to provide the signals to the appropriate location at the proper timing. Moreover, it is obvious that both proliferation and differentiation signals should follow the migration of CTPs and EPCs into the repairing site. Thus, the fine-tuning of sequential signals for mobilization, homing, proliferation, migration, and differentiation of CTPs and EPCs will likely be required to obtain the best outcome.

32.5 Conclusions

Targeting native tissue-specific connective tissue progenitors (CTPs) in situ is a rational strategy for enhancing connective tissue repair. This chapter has provided an overview of a more nuanced strategy of enhancing mobilization of CTPs into circulation and subsequently their homing to sites of tissue repair, regeneration, or remodeling. This approach has the potential to overcome some of the limitations of cell-based strategies that rely on ex vivo expansion to generate transplantable populations, such as mesenchymal stromal cells or other populations. In particular, these problems include (1) the time-consuming expensive culture expansion process, (2) difficulty to preservation of repeatable and reproducible multipotency in extended expansion, and (3) the frequent failure of survival and differentiation of culture expanded MSCs in vivo.

The mobilization and homing strategy can bypass these problems. Mobilization of intrinsic CTPs and other stem and progenitor populations may be accomplished using systemic pharmacological agents with minimal morbidity. In fact, many of these mechanisms may already be activated as part of the natural response to major trauma. Preclinical studies using fracture models provide a good proof that CTPs that transit through blood will home to sites of injury or active tissue repair. Some of the mechanisms responsible for this homing may be enhanced. However, further investigations are required to (1) identify and refine minimally invasive methods for delivery or activation of optimal homing signals at the site where tissue regeneration and repair are desired and (2) to fine our understanding of possible homing strategies for non-bone tissues and other chronic orthopedic diseases (e.g., osteoporosis and osteoarthritis), particularly those involving hypovascular tissues (e.g., cartilage). If successful, this new cell therapy approach targeting stem and progenitor cell populations in situ will open new avenues to orthopedic treatments, accelerate tissue repair and regeneration, and shorten recovery times and period of immobilization.

> **Take-Home Messages**
>
> - Conventional cell-based therapy requires the time-consuming expensive in vitro cell expansion process, which results in treatment delays, especially in the case of acute injuries.
> - Culture-expanded cells represent a highly selected and behaviorally altered population that responds differently in vivo.
> - Targeting native tissue-specific connective tissue progenitors (CTPs) in situ can be a rational strategy for enhancing connective tissue repair that bypasses these drawbacks of culture-expanded cell-based therapy.
> - Tissue repair processes such as mobilization, recruitment, and differentiation of CTPs can be enhanced in vivo by using systemic pharmacological agents. Preclinical studies using fracture models provide good proof of this new strategy.
> - While signals related to the mobilization and homing of CTPs may be common among tissues, signals driving differentiation are likely to be specific to each tissue. Thus, further investigations are required to develop optimal treatment regimens for each tissue.

References

1. Muschler GF, Nakamoto C, Griffith LG. Engineering principles of clinical cell-based tissue engineering. J Bone Joint Surg Am. 2004;86(7):1541–58.
2. Kwee E, Saidel G, Powell K, Heylman C, Boehm C, Muschler G. Quantifying proliferative and surface marker heterogeneity in colony-founding connective tissue progenitors and their progeny using time-lapse microscopy. J Tissue Eng Regen Med. 2019;13(2):203–16.
3. Mantripragada VP, Piuzzi NS, Bova WA, Boehm C, Obuchowski NA, Lefebvre V, et al. Donor-matched comparison of chondrogenic progenitors resident in human infrapatellar fat pad, synovium, and periosteum—implications for cartilage repair. Connect Tissue Res. 2019;60(6):597–610.
4. Muschler GF, Midura RJ, Nakamoto C. Practical modeling concepts for connective tissue stem cell and progenitor compartment kinetics. J Biomed Biotechnol. 2003;2003(3):170–93.
5. Krafts KP. Tissue repair. Organogenesis. 2010;6(4):225–33.
6. Liu J, Saul D, Böker KO, Ernst J, Lehman W, Schilling AF. Current methods for skeletal muscle tissue repair and regeneration. Biomed Res Int. 2018;2018:1–11.
7. Miller LW. New strategies to enhance stem cell homing for tissue repair. Stem Cell Gene Ther Cardiovasc Dis. 2016:485–96.
8. Szczesny SE, Lee CS, Soslowsky LJ. Remodeling and repair of orthopedic tissue: role of mechanical loading and biologics. Am J Orthop. 2018;13
9. Tsiridis E, Upadhyay N, Giannoudis P. Molecular aspects of fracture healing: which are the important molecules? Injury. 2007;38(1):S11–25.
10. Berebichez-Fridman R, Gómez-García R, Granados-Montiel J, Berebichez-Fastlicht E, Olivos-Meza A, Granados J, et al. The holy grail of orthopedic surgery: mesenchymal stem cells—their current uses and potential applications. Stem Cells Int. 2017;2017:1–14.
11. Pittenger MF, Discher DE, Péault BM, Phinney DG, Hare JM, Caplan AI. Mesenchymal stem cell perspective: cell biology to clinical progress. NPJ Regen Med. 2019;4(1):22.
12. Caplan Arnold I, Correa D. The MSC: an injury drugstore. Cell Stem Cell. 2011;9(1):11–5.
13. Keating A. Mesenchymal stromal cells: new directions. Cell Stem Cell. 2012;10(6):709–16.
14. Chen F-M, Wu L-A, Zhang M, Zhang R, Sun H-H. Homing of endogenous stem/progenitor cells for in situ tissue regeneration: promises, strategies, and translational perspectives. Biomaterials. 2011;32(12):3189–209.
15. Ito H. Chemokines in mesenchymal stem cell therapy for bone repair: a novel concept of recruiting mesenchymal stem cells and the possible cell sources. Mod Rheumatol. 2011;21(2):113–21.
16. Eghbali-Fatourechi GZ, Nagel D. Circulating osteoblast-lineage cells in humans. N Engl J Med. 2005;8
17. Kuznetsov SA, Mankani MH, Gronthos S, Satomura K, Bianco P, Robey PG. Circulating skeletal stem cells. J Cell Biol. 2001;153(5) 1133–40.
18. Zvaifler NJ, Marinova-Mutafchieva L, Adams G, Edwards CJ, Moss J, Burger JA, et al. Mesenchymal precursor cells in the blood of normal individuals. Arthritis Res Ther. 2000;2(6):477.
19. Kumagai K, Vasanji A, Drazba JA, Butler RS, Muschler GF. Circulating cells with osteogenic potential are physiologically mobilized into the fracture healing site in the parabiotic mice model. J Orthop Res. 2008;26(2):165–75.
20. Otsuru S, Tamai K, Yamazaki T, Yoshikawa H, Kaneda Y. Circulating bone marrow-derived osteoblast progenitor cells are recruited to the bone-forming site

by the CXCR4/stromal cell-derived factor-1 pathway. Stem Cells. 2008;26(1):223–34.

21. Einhorn TA, Gerstenfeld LC. Fracture healing: mechanisms and interventions. Nat Rev Rheumatol. 2015;11(1):45–54.

22. Marsell R, Einhorn TA. The biology of fracture healing. Injury. 2011;42(6):551–5.

23. Phillips AM. Overview of the fracture healing cascade. Injury. 2005;36(3):S5–7.

24. Shirley D, Marsh D, Jordan G, McQuaid S, Li G. Systemic recruitment of osteoblastic cells in fracture healing. J Orthop Res. 2005;23(5):1013–21.

25. Yellowley C. CXCL12/CXCR4 signaling and other recruitment and homing pathways in fracture repair. Bonekey Rep. 2013;2:300.

26. Gunawardene P, Al Saedi A, Singh L, Bermeo S, Vogrin S, Phu S, et al. Age, gender, and percentage of circulating osteoprogenitor (COP) cells: the COP study. Exp Gerontol. 2017;96:68–72.

27. Pignolo RJ, Kassem M. Circulating osteogenic cells: implications for injury, repair, and regeneration. J Bone Miner Res. 2011;26(8):1685–93.

28. Rochefort GY, Delorme B, Lopez A, Hérault O, Bonnet P, Charbord P, et al. Multipotential mesenchymal stem cells are mobilized into peripheral blood by hypoxia. Stem Cells. 2006;24(10):2202–8.

29. Crane GM, Jeffery E, Morrison SJ. Adult haematopoietic stem cell niches. Nat Rev Immunol. 2017;17(9):573–90.

30. Tay J, Levesque J-P, Winkler IG. Cellular players of hematopoietic stem cell mobilization in the bone marrow niche. Int J Hematol. 2017;105(2):129–40.

31. Pessach I, Resnick I, Shimoni A, Nagler A. G-CSF-primed BM for allogeneic SCT: revisited. Bone Marrow Transplant. 2015;50(7):892–8.

32. Broxmeyer HE, Orschell CM, Clapp DW, Hangoc G, Cooper S, Plett PA, et al. Rapid mobilization of murine and human hematopoietic stem and progenitor cells with AMD3100, a CXCR4 antagonist. J Exp Med. 2005;201(8):1307–18.

33. Katayama Y, Battista M, Kao W-M, Hidalgo A, Peired AJ, Thomas SA, et al. Signals from the sympathetic nervous system regulate hematopoietic stem cell egress from bone marrow. Cell. 2006;124(2):407–21.

34. Lévesque J-P, Hendy J, Takamatsu Y, Simmons PJ, Bendall LJ. Disruption of the CXCR4/CXCL12 chemotactic interaction during hematopoietic stem cell mobilization induced by GCSF or cyclophosphamide. J Clin Investig. 2003;111(2):187–96.

35. Tzeng Y-S, Li H, Kang Y-L, Chen W-C, Cheng W-C, Lai D-M. Loss of Cxcl12/Sdf-1 in adult mice decreases the quiescent state of hematopoietic stem/progenitor cells and alters the pattern of hematopoietic regeneration after myelosuppression. Blood. 2011;117(2):429–39.

36. Pitchford SC, Furze RC, Jones CP, Wengner AM, Rankin SM. Differential mobilization of subsets of progenitor cells from the bone marrow. Cell Stem Cell. 2009;4(1):62–72.

37. Toupadakis CA, Granick JL, Sagy M, Wong A, Ghassemi E, Chung D-J, et al. Mobilization of endogenous stem cell populations enhances fracture healing in a murine femoral fracture model. Cytotherapy. 2013;15(9):1136–47.

38. Kumar S, Ponnazhagan S. Mobilization of bone marrow mesenchymal stem cells in vivo augments bone healing in a mous e model of segmental bone defect. Bone. 2012;50(4):1012–8.

39. Meeson R, Sanghani-Keri A, Coathup M, Blunn G. VEGF with AMD3100 endogenously mobilizes mesenchymal stem cells and improves fracture healing. J Orthop Res. 2019;37(6):1294–302.

40. Bianchi ME, Crippa MP, Manfredi AA, Mezzapelle R, Querini PR, Venereau E. High-mobility group box 1 protein orchestrates responses to tissue damage via inflammation, innate and adaptive immunity, and tissue repair. Immunol Rev. 2017;280(1):74–82.

41. Palumbo R, Galvez BG, Pusterla T, De Marchis F, Cossu G, Marcu KB, et al. Cells migrating to sites of tissue damage in response to the danger signal HMGB1 require NF-κB activation. J Cell Biol. 2007;179(1):33–40.

42. Qiu J, Nishimura M, Wang Y, Sims JR, Qiu S, Savitz SI, et al. Early release of HMGB-1 from neurons after the onset of brain ischemia. J Cereb Blood Flow Metab. 2008;28(5):927–38.

43. Straino S, Di Carlo A, Mangoni A, De Mori R, Guerra L, Maurelli R, et al. High-mobility group box 1 protein in human and murine skin: involvement in wound healing. J Investig Dermatol. 2008;128(6):1545–53.

44. Tamai K, Yamazaki T, Chino T, Ishii M, Otsuru S, Kikuchi Y, et al. PDGFR-positive cells in bone marrow are mobilized by high mobility group box 1 (HMGB1) to regenerate injured epithelia. Proc Natl Acad Sci. 2011;108(16):6609–14.

45. Chavakis E, Hain A, Vinci M, Carmona G, Bianchi ME, Vajkoczy P, et al. High-mobility group box 1 activates integrin-dependent homing of endothelial progenitor cells. Circ Res. 2007;100(2):204–12.

46. Chavakis E, Urbich C, Dimmeler S. Homing and engraftment of progenitor cells: a prerequisite for cell therapy. J Mol Cell Cardiol. 2008;45(4):514–22.

47. Palumbo R, Bianchi ME. High mobility group box 1 protein, a cue for stem cell recruitment. Biochem Pharmacol. 2004;68(6):1165–70.

48. Aikawa E, Fujita R, Kikuchi Y, Kaneda Y, Tamai K. Systemic high-mobility group box 1 administration suppresses skin inflammation by inducing an accumulation of PDGFRα+ mesenchymal cells from bone marrow. Sci Rep. 2015;5(1):11,008.

49. Goto T, Miyagawa S, Tamai K, Matsuura R, Kido T, Kuratani T, et al. High-mobility group box 1 fragment suppresses adverse post-infarction remodeling by recruiting PDGFRα-positive bone marrow cells. PLoS One. 2020;15(4):e0230392.

50. Kido T, Miyagawa S, Goto T, Tamai K, Ueno T, Toda K, et al. The administration of high-mobility group box 1 fragment prevents deterioration of cardiac performance by enhancement of bone

marrow mesenchymal stem cell homing in the delta-sarcoglycan-deficient hamster. PLoS One. 2018;13(12):e0202838.

51. Lin F, Xue D, Xie T, Pan Z. HMGB1 promotes cellular chemokine synthesis and potentiates mesenchymal stromal cell migration via Rap1 activation. Mol Med Rep. 2016;14(2):1283–9.

52. Lin F, Zhang W, Xue D, Zhu T, Li J, Chen E, et al. Signaling pathways involved in the effects of HMGB1 on mesenchymal stem cell migration and osteoblastic differentiation. Int J Mol Med. 2016;37(3):789–97.

53. Meng E, Guo Z, Wang H, Jin J, Wang J, Wang H, et al. High mobility group box 1 protein inhibits the proliferation of human mesenchymal stem cells and promotes their migration and differentiation along osteoblastic pathway. Stem Cells Dev. 2008;17(4):805–14.

54. Lv Y, Lin C. High mobility group box 1-immobilized nanofibrous scaffold enhances vascularization, osteogenesis and stem cell recruitment. J Mater Chem B. 2016;4(29):5002–14.

55. Xue D, Zhang W, Chen E, Gao X, Liu L, Ye C, et al. Local delivery of HMGB1 in gelatin sponge scaffolds combined with mesenchymal stem cell sheets to accelerate fracture healing. Oncotarget. 2017;8(26):42,098–115.

56. Xie H-L, Zhang Y, Huang Y-Z, Li S, Wu C-G, Jiao X-F, et al. Regulation of high mobility group box 1 and hypoxia in the migration of mesenchymal stem cells. Cell Biol Int. 2014;38(7):892–7.

57. Hong HS, Lee J, Lee E, Kwon YS, Lee E, Ahn W, et al. A new role of substance P as an injury-inducible messenger for mobilization of CD29+ stromal-like cells. Nat Med. 2009;15(4):425–35.

58. Onuoha GN. Circulating sensory peptide levels within 24 h of human bone fracture. Peptides. 2001;22(7):1107–10.

59. Onuoha GN, Alpar EK. Elevation of plasma CGRP and SP levels in orthopedic patients with fracture neck of femur. Neuropeptides. 2000;34(2):116–20.

60. Zhang Y, An S, Hao J, Tian F, Fang X, Wang J. Systemic injection of substance P promotes murine calvarial repair through mobilizing endogenous mesenchymal stem cells. Sci Rep. 2018;8(1):12996.

61. Hausman MR, Schaffler MB, Majeska RJ. Prevention of fracture healing in rats by an inhibitor of angiogenesis. Bone. 2001;29(6):560–4.

62. Lu C, Saless N, Wang X, Sinha A, Decker S, Kazakia G, et al. The role of oxygen during fracture healing. Bone. 2013;52(1):220–9.

63. Stegen S, van Gastel N, Carmeliet G. Bringing new life to damaged bone: the importance of angiogenesis in bone repair and regeneration. Bone. 2015;70:19–27.

64. Atesok K, Li R, Stewart DJ, Schemitsch EH. Endothelial progenitor cells promote fracture healing in a segmental bone defect model. J Orthop Res. 2010;28(8):1007–14.

65. Asahara T, Takahashi T, Masuda H, Kalka C, Chen D, Iwaguro H, et al. VEGF contributes to postnatal neovascularization by mobilizing bone marrow-derived endothelial progenitor cells. EMBO J. 1999;18(14):3964–72.

66. Keramaris NC, Calori GM, Nikolaou VS, Schemitsch EH, Giannoudis PV. Fracture vascularity and bone healing: a systematic review of the role of VEGF. Injury. 2008;39:S45–57.

67. Street J, Bao M, de Guzman L, Bunting S, Peale FV, Ferrara N, et al. Vascular endothelial growth factor stimulates bone repair by promoting angiogenesis and bone turnover. Proc Natl Acad Sci. 2002;99(15):9656–61.

68. Jujo K, Ii M, Sekiguchi H, Klyachko E, Misener S, Tanaka T, et al. CXC-chemokine receptor 4 antagonist AMD3100 promotes cardiac functional recovery after ischemia/reperfusion injury via endothelial nitric oxide synthase–dependent mechanism. Circulation. 2013;127(1):63–73.

69. Petit I, Jin D, Rafii S. The SDF-1–CXCR4 signaling pathway: a molecular hub modulating neo-angiogenesis. Trends Immunol. 2007;28(7):299–307.

70. Asahara T, Murohara T, Sullivan A, Silver M, van der Zee R, Li T, et al. Isolation of putative progenitor endothelial cells for angiogenesis. Science. 1997;275(5302):964–6.

71. Stroncek DF, Clay ME, Herr G, Smith J, Jaszcz WB, Ilstrup S, et al. The kinetics of G-CSF mobilization of CD34+ cells in healthy people. Transfus Med. 1997;7(1):19–24.

72. Herrmann M, Zeiter S, Eberli U, Hildebrand M, Camenisch K, Menzel U, et al. Five days granulocyte Colony-stimulating factor treatment increases bone formation and reduces gap size of a rat segmental bone defect: a pilot study. Front Bioeng Biotechnol. 2018;6:5.

73. Kuroda R, Matsumoto T, Kawakami Y, Fukui T, Mifune Y, Kurosaka M. Clinical impact of circulating CD34-positive cells on bone regeneration and healing. Tissue Eng Part B Rev. 2014;20(3):190–9.

74. Kuroda R, Matsumoto T, Miwa M, Kawamoto A, Mifune Y, Fukui T, et al. Local transplantation of G-CSF-mobilized CD34+ cells in a patient with tibial nonunion: a case report. Cell Transplant. 2011;20(9):1491–6.

75. Kuroda R, Matsumoto T, Niikura T, Kawakami Y, Fukui T, Lee SY, et al. Local transplantation of granulocyte colony stimulating factor-mobilized CD34+ cells for patients with femoral and tibial nonunion: pilot clinical trial. Stem Cells Transl Med. 2014;3(1):128–34.

76. Karp JM, Leng Teo GS. Mesenchymal stem cell homing: the devil is in the details. Cell Stem Cell. 2009;4(3):206–16.

77. Hu C, Yong X, Li C, Lü M, Liu D, Chen L, et al. CXCL12/CXCR4 axis promotes mesenchymal stem cell mobilization to burn wounds and contributes to wound repair. J Surg Res. 2013;183(1):427–34.

78. Iinuma S, Aikawa E, Tamai K, Fujita R, Kikuchi Y, Chino T, et al. Transplanted bone marrow–derived circulating PDGFRα + cells restore type VII collagen

in recessive dystrophic epidermolysis bullosa mouse skin graft. J Immunol. 2015;194(4):1996–2003.

79. Kitaori T, Ito H, Schwarz EM, Tsutsumi R, Yoshitomi H, Oishi S, et al. Stromal cell–derived factor 1/ CXCR4 signaling is critical for the recruitment of mesenchymal stem cells to the fracture site during skeletal repair in a mouse model. Arthritis Rheum. 2009;60(3):813–23.

80. Xu J, Chen Y, Liu Y, Zhang J, Kang Q, Ho K, et al. Effect of SDF-1/Cxcr4 signaling antagonist AMD3100 on bone mineralization in distraction osteogenesis. Calcif Tissue Int. 2017;100(6):641–52.

81. Ponte AL, Marais E, Gallay N, Langonné A, Delorme B, Hérault O, et al. The in vitro migration capacity of human bone marrow mesenchymal stem cells: comparison of chemokine and growth factor chemotactic activities. Stem Cells. 2007;25(7):1737–45.

82. Toupadakis CA, Wong A, Genetos DC, Chung D-J, Murugesh D, Anderson MJ, et al. Long-term administration of AMD3100, an antagonist of SDF-1/ CXCR4 signaling, alters fracture repair. J Orthop Res. 2012;30(11):1853–9.

83. Higashino K, Viggeswarapu M, Bargouti M, Liu H, Titus L, Boden SD. Stromal cell-derived factor-1 potentiates bone morphogenetic protein-2 induced bone formation. Tissue Eng Part A. 2011;17(3–4):523–30.

84. Shinohara K, Greenfield S, Pan H, Vasanji A, Kumagai K, Midura RJ, et al. Stromal cell-derived factor-1 and monocyte chemotactic protein-3 improve recruitment of osteogenic cells into sites of musculoskeletal repair. J Orthop Res. 2011;29(7):1064–9.

85. Roberts SJ, Ke HZ. Anabolic strategies to augment bone fracture healing. Curr Osteoporos Rep. 2018;16(3):289–98.

86. Govender S, Csimma C, Genant HK, Valentin-Opran A, Amit Y, Arbel R, et al. Recombinant human bone morphogenetic protein-2 for treatment of open tibial fractures: a prospective, controlled, randomized study of four hundred and fifty patients. J Bone Joint Surg Am. 2002;84(12):2123–34.

87. Swiontkowski MF. Recombinant human bone morphogenetic protein-2 in open tibial fractures: a subgroup analysis of data combined from two prospective randomized studies. Yearbook Orthop. 2007;2007:89.

88. Wei S, Cai X, Huang J, Xu F, Liu X, Wang Q. Recombinant human BMP-2 for the treatment of open tibial fractures. Orthopedics. 2012;35(6):e847–54.

89. Kawaguchi H, Oka H, Jingushi S, Izumi T, Fukunaga M, Sato K, et al. A local application of recombinant human fibroblast growth factor 2 for tibial shaft fractures: a randomized, placebo-controlled trial. J Bone Miner Res. 2010;25(12):2735–43.

90. Aspenberg P, Genant HK, Johansson T, Nino AJ, See K, Krohn K, et al. Teriparatide for acceleration of fracture repair in humans: a prospective, randomized, double-blind study of 102 postmenopausal women with distal radial fractures. J Bone Miner Res. 2010;25(2):404–14.

91. Peichl P, Holzer LA, Maier R, Holzer G. Parathyroid hormone 1-84 accelerates fracture-healing in pubic bones of elderly osteoporotic women. J Bone Joint Surg Am. 2011;93(17):1583–7.